INTERNATIONAL cell culture congress - Cell Culture & it's Application.

CELL CULTURE AND ITS APPLICATION

ORGANIZING COMMITTEE

Ronald T. Acton, University of Alabama in Birmingham, Chairman
J. Daniel Lynn, University of Alabama in Birmingham, Co-Chairman
John Papaconstantinou, Oak Ridge National Laboratory, Program Director
Kevin Bilson, University of Alabama in Birmingham, Registration
Georgia Cosmas, University of Alabama in Birmingham, Treasurer
Gloria Goldstein, University of Alabama in Birmingham, Publicity
Alvin Bridges, KC Biologicals, Inc., Exhibits
Carol Bishop, Kahler Plaza Hotel, Reservations
Susan Stancil, University of Alabama in Birmingham, Secretary
Jackie Morris, University of Alabama in Birmingham, Secretary

ADVISORY COMMITTEE

Herman Lewis, National Science Foundation
Jerry J. Callis, Plum Island Animal Disease Center
Leonard Keay, Washington University Medical School

THE FIRST INTERNATIONAL CELL CULTURE CONGRESS,
Birmingham, Alabama, September 21–25, 1975

CELL CULTURE AND ITS APPLICATION

EDITED BY

Ronald T. Acton
J. Daniel Lynn

Department of Microbiology
University of Alabama in Birmingham
Birmingham, Alabama

ACADEMIC PRESS
New York San Francisco London 1977
A Subsidiary of Harcourt Brace Jovanovich, Publishers

ACADEMIC PRESS RAPID MANUSCRIPT REPRODUCTION

ACADEMIC PRESS, INC.
111 Fifth Avenue, New York, New York 10003

United Kingdom Edition published by
ACADEMIC PRESS, INC. (LONDON) LTD.
24/28 Oval Road, London NW1

Library of Congress Cataloging in Publication Data

International Cell Culture Congress, 1st, Birmingham, Ala., 1975.
Cell culture and its application.

1. Cell culture—Congresses. I. Acton, Ronald T. II. Lynn, J. Daniel. III. Title. [DNLM: 1. Cells, Cultures—Congresses. 2. Cell differentiation—Congresses. 3. Cytological technics—Congresses. W3 IN1242M 1st 1975c / QH573 I51 1975c]
QH585.I534 1975 574.8′028 77-14087
ISBN 0-12-043050-9

PRINTED IN THE UNITED STATES OF AMERICA

CONTENTS

SYMPOSIUM III. LARGE-SCALE PRODUCTION

SYMPOSIUM IV. CELL CULTURE FOR THE STUDY OF DISEASE

PLENARY SESSIONS

WORKSHOP I. DIFFERENTIATION

WORKSHOP II. NUTRITIONAL REQUIREMENTS FOR CELL GROWTH

WORKSHOP III. PROBLEMS OF LARGE-SCALE CELL CULTURE AND QUALITY CONTROL

WORKSHOP IV. CELL CULTURE FOR THE STUDY OF DISEASE

LIST OF CONTRIBUTORS

Numbers in parentheses refer to the pages on which authors' contributions begin.

Acton, Ronald T. (129), Department of Microbiology and Diabetes Research and Training Center, Univeristy of Alabama in Birmingham, 1808 Seventh Avenue South, University Station, Birmingham, Alabama 35294

Allen, H. J. (689), Department of Gynecology, Roswell Park Memorial Institute, 666 Elm Street, Buffalo, New York 14263

Amborski, Grace F. (653), Department of Veterinary Microbiology and Parasitology, Louisiana State University, Baton Rouge, Louisiana 70803

Amborski, Robert L. (653), Department of Microbiology, Louisiana State University, Baton Rouge, Louisiana 70803

Ames, Gregory (637), Biology and Chemistry Departments, Seton Hall University, South Orange, New Jersey 07079

Bachrach, H. L. (603), Plum Island Animal Disease Center, USDA, ARS, NER, P.O. Box 848, Greenport, New York 11944

Bag, Jnanankur (389), Department of Neurology, 20 Staniford Street, Harvard Medical School, Boston, Massachusetts 02114

Balciunas, Rytis (637), Biology and Chemistry Departments, Seton Hall University, South Orange, New Jersey 07079

Bando, B. M. (667), U.S. Department of Agriculture, Agricultural Research Service, Arthropod-borne Animal Disease Research Laboratories, Denver, Colorado 80225

Barile, Michael F. (291), DHEW, Food and Drug Administration, Bureau of Biologics, Division of Bacterial Products, Mycoplasma Branch, Bethesda, Maryland 20014

Barlow, J. J. (689), Department of Gynecology, Roswell Park Memorial Institute, 666 Elm Street, Buffalo, New York 14263

Barstad, Paul A. (129), Department of Microbiology and Diabetes Research and Training Center, University of Alabama in Birmingham, 1808 Seventh Avenue South, University Station, Birmingham, Alabama 35294

Benton, Charles V. (559, 571), National Cancer Institute, Frederick Cancer Research Center, Frederick, Maryland 21701

BIRCH, J. R. (503), Searle Research Laboratories, P.O. Box 53, Lane End Road, High Wycombe, Bucks., HP12 4HL, England

BODMER, WALTER F. (247), Department of Biochemistry, Genetics Laboratory, University of Oxford, Oxford, England

BRONSON, D. L. (711), Department of Urologic Surgery, University of Minnesota Health Sciences Center, 412 Union Street Southeast, Minneapolis, Minnesota 55455

BRUMBAUGH, J. A. (491), School of Life Sciences, University of Nebraska, Lincoln, Nebraska 68508

BUONINCONTRI, LOIS (637), Biology and Chemistry Departments, Seton Hall University, South Orange, New Jersey 07079

CALLIS, J. J. (217), Plum Island Animal Disease Center, USDA, ARS, NER, P.O. Box 848, Greenport, New York 11944

CARPENTER, GRAHAM (83), Department of Biochemistry, Vanderbilt University, Nashville, Tennessee 37232

CERVENKA, J. (711), Department of Urologic Surgery, University of Minnesota Health Sciences Center, 412 Union Street Southeast, Minneapolis, Minnesota 55455

CLEVELAND, P. H. (711), Department of Urologic Surgery, University of Minnesota Health Sciences Center, 412 Union Street Southeast, Minneapolis, Minnesota 55455

COFFINO, PHILIP (49), Department of Microbiology, University of California, San Francisco, San Francisco, California 94143

COHEN, STANLEY (83), Department of Biochemistry, Vanderbilt University, Nashville, Tennessee 37232

COX, R. MICHAEL (129), Department of Microbiology, University of Alabama in Birmingham, University Station, Birmingham, Alabama 35294

CRISTOFALO, V. J. (223), The Wistar Institute, 36th Street at Spruce, Philadelphia, Pennsylvania 19104

CRUMPTON, MICHAEL J. (247), National Institute for Medical Research, Mill Hill, London NW7 1AA, England

ELLIOTT, A. Y. (711), Department of Urologic Surgery, University of Minnesota Health Sciences Center, 412 Union Street Southeast, Minneapolis, Minnesota 55455

FRALEY, E. E. (711), Department of Urologic Surgery, University of Minnesota Health Sciences Center, 412 Union Street Southeast, Minneapolis, Minnesota 55455

GIESING, M. (417), Institute for Physiological Chemistry, University of Bonn, 53 Bonn, GFR

GILDEN, RAYMOND V. (335), Division of Viral Oncology, Frederick Cancer Research Center, Frederick, Maryland 21701

GIRARD, H. C. (111), Foot and Mouth Disease Institute, Ankara, Turkey

GOLDSTEIN, GIDEON (39), Memorial Sloan-Kettering Cancer Center, New York, New York 10021

GOODFELLOW, PETER (247), Department of Biochemistry, Genetics Laboratory, University of Oxford, Oxford, England

GOSPODAROWICZ, DENIS (55), The Salk Institute for Biological Studies, P.O. Box 1809, San Diego, California 92112

GROVE, G. L. (223), The Wistar Institute, 36th Street at Spruce, Philadelphia, Pennsylvania 19104

GURLEY, LAWRENCE R. (5), Cellular and Molecular Biology Group, Los Alamos Scientific Laboratory, University of California, Los Alamos, New Mexico 87544

HAM, RICHARD G. (533), Department of Molecular, Cellular, and Developmental Biology, University of Colorado, Boulder, Colorado 80309

HATGI, JOHN (571), National Cancer Institute, Frederick Cancer Research Center, Frederick, Maryland 21701

HILDEBRAND, CARL E. (5), Cellular and Molecular Biology Group, Los Alamos Scientific Laboratory, University of California, Los Alamos, New Mexico 87544

HOHMANN, PHILIP G. (5), Cellular and Molecular Biology Group, Los Alamos Scientific Laboratory, University of California, Los Alamos, New Mexico 87544

HOLYK, NESTOR (637), Biology and Chemistry Department, Seton Hall University, South Orange, New Jersey 07079

HOPKINS, D. W. (503), Searle Research Laboratories, P.O. Box 53, Lane End Road, High Wycombe, Bucks., HP12 4HL, England

HU, FUNAN (449), Departments of Cutaneous Biology and Pathology, Oregon Regional Primate Research Center, 505 Northwest 185th Avenue, Beaverton, Oregon 97005

JENSEN, MONA D. (589), Instrumentation Laboratory, Inc., Lexington, Massachusetts 02173

JOHNSON, E. A. Z. (689), Department of Gynecology, Roswell Park Memorial Institute, 666 Elm Street, Buffalo, New York 14263

JOHNSON, ROGER W. (559, 571), National Cancer Institute, Frederick Cancer Research Center, Frederick, Maryland 21701

JONES, R. H. (667), U.S. Department of Agriculture, Agricultural Research Service, Arthropod-borne Animal Disease Research Laboratory, Denver, Colorado 80225

JONES, W. I. (559), National Cancer Institute, Frederick Cancer Research Center, Frederick, Maryland 21701

KAZAMA, NORIMOTO (347), Division of Reproductive Biology, C.S. Mott Center for Human Growth and Development, Wayne State University, Detroit, Michigan

KEAY, LEONARD (513), Department of Microbiology and Immunology, Division of Biological and Medical Sciences, Washington University School of Medicine, St. Louis, Missouri 63110

KLETZIEN, ROLF F. (379), Sidney Farber Cancer Center, 35 Binney Street, Boston, Massachusetts 02115

LACEY, JAMES C., JR. (541), Laboratory of Molecular Biology, University of Alabama in Birmingham, University Station, Birmingham, Alabama 35294

LEE, HAROLD H. (347), Department of Biology, University of Toledo, Toledo, Ohio 43606

LEIBOVITZ, ALBERT (675), Scott and White Clinic, Temple, Texas 76501

LEVINE, D. W. (191), The Cell Culture Center, Department of Nutrition and Food Science, Massachusetts Institute of Technology, Cambridge, Massachusetts

LEVINS, PETER (637), Biology and Chemistry Departments, Seton Hall University, South Orange, New Jersey 07079

LEWIS, HERMAN W. (1), National Science Foundation, Cellular Biology Section, Washington, D.C. 20550

LIANG-TANG, LORETTA (347), Department of Obstetrics and Gynecology, University of Rochester Medical School, Rochester, New York

LIM, RAMON (461), Division of Chicago Hospitals, 950 East 59th Street, Chicago, Illinois 60637

LYNN, J. DANIEL (129), Department of Microbiology, University of Alabama in Birmingham, University Station, Birmingham, Albama 35294

MABRY, NANCY D. (675), Scott and White Clinic, Temple, Texas 76501

MCCOMBS, WILLIAM B. III (675), Scott and White Clinic, Temple, Texas 76501

MCCOY, CAMERON E. (675), Scott and White Clinic, Temple, Texas 76501

MALEMUD, CHARLES J. (481), Department of Pathology, Health Science Center, State University of New York at Stony Brook, Stony Brook, New York

MARKS, MICHAEL J. (23), Department of Biophysics and Genetics, University of Colorado Medical Center, Denver, Colorado 80220

MAZUR, KENNETH C. (675), Scott and White Clinic, Temple, Texas 76501

MITSUNOBU, KATSUSUKE (461), Department of Neuropsychiatry, University of Okayama, Okayama City, Japan

MORAN, JOHN S. (55), The Salk Institute for Biological Studies, P.O. Box 1809, San Diego, California 92112

MUSTAFA, S. JAMAL (729), Department of Pharmacology, University of South Alabama, College of Medicine, Mobile, Alabama 36688

NAKEFF, ALEXANDER (433), Section of Cancer Biology, Mallinckrodt Institute of Radiology, Washington University School of Medicine, 510 South Kingshighway, St. Louis, Missouri 63110

NORBY, DAVID P. (481), Department of Pathology, Health Science Center, State University of New York at Stony Brook, Stony Brook, New York

NYIRI, LASZLO K. (161), Fermentation Design, Inc., Division of New Brunswick Scientific Co., Inc., Bethlehem, Pennsylvania 18015

ORR, MARY FAITH (363), Department of Anatomy, Northwestern University, Medical and Dental Schools, 303 East Chicago Avenue, Chicago, Illinois 60611

ORSI, ERNEST V. (637), Biology and Chemistry Departments, Seton Hall University, South Orange, New Jersey 07079

PARIZA, MICHAEL W. (379), McArdle Laboratory for Cancer Research, Medical School, University of Wisconsin, Madison, Wisconsin 53706

PASZTOR, LINDA M. (449), Departments of Cutaneous Biology and Pathology, Oregon Regional Primate Research Center, 505 Northwest 185th Avenue, Beaverton, Oregon 97005

PERRY, ALBERT (559, 571), National Cancer Institute, Frederick Cancer Research Center, Frederick, Maryland 21701

POLATNICK, J. (603), Plum Island Animal Disease Center, USDA, ARS, NER, P.O. Box 848, Greenport, New York 11944

POTTER, VAN R. (379), McArdle Laboratory for Cancer Research, Medical School, University of Wisconsin, Madison, Wisconsin 53706

RAMIREZ, GALO (23), Department of Biophysics and Genetics, University of Colorado Medical Center, Denver, Colorado 80220

RYAN, J. M. (223), The Wistar Institute, 36th Street at Spruce, Philadelphia, Pennsylvania 19104

SARKAR, SATYAPRIYA (389), Department of Muscle Research, Boston Biomedical Research Institute, Harvard Medical School, Boston, Massachusetts 02114

SATO, GORDON (107), Department of Biology, Muir College, University of California, San Diego, La Jolla, California 92093

SCHALL, D. G. (491), School of Life Sciences, University of Nebraska, Lincoln, Nebraska 68508

SEEDS, NICHOLAS W. (23), Department of Biophysics and Genetics, University of Colorado Medical Center, Denver, Colorado 80220

SEGREST, JERE P. (283), Departments of Pathology, Biochemistry, and Microbiology, Comprehensive Cancer Center, Institute of Dental Research, University of Alabama in Birmingham, University Station, Birmingham, Alabama 35294

SHIBLEY, GEORGE P. (559, 571), National Cancer Institute, Frederick Cancer Research Center, Frederick, Maryland 21701

SLY, WILLIAM S. (267), The Edward Mallinckrodt Department of Pediatrics, Washington University School of Medicine, The Division of Medical Genetics, St. Louis Children's Hospital, St. Louis, Missouri 63110

SNARY, DAVID (247), National Institute for Medical Research, Mill Hill, London NW7 1AA, England

STEPHENS, DANIEL P., JR. (541), Laboratory of Molecular Biology, University of Alabama in Birmingham, University Station, Birmingham, Alabama 35294

STINSON, JAMES C. (675), Scott and White Clinic, Temple, Texas 76501

STROBEL, JANNA D. (541), Laboratory of Molecular Biology, University of Alabama in Birmingham, University Station, Birmingham, Alabama 35294

SULLIVAN, PATRICIA M. (533), Department of Molecular, Cellular, and Developmental Biology, University of Colorado, Boulder, Colorado 80309

THILLY, W. G. (191), The Cell Culture Center, Department of Nutrition and Food Science, Massachusetts Institute of Technology, Cambridge, Massachusetts

TOBEY, ROBERT A. (5), Cellular and Molecular Biology Group, Los Alamos Scientific Laboratory, University of California, Los Alamos, New Mexico 87544

TOTH, G. M. (617), Fermentation Design, Inc., Division of New Brunswick Scientific Co., Inc., Bethlehem, Pennsylvania 18017

TROY, SHUANG S. (461), Brain Research Institute, University of Chicago, Chicago, Illinois 60637

TURRIFF, DAVID E. (461), Department of Biochemistry, University of Chicago, Chicago, Illinois 60637

VALERIOTE, FREDERICK A. (433), Section of Cancer Biology, Mallinckrodt Institute of Radiology, Washington University School of Medicine, 510 South Kingshighway, St. Louis, Missouri 63110

WALTERS, RONALD A. (5), Cellular and Molecular Biology Group, Los Alamos Scientific Laboratory, University of California, Los Alamos, New Mexico 87544

WANG, D. I. C. (191), The Cell Culture Center, Department of Nutrition and Food Science, Massachusetts Institute of Technology, Cambridge, Massachusetts

WISE, KIM S. (129), Department of Microbiology and Diabetes Research and Training Center, University of Alabama in Birmingham, 1808 Seventh Avenue South, University Station, Birmingham, Alabama 35294

ZILLIKEN, F. (417), Institute for Physiological Chemistry, University of Bonn, 53 Bonn, GFR

ZWERNER, ROBERT K. (129), Department of Microbiology and Diabetes Research and Training Center, University of Alabama in Birmingham, 1808 Seventh Avenue South, University Station, Birmingham, Alabama 35294

PREFACE

The growth and manipulation of animal cells in culture has become a vital addition to the arsenal of procedures for many studies in biology today. There is little doubt that early workers in the field could not possibly have imagined the widespread use of this technique. Since Wilhelm Roux attempted to grow tissues from a chicken embryo in 1885, numerous types of cells have been induced to grow *in vitro*. The ability to grow cells in culture has advanced to the point where many investigators feel that the mysteries concerning differentiation and function *in vivo* might be unraveled by the use of cell culture. Moreover, many individuals interested in both basic and applied science view cells as factories that can be induced to generate a number of components such as viruses and a variety of substances involved in modulating cellular activities. Many believe that these latter substances may be the key for controlling various disease states in the future. The impetus to organize this first International Cell Culture Congress was not due entirely to the ferment of activity surrounding the use of cells in culture. It was due in part to the foresight and wisdom of Dr. Herman Lewis, head of the Cellular Biology Section of the National Science Foundation, and his panel of *ad hoc* advisors, who felt that emphasis should be placed on the establishment of facilities for the cultivation of mammalian cells, which in turn would hopefully enhance basic research in cell biology. The Cell Culture Center within the Department of Microbiology at the University of Alabama in Birmingham was the second such facility funded (the MIT Cell Culture Center was inaugurated March 28, 1975) by the NSF with this goal in mind. Our major goal was to design a facility in which the large-scale propagation of mammalian cells in suspension could be conducted and their products obtained in an economical and efficient manner. It was felt that these materials would be valuable resources to investigators working in the area of cell biology and greatly facilitate the acquisition of new knowledge. Due to the paucity of techniques available for the large-scale production of cells and their products, the center was designed in a manner to allow innovative approaches to these problems. It was felt by us and the NSF that the development of this technology should involve the collaboration of investigators from academic institutions as well as industry. This Cell Culture Congress was organized to provide a forum for the discussion of how this technology could impact on cell biology. As judged by the quality of the plenary and symposia speakers as well as the lively workshop sessions that ensued, this goal was largely accomplished. Many left the congress pondering the possibility of new approaches to old problems made possible by the ability to generate a variety of cell types in large quantities and to harvest subcellular organelles as well as various secreted components.

The Organizing Committee of this congress elected to accept for publication all manuscripts submitted by the speakers. Insofar as this congress was mainly organized in order to disseminate information, the Committee felt this policy would best serve that mission. We, on behalf of all the organizers of the congress, would like to express appreciation to the Department of Microbiology, University of Alabama in Birmingham, and in addition, to other basic science and clinical departments who so willingly provided support for this meeting. A special tribute is owed also to the various vendors of scientific equipment and biological reagents who not only provided monetary support but their perspective on various approaches to cell culture. Finally, gratitude is expressed to the more than 300 individuals who chose to attend this first congress and who contributed to its success.

OPENING REMARKS

HERMAN W. LEWIS

National Science Foundation
Washington, D. C.

It is a pleasure for me to add words of welcome to you on behalf of the NSF. It is a primary function of the NSF to foster the growth, development and good health of science in the U.S. and the purpose that brings us together on this occasion serves well this NSF objective. I shall try to put this meeting into perspective, as it relates to NSF interests.

The second stage of the revolution in biological research has begun. It is in this stage, that of modern cell biology, that the full ramifications will be seen of the revolution started 20 years ago. The first stage of the revolution in biological research was the discoveries and development of concepts of molecular biology carried out with the simplest of living organisms: bacteria and the viruses that infect them. The aim was to get an understanding of the structure, regulation and function of important biological macromolecules. The second stage of the revolution in biological research that we have now entered is an extension of interest in molecular processes to the cellular level and the focus is on mammalian or human cells. The possibility of understanding the interactions and regulation of molecular processes in complex cell systems is challenging, exciting, and attractive to large numbers of bright, young as well as established investigators, many of whom are present today.

The new burst of activity in this area of research and the blossoming of the new cell biology depends on a number of technical advances, the most important being the development of techniques for growing a variety of mammalian cells in culture. Future progress in cell biology may well depend on further development of this technology.

This cell culture congress that is now convened may be a benchmark event in the history of cell biology. To the best of my knowledge this is the first meeting of this type and is most timely. The organizers recognize the interlocking relationship of technology and the design of biological experiments and I am pleased with the good blend of cell culture technology and substantive cell biology on the program. The format of interspersed plenary sessions and workshops offers good opportunity not only to summarize and exchange information about current knowledge, but also encourages cross fertilization and grappling with problems in need of solution.

One area that will receive special attention during this meeting will be large scale cell culture technology and one of the opportunities we will have during the week will be to tour the cell culture center of our host institution. The NSF, with the University of Alabama Medical Center, has established this cell center and we believe it is innovative, well designed and will be very useful in enhancing basic research in cell biology. I hope you will all inspect it and make inquiries about its services. I look forward to a very stimulating week and I hope everyone present benefits from the meeting, as I am sure I shall.

Symposium I

CELL CULTURE FOR THE STUDY OF DIFFERENTIATION AND CELL CYCLE

SEQUENTIAL BIOCHEMICAL EVENTS IN THE CELL CYCLE

Robert A. Tobey
Ronald A. Walters
Philip G. Hohmann
Carl E. Hildebrand
Lawrence R. Gurley

Cellular and Molecular Biology Group
Los Alamos Scientific Laboratory
University of California
Los Alamos, New Mexico

I. INTRODUCTION

The detection of specific sequences of biochemical processes as cells prepare for DNA replication and division (Baserga, 1968; Mitchison, 1971; Tobey *et al.*, 1974, 1975) suggests the temporally ordered expression of genetic information during the cell cycle. Although little is known about the underlying regulatory mechanisms, it has been suggested that changes in chromatin structure may be involved in control of cycle-specific events (Mazia, 1963). One possible avenue for altering chromatin structure involves the reversible modification of histones (Allfrey, 1971; Bradbury and Crane-Robinson, 1971). Consequently, over the past few years we have utilized CHO cells to examine the relationship between one such modification, *phosphorylation of histones*, and changes in the structure of chromatin during the cell cycle. Since the program is part of a long-term study, this communication, in effect, represents a progress report in which we summarize our current views of this rapidly developing area of research.

II. CELL SYNCHRONIZATION AND ANALYSIS

Biochemical dissection of the cell cycle is greatly facilitated by use of populations synchronized at different stages in the cycle (Petersen *et al.*, 1969). Cells synchronized in one portion of the cell cycle are generally unsuitable for studies in other phases due to synchrony decay, a naturally occurring loss of synchrony arising from variations in the rate of cycle progression by individual cells in the population (Engelberg, 1964; Petersen *et al.*, 1969). In particular, cells synchronized in mitosis or early G_1 should never be utilized to study "G_2 events," since the G_2 cells are badly contaminated with cells from other phases of the cell cycle.

For example, Fig. 1A summarizes the kinetic response of CHO cells following mitotic selection (Tobey *et al.*, 1967). Essentially identical results are obtained in cultures released from G_1-arrest in isoleucine-deficient medium (Tobey and Ley, 1971; Tobey, 1973). Although the initial mitotic population possesses an Engelberg synchrony index (Engelberg, 1964) of 0.97 (where 1.0 is perfect synchrony), note that the maximum proportion of G_2 cells is only 30% - the result of synchrony decay. Therefore, it is apparent that cells prepared by mitotic selection or by the isoleucine-deficiency method are useful for studying early interphase events, but much less satisfactory for investigating mid- to late-cycle processes.

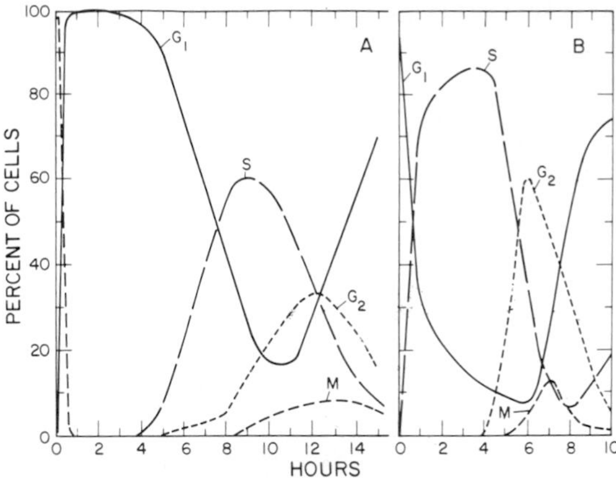

FIGURE 1 *Cell-cycle kinetics following synchronization of CHO cells. (A) Cells synchronized by mitotic selection (Tobey et al., 1967) were placed in suspension culture, and aliquots were removed at intervals thereafter for determination of cell-cycle distribution. (B) Cells released from isoleucine-mediated G_1-arrest (Tobey and Ley, 1971; Tobey, 1973) were treated for 9 hr with 10^{-3} M hydroxyurea as they were traversing G_1, resulting in an accumulation of cells at the G_1/S boundary (Tobey and Crissman, 1972a). The cells were then washed and resuspended in drug-free medium (t = 0). Aliquots were removed at intervals thereafter to allow determination of the cell-cycle distribution. Cell-cycle distributions were determined by a combination of flow microfluorometry, thymidine-^{3}H autoradiography, and cell number enumeration, as described previously (Tobey and Crissman, 1972b).*

In order to examine events in mid/late interphase, we find it necessary to treat cells traversing G_1 (following either mitotic selection or release from isoleucine-deficient medium) with 10^{-3} M hydroxyurea, which causes the cells to accumulate near the G_1/S border (Tobey and Crissman, 1972a) due to an inhibition of synthesis of dATP (Walters *et al.*, 1973). After removing hydroxyurea, the cells proceed synchronously through the cycle as shown in Fig. 1B. These cultures are especially well suited for investigating events at the G_1/S boundary and in S phase, but they may also be used for examining later events in view of the 60-65% value of G_2 cells obtained approximately 6 hr after drug removal. Similar

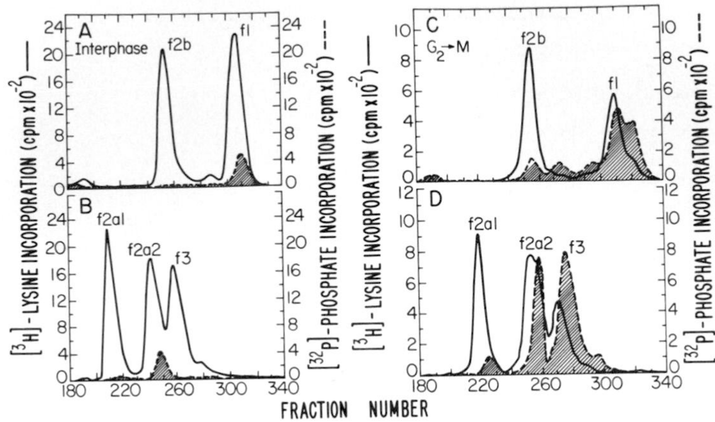

FIGURE 2 *Preparative polyacrylamide gel electrophoresis profiles of histone phosphorylation in CHO cells from interphase cultures (A and B) and from cultures traversing from G_2 into mitosis (C and D). The cells were prelabeled for 52 hr with 50 μCi lysine-^3H per liter of culture prior to exposure for a 2-hr period to 20 mCi of carrier-free $^{32}PO_4$ per liter of culture. Specific details involving cell synchronization and isolation and fractionation of histones may be found elsewhere (Gurley et al., 1974a).*

kinetic data are obtained in cultures synchronized by double-thymidine blockade or isoleucine-thymidine blockade (Kraemer and Tobey, 1972).

Once synchronization techniques are developed, it is absolutely essential to know at all stages of the experiment *precisely* where in the cycle the cells are located. In this laboratory, we utilize a combination of flow microfluorometry, cell number enumeration, and autoradiography to monitor population kinetics (Tobey and Crissman, 1972b). Thus, each time a sample is removed for biochemical analysis, a duplicate sample is removed for determination of the cell-cycle distribution, thereby ensuring that we are measuring the response of a known population.

III. CELL-CYCLE-SPECIFIC PHOSPHORYLATION OF HISTONES

Our early studies utilized preparative polyacrylamide gel electrophoresis to examine the patterns of phosphorylation of the various histones at different stages in the CHO cell cycle. In G_1-arrested populations maintained 38 hr in isoleucine-deficient medium, only histone f2a2 was significantly phosphorylated during a 1-hr exposure to $^{32}PO_4$ (Gurley *et al.*, 1973a).

In contrast, examination of interphase populations (i.e., cycle-traversing cells in G_1, S, and G_2) revealed that two histones were phosphorylated: f2a2 and f1 (Fig. 2A). As cells progressed from G_2 into M, three histones were phosphorylated to an appreciable extent: histones f2a2, f1, and f3, with evidence for more than one phosphorylated site on f1 (Fig. 2B) (Gurley *et al.*, 1973b, 1974a). The f1 and f3 fractions were specifically phosphorylated as cells entered mitosis and were rapidly dephosphorylated as cells reentered G_1 (Gurley *et al.*, 1974a).

Cell-cycle kinetic studies revealed that interphase-specific phosphorylation of histone f1 (also measured by preparative gel electrophoresis) commenced 2 hr prior to entry into S phase (Gurley *et al.*, 1974a, b). This observation allowed us to conclude that the events involved phosphorylation of old, preexistent f1 molecules and provided an early indication that the cells had recovered from the nonproliferating state and were preparing to replicate DNA (Gurley *et al.*, 1974a, b).

To obtain more detailed information concerning the phosphorylation events associated with cell-cycle progression, it was necessary to utilize an additional analytical technique. Patterns of phosphorylation of the various subfractions of histone f1 were examined by ion exchange column chromatography on Bio-Rex 70. As shown in Fig. 3A, this method resolves at least four distinct f1 subfractions from CHO cells. In cultures arrested in late G_1 following release of isoleucine-deficient cells into complete medium containing hydroxyurea, a single form of phosphorylated f1 was

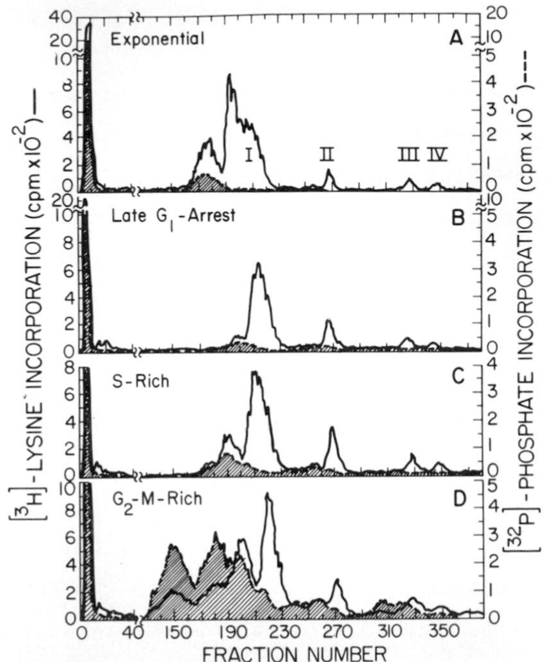

FIGURE 3 *Bio-Rex 70 ion exchange column chromatography of histone f1 from exponential and synchronized cultures of CHO cells. Cells prelabeled with lysine-^3H, as in Fig. 2 above, were pulse-labeled for 1-hr periods with 20 mCi of carrier-free $^{32}PO_4$ per liter of culture. Four different unphosphorylated f1 subfractions are indicated by Roman numerals. Specific details involving cell synchronization and isolation and fractionation of histones are presented elsewhere (Gurley et al., 1975).*

found to exist for each of the four subfractions (Fig. 3B) (Gurley *et al.*, 1975); this G_1 phosphorylation event was designated $f1_{G1}$.

When the cells in Fig. 3B entered S phase after removal of hydroxyurea, a second, more highly phosphorylated form of f1 ($f1_S$) was observed as a faster eluting shoulder in the phosphorylated $f1_{G1}$ peak (Fig. 3C). The ^{32}P:^3H ratio of the $f1_S$ peak was twice that of the $f1_{G1}$ peak (Gurley *et al.*, 1975).

Finally, in G_2/M-rich cultures, two new forms of phosphorylated f1 were observed which had ^{32}P:^3H ratios approximately four times greater than that observed in the phosphorylated $f1_{G1}$ peak

(Fig. 3D). This mitotic-specific phosphorylation event was designated $f1_M$ (Gurley *et al.*, 1975).

Detailed kinetic analysis of a number of experiments similar to those described in Fig. 3 allowed us to draw the following conclusions regarding phosphorylation events in the CHO cell (Gurley *et al.*, 1975)

(a) Two hours before initiation of DNA replication, approximately 20% of the "old" f1 histone molecules were phosphorylated ($f1_{G1}$).

(b) A second phosphorylation event ($f1_S$) began as cells entered S phase. This event involved the phosphorylation of a second site in addition to the $f1_{G1}$ phosphorylation site. Whereas the amount of phosphorylated $f1_{G1}$ accumulated in chromatin continued to increase throughout the entire cell cycle, the amount of phosphorylated $f1_S$ did not increase as cells progressed through S and G_2. Maximally, only 10% of the f1 molecule exhibited this $f1_S$ type of phosphorylation at any one time.

(c) A third phosphorylation event ($f1_M$) resulted from a burst of phosphorylation as cells entered mitosis, adding two new chromatographic fractions whose $^{32}P:^{3}H$ ratios were four-fold greater than that of phosphorylated $f1_{G1}$. In mitotic cells, all f1 molecules were phosphorylated, and $f1_M$ was specifically dephosphorylated as cells reentered G_1.

(d) In CHO cells, each of the f1 phosphorylation events associated with specific times in the cell cycle ($f1_{G1}$, $f1_S$, and $f1_M$) occurred on all four histone subfractions (I-IV).

Similarly, from studies utilizing Triton X-100 polyacrylamide gel electrophoresis, it was concluded that the mitotic specific

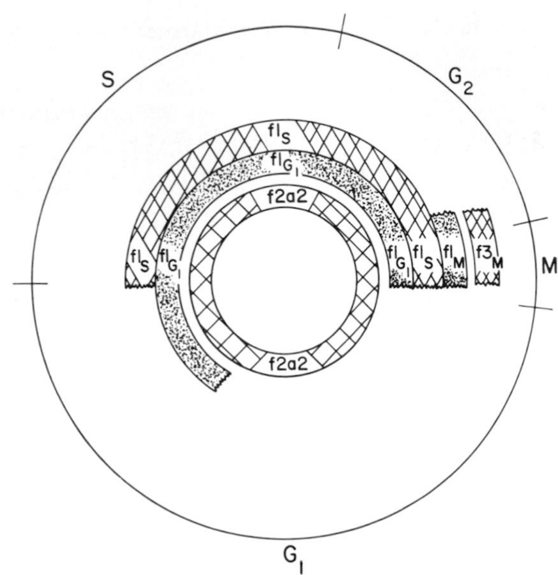

FIGURE 4 *Relationship of histone phosphorylation to the cell cycle of line CHO Chinese hamster cells. The shaded bands indicate the times in the cell cycle that histones f2a2, f1, and f3 are phosphorylated.*

$f3_M$ phosphorylation occurred on all f3 subfractions (Gurley *et al.*, 1975).

The relationship between histone phosphorylation and the cell cycle is depicted diagrammatically in Fig. 4. Whereas phosphorylation of f3 is confined to the mitotic period, histone f1 is phosphorylated at three different stages in the cell cycle, and f2a2 occurs throughout the entire cycle and also in G_1-arrested cells maintained in isoleucine-deficient medium (Gurley *et al.*, 1973a).

IV. PHOSPHORYLATION OF SPECIFIC REGIONS WITHIN THE f1 MOLECULE DURING CELL-CYCLE PROGRESSION

The phosphorylation events which occur on histone f1 during late G_1, S phase, and mitosis suggest an ordered, sequential modification of the histones as cells progress through the cell cycle. By analyzing ^{32}P-labeled phosphopeptides from purified f1 histone in synchronized CHO cells, we were able to demonstrate that distinct regions and amino acids (Hohmann *et al.*, 1975) within the f1 molecule were phosphorylated at different stages in the cell cycle.

In these studies, the f1 histone was first digested with trypsin, and the phosphopeptides released were resolved into four fractions (I-IV) by high-voltage paper electrophoresis (Hohmann *et al.*, 1975). Two of these fractions (II and III) were further resolved by a second electrophoretic step (Hohmann *et al.*, 1975). Figure 5 demonstrates that fraction II from mitotic CHO cells contains five phosphopeptides, while only one of these peptides (II_4) is found in interphase cells. Therefore, these results confirm the general conclusions reached above that phosphorylation of $f1_M$ is more complex than phosphorylation of $f1_{G1}$ or $f1_S$.

Experiments were then performed which allowed us to assign the phosphopeptides shown in Fig. 5 to distinct regions within f1 molecule. The f1 histone was first cleaved at the single tyrosine residue using N-bromosuccinimide (Bustin and Cole, 1969; Bustin *et al.*, 1969), and the N-terminal and C-terminal fragments were separated by exclusion chromatography. These two large fragments were then digested with trypsin, and the various peptides were fractionated by electrophoresis. All the peptides in fraction II were assigned to the N-terminal and C-terminal regions of the f1 molecule, as shown in Fig. 5. Other experiments (not shown) allowed us to assign a total of 16 phosphopeptides to either the N-terminal or C-terminal fragment of the f1 molecule (Hohmann, Tobey, and Gurley, in preparation). Due to ambiguous tryptic

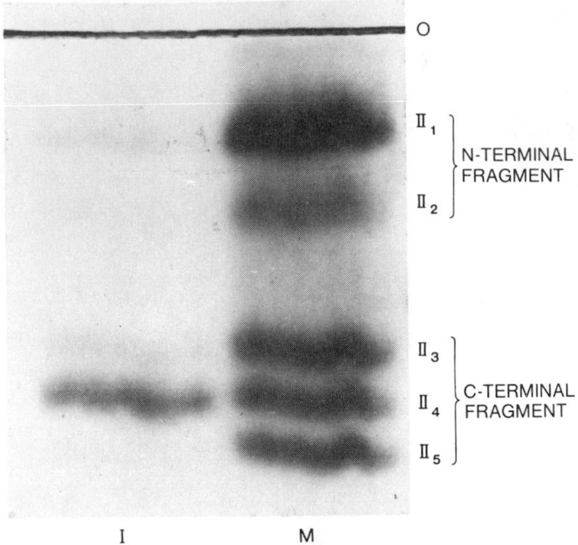

FIGURE 5 *Autoradiograph depicting histone f1 tryptic fraction II phosphopeptides found in interphase (I) and mitotic (M) cultures of CHO cells. Cells were pulse-labeled for a 2-hr period with 50 μCi of carrier-free* $^{32}PO_4$ *per liter of culture. Following purification of f1 by ion exchange chromatography, tryptic fragments were prepared and resolved into four fragments by high-voltage paper electrophoresis. One of these tryptic fragments (fraction II) (Hohmann et al., 1975) was further resolved into five phosphopeptides* (II_{1-5}) *by an additional electrophoretic step, as described previously (Hohmann et al., 1975). Localization of tryptic fraction II phosphopeptides in the N-terminal or C-terminal portion of the f1 molecule was accomplished in the manner described in the text.*

cleavages, it is possible that many of the 16 phosphopeptides are related to each other (i.e., contain the same site). Thus, the number of phosphopeptides released is *not* a true indication of the number of sites. Precise determination of the number of sites must be derived ultimately from sequence analysis.

Our previous studies with CHO cells (Hohmann *et al.*, 1975) demonstrated that threonine was phosphorylated more rapidly than serine in mitotic cells, while the reverse was true in interphase cells. Thus, the phosphopeptides from histone f1 can be distinguished not only by their position in the f1 molecule and the time in the cell cycle when they appear but also by their content

TABLE I Summary of Phosphorylation Events During the CHO Cell Cycle

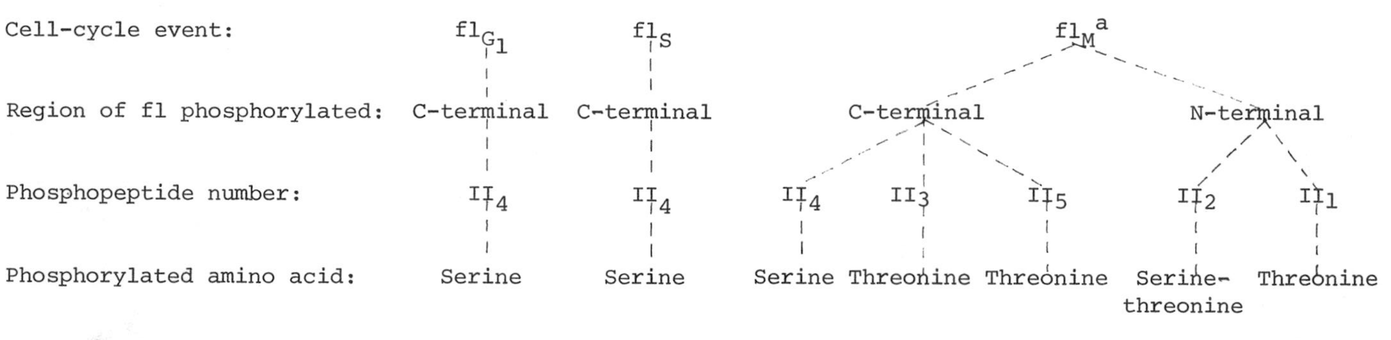

Cell-cycle event:	fl_{G_1}	fl_S	fl_M[a]				
Region of fl phosphorylated:	C-terminal	C-terminal	C-terminal			N-terminal	
Phosphopeptide number:	II_4	II_4	II_4	II_3	II_5	II_2	II_1
Phosphorylated amino acid:	Serine	Serine	Serine	Threonine	Threonine	Serine-threonine	Threonine

[a]Other peptides not shown in Fig. 5 support this assignment.

of phosphoserine and phosphothreonine. Since tryptic fraction II is representative of all these parameters, a summary of our general conclusions regarding the phosphorylation of distinct regions and amino acids in CHO f1 histone as a function of the cell cycle is shown in Table I.

Phosphorylation of f1 begins in G_1 ($f1_{G1}$) in the C-terminal part of the molecule on serine (phosphopeptide II_4). This event ($f1_{G1}$) continues into S phase. Peptides corresponding to the $f1_S$ event have not been positively identified, presumably due to the low level of this phosphorylation event. As the cells enter mitosis, phosphorylation occurs in the C-terminal region on both serine and threonine, *and*, for the first time, in the N-terminal region on both serine and threonine.

These results indicate that phosphorylation of histone f1 in interphase CHO cells involves serine in the C-terminal fragment. In contrast, serine *and* threonine residues are phosphorylated in mitotic cells in *both* C-terminal and N-terminal fragments. Thus, interphase and mitotic phosphorylation events are both qualitatively and quantitatively different.

V. CELL-CYCLE-SPECIFIC CHANGES IN CHROMATIN ORGANIZATION

Once it was demonstrated that the pattern of histone phosphorylation changed during the cell cycle, the intimate association of histones and DNA in chromatin suggested that there might be corresponding cell-cycle-specific changes in the structure of chromatin during interphase as well as mitosis. To explore this notion, we initiated a study with the chemical agent, heparin (a natural polyanion), which interacts predominantly with the histone component of chromatin (Arnold *et al.*, 1972). Treatment of chromatin or isolated nuclei with heparin causes histones to be removed in a well defined sequence (lysine-rich f1 is released preferentially) (Hildebrand *et al.*, 1975). Concomitant with the

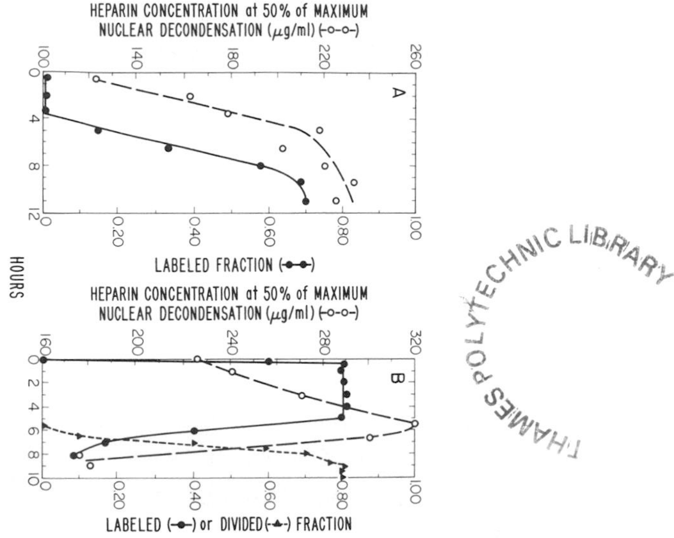

FIGURE 6 *Cell-cycle dependence of heparin-mediated release from nuclei of prelabeled DNA. (A) Early interphase: cells synchronized by mitotic selection, and (B) late interphase: cells synchronized by mitotic selection/hydroxyurea treatment. Experimental details are described elsewhere (Hildebrand and Tobey, 1975).*

removal of histones with heparin, the DNA is released in a highly decondensed state (Arnold *et al.*, 1972). Thus, changes in the amount of heparin which will decondense a given amount of DNA in a specific time reflect alterations in the organization of chromatin related to the arrangement of histones on the DNA.

The cell-cycle kinetics of chromatin structural changes were measured in CHO cells traversing early interphase by determining the concentration of heparin required to release 50% of the DNA from nuclei during a 5-min incubation period (Fig. 6A) (Hildebrand and Tobey, 1975). The resistance of chromatin to heparin-mediated DNA decondensation increased as the cells progressed through G_1 into S. Similarly, this resistance continued increasing through S and G_2 but then fell abruptly when the cells divided (i.e., reentered G_1) (Fig. 6B). Taken together, the data in Fig. 6 suggest a continuing change throughout the cell cycle in the susceptibility of chromatin-bound histones to interaction with heparin.

VI. STATE OF THE CHO CELL CYCLE: 1975

When the results presented in this report are combined with data from our previous studies (Tobey *et al.*, 1974, 1975), it is possible to construct a highly detailed map illustrating the temporal arrangement of biochemical events during the cell cycle (Fig. 7). Our results strongly suggest that, in our cultured cell system, proliferation capacity depends on a complex series of coordinated events. That is, even cells grown in the artificial environment of tissue culture are subject to specific regulatory processes which ensure an orderly sequence of operations in preparation for DNA replication and cell division.

Inhibition of DNA replication does not necessarily prevent initiation of other preparative events, since the phosphorylation of histone f1 (Gurley *et al.*, 1974b), the increase in lipoprotein-associated DNA (Hildebrand and Tobey, 1973), and the elevation in level of three of the deoxyribonucleoside triphosphates (Walters *et al.*, 1973) are initiated sequentially in cultures traversing G_1 in the presence of hydroxyurea. That is, those processes normally prerequisite to initiation of genome replication are "switched on" in proper order at the appropriate time, even when DNA synthesis is inhibited. Detailed discussions of the interrelationships between events in the CHO cell cycle (Fig. 7) are presented elsewhere (Tobey *et al.*, 1974, 1975).

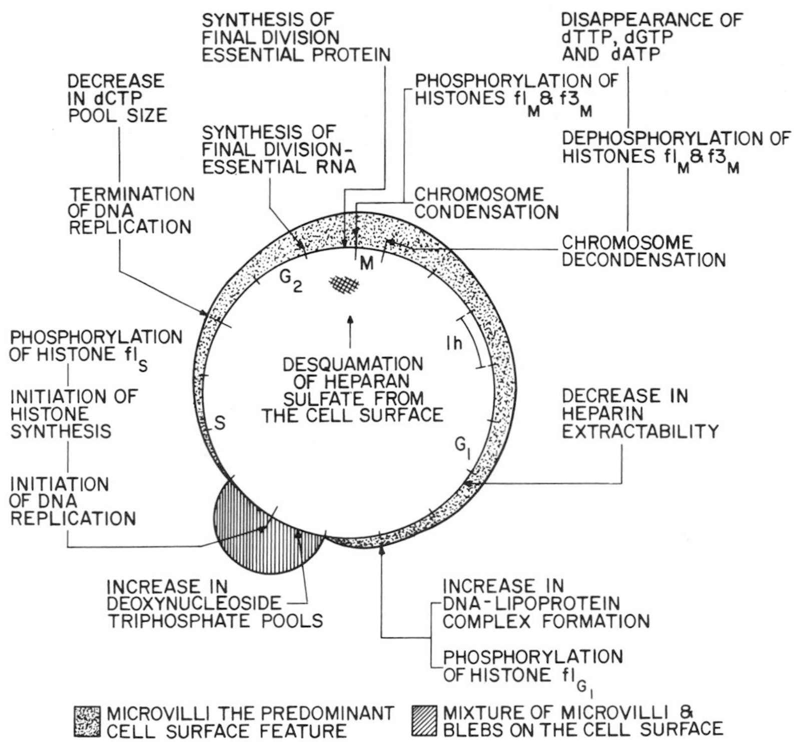

FIGURE 7 *Temporal sequence of biochemical events in the CHO cell cycle.*

VII. IMPLICATIONS

In 1963, Mazia (1963) speculated that the condensation of chromosomes observed during mitosis represented only a small, readily visible portion of a dynamic "chromosome cycle" which extended throughout both interphase and mitosis. Recent studies utilizing chemical probes such as actinomycin and deoxyribonuclease (Pederson, 1972; Pederson and Robbins, 1972), sarcosyl-Mg^{++} crystals (Hildebrand and Tobey, 1973; Yamada and Hanoaka, 1973), heparin (Hildebrand and Tobey, 1975; Hildebrand *et al.*, 1975), and ethidium bromide (Nicolini *et al.*, 1975) all tend to support the concept of a nonstatic structure of chromatin which changes as cells progress around the cell cycle. If reversible phosphorylation of histones represents a major mechanism for altering chromatin structure (Allfrey, 1971; Bradbury and Crane-Robinson, 1971), then our studies demonstrating distinct cell-cycle-specific patterns of phosphorylation suggest that this histone modification may be responsible for introducing at least some of the observed cell-cycle-dependent variation in chromatin structure discussed above.

Support for the notion that histone phosphorylation represents a mechanism for altering chromatin structure comes from the studies of Louie and Dixon (1973) who correlated specific histone modifications with chromatin structural changes during spermatogenesis in trout. Also, the model studies of Adler *et al.* (1971, 1972) support this concept by demonstrating that specific multiple phosphorylations of f1 alter the ability of f1 to change the conformation of DNA.

Our findings agree with this concept by demonstrating that changes in chromatin structure occur during interphase and mitosis and that, coincident with these changes, histones are phosphorylated in a highly ordered manner, exhibiting cell-cycle specificity for both the amino acid residues and the sites phosphorylated. Similar correlations have been made in studies by Marks *et al.* (1973) and by Bradbury and his associates (1973, 1974) regarding histone f1 phosphorylation and changes in chromatin structure. In view of these studies and our own investigations, we consider the orderly progression of specific histone modifications to play an essential role in the structure and function of chromatin as cells progress through the proliferation cycle.

REFERENCES

Adler, A. J., Schaffhausen, B., Langan, T. A., Fasman, G. D., (1971), *Biochemistry 10*, 909-913.

Adler, A. J., Langan, T. A., Fasman, G. D., (1972), *Arch. Biochem. Biophys. 153,* 769-777.

Allfrey, V. G., (1971), *in* "Histones and Nucleohistones," (D. M. Phillips, ed.), pp. 241-294, Plenum Publishing Co., New York.

Arnold, E. A., Yawn, D. H., Brown, D. G., Wyllie, R. C., Coffey, D. S., (1972), *J. Cell Biol. 53,* 737-757.

Baserga, R., (1968), *Cell Tissue Kinet. 1,* 167-191.

Bradbury, E. M., Crane-Robinson, C., (1971), *in* "Histones and Nucleohistones," (D. M. Phillips, ed.), pp. 85-134, Plenum Publishing Co., New York.

Bradbury, E. M., Carpenter, B. G., Rattle, H. W. E., (1973), *Nature 241,* 123-126.

Bradbury, E. M., Inglis, R. J., Matthews, H. R., (1974), *Nature 247,* 257-261.

Bustin, M., Cole, R. D., (1969), *J. Biol. Chem. 244,* 5291-5294.

Bustin, M., Rall, S. C., Stellwagen, R. H., Cole, R. D., (1969), *Science 163,* 391-393.

Engelberg, J., (1964), *Exp. Cell Res. 36,* 647-662.

Gurley, L. R., Walters, R. A., Tobey, R. A., (1973a), *Arch. Biochem. Biophys. 154,* 212-218.

Gurley, L. R., Walters, R. A., Tobey, R. A., (1973b), *Biochem. Biophys. Res. Commun. 50,* 744-750.

Gurley, L. R., Walters, R. A., Tobey, R. A., (1974a), *J. Cell Biol. 60,* 356-364.

Gurley, L. R., Walters, R. A., Tobey, R. A., (1974b), *Arch. Biochem. Biophys. 164,* 469-477.

Gurley, L. R., Walters, R. A., Tobey, R. A., (1975), *J. Biol. Chem. 250,* 3936-3944.

Hildebrand, C. E., Tobey, R. A., (1973), *Biochim. Biophys. Acta 331,* 165-180.

Hildebrand, C. E., Tobey, R. A., (1975), *Biochem. Biophys. Res. Commun. 63,* 134-139.

Hildebrand, C. E., Gurley, L. R., Tobey, R. A., Walters, R. A., (1975), *Federation Proc. 34,* 581 (Abstract #2049).

Hohmann, P., Tobey, R. A., Gurley, L. R., (1975), *Biochem. Biophys. Res. Commun. 63,* 126-133.

Kraemer, P. M., Tobey, R. A. (1972). *J. Cell Biol. 55,* 713-717.

Louie, A. J., Dixon, G. H., (1973), *Nature 243,* 164-168.

Marks, D. B., Paik, W. K., Borun, T. W., (1973), *J. Biol. Chem. 248,* 5660-5667.

Mazia, D., (1963), *J. Cell. Comp. Physiol. 62* (Suppl. 1), 123-140.

Mitchison, J. M., (1971), "The Biology of the Cell Cycle," Cambridge University Press, London.

Nicolini, C., Ajiro, K., Borun, T. W., Baserga, R., (1975), *J. Biol. Chem. 250,* 3381-3385.

Pederson, T., (1972), *Proc. Natl. Acad. Sci. U.S.A. 69,* 2224-2228.

Pederson, T., Robbins, E., (1972), *J. Cell Biol. 55,* 322-327.

Petersen, D. F., Tobey, R. A., Anderson, E. C., (1969), *Federation Proc. 28,* 1771-1778.

Tobey, R. A., (1973), *in* "Methods in Cell Biology," (D. M. Prescott, ed.), Vol. VI, pp. 67-112, Academic Press, New York.

Tobey, R. A., Crissman, H. A., (1972a), *Exp. Cell Res. 75,* 460-464.

Tobey, R. A., Crissman, H. A., (1972b), *Cancer Res. 32,* 2726-2732.

Tobey, R. A., Ley, K. D., (1971), *Cancer Res. 31,* 46-51.

Tobey, R. A., Anderson, E. C., Petersen, D. F., (1967), *J. Cell. Physiol. 70,* 63-68.

Tobey, R. A., Gurley, L. R., Hildebrand, C. E., Ratliff, R. L., Walters R. A., (1974), *in* "Control of Proliferation in Animal Cells," (B. Clarkson and R. Baserga, eds.), pp. 668-679, Cold Spring Harbor Press, New York.

Tobey, R. A., Gurley, L. R., Hildebrand, C. E., Kraemer, P. M., Ratliff, R. L., Walters, R. A., (1975), *in* "Mammalian Cells: Probes and Problems," (C. R. Richmond, D. F. Petersen, P. F. Mullaney, E. C. Anderson, eds.), pp. 152-167, National Technical Information Service, Springfield, Virginia.

Walters, R. A., Tobey, R. A., Ratliff, R. L., (1973), *Biochim. Biophys. Acta 319,* 336-347.

Yamada, M., Hanoaka, F., (1973), *Nature (New Biology) 243,* 227-230.

AGGREGATE CULTURES: A MODEL FOR STUDIES OF BRAIN DEVELOPMENT

Nicholas W. Seeds
Galo Ramirez
Michael J. Marks

Department of Biophysics and Genetics
University of Colorado Medical Center
Denver, Colorado

I. INTRODUCTION

Brain is a very complex and heterogenous tissue whose morphogenesis reflects extensive cell migration and specific cell associations. During development brain cells must possess both self-recognition properties for interaction with homologous cells and complementary recognition mechanisms for specific cell interactions with heterologous cell populations. These cell recognition capabilities display temporal alterations during development suggesting concomitant changes in the cell surface (Gottlieb *et al.*, 1974; Moscona, 1974; Seeds, 1975). Furthermore normal brain function requires that specialized intercellular associations, such as those related to synaptogenesis and myelination, be formed in a very selective manner. Thus these recognition mechanisms regulated by the cell genome (Sidman, 1974) and the cell surface (Moscona, 1974) are necessary developmental events.

Ideally a culture system for studying the cellular and molecular aspects of brain development should retain the specificity of cell interaction found in *situ* and display morphological, biochemical and electrical differentiation similar to normal brain maturation. Furthermore, it should be possible to manipulate the numbers and types of cells in a rigorously controlled culture environment.

II. AGGREGATION

Embryonic cells possess a very selective adhesiveness that allows them to distinguish like from unlike cell types (Moscona, 1973). This property of fetal cells permits dissociated single cells to reaggregate into multicellular structures where specific cell-cell contacts can be reestablished. Formation of these aggregates is enhanced by gently rotating suspensions of trypsin dissociated cells to increase the number of collisions between sticky cells while minimizing the shearing forces. Aggregation is influenced by the tissue of origin, the state of differentiation, cell specific aggregation factors, as well as environmental factors including rotation speed, culture medium, temperature and serum proteins (Moscona, 1973; Seeds, 1973). Aggregation is a rapid process being completed in a matter of hours. The initial aggregates are composed of loosely packed and randomly dispersed cells (Fig. 1a). Numerous small cellular extensions are present by one day of culture and after several days there are extensive cellular outgrowths that serve to stabilize the complex.

During the first week of culture the cells undergo extensive migration and segregate into clusters of similar cell types (Fig. 1b). In addition reaggregates prepared from specific brain regions show distinctive patterns of aggregation. Cerebral cortex aggregates display a cell poor peripheral region composed largely of outwardly directed cellular extensions from cells whose peri-

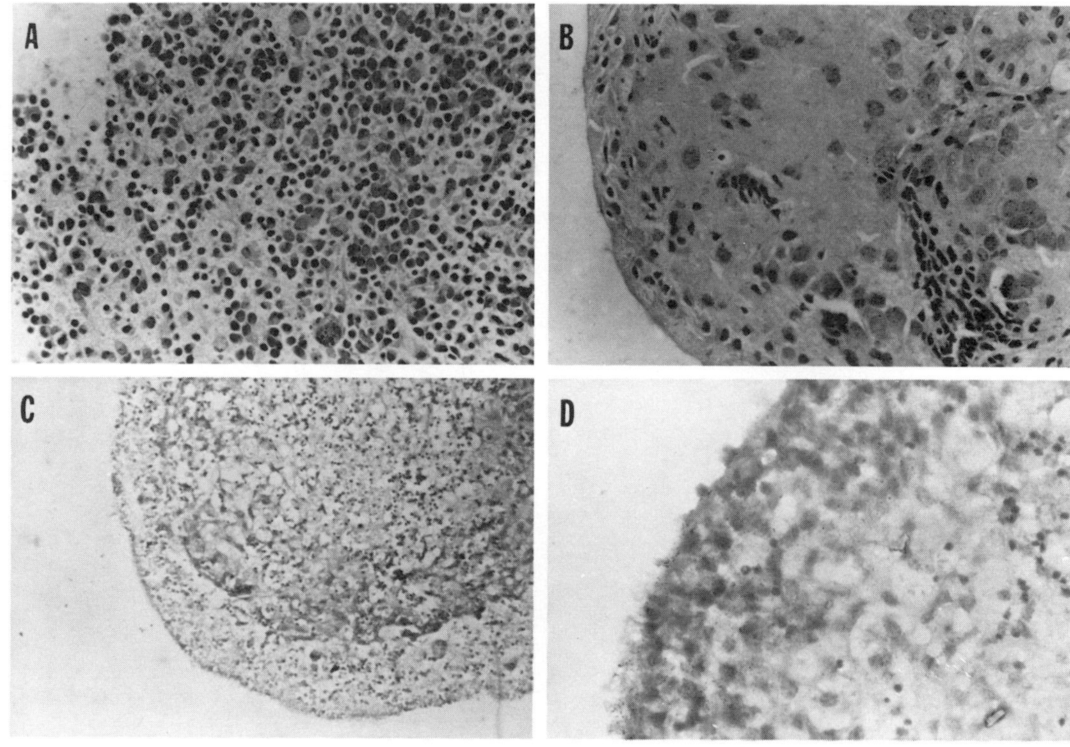

FIGURE 1 *Histogenesis in Aggregate Cultures. Light microscopic sections of whole brain reaggregate cultures at one day (a) and twelve days (b) of culture. Similar sections were prepared from cerebral cortex (c) and cerebellar cortex reaggregates (d) after 8 days of culture. The magnification of 1c is only 40% that of the others.*

karyon is more centrally located (Fig. 1c). In contrast cerebellar aggregates have a cell dense periphery of large cells whose extensions are directed internally or laterally (Fig. 1d). The identity of these exterior cells in cerebellar aggregates is not complete; however, many of these cells can concentrate ^{3}H-GABA from the medium and fetal cells that undergo their last nuclear division on gestational day 13, namely Purkinje and Golgi II neurons, are only found in this region; whereas, later developing cells principally occupy more internal positions. More impressive is the histotypic pattern formation demonstrated by DeLong (1970) for pyramidal neurons in hippocampal aggregates and the lack of this cellular alignment in aggregates of cells from the "reeler" mutant mouse (DeLong and Sidman, 1970).

III. MORPHOLOGICAL DEVELOPMENT

The ability of dissociated cells to reproduce histotypic cellular patterns suggested that these cultures may show ultrastructural and biochemical differentiation similar to that of developing mouse brain. The two most characteristic features of brain maturation are synaptogenesis and myelination. Both of these events occur late in brain differentiation; most mouse brain synapses form during the third postnatal week and myelination is most active during the fourth week (Aghajanian and Bloom, 1967; Karlsson, 1967). The tissue we most often use to prepare aggregate cultures, 16 day old fetal mouse brain, has very few synapses and these are generally restricted to the spinal cord (Crain *et al.*, 1968).

The initial reaggregates are loosely associated and lack any membrane specialization suggestive of synaptic-like structures. After four days of culture numerous symmetrical intercellular complexes or desmosomes are found (Seeds and Haffke, 1976). The first synapses are seen at 7 to 9 days of culture and they both

TABLE I Synaptogenesis in Aggregate Cultures[a]

Days in culture	Synapses/field
1	0
7	0
9	3.7
15	4.3
21	6.7
25	10
33	11
49	8.3

[a]Aggregate cultures were prepared from dissociated brain cells of 16 day mouse fetuses. Aggregates were fixed, embedded and sectioned for electron microscopy at several day intervals. The sections were observed at a constant magnification (43 μ^2 field) and photographs taken of fields containing at least one synapse. The number of synapses per field was determined from about ten photographs at each time point.

increase in number and mature in appearance during subsequent culture (Seeds and Vatter, 1971). Synaptogenesis reaches a peak at 4-5 weeks of culture (Table I). A typical mature synaptic complex containing many clear vesicles and encircled by numerous electron opaque glial cells with abundant glial filaments and glycogen granules is shown in Fig. 2.

During culture the reaggregated cells show increased amounts of rough endoplasmic reticulum and golgi complexes. Myelinated axons are present by the fourth week of culture (Seeds and Vatter, 1971). Furthermore, studies in our laboratory by Schmidt (1975) have demonstrated increased sulfatide synthesis during chick brain aggregate development.

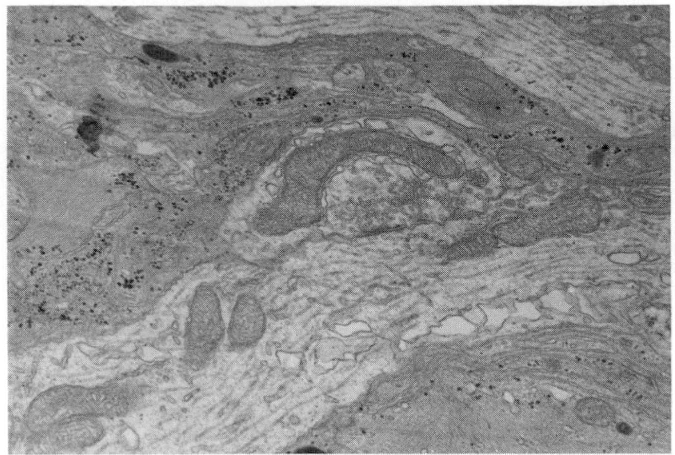

FIGURE 2 *Ultrastructure of a Brain Aggregate. A clear vesicle containing synaptic complex surrounded by more electron dense glial processes is present.*

IV. BIOCHEMICAL DEVELOPMENT

A. Neuronal Development

One of the main reasons for choosing the mouse as the source of tissue was the limited development of biochemical activities related to neurotransmission in the fetal mouse brain. The greatest development occurs postnatally and the 16 day fetal brain cells are relatively undifferentiated. However, when cultured as reaggregates these cells show marked increases in specific activities of three neuronal enzymes: choline acetyltransferase (EC 2.3.1.6), acetylcholinesterase (EC 3.1.1.7) and glutamate decarboxylase (EC 4.1.1.15) (Fig. 3). The greatest increase in these activities occurs during the second week of culture, thus resembling the *in situ* development in both rate and extent (Seeds, 1973). Cultures maintained for as long as three months possess high levels of enzyme activity often surpassing the specific activities found in adult mouse brain. Furthermore, these enzymes appear

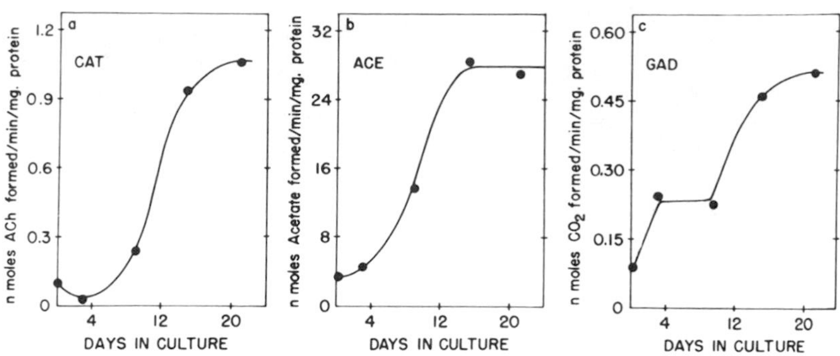

FIGURE 3 *Biochemical Differentiation in Aggregate Culture. The development of choline acetyltransferase (a), acetylcholinesterase (b), and glutamate decarboxylase (c) are shown as a function of time in culture. The starting undissociated brain tissue has activities of 0.18, 11, and 0.13 respectively (from Seeds, 1971).*

functional in the aggregates, since incubation with radiolabeled choline and glutamate leads to accumulation in the aggregates of acetylcholine and γ-aminobutyrate (Fig. 4).

In addition to the presynaptic cholineacetyltransferase activity, we have examined the postsynaptic component of cholinergic development. Development of the muscarinic receptor in reaggregates has been followed by specific binding of ^{3}H-guinuclidinyl benzilate (Yamamura and Synder, 1974) to particulate fractions of reaggregate homogenates. The developmental profile of the muscarinic receptor is similar to that of cholineacetyltransferase activity but somewhat slower in rate of appearance, reaching a maximum expression of receptor after 24 days in culture (Seeds and Haffke, 1976). Thus cholinergic development in these reaggregate cultures reaches maximum levels prior to synaptogenesis; however, since we cannot at present distinguish microscopically the cholinergic synapses, it is possible that cholinergic synapses may be formed first and the slower development reflects other types of synapses.

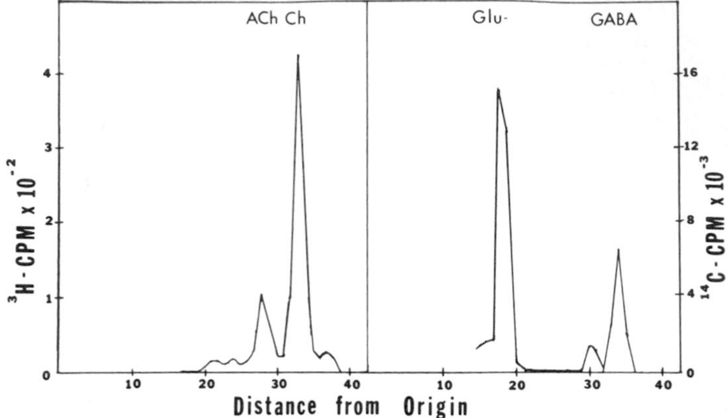

FIGURE 4 *Neurotransmitter Synthesis in Reaggregate Cultures. Homogenates from cultures incubated with ^{3}H-choline or ^{14}C-glutamate were subjected to high voltage electrophoresis at pH 1.9 to separate acetylcholine (ACh) and γ-aminobutyrate (GABA) from their precursors.*

B. Brain Development

Reaggregated brain cell cultures also acquire enzymatic activities found in cells other than neurons (Seeds, 1975a: Marks and Seeds, 1975). Some activities such as monoamine oxidase (EC 3.6.1.3) increase 3 to 10 fold during brain development; whereas, other enzymes such as catechol-O-methyltransferase (EC 2.1.1.6) show little change. Similar developmental increases are found in reaggregate cultures (Table II). There is a shift in the relative amount of lactate dehydrogenase isoenzymes during brain development; this transition is also seen in the aggregates (Seeds, 1975a). In addition, the cultures express the brain specific S-100 protein that is found relatively late in development and is concentrated in glial cells. However, the maximal level of S-100 is only 25% that found in adult brain, and possibly reflects the limited (<5%) division of cells in reaggregate cultures.

TABLE II Development of Biochemical Activities[a]

Days in culture	Specific activity MAO	LDH	COMT	Na^+, K^+ ATPase	S-100
0	0.34	367	0.065	5.0	<0.01
10	1.05	658	0.080	24.7	1.9
20	1.22	932	0.075	61.8	1.8

[a]Reaggregated brain cell cultures from 16 day fetal mice were maintained in basal Eagle's medium with 10% fetal calf serum. Enzyme specific activities are expressed as n moles product/min/mg protein. S-100 protein is expressed as μg S-100 per mg soluble protein.

TABLE III Biochemical Development versus Aggregate Size[a]

	Cells	ACE	CAT	MAO	LDH	S-100
	Dissociated cells (5-50 μ)	2.5	0.08	0.30	367	<0.01
I	Big aggregates (1050 μ diam)	36.0	0.84	1.05	985	-
I	Small aggregates (365 μ diam)	25.0	0.38	1.37	878	-
II	Big aggregates (1066 μ diam)	25.0	0.54	-	-	0.7
II	Small aggregates (510 μ diam)	11.0	0.22	-	-	0.6

[a]The biochemical activity of reaggregated brain cells cultured for 15 days in basal Eagle's medium on gyrotary shaker baths is compared to that of the initial dissociated cells. The speed of rotation was 70 or 90 rpm to produce aggregates of two distinct size classes, with the average aggregate diameter given in parenthesis.

TABLE IV Potassium Induction of Na^+, K^+-ATPase[a]

Days in culture	Specific activity	
	5mM	25mM
2	11	7
4	10	8
5	17	21
7	19	28
9	25	36
11	29	39
17	51	60
21	59	69

[a]Aggregate brain cell cultures from 16 day fetal mice maintained in basal Eagle's medium (5.4mM K^+) or in medium where some of the $NaHCO_3$ had been replaced with $KHCO_3$ to 25mM K^+. The enzyme specific activity is expressed as n moles Pi released/min/mg protein and standard deviations were 3-8%.

The appearance of neurohormone receptors on the surface of brain cells is a characteristic feature of development. One such molecule is the β-adrenergic receptor that is closely linked to adenyl cyclase and appears between postnatal day 4 and 10 (Schmidt *et al.*, 1970). Binding of norepinephrine to this receptor leads to a rapid 5 fold increase of cyclic AMP levels in brain slices. Although newly formed aggregates show no response to catecholamines, one week old aggregate cultures show a similar 4-6 fold response in cellular cAMP levels (Seeds and Gilman, 1971). This response probably represents the formation or activation of cell surface β-adrenergic receptors.

The biochemical development of these aggregate cultures is sensitive to a variety of environmental factors, including ion levels, drugs and culture medium composition (Marks and Seeds, 1977; Seeds, 1976c). Aggregate size is also an important variable (Table III). Small aggregates (200-500 μ) show greatly reduced levels of choline acetyltransferase and acetylcholinesterase activity compared to normal (900-1100 μ) size aggregates. Furthermore, only the neuronal activities are affected, monoamine oxidase, lactate dehydrogenase and S-100 develop normally, thus suggesting that a critical cell mass or cell interactions are necessary for maximal neuronal differentiation; whereas, glial development may not be so restrictive.

C. ATPase Induction

One of the major advantages for studying development in cell culture is the opportunity to control the environment. Regulation of the ionic balance between intra- and extracellular fluids is critical for functional activity of the nervous system and this regulation is performed by the energy dependent Na^+, K^+ ATPase in brain, we have examined the influence of extracellular ions on this enzyme during development (Marks and Seeds, 1977). Chronic exposure to elevated K^+ during culture leads to an induction of the Na^+, K^+-ATPase activity. Table IV shows the development of the enzyme activity in normal medium, 5.4mM K^+ and in medium containing 25 mM K^+. After a lag of several days there is a 50% increase in the activity by day 7 and this increment is retained during further development. Furthermore, this stimulation increases linearly as the K^+ is raised.

Additional studies using ouabain to inhibit the Na^+, K^+-ATPase in the aggregate cultures show a similar induction of enzyme activity with exponential increases between 10^{-5} and 2×10^{-4}M ouabain. Collectively these findings suggest that changes in membrane potential may regulate the synthesis of specific cell proteins necessary for controlling the membrane potential; thus, increased bioelectric activity during brain maturation may promote the observed developmental increases in Na^+, K^+-ATPase of brain.

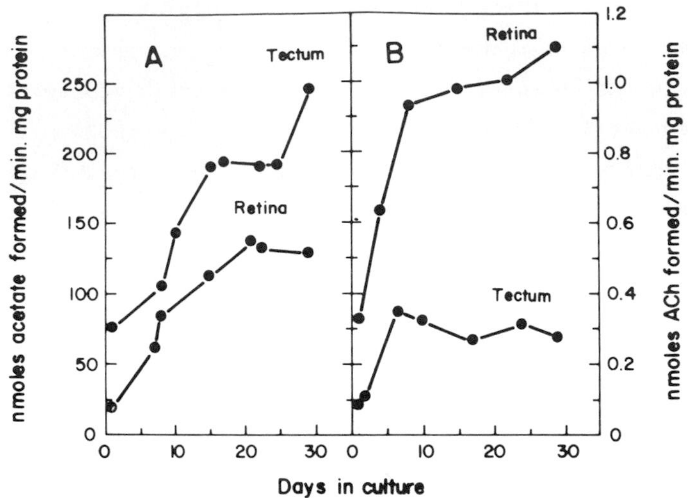

FIGURE 5 *Development of Acetylcholinesterase (A) Choline Acetyltransferase (B) Activity in Tectum and Retina Aggregates. Aggregate cultures were prepared from seven day old chick embryos and enzyme activity was determined at periodic intervals (from Ramirez and Seeds, 1977).*

D. Cell Interaction

The specificity of cell recognition displayed by retina and optic tectum cells of the chick is well documented (Barbera *et al.*, 1973; Merrell and Glaser, 1973; Gottlieb *et al.*, 1974). These tissues show both regional and temporal specificity in their interaction. For these reasons the retino-tectal system is an attractive candidate for studying the relationship of cell recognition events and biochemical development (Ramirez and Seeds, 1977). Both chick retina and optic tectum show large increases in choline acetyltransferase and acetylcholinesterase specific activities from fetal day 7 to ten days after hatching. The specific activity of acetylcholinesterase in aggregate cultures prepared from 7 day chick embryo retina and tectum increases 3-6 fold during four weeks of culture (Fig. 5A). The maximal activity is 55-75% of the *in situ* activity. In retina aggregates the choline acetyltransferase activity increases 4 fold to 55% of the

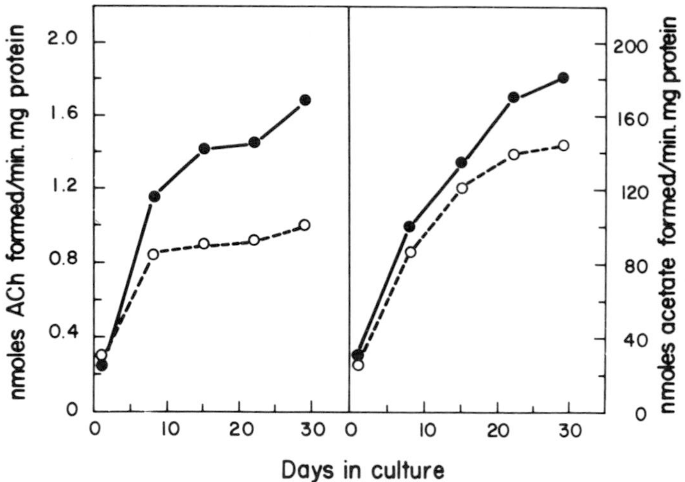

FIGURE 6 *Biochemical Development of Retinal-Tectal CoAggregates. Seven day old chick embryo retina and tectum cells were mixed and allowed to coaggregate, then assayed for choline acetyltransferase and acetylcholinesterase activity. The culture contained 87% retina and 13% tectum cells. The dashed line (open circle) represents the expected developmental values for the above cell ratio determined from Figure 5 (from Ramirez and Seeds, 1977).*

in situ levels; however, optic tectum aggregates show only a slight increase, 7% of the activity in a 10 day old chick tectum (Fig. 5B).

The influence of cell recognition events on development of these cholinergic activities was examined by preparing retinotectal coaggregates from dissociated retina and tectum cells. The development of these coaggregates is shown in Fig. 6. There is a 100% synergistic effect on choline acetyltransferase development and a 30% stimulation of acetylcholinesterase specific activity. In addition, this synergism requires cell-cell contact since coculture of retinal and tectal aggregates does not lead to an increase in enzyme activity. Furthermore, the synergistic effect shows temporal specificity, requiring the two cell populations to be of identical embryonic age (Ramirez and Seeds, 1977). These results suggest that cell recognition events in aggregate cultures and the resulting cell-cell interaction can lead to specific biochemical changes in the participating cells. The possibility that this synergism reflects the formation of specific retino-tectal synapses in the coaggregates is currently being investigated.

Thus, reaggregates offer a useful cell culture system for studies of development, especially where three dimensional cell-cell interactions may be required. Hopefully these cultures will provide new information on the role of cell recognition and cell-cell interaction in the regulation of morphogenesis and biochemical differentiation.

ACKNOWLEDGMENTS

We gratefully acknowledge the excellent technical assistance of Susan Haffke and Barbara Fonda, and wish to thank Sally Planalp for preparing the manuscript. The studies reported here were supported in part by U.S. Public Health Service Grants NS-09818 and CA-15549 and a Career Development Award K4-GM40170.

REFERENCES

Aghanjanian, G. K., Bloom, F. E., (1967), *Brain Res. 6,* 716-727.

Barbera, A. J., Marchese, R. B., Roth, S., (1973), *Proc. Nat. Acad. Sci. U.S.A. 70,* 2482-2486.

Crain, S. M., Peterson, E. R., Bornstein, M. B., (1968), *in* "Growth of the Nervous System," Ciba Symposium, pp. 13-40, Little, Brown, Boston.

DeLong, G. R., (1970), *Develop. Biol. 22,* 563-583.

DeLong, G. R., Sidman, R. L., (1970), *Develop. Biol. 22,* 584-600.

Gottlieb, D. E., Merrell, R., Glaser, L., (1974), *Proc. Nat. Acad. Sci. U.S.A. 71,* 1800-1802.

Karlsson, U., (1967), *J. Ultrastruct. Res. 17,* 158-175.

Marks, M. J., Seeds, N. W., (1977), *J. Biol. Chem. (in press).*

Merrell, R., Glaser, L., (1973), *Proc. Nat. Acad. Sci. U.S.A. 70,* 2794-2798.

Moscona, A. A., (1973), *in* "Cell Biology in Medicine," (E. Bittar, ed.), pp. 571-591, J. Wiley, New York.

Moscona, A. A., (1974), *in* "The Cell Surface in Development," (A. A. Moscona, ed.), pp. 67-99, J. Wiley, New York.

Ramirez, G., Seeds, N. W., (1977), *Nature* (in press).

Schmidt, G. L., (1975), *Brain Res. 87,* 110-113.

Schmidt, J. J., Palmer, E. C., Dettbarn, W. D., Robison, G. A., (1970), *Develop. Biol. 3,* 53-67.

Seeds, N. W., (1971), *Proc. Nat. Acad. Sci. U.S.A. 68,* 1858-1861.

Seeds, N. W., (1973), *in* "Tissue Culture of the Nervous System," (G. Sato, ed.), pp. 35-53, Plenum Press, New York.

Seeds, N. W., (1975a), *J. Biol. Chem. 250,* 5455-5458.

Seeds, N. W., (1975b), *Proc. Nat. Acad. Sci. U.S.A. 72,* 4110-4114.

Seeds, N. W., (1976), *in* "Molecular and Cellular Analysis of Mental Disorders," (H. Matthaei, ed.), Springer-Verlag, Berlin, (in press).

Seeds, N. W., Gilman, A. G., (1971), *Science 174,* 292.

Seeds, N. W., Haffke, S. C., (1976), submitted for publication.

Seeds, N. W., Vatter, A. E., (1971), *Proc. Nat. Acad. Sci. U.S.A. 68,* 3219-3222.

Sidman, R. L., (1974), *in* "The Cell Surface in Development," (A. Moscona, ed.), pp. 221-253, J. Wiley, New York.

Yamamura, H. I., Synder, S. H. (1974), *Proc. Nat. Acad. Sci. U.S.A. 71,* 1725-1729.

DIFFERENTIATION OF T CELLS *IN VITRO*

GIDEON GOLDSTEIN

Memorial Sloan-Kettering Cancer Center
New York, New York

Thymus derived lymphocytes (T cells) represent a population of cells that is replenished by differentiation of precursor cells, even in the adult organism. Thus a lineage has been traced whereby thymocyte precursors originate in the hemopoetic tissue and migrate to thymus, differentiating therein to thymocytes. Thymocytes in turn give rise to a number of subclasses of more mature T cells which can be detected in the thymic medulla (Cantor and Boyse, 1975), and which leave the thymus to form the T cell population of the lymphoid tissues, blood and lymphatics.

The differentiation of thymocytes from precursor cells can be induced *in vitro* (Komuro and Boyse, 1973) and is monitored by the rapid appearance on the cell surface of differentiation alloantigens. Specific alloantisera generated between inbred or congenic mouse strains identify differentiation alloantigens on the T cell surface, the reaction being measured by the cytotoxicity test. The cell surface of the thymocyte can be characterized in appropriate inbred strains as having the following phenotype: TL^+, $Thy\text{-}1^+$, $Ly\text{-}1^+$, $Ly\text{-}2^+$, $Ly\text{-}3^+$, $Ly\text{-}5^+$ and G_{IX}^+.

These antigens are not present on cells found in low density fractions of murine spleen or bone marrow, and Komuro and Boyse (1973) showed that a proportion of this population of cells could be induced to display these antigens after a brief incubation period *in vitro* with certain tissue extracts. This constitutes the Komuro-Boyse assay and is a model system for T cell differentiation *in vitro*.

I. THE KOMURO-BOYSE ASSAY

An enriched population of inducible cells was prepared by centrifuging spleen cells in a discontinuous density gradient of bovine serum albumin. Inducible cells, which have a lower density than the majority of spleen cells, accumulated in a less dense fraction termed the B layer. B layer cells were incubated with an inducing agent (see below); after two hours approximately 25% of the cells developed T cell alloantigens on their surface, as detected by the cytotoxicity test.

Studies utilizing this assay of *in vitro* differentiation of T cells have revealed a number of important points. First, the nature of the alloantigen expressed depends on the genetic background of the mouse from which the inducible cells were obtained. Thus, as stated above, the cells from an appropriate strain will manifest TL, Thy-1, Ly-1, Ly-2, Ly-3 and G_{IX}. However, cells from a TL^- mouse, for example, cannot be induced to develop TL. Thus the inducing agent cannot be responsible for the genetic information incorporated in the manifested antigens; rather it must act as a trigger to initiate manifestation of a pre-existing genetic program within the precursor cell.

The action of inducing agents is likely mediated by a cyclic AMP second message (Scheid *et al.*, 1975a). Cyclic AMP itself is an effective inducing agent and drugs such as theophylline and immidazole, which modify intra-cellular levels of cyclic AMP, appropriately effect the action of other inducing agents. The finding that cyclic AMP mediates induction of T cell differentiation further emphasizes that the inducing agent must act as a trigger and does not itself carry instructions for details of the induced differentiation program, which is contained within the inducible cells.

Many polypeptide hormone actions are mediated by a cyclic AMP second message. In these cases the action of the physiological hormone can be mimicked by other agents which elevate intracellular cyclic AMP. Thus it would not be surprising to find that the action of the physiological hormone which induces T cell differentiation could be similarly mimicked, and this has indeed been found to be the case. Scheid *et al.* (1973) reported that a large number of substances were capable of inducing T cell differentiation *in vitro* in the Komuro-Boyse assay. These included epinephrine, isoproterenol, poly AU and bacterial endotoxin. The effect of a putative thymic hormone preparation termed thymosin fraction V (A. L. Goldstein *et al.*, 1972) was of particular interest. This extract was active in inducing T cell differentiation but similarly prepared extracts of spleen or muscle were also effective. Clearly inducing activity in the Komuro-Boyse assay was not a sufficient criterion for defining a putative thymic hormone. A further criterion was formulated; a thymic hormone should be active in the Komuro-Boyse assay but should not trigger differentiation of a cyclic AMP mediated inductive event in another cell type.

A parallel assay of B cell differentiation was developed to serve as a control. This involved the differentiation of B cells lacking complement receptors to cells bearing complement receptors (termed CR^+ B cells) (Goldstein *et al.*, 1975, Scheid *et al.*, 1975 a, b). This B cell differentiation *in vitro* was similar to the T cell differentiation in that it was also mediated by cyclic

AMP and developed with two hours incubation *in vitro*. The various non-thymus-related substances capable of inducing T cell differentiation *in vitro* were also found to induce this B cell differentiation *in vitro*. The one substance that has been found to be specific for T cell differentiation *in vitro* and not to also induce CR^- to CR^+ B cell differentiation is thymopoietin (see below).

II. INDUCING AGENTS

Thymopoietin is a 5,562 molecular weight polypeptide that was isolated from bovine thymus (Goldstein, 1974). Thymopoietin was isolated by its presumably secondary effect on neuromuscular transmission and not by its effect on T cell differentiation. The neuromuscular effect was detected in the course of experiments related to the human disease myasthenia gravis in which thymic disease is regularly associated with a deficit of neuromuscular transmission. The neuromuscular lesion was shown to be caused by a substance present in thymus and this proved on isolation to be thymopoietin. Purified thymopoietin is active in nanogram amounts in causing a detectable neuromuscular deficit in mice (Goldstein, 1974). Of most interest, however, was the finding that purified thymopoietin was effective in concentrations down to 20 picogram/ml in inducing the *in vitro* differentiation of T cells in the Komuro-Boyse assay (Basch and Goldstein, 1974). Furthermore, thymopoietin was shown to be selective in its action in that it induced T cell differentiation and not the differentiation of CR^- to CR^+ B cells (see above) (Goldstein *et al.*, 1975). Thus we believe thymopoietin to be the thymic hormone acting physiologically *in vivo* to induce T cell differentiation in the thymus and the evidence for this is both specificity of isolation (that is the neuromuscular effect, which is unambiguous, can only be detected in extracts of thymus and not other organs) (Goldstein, 1968, 1975) and also selectivity of target action in that thymopoietin induces T cell differentiation but does not induce CR^- to CR^+ B cell differentiation (Goldstein *et al.*, 1975).

The complete 49 amino acid sequence of thymopoietin has been determined (Schlesinger and Goldstein, 1975a) and a 13 amino acid fragment (tridecapeptide) corresponding to positions 29 to 41 has been synthesized (Schlesinger *et al.*, 1975a); this tridecapeptide retains the selective activity of the parent thymopoietin molecule on *in vitro* differentiation of T cells (Schlesinger *et al.*, 1975a) and also causes the neuromuscular effect of thymopoietin (Goldstein and Schlesinger, 1975).

Ubiquitin is a 8,451 molecular weight polypeptide that was first isolated from thymus extracts because it was the most copious polypeptide present in the molecular size range of thymopoietin (Goldstein *et al.*, 1975). Purified ubiquitin did not effect neuromuscular transmission and was therefore discarded as a candidate for the neuromuscular-blocking hormone in thymus. However purified ubiquitin was found to have some interesting effects in the *in vitro* differentiation systems. In concentrations ranging from 1 to 100 nanogram/ml ubiquitin induced not only the differentiation of T cells but also the differentiation of CR^- to CR^+ B cells. Thus this polypeptide isolated from thymus extracts could mimic the action of the thymus hormone thymopoietin but also had non-thymus-related effects such as the differentiation of CR^- to CR^+ B cells.

That ubiquitin is not related to a thymic hormone is further emphasized by its widespread distribution; it is found in all tissues by polyacrylamide gell electrophoresis and this distribution is confirmed by the findings with a sensitive and specifice radioimmunoassay; ubiquitin is not only present in all mammalian tissues tested but also in extracts of tissues from fish, squid, yeast, bacteria and even higher plants such as celery. This extraordinary conservation and widespread distribution prompted us to name this polypeptide ubiquitin. Small amounts of ubiquitin have been prepared from E coli and celery and preliminary sequence data indicate strong conservation of amino acid sequence and function of ubiquitin from phylogenetically distant celery on murine T cell differentiation *in vitro*.

The complete 74 amino acid sequence of ubiquitin has been determined for cattle and man, being identical in both (Schlesinger, Goldstein and Niall, 1975, Schlesinger and Goldstein, 1975b) and a 16 amino acid fragment (hexadecapeptide) has been synthesized (Schlesinger *et al.*, 1975b) which retains the non-selective action of the parent molecule in inducing both T cell and CR^- to CR^+ B cell differentiation *in vitro* (this is in contrast with the selective action of the tridecapeptide fragment based on the sequence of thymopoietin).

The inductive effects of ubiquitin are blocked by the β-adrenergic blocking substance propranolol (which does not effect the inductive action of thymopoietin) (Goldstein *et al.*, 1975). Thus we infer that ubiquitin engages widespread β-adrenergic receptors linked to adenylate cyclase and thus triggers induction of differentiation in a number of precommitted precursor cells, each type manifesting its precommitted program. Thymopoietin receptors, which would be discreet from the β-adrenergic receptors engaged by ubiquitin, would likely be restricted to the prothymocyte, and thus thymopoietin would selectively induce elevations of intracellular cyclic AMP in this class of precommitted precursor cells.

These studies clearly establish that ubiquitin has no physiological relationship to the action of thymic hormones yet they emphasize the difficulties of assaying for an organ-specific inductive effect in the presence of large amounts of a substance capable of mimicry. Ubiquitin is present in virtually all tissue extracts. In the light of these findings we were fortunate that thymopoietin was isolated by the unambiguous neuromuscular assay which was dependent on a secondary effect of the thymopoietin molecule. These considerations would probably apply in other *in vitro* inductive assays of differentiation and may well account for the previous findings that although embryonic inductions could be observed with tissue extracts these were found, in general, to be not specific for the appropriate inductor tissue, being mimicked by inappropriate tissue extracts (Boyse and Abbot, 1974). Such problems may be worth re-investigating in the light of our new-found knowledge of the presence and effect of ubiquitin, since ubiquitin effects in tissue extracts can be selectively prevented.

III. MOLECULAR MECHANISMS OF DIFFERENTIATION

The rapid *in vitro* differentiation of T cells induced by thymopoietin presents a valuable model for the study of molecular mechanisms of differentiation. Thymopoietin, a chemically defined inducing agent, induces, within 2 hours, the manifestation of serologically defined molecules on the cell surface which represent the products of at least 6 unlinked genes.

The initial steps of induction involve the elevation of intra-cellular cyclic AMP (see above). What are the ensuing molecular events leading to the expression of cell surface antigens? Our preliminary studies with drugs effecting macromolecular synthesis show that DNA replication (corresponding to cell division) is not necessary (Storrie *et al.*, 1975). Thus cytosine arabinoside or hydroxyurea, in amounts that inhibited DNA replication, did not effect the induction by thymopoietin of TL antigen on the cell surface of prothymocytes. However, drugs affecting the production of messenger RNA transcripts (actinomycin D, camptothecin and cordycepin) *did* block induction of TL and Thy-1, as did drugs effecting the translation of messenger RNA (puromycin and cycloheximide).

These preliminary studies suggest then that transcription and translation of messenger RNA is involved in T cell differentiation induced *in vitro* by thymopoietin. Whether this actually involves the transcription and translation of the structural information for the antigens themselves, or whether regulatory molecules are involved, has yet to be answered. It should be emphasized that the cell surface changes we are observing in the Komuro-Boyse assay may represent an early phase of a more prolonged differentiation program. While replication is clearly not necessary for the cell surface changes observed it may be necessary for the full development of the differentiation program and production of later, perhaps functional, T cells. These studies of T cells differentiation *in vitro* are only just beginning. We believe that they offer an unusual opportunity for analyzing the molecular mechanisms involved in the fulfillment of a precommitted differentiation program within a eukaryotic cell.

ACKNOWLEDGMENTS

I thank my colleagues and collaborators, especially Drs. E. A. Boyse, R. S. Basch, M. Scheid and D. H. Schlesinger.

Supported by U. S. Public Health Service Grants CA-08748, Al-12487, CA-17085 and Contract CB-53868 from the National Cancer Institute.

REFERENCES

Basch, R. S. and Goldstein, G. (1974). *Proc. Nat. Acad. Sci. U.S.A. 71,* 1474-1478.

Boyse, E. A. and Abbott, J. (1975). *Federation Proceedings 34,* 24-27.

Cantor, H. and Boyse, E. A. (1975). *J. Exp. Med. 141,* 1376-1389.

Goldstein, G. (1968). *Lancet 2,* 119-122.

Goldstein, G. (1974). *Nature 247,* 11-14.

Goldstein, G. (1975). *Ann. N.Y. Acad. Sci. 249,* 177-183.

Goldstein, A. L., Guha, A., Zatz, M. M., Hardy, M. A. and White,

A. (1972). *Proc. Nat. Acad. Sci. U.S.A. 69*, 1800.

Goldstein, G., Scheid, M., Hammerling, U., Boyse, E. A., Schlesinger, D. H. and Niall, H. D. (1975). *Proc. Nat. Acad. Sci. U.S. A. 72*, 11-15.

Goldstein, G. and Schlesinger, D. H. (1975). *Lancet 2*, 256-259.

Komuro, K., Boyse, E. A. and Old, L. J. (1973). *J. Exp. Med. 137*, 533-536.

Scheid, M., Goldstein, G., Hammerling, Y. and Boyse, E. A. (1975). *Ann. N.Y. Acad. Sci. 249*, 531-540.

Scheid, M. P., Goldstein, G., Hammerling, U. and Boyse, E. A. (1975). *In* "Membrane Receptors of Lymphocytes" North Holland Publishing Company. p. 353-359.

Scheid, M. P., Hoffmann, M. K., Komuro, K., Hammerling, U., Abbott, J., Boyse, E. A., Cohen, G. H., Hooper, J. A., Schulof, R. S. and Goldstein, A. L. (1973). *J. Exp. Med. 138*, 1027-1032.

Schlesinger, D. H. and Goldstein, G. (1975). *Nature 255*, 423-424.

Schlesinger, D. H. and Goldstein, G. (1975). *Cell 5*, 361-365.

Schlesinger, D. H., Goldstein, G. and Niall, H. D. (1975). *Biochemistry 14*, 2214.

Schlesinger, D. H., Goldstein, G., Scheid, M. P. and Boyse, E. A. (1975). *Cell 5*, 367-370.

Schlesinger, D. H., Goldstein, G., Scheid, P., Boyse, E. A. and Tregear, G. (1975). *Fed. Proc.* 34, 551.

Storrie, B., Goldstein, G., Boyse, E. A. and Hammerling, U. Submitted for publication.

SUMMARY REMARKS OF CHAIRMAN

PHILIP COFFINO

Department of Microbiology
University of California
San Francisco, California

Tissue culture systems have been preeminently useful in study of the growth cycle of animal cells. This is so partly for technical reasons: the cells are accessible for analysis and manipulation. Mitosis can be readily observed and morphologically less dramatic events followed by labeling with radioisotopes. Synchronized cell populations may be induced with inhibitors or obtained by selective procedures. The work Dr. Tobey has presented here is indicative of the degree of sophistication and precision possible in studies of the biochemical events that mark the cell cycle. He has demonstrated that at each phase of the cycle, highly specific histone phosphorylations occur. Students of the cell cycle find themselves in such a pleasant experimental position for two reasons. First, growth is a property of individual cells. Under suitable conditions, an isolated cell in tissue culture is capable of division. Hence, for many kinds of studies, cell-cell interactions can be ignored. Second, cultured cells consist of a homogeneous population. Therefore, properties of individual cells can be inferred from the biochemical analysis of cell populations.

In contrast, differentiation can be viewed as a phenomena of interacting, inductive, inhomogeneous cell populations. It is not surprising then that such relatively complex events have only recently become amenable to experimental analysis in tissue culture systems. We have had examples of three distinct strategies for coping with these problems in this session. Dr. Seeds has devised a means for culturing cell aggregates, thereby promoting the cell interaction that occurs in developing brain. Under these conditions, one finds expression of many of the biochemical markers associated with control nervous system maturation.

Dr. Goldstein has used a different approach. T cell maturation *in vivo* appears to require that precursor cells interact with thymic epithelium. Dr. Goldstein has demonstrated that thymopoietin, a protein extracted from the thymus and now purified to homogeneity, induces in T cell precursors membrane changes associated with maturation. This implies that at least certain thymic functions are mediated by a hormone, and has made possible investigation of the biochemical basis of the differentiation process.

Dr. Papaconstantinou induces hemoglobin synthesis by treating Friend leukemia cells with dimethylsulfoxide. The mechanism whereby this simple chemical produces complex changes in gene expression is not known. While dimethylsulfoxide is certainly not the physiologic "messenger" for differentiation, it is convenient to use, reproducible in its effects, and productive of genuine expression of differentiated function. The system is under intensive investigation in a number of laboratories. This work, in effect, holds in abeyance the study of the physiologic initiator of differentiation to concentrate on subsequent events.

The kinetics of the response of T cell precursors to thymopoietin and Friend leukemia cells to dimethylsulfoxide stand in sharp contrast. T cell specific surface antigens appear on pre-T cells in hours, but several rounds of cell replication must supervene before hemoglobin synthesis can be demonstrated in the leukemia cells. Pre-T cells thus appear fully committed to a particular pathway of differentiation, awaiting only an inductive signal. The signal, as in other systems where expression of differentiated function is triggered, appears to act through modulation of cyclic AMP levels. There is much evidence that cyclic nucleotides, in turn, function by regulating the activity of protein phosphokinase. Recent work using kinase mutants has helped to confirm this (Coffino *et al.*, 1976). Perhaps it will prove true that phosphorylation and dephosphorylation events act as a general switch to regulate pre-existing cellular mechanisms, while more complex (and less readily reversible?) biochemical events, sometimes requiring cell division, occur when commitment to differentiation is induced in more pluripotent cells.

REFERENCES

1. Coffino, P., Bourne, H. R., Friedrich, U., Hochman, J., Insel, P. A., LeMaire, I., Melmon, K. L. and Tomkins G. M. *In* Recent Progress in Hormone Research (R. O. Greep, ed.), Vol. 32, pp. 669-684, 1976.

Symposium II

IN VITRO STIMULATORS OF GROWTH

FIBROBLAST GROWTH FACTOR: EFFECT ON GROWTH OF FIBROBLASTS AND RELATED CELL TYPES

DENIS GOSPODAROWICZ *
JOHN S. MORAN

The Salk Institute for Biological Studies
San Diego, California

I. INTRODUCTION

Elucidation of the internal mechanisms controlling cell division requires the identification of exogenous agents that can stimulate cell division. Since eukaryotic cells are extremely sensitive to perturbations of their plasma membrane, a variety of non-specific agents, such as lectins (Aubery, *et al.*, 1975) and proteases (Burger, 1970), can induce cell division in a portion of a resting cell population under certain culture conditions. However, these agents may act by a mechanism unrelated to the way physiological molecules involved in the normal control of proliferation act, and the information gathered with such unspecific agents may, therefore, be misleading and thus of limited value.

*Present address: University of California San Francisco Medical School Cancer Research Institute, San Francisco, California 94143.

It seems reasonable to assume that a mitogen with a physiological role should meet two criteria: First, it should be obtainable from the organism in which it is active (unlike plant lectins). Second, its primary action should be to stimulate division; that is, it should be more potent for producing a mitogenic effect than for producing some secondary effect, not directly related to mitogenesis (unlike proteases). In addition, it should act directly, rather than by potentiating the effect of another mitogen (permissive effect). In looking for such a mitogen, we have investigated two likely sources: serum and tissue extracts. Purification of mitogens from serum has been shown by others to be a very difficult task. Difficulties encountered are due to two reasons: First, mitogens are present in serum in low concentrations. Second, target cells may be sensitive to more than one mitogen, thus causing mitogenic activity to appear in a wide range of serum fractions that are quite different chemically (Gospodarowicz and Moran, 1976).

In view of the above considerations, we chose to purify mitogens from tissue. Our attention has focused on two promising sources of growth factors: the pituitary, since it has been reported that this organ is a potential source of growth factor(s) (Holley and Kiernan, 1968; Armelin, 1973; Corvol, *et al.*, 1972), and neural tissue, since nerves are clearly necessary for regeneration in lower vertebrates (Singer, 1974).

From both these sources (bovine pituitary--Gospodarowicz, 1975, and bovine brain--Gospodarowicz, *et al.*, 1976a), we have isolated a mitogen that we call Fibroblast Growth Factor (FGF).

While FGF was given its name because of its effect on fibroblasts, later studies have shown that it is a potent mitogen for a variety of mesoderm-derived cells, such as adrenal cells (Gospodarowicz and Handley, 1976), chondrocytes (Jones and Allison, 1975; Gospodarowicz, *et al.*, 1976a), myoblasts at high cell density (Gospodarowicz, *et al.*, 1976b), glial cells[1], smooth muscle cells[2], endothelial cells (Gospodarowicz, *et al.*, 1976c,d, 1977a), and for cells of undetermined type obtained from amniotic fluid (Gospodarowicz, *et al.*, 1976).

In this review we discuss only the mitogenic effect of FGF on an established fibroblastic cell line, on diploid fibroblasts in early passage cultures, and on chondrocytes. We also review what is known about the mechanism of action of FGF. As far as the physiological role of FGF is concerned, the reader is referred to a recent publication describing the effect of FGF on the regeneration of limbs (Gospodarowicz, *et al.*, 1975b), and to a recent report on the possible role of FGF in wound-healing in higher vertebrates (Gospodarowicz, *et al.*, 1976a).

II. RESULTS

A. Physicochemical Properties of FGF

Brain and pituitary FGF are polypeptides with apparent molecular weights between 13,000 and 13,400. They do not contain carbohydrate. They have an isoelectric point of 9.5-9.7 and give a single band when analyzed by polyacrylamide electrophoresis at pH 4.5. They are rich in lysine and arginine, reflecting their basic nature. The mitogenic activity of both brain and pituitary FGF is destroyed by heat (60° at ph 7 for 5 minutes) or acid (below ph 3.5) treatment (Gospodarowicz, 1975; Gospodarowicz, *et al.*, 1976a).

FGF from both brain and pituitary is devoid of Nerve Growth Factor activity in the chick dorsal root ganglion fiber outgrowth assay; however, it is 60% as potent as the Nerve Growth Factor

[1]Jan Ponten, personal communication.
[2]Russell Ross, personal communication.

when the attachment of neurons is used as an assay[1]. It does not have any significant sulfation factor activity[2], and it does not compete for the receptor site for Epidermal Growth Factor[3].

No esterase activity was detected, even at FGF concentrations as high as 1 mg/ml with either denatured bovine serum albumin or N α -benzoyl-L-arginine ethyl ester as substrates. This indicates that if any trypsin-like activity is present in FGF, it is less than 0.005%. Treatment with proteolytic enzymes such as trypsin, chymotrypsin, pepsin, or proteases destroyed 90% of the FGF activity (Gospodarowicz, *et al.*, 1976a).

B. Effect of Brain and Pituitary FGF on the Proliferation of 3T3 Cells

The addition of brain or pituitary FGF to resting populations of Balb/c 3T3 cells results in the initiation of DNA synthesis (Gospodarowicz, 1974). The minimal effective dose is approximately 0.01 ng/ml. Plateau values are obtained at 0.5 to 1 ng/ml for pituitary FGF. Brain FGF gives a greater effect at concentrations as high as 10 ng/ml (Fig. 1). The initiation of DNA synthesis stimulated by FGF is only 30 to 50% of that observed after the addition of an optimal concentration of serum. In the presence of a glucocorticoid, such as dexamethasone, FGF has a stimulatory effect on the initiation of DNA synthesis comparable to that of serum (Fig. 2).

FGF with glucocorticoids also induces cell proliferation. The effect of FGF is more pronounced when cells are maintained in the presence of defibrinated plasma than in serum because plasma can substitute for serum for maintaining the cells in good health, while at the same time plasma has a lower mitogenic content than serum (Gospodarowicz, *et al.*, 1975a). The addition of FGF to sparse (7 cells/mm^2), resting populations of 3T3 cells in the

[1]Sig Norr, personal communication.
[2]Judson Van Wyk, personal communication.
[3]Stanley Cohen, personal communication.

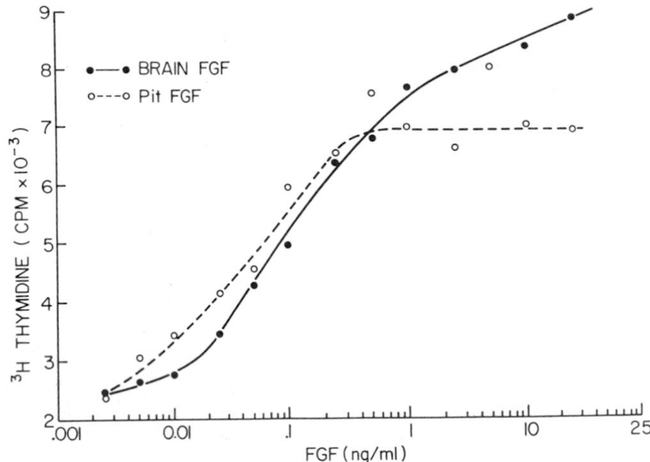

FIGURE 1 *Stimulation of DNA synthesis in 3T3 cells by various concentrations of brain (● - ●) and pituitary FGF (o - o). 3T3 cells (2 x 10^4 per 35 mm dish) were plated in Dulbecco's Modified Eagle's medium with 5% calf serum. After 24 hours, the medium was replaced with 2 ml of medium containing 0.4% calf serum. Twenty-six hours later, FGF was added, followed 16 hours later by the addition of [^{3}H] thymidine. After a 12-hour labelling period, the incorporation of [^{3}H] thymidine into DNA was determined as described by Gospodarowicz and Moran, 1974a. [^{3}H] thymidine in corporation was 110 cpm per dish in controls, and 15,200 cpm when 10% calf serum was added.*

presence of 2.5% plasma results in a final cell density (400 cells/mm^2) which is similar to that observed when an optimal concentration of serum is added to the cultures.

The mitogenic effect of FGF is not limited to sparse populations of 3T3. FGF causes a marked overgrowth of "contact inhibited" 3T3 cells maintained in 10% serum (Gospodarowicz and Moran, 1974a, b). The addition of FGF results in the cells losing their polygonal shape and becoming elongated and highly refractile. The culture loses its cobblestone appearance and takes on the appearance of a bird's nest as the cells grow in multiple layers (Fig. 3). The addition of glucocorticoid potentiates the effect of FGF, and the final density reached is 4 to 5 times that observed with controls to which fresh serum is added. The

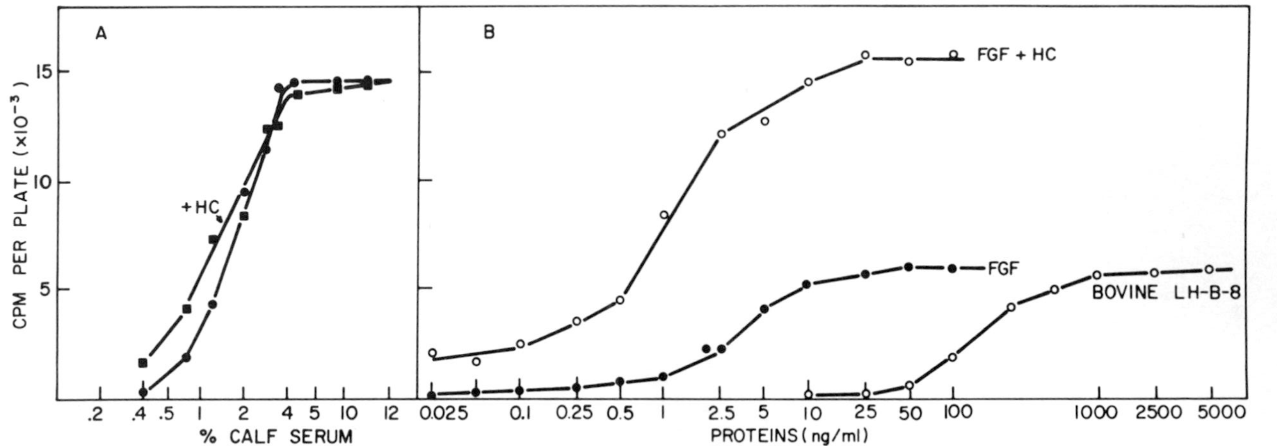

FIGURE 2 *Stimulation of DNA synthesis in 3T3 cells by serum and FGF in the presence and absence of hydrocortisone. 3T3 cells (2.4 x 10^4 cells per 60 mm dish) were plated and maintained as described in Fig. 1. (A) [^3H] thymidine incorporation in response to various concentrations of serum in the presence (■) or absence (●) of 0.1 µg/ml of hydrocortisone. The incorporation of [^3H] thymidine was 300 cpm per dish in controls. (B) [^3H] thymidine incorporation in response to various concentrations of FGF in the presence (○) or absence (●) of 0.1 µg/ml hydrocortisone. The stimulatory effect of NIH bovine LH-B-8 (bovine luteinizing hormone) where the FGF is present as a contaminant is shown by the lowest line. The purified FGF is about 200-500 times as potent as that preparation. The stimulatory effect of FGF on the initiation of DNA synthesis varied with the length of the time the cells were left resting in low serum and on the size of the dish in which the cells were cultured. Cells maintained in low serum for 26-hours were more responsive than cells maintained longer. Cells tested in 2 ml of medium in 35 mm dishes were more responsive than cells tested in 5 ml of medium in 60 mm dishes.*

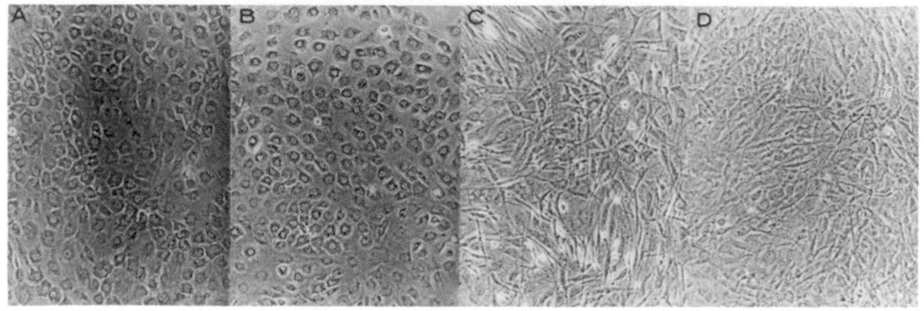

FIGURE 3 *Effect of FGF and glucocorticoid on "contact inhibited" Balb/c 3T3 in the presence of 10% serum. Balb/c 3T3 cells were grown in Dulbecco's modified Eagle's medium with 10% calf serum. Two days after reaching confluency (A), either 10% serum (B), 25 ng/ml of FGF (C), 25 ng/ml of FGF plus 1 μg/ml of dexamethasone (D), were added to the cells. FGF or FGF plus dexamethasone were added daily; serum was added only once. Four days later, photographs of living cells were taken with phase contrast optics. While the addition of serum did not change the morphology of the cells, the addition of FGF alone to cultures grown in 10% serum caused the cells to become elongated, density inhibition was overcome, and the cells lost their orientation and grew in multiple layers. When FGF plus dexamethasone was added, the cells were flatter but grew to an even higher density than with FGF alone (Gospodarowicz and Moran, 1974 a, b). When the cells were trypsinized, replated, and grown in the presence of serum, they resumed their usual morphology, thus demonstrating that the transformation observed in the presence of FGF is reversible.*

addition of insulin to cells in 10% serum or to cells in 10% serum plus FGF and dexamethasone does not have any effect.

The mitogenic effect of FGF is not restricted to the particular line of Balb/c 3T3 maintained in this laboratory. It is also observed with other 3T3 cell lines (Table I) and with mouse fibroblasts in early passages (Gospodarowicz and Moran, 1975).

C. Action of FGF on Transformed Cells

A particularly interesting aspect of FGF's action is its mitogenic effect on thermosensitive mutants of polyoma transformed Balb/c 3T3 cells (Rudland, *et al.*, 1974b). Chemical or viral transformation of cells in tissue culture results in morphologi-

TABLE I FGF and Different Mouse Cells[a]

Cell type	DNA synthesis ([^{3}H] thymidine cpm/10 cells)			% Labeled nuclei		Cell number x 10^{-5} (48 hr after additions)	
	[b]	+FGF	+serum	[b]	+FGF	[b]	+FGF
BALB/c 3T3a[c]	0.2	13	14	0.4	95	3.9	7.4
BALB/c 3T3b	0.1	12	13	0.5	100	7.0	13
BALB/c 3T3c	0.1	8.2	12	0.3	65	6.8	11
Swiss 3T3-4A	0.8	4.9	7.6	0.5	64	7.5	12
2nd embryos	0.1	4.9	5.1	0.4	81	16	28
BMK: fibroblasts	0.3	3.5	4.5	0.4	89	--	--
epithelial				0.6	1.0	--	--
Py3T3	3.2	2.9	7.4	--	--	10	9.0
SV3T3	2.1	1.8	8.4	30	26	5.1	4.2

[a]Cultures were "resting" for 3 days (or 2 days for SV3T3) in serum-free medium as described by Rudland, *et al.*, (1974a) before additions of saturating amounts of FGF (50 ng/ml except for 2nd embryos, which was 100 ng/ml). 0.5% or less of all cultures were radioactively labeled with [^{3}H] thymidine during the subsequent 24 hours except 30% for SV and Py 3T3.

[b]Controls, no additions.

[c]Abbreviations: BALB/c 3T3a, 3T3b, 3T3c sublines of BALB/c 3T3 (3T3a was from D. Gospodarowicz, 3T3b and c were from Dr. S. Aaronson); Swiss 3T3-4A is a line of Swiss 3T3 from Dr. H. Green; 2nd embryos, secondary mouse embryos; BMK, baby mouse kidney cultures containing both fibroblast and epithelial cells; Py3T3, polyoma-transformed BALB/c 3T3b; SV3T3, simian virus 40-transformed 3T3-4A.

cal changes, surface alterations, and an increased ability to grow under various conditions. The changed pattern of growth control of many transformed cells is manifested *in vitro* by their ability to grow to higher densities in monolayer cultures and by their decreased requirement for serum for growth.

A comparative biochemical study of normal and transformed cells in tissue culture is complicated by the fact that often two very different cell types are being compared (for example, there are chromosomal differences between SV 3T3 and 3T3). The observed differences can then arise as a result of the transformation event itself or as a consequence of a selection for a particularly stable transformant. With cells possessing thermosensitive lesions in the expression of the transformed phenotype, however, the latter difficulty is obviated since the same cell is being used as a transformed cell and as its control.

We investigated the ability of FGF to promote the growth of mutant Balb/c 3T3 cells which show temperature dependent changes in many properties characteristic of the transformed state. We find that FGF stimulates the initiation of DNA synthesis and cell proliferation only at the non-permissive temperature (normal state) but fails to stimulate DNA synthesis or promote cell proliferation at permissive temperature (transformed state), Table II.

Since transformation is most strikingly reflected by changes in the plasma membrane, the lack of effect of FGF at the permissive temperature suggests that FGF acts at the plasma membrane level.

D. Mode of Action of FGF

Most cells in tissue culture exist in one of two reservible growth states--one of rapid proliferation and one of relative quiescence. Transition between the two stages is believed to be accomplished by a sequential series of regulated steps which constitutes a "pleiotypic" response (Hershko, *et al.*, 1971). These

TABLE II Additions to Growing Cultures[a]

Additions	3T3		ts 3T3 Cl.1		Py3T3	
	a,cpm	b,cells	a,cpm	b,cells	a,cpm	b,cells
32° C						
None	1.4	4.4	2.4	4.2	2.2	6.5
FGF	6.1	10.8	2.2	4.3	2.3	6.7
Serum	7.1	12.8	3.9	6.9	4.1	10.4
39° C						
None	1.6	4.0	2.0	4.1	6.5	2.1
FGF	7.6	6.6	9.4	6.6	5.6	2.0
Serum	6.0	7.1	9.7	6.7	9.3	2.9

[a]BALB/c 3T3 cells, polyoma transformed BALB/c cells (Py3T3), and temperature-sensitive transformed BALB/c 3T3 cells (ts 3T3 Cl.1) were plated at 10^5 per 5 cm Petri dish and grown in Dulbecco's Modified Eagle's Medium (5 ml) in 2% calf serum (Colorado) at 32° or 39° in a 10% CO_2 atmosphere. The CO_2 pressure was slightly varied to maintain a constant pH in the medium at the two temperatures. After 2 or 3 d when the final cell density was about 3×10^5 per plate (3T3: 3.0 and 3.1; ts 3T3: 2.9, 3.0; and Py3T3: 1.6 and 4.5×10^5 cells at 39°C or 32°C, respectively) 50 ng/ml of FGF or 10% serum (6 mg/ml) were added to the growing cultures. Autoradiography of control dishes without additions showed that more than 70% of all cell nuclei became radioactively labeled during the first 12 hours after the time of addition of [^{3}H] thymidine. Cultures were radioactively labeled with 3 μCi ml^{-1} [^{3}H] methyl thymidine at 3 μM from 7 until 30 hours after additions, or the cells were counted in a Coulter Counter 48 hours afterwards. The cpm of [^{3}H] thymidine incorporated into DNA per 10 cells (a) or the number of cells per 5 cm Petri dish $\times 10^{-5}$ were recorded (b).

sequential changes include stimulation of cellular transport systems, polyribosome formation, protein synthesis, ribosomal and tRNA synthesis, and eventually DNA synthesis followed by cell division. We have compared the effect of FGF plus hydrocortisone to serum for delivery of the full "pleiotypic" and mitogenic responses in lines of Balb/c 3T3 cells.

The addition of serum to confluent, resting (less than 0.5% radioactive labeled nuclei after a 24-hour exposure to [^{3}H] thymidine) cultures of Balb/c 3T3 cells leads to an increased uptake of macromolecular precursors. Uridine, total amino acids, and thymidine uptake was measured after the addition of either 10% serum or saturating amounts of FGF (25 ng/ml). The time course of the increase in net incorporation into the cell was almost identical for both FGF and serum additions. Amino acid uptake increased immediately, there was a 15-minute lag for uridine, and approximately a 3 to 5-hour lag for thymidine uptake (Fig. 4A). The rate of protein synthesis increased (about 300%) within 5 hours, that of mRNA (about 15-20%) within 5 hours, and DNA (60 to 100-fold) within 24 hours. These increases were almost identical for cultures activated with FGF or serum, both in the kinetics and extent of the increases (Fig. 4B). Finally, cell division occurred from about 25 to 32 hours in both cases. The dose response of varying concentrations of FGF for the stimulation of protein synthesis at 6 hours, ribosomal RNA synthesis at 14 hours, and the eventual induction of DNA synthesis were roughly parallel; maximal stimulations were achieved at 5-20ng/ml in all cases when 10 cm dishes were used; lower concentrations of FGF were required with smaller dishes.

The abilities of FGF and serum to induce polysome reformation measured 6 hours after additions were virtually identical (Fig. 5). The m RNA location in resting cultures was approximately 28% in polysomes and 72% in the 30S-80S messenger ribonucleoprotein complex. Addition of either FGF or serum results in greater than 85% of the mRNA being located in the polysomal region of the sucrose gradient. Addition of hydrocortisone alone to quiescent cultures in the absence of serum induces no detectable changes compared with control cultures except for a small increase in polyribosome formation (Fig. 5a). The extent of the increase would probably be too small (about 10%) to detect a net increase in overall protein synthesis. Without hydrocortisone and with

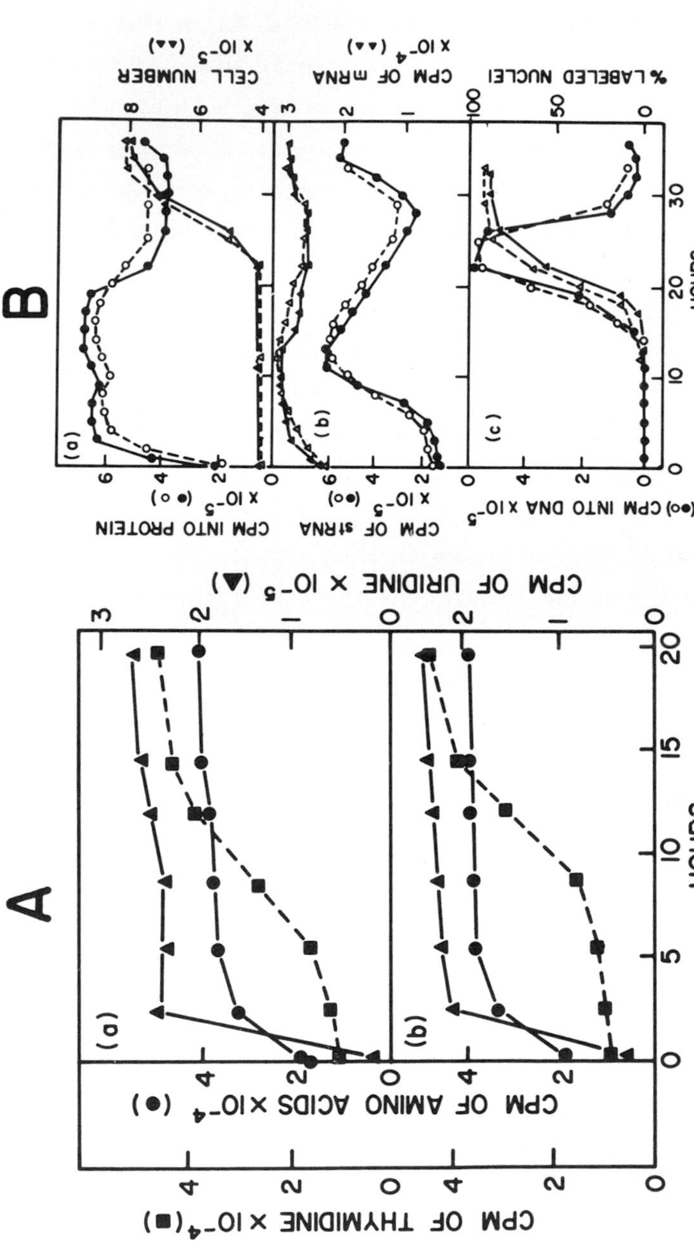
A
B
CPM OF THYMIDINE × 10⁻⁴ (■)
CPM OF AMINO ACIDS × 10⁻⁴ (●)
CPM OF URIDINE × 10⁻⁵ (▲)
HOURS
CPM INTO PROTEIN × 10⁻⁵ (● ○)
CELL NUMBER × 10⁻⁵ (▲△)
CPM OF sRNA × 10⁻⁵ (●○)
CPM OF mRNA × 10⁻⁴ (▲△)
(●○) CPM INTO DNA × 10⁻⁵
% LABELED NUCLEI
HOURS

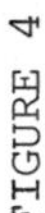

FIGURE 4

FIGURE 4 *(a) Uptake changes in some macromolecular precursors. Balb/c 3T3 cultures were prepared as described by Rudland, et al., (1974a). Either (a) 20 ng/ml FGF or (b) 10% dialyzed serum (6 mg/ml) was added to the resting cultures (see Table I for autoradiography data) at time zero. Either [^3H] aminoacids (●), 5 μCi/ml of [^3H] uridine at 2.5 μM (▲), or [^3H] thymidine (■) was added at various times and the cultures were isolated 30 minutes later. The cpm of each isotope incorporated into trichloroacetic acid-soluble material per 30 min/mg of cell protein is recorded. Time-points indicate the mid-point of the radioactive labeling time. Approximately 80-90% of the cells' nuclei became radioactively labeled with [^3H] thymidine after FGF or serum additions. (b) Sequentially induced macromolecular changes. Cells were prepared as described by Rudland, et al., (1974a). Either 25 ng/ml of FGF (solid line) or 10% dialyzed serum (6 mg/ml) (broken line) was added. Parallel cultures were then pulse-labeled for 2 hours with (a) [^3H] aminoacids for determination of protein synthesis (●, o); or (b) [^3H] uridine for stable RNA (stRNA = rRNA and tRNA) synthesis (●, o) or messenger RNA (mRNA) synthesis (▲, Δ); or (c) [^3H] thymidine for DNA synthesis (●, o). Results are expressed in corrected cpm of isotope incorporated per 2 hr/mg of cell protein. Time-points indicate the mid-point of the pulse-label, except at zero time when cultures were radioactively labeled for 2 hours and then isolated immediately before the additions. In addition, (a) cell numbers (▲, Δ), and (c) the percentage (%) of [^3H] thymidine-labeled cell nuclei from cultures labeled from zero time until their isolation time (▲, Δ) are shown.*

only FGF present, all uptake and macromolecular synthesis rates studied were approximately 40-50% and 20-25% respectively, of the values measured with hydrocortisone present.

Studies to determine how long FGF and glucocorticoid must be present in the medium to induce cells to divide have indicated that when FGF or FGF plus glucocorticoid was present for only 2.3 hours and the medium was then changed to fresh 0.1% serum, a small proportion of the cells (10% over controls) were induced to make DNA. When FGF or FGF plus glucocorticoid was removed after 6.3 hours, 50% of the cells were already committed to make DNA, while at 16 hours, 100% were committed (Fig. 6A). Since most of the macromolecular changes take place within 5 hours (increased

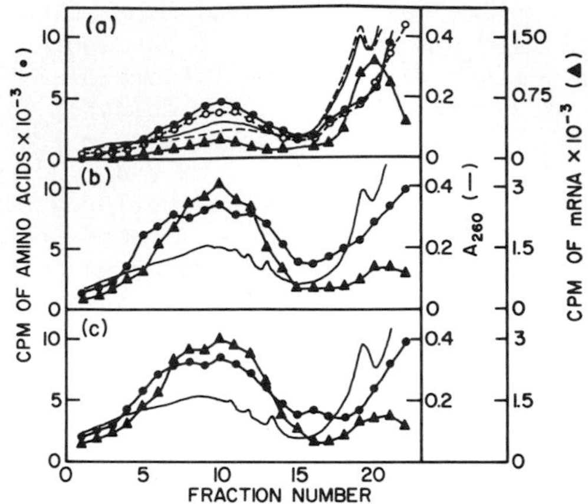

FIGURE 5 *Formation of polyribosomes. Sucrose gradients of polyribosomes isolated from (a) resting cell cultures without hydrocortisone (broken line) and with hydrocortisone (solid line); (b) cells stimulated by FGF; or (c) cells stimulated by serum. Incorporation of* $[^3H]$ *aminoacids into nascent peptides* (•, o), A_{260} *(solid line), and mRNA* (▲) *are indicated. Cells were grown in 9-cm petri dishes in 10 ml of medium and allowed to become quiescent in 0.1 ng/ml of FGF and 250 μg/ml of bovine serum albumin with 0.5 μg/ml of hydrocortisone (solid line) or without (broken line) hydrocortisone. After 3 days, polysomes were isolated from cultures 6 hours after (a) no additions, (b) addition of 25 ng/ml of FGF, or (c) addition of 10% dialyzed serum (6 mg/ml). Cultures were radioactively labeled for 2 minutes with 250 μCi/ml of* $[^3H]$ *aminoacids in modified Eagle's medium containing one-fifth the normal concentration of amino acids* (•, o) *or for 6 hours immediately before isolation with* $[^3H]$ *uridine. Two petri dish cultures were isolated for each isotope and four* (4×10^5 *cells) for nonradioactive cultures.*

uptake of macromolecular precursors and protein synthesis), it is not surprising that a 6-hour exposure of the cells to FGF commits 50% of their population to divide.

Since the presence of glucocorticoid is required to permit the cells to respond to FGF to the same extent as they respond to serum, cells were exposed to FGF or FGF plus glucocorticoid and the medium was then changed to fresh 0.1% calf serum plus glucocorticoid. Under these circumstances, at 2.3 hours 50% of the cells were committed to make DNA when they had been exposed to FGF or FGF plus glucocorticoids. At 6.3 hours, 100% of them were

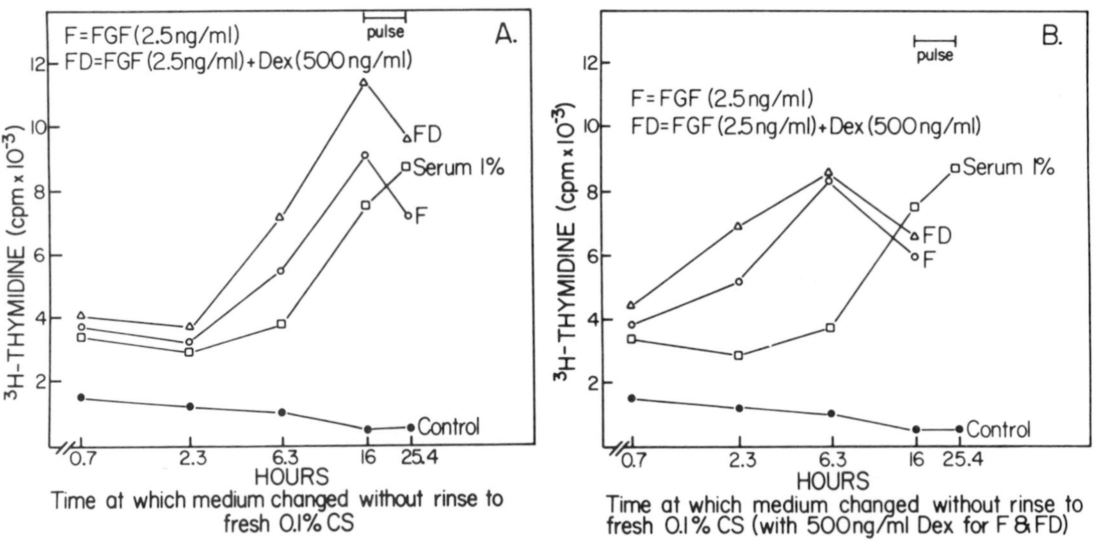

FIGURE 6 *Stimulation of DNA synthesis in 3T3 cells after exposure to FGF or serum for various lengths of time. Cells were plated in Dulbecco's modified Eagle's medium with 5% calf serum at a density of 80,000 cells per 6 cm dish. After 24 hours, the medium was replaced with fresh medium containing 0.1% calf serum. At time 0, either 1% serum, 2.5 ng/ml FGF or 2.5 ng/ml FGF plus 500 ng/ml dexamethasone were added to the dishes. At the times indicated on the abscissa, the media were removed and replaced with medium containing either 0.1% serum (A, in "control" and "serum 1%" in B) or 0.1% serum plus 500 ng/ml dexamethasone ("F" and "FD" in B). The amount of [^3H] thymidine incorporated into DNA (cpm/dish) during a 9.4-hour labeling period, starting 16 hours after sample addition, is shown on the ordinate. Less than optimal amounts of FGF and serum were added to the dishes so that no significant amount of FGF or serum would be remaining in the dish after the medium was replaced with fresh medium.*

committed (Fig. 6B). This points out that cells committed by FGF plus glucocorticoid to make DNA required the continued presence of glucocorticoid to carry through with their commitment.

E. Effect of FGF on Human Foreskin Fibroblasts and Chondrocytes

1. *Human Foreskin Fibroblasts*

Since Todaro and Green (1963) describe the establishment of the 3T3 cell line as a process of selection for mutant, usually heteroploid, cells adapted to *in vitro* culture, it cannot be assumed that growth control in these cells is the same as that in the population from which they originated. Therefore, in order to further ascertain the role of FGF in the control of cell division, we have examined the mitogenic effect of FGF on cells in early passage derived from normal (non-neoplastic) tissue from human foreskin (Gospodarowicz and Moran, 1975). Quiescent cultures of human foreskin fibroblasts maintained in 5% serum were very responsive to FGF. When calf serum or FGF was added, DNA synthesis was initiated (Fig. 7A). The minimal effective dose of FGF was 0.5 ng/ml (4×10^{-11}M). At 10 ng/ml, FGF stimulated DNA synthesis as effectively as did the optimal amount of added serum. When 100 ng/ml of FGF was added, [^{3}H] thymidine incorporation was 11 times the control value and 47% higher than the maximal incorporation with serum.

In order to determine if the greater [^{3}H] thymidine incorporation in response to FGF as compared with serum was due to a larger proportion of the cells initiating DNA synthesis, autoradiography of cultures in 5% calf serum was performed. In controls, less than 3% of the nuclei incorporated [^{3}H] thymidine in a 12-hour labeling period. When 10% calf serum was added to the cultures (to result in a serum concentration of 15%), 17% of the nuclei were labeled. However, when 50 ng/ml FGF was added, 58%

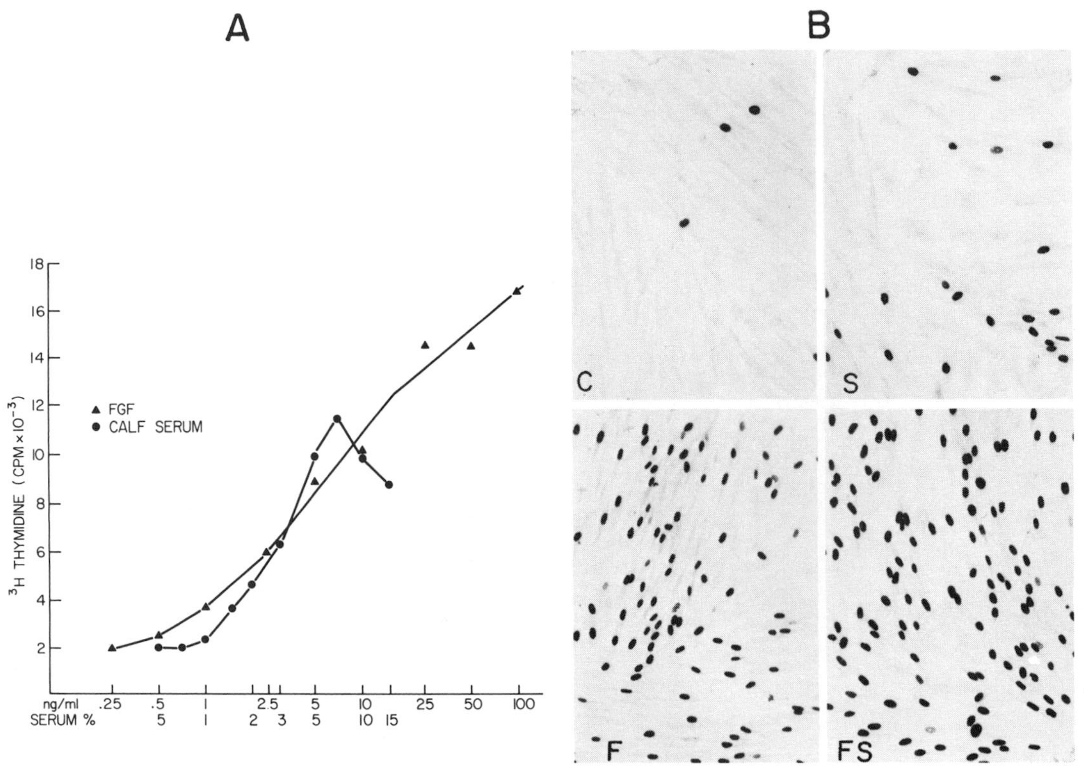
A
B
▲ FGF
● CALF SERUM
^{3}H THYMIDINE (CPM × 10^{-3})
18
16
14
12
10
8
6
4
2
ng/ml .25 .5 1 2.5 5 10 25 50 100
SERUM % 5 1 2 3 5 10 15
C
S
F
FS

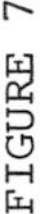

FIGURE 7

FIGURE 7 *(a) DNA synthesis in subconfluent human (HF) cells in response to various concentrations of serum or FGF. Serum and FGF were added to subconfluent (2.8×10^4 cells/cm^2) cultures that had been made quiescent by maintaining them for 4 days in 5% calf serum (5 ml of medium/6-cm dish). DNA was labeled with [^3H] thymidine for a 24-hour period beginning 12 hours after sample addition. Ordinate, ^3H thymidine for a 24-hour period beginning 12 hours after sample addition. Ordinate, [^3H] cpm/dish in cold trichloroacetic acid-precipitable material. Abscissa, nanograms/milliliters FGF added (▲ - ▲) or percent serum added (● - ●). Incorporation was 1,550 cpm/dish in controls. Standard errors did not average more than 10% of the mean. (b) DNA synthesis in subconfluent human (HF) cells in response to serum or FGF. Samples were added to cultures prepared as described in A. The cells were exposed to [^3H] thymidine for a 12-hour period beginning 12 hours after sample addition. Autoradiographs show DNA synthesis after the following additions: C - none; S - 10% calf serum; F - 100 ng/ml FGF; FS - 10% calf serum plus 100 ng/ml FGF.*

of the nuclei were labeled during the same period, and the additiontion of 10% serum plus 50 ng/ml FGF resulted in more than 85% labeling (Fig. 7B).

To see whether or not the initiation of DNA synthesis described in Fig. 7 was followed by cell division, the number of cells per dish was determined 63 hours after sample additions. The cell densities were 4.5×10^4 when 25 ng/ml FGF had been added, 6.8×10^4 when both 10% serum and 25 ng/ml FGF had been added. The cell density in control dishes was 3.0×10^4 cells/cm^2.

Since the initiation of DNA synthesis in response to FGF does not prove that FGF will induce mitosis in sparse cultures, the increases in cell number in cultures maintained in the presence of FGF, 10% serum, or 10% serum plus FGF were compared. When 10% calf serum was added to sparse, quiescent cultures maintained in 0.2% serum, the number of cells increased exponentially with an apparent doubling time of approximately 48 hours, and the final density was 3×10^4 cells/cm^2 (Fig. 8D). Culture growth followed a similar pattern in 20% fetal calf serum. When FGF (25 ng/ml added daily) was added to quiescent cultures in 0.2% serum, the cell number doubled. This result is in accord with the observa-

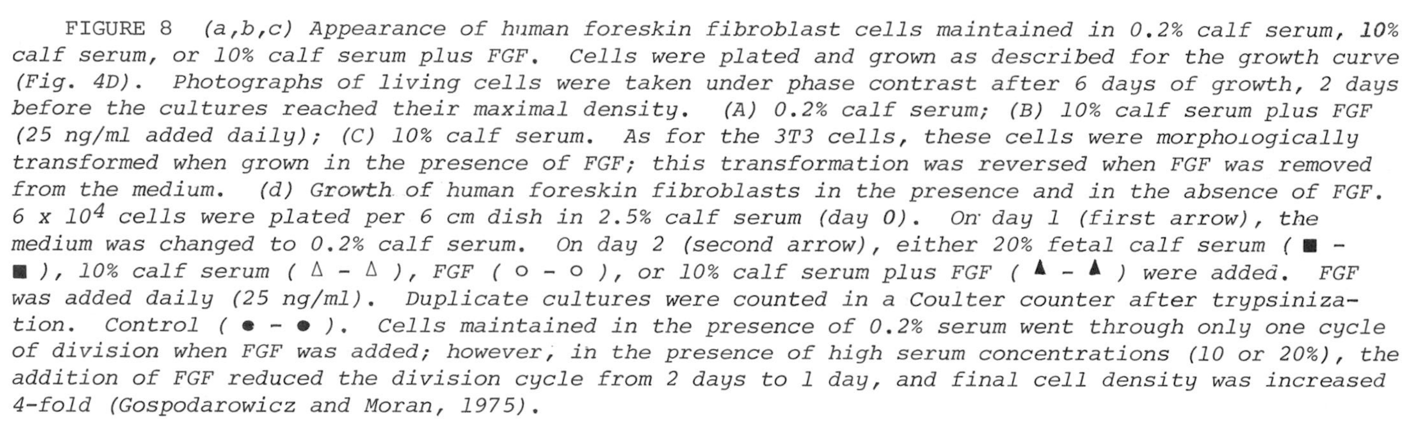

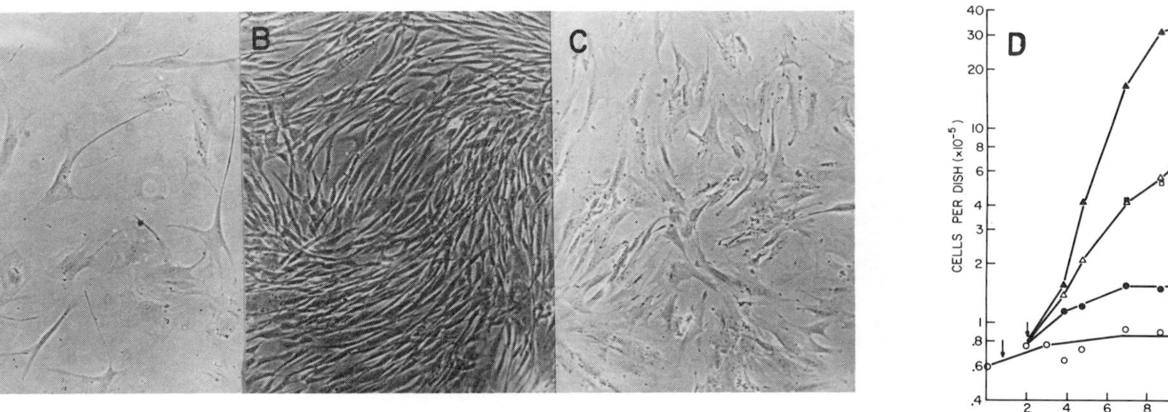

FIGURE 8 *(a,b,c) Appearance of human foreskin fibroblast cells maintained in 0.2% calf serum, 10% calf serum, or 10% calf serum plus FGF. Cells were plated and grown as described for the growth curve (Fig. 4D). Photographs of living cells were taken under phase contrast after 6 days of growth, 2 days before the cultures reached their maximal density. (A) 0.2% calf serum; (B) 10% calf serum plus FGF (25 ng/ml added daily); (C) 10% calf serum. As for the 3T3 cells, these cells were morphologically transformed when grown in the presence of FGF; this transformation was reversed when FGF was removed from the medium. (d) Growth of human foreskin fibroblasts in the presence and in the absence of FGF. 6 x 10^4 cells were plated per 6 cm dish in 2.5% calf serum (day 0). On day 1 (first arrow), the medium was changed to 0.2% calf serum. On day 2 (second arrow), either 20% fetal calf serum (■ - ■), 10% calf serum (△ - △), FGF (○ - ○), or 10% calf serum plus FGF (▲ - ▲) were added. FGF was added daily (25 ng/ml). Duplicate cultures were counted in a Coulter counter after trypsinization. Control (● - ●). Cells maintained in the presence of 0.2% serum went through only one cycle of division when FGF was added; however, in the presence of high serum concentrations (10 or 20%), the addition of FGF reduced the division cycle from 2 days to 1 day, and final cell density was increased 4-fold (Gospodarowicz and Moran, 1975).*

tion that FGF provoked only a slight (two-fold) increase in [^{3}H] thymidine incorporation in cultures maintained in low (0.05%) serum (Gospodarowicz and Moran, 1975). When FGF was added along with serum, the apparent doubling time was half that observed with 10% calf serum or 20% fetal calf serum alone, and the final density was 10^5 cells/cm^2, 3.3 times as great as with serum alone.

The appearance of the cells in the presence of FGF was markedly different from that of cells in either 0.2% or 10% serum alone. In the presence of FGF, the cells were elongated and not so flat as cells in serum alone. This morphological change took place within 2 days of the addition of FGF and was not dependent on cell density (Fig. 8 A-C).

When cells that had been grown to a high density in the presence of FGF were replated, their growth characteristics and morphology in 0.2 and 10% serum were the same as those of cells that had never been exposed to FGF.

2. *Early Passage Cultures of Rabbit Ear Chondrocytes*

The effect of FGF on cartilage-producing cells obtained by trypsinization of rabbit ear has been investigated. This cell type was selected because it is derived from the mesoderm and has many characteristics in common with fibroblasts. Furthermore, with cartilage cells, not only proliferation can be studied, but also differentiation. Colonies of rabbit-ear cartilage can be detected with the naked eye by their three-dimensional, nearly hemispherical appearance (Fig. 9). Also, differential staining demonstrates areas with cartilage-like matrix (Fig. 10). As shown in Fig. 10 A-C, the addition of FGF, from either brain or pituitary, induced cartilage cells plated at low density (10 cells per 60 mm dish) to proliferate rapidly, forming well-differentiated colonies by 10 days. In contrast, no or little proliferation took place in controls maintained in the presence of 2.5% serum.

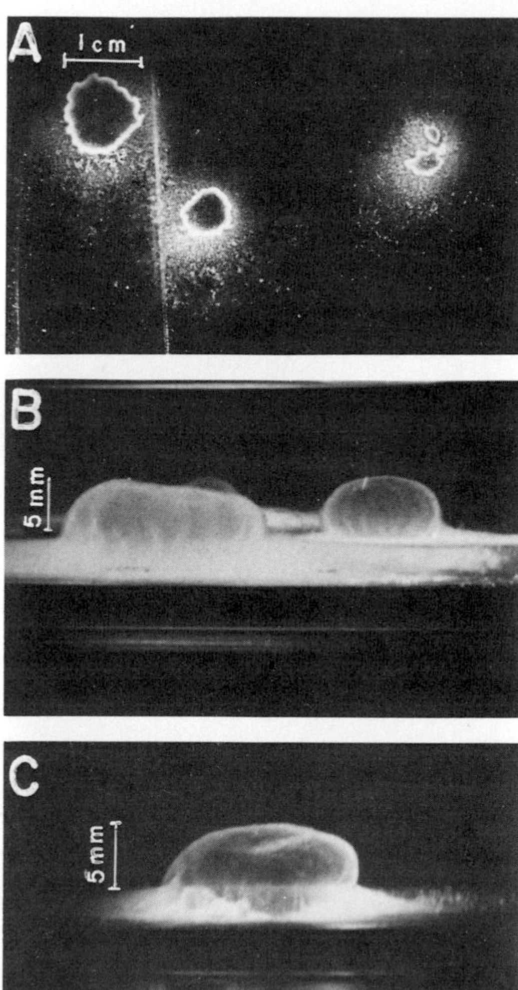

FIGURE 9 *Appearance of chondrocyte colonies after long-term culture. Chondrocytes obtained from rabbit ear by collangenase dissociation were plated at a density of 5 cells per 15 cm dish in the presence of 5% fetal calf serum and 100 ng/ml FGF in F 12 medium. Cultures were fixed in 10% formalin and photographs were taken after 3 weeks of culture. (A) Top view of three colonies. The dark centers of the colonies are cartilage; the light perimeters are chondrocytes. (B) Side view of the two colonies shown to the left in A. Cartilage projects 5 mm from the surface of the culture dish. (C) Side view of another colony.*

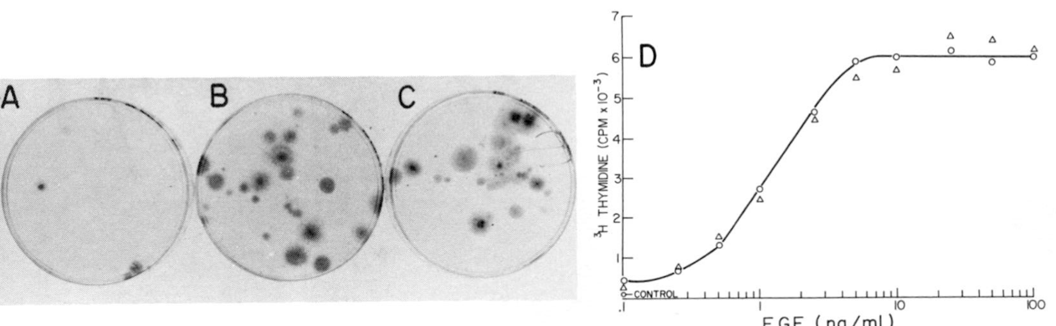

FIGURE 10 *(a,b,c) Comparison of the effect of brain and pituitary FGF on the clonal growth of chondrocytes plated at low density. Chondrocytes obtained by collagenase dissociation of rabbit-ear cartilage were plated at a concentration of 50 cells per 6 cm dish in the presence of 5 ml of F-12 medium with 2.5% fetal calf serum. The following additions were made. (A) none; (B) pituitary FGF (100 ng/ml); (C) brain FGF (100 ng/ml). Media were renewed once every five days. FGF was added every other day with each media change. After 12 days, the plates were washed and fixed with 10% formalin pH 7.2, and the cells were stained with alcian green to stain the cartilage in green and counterstained with metanil yellow which stains the chondrocytes in yellow--differentiated colonies appear as yellow with a green spot in the center (Ham, 1972) (on the figure, colonies are gray with dark spots indicating deep green staining). (d) Effect of increasing concentrations of brain and pituitary FGF on the initiation of DNA synthesis in rabbit-ear chondrocytes. The experiment was done as described by Gospodarowicz and Moran (1974a), using rabbit-ear chondrocytes instead of 3T3 cells with 4 × 10^4 cells per 35 mm dish. Each point represents the average of three plates. Brain FGF (o - o); pituitary FGF (Δ - Δ).*

The effect of FGF on the initiation of DNA synthesis in cultures of chondrocytes maintained in low serum was also measured. Two to 5 ng/ml of FGF stimulated DNA synthesis (Fig. 10D).

This proves that FGF is among the most potent known mitogenic agents for chondrocytes.

III. DISCUSSION

In view of the strong mitogenic effect of FGF on cells maintained in tissue culture, one may wonder what its physiological role could be and how our observations obtained using tissue culture techniques can be related to *in vivo* situations.

Two observations made in tissue culture can serve as a basis to suggest that FGF could be involved in the wound-healing process in higher vertebrates and in one of the early steps in regeneration in lower vertebrates. Although FGF has been isolated from the brain and pituitary, and we have not shown that it is present in the blood (We have not been able to look for it in the blood stream due to our inability to obtain antibody against it), we have been able to show that for cells which can grow in serum, but not in plasma, FGF can replace platelet extract. This suggests that FGF could be similar to platelet growth-promoting factors which have been implicated in the development of arteriosclerosis (Ross, *et al.*, 1974) and the wound-healing process.

As far as the wound-healing process is concerned, we have shown that FGF, from either brain or pituitary, is most active on cell types which are mesoderm-derived. Those cells are chondrocytes (Gospodarowicz, *et al.*, 1976a; Jones and Allison, 1975), myoblasts (Gospodarowicz, *et al.*, 1976b), fibroblasts (Gospodarowicz, 1974; Holley and Kiernan, 1974; Gospodarowicz and Moran, 1974a,b, 1975), endothelial cells (Gospodarowicz, *et al.*, 1976c,d, 1977a), adrenal cells from tumors (Gospodarowicz and Handley, 1975), as well as normal tissue (Gospodarowicz, *et al.*, 1976a,

1977b). FGF can, in these cell types, either replace serum (chondrocytes, myoblasts, 3T3) or have an additive effect over serum (endothelial cells, human foreskin fibroblasts). Out of all these cell types, two (endothelial cells and fibroblasts) are known to play an important role in the formation of the granulation tissue which is the first step of wound-healing. The effect of FGF on endothelial cells and smooth muscle of the arterial wall also calls our attention to its role in the formation of new capillaries which are formed in granulation tissue by endothelial cells.

The second observation relating FGF to the wound-healing process is that mitogenic activity similar to FGF from the pituitary can be obtained from neural tissue (Gospodarowicz, *et al.*, 1976a). The importance of nerves in the development of the blastema, one of the early steps in regeneration in lower vertebrates, was observed long ago, and it has been speculated that axons could release growth factor(s) involved in the proliferation of blastema cells (Singer, 1974). Furthermore, it has recently been shown that sensory axons release mitogenic factor(s) for Schwann cells (Wood and Bunge, 1975), and it has been observed that the mitogenic capacities of the axon may explain the *in vivo* observation of Thomas (1970) that repeated crushing of a peripheral nerve leads to a remarkable proliferation of Schwann cells in the distal nerve stump to such an extent that these cells are applied in multiple layers around a single axon. The importance of the axon in provoking Schwann cell proliferation is also indicated by observations on the proliferation of Schwann cells that occurs after a crush injury to normal (Logan, *et al.*, 1953) or dystrophic nerve (Stirling, 1975).

Since FGF has been shown by us (Gospodarowicz, *et al.*, 1975b) to induce the early steps of regeneration in frog limbs, namely the blastema formation, it could qualify as a neurotrophic factor. Experiments in tissue culture using the Schwann cell as a target cell may further reinforce the concept that brain FGF could be a neurotrophic factor.

ACKNOWLEDGMENTS

We thank J. Weseman, G. Greene, H. Bialecki, and D. Goldminz for excellent technical assistance. We thank Macmillan Journals, Ltd. and The Rockefeller University Press for permission to reproduce figures. Since this paper was limited in length, we have presented mostly our work; contributions by others to the field of growth factors are acknowledged in the references of the publications to which the reader is referred.

This work was supported by National Institutes of Health Grants (No. HD 07651 and HD 081801) and the American Cancer Society (BC 152 and VC 196).

REFERENCES

Armelin, H. (1973) *Proc. Nat. Acad. Sci. U.S.A. 70,* 2702-2706.

Aubery, M.. Bourrillon, R., and O'Neill, C.H. (1975) *Exptl. Cell Res. 93,* 47-54.

Burger, M. M. (1970) *Nature 227,* 170-171.

Corvol, M., Malemud, C.J., and Sokoloff, L. (1972) *Endocrinol. 90,* 262-271.

Gospodarowicz, D. (1974) *Nature 249,* 123-127.

Gospodarowicz, D. (1975) *J. Biol. Chem 250,* 2515-2520.

Gospodarowicz, D., and Handley, H. (1975) *Endocrinol. 97,* 102-107.

Gospodarowicz, D., and Moran, J. (1974a) *Proc. Nat. Acad. Sci. U.S.A. 71,* 4584-4588.

Gospodarowicz, D., and Moran, J. (1974b) *Proc. Nat. Acad. Sci. U.S.A. 71,* 4648-4652.

Gospodarowicz, D., and Moran, J. (1975) *J. Cell Biol. 66,* 451-457.

Gospodarowicz, D., and Moran, J. (1976) *Ann. Rev. Biochem., 45,* 531-558.

Gospodarowicz, D., Greene, G., and Moran, J. (1975a) *Biochem. Biophys. Res. Com. 65,* 779-787.

Gospodarowicz, D., Rudland, P., Lindstrom, J., and Benirschke, K. (1975b) *Advances in Metab. Dis. 8*, 302-335.

Gospodarowicz, D., Moran, J., and Bialecki, H. (1976a) *Third Int. Symp. on Growth Hormone and Related Peptides. Excerpta Medica.* Elsevier, New York, 141-156.

Gospodarowicz, D., Weseman, J., Moran, J., and Lindstrom, J. (1976b) *J. Cell Biol. 70,* 395-406.

Gospodarowicz, D., Moran, J., Braun, D., and Birdwell, C. R., (1976c) *Proc. Nat. Acad. Sci. U.S.A. 73,* 4120-4124.

Gospodarowicz, D., (1976d) *In* Progr. In Clin. and Biol. Res-Membranes and Neoplasia. New Approaches and Strategies (V. T. Marchesi, ed.) A. R. Liss Inc., New York, p. 1-19.

Gospodarowicz, D., Moran, J., Braun, D., *J. Cell. Physiol.* (1977a) in press.

Gospodarowicz, D., Ill, C. R., Hornby, P., Gill, G., (1977b) *Endocrinology,* April.

Ham, R.G. (1972) in *Methods in Cell Physiology 5,* 37-74.

Hershko, A., Mamont, P., Shields, R., Tomkins, G. (1971) *Nature New Biol. 232,* 206-211.

Holley, R.W., and Kiernan, J.A. (1968) *Proc. Nat. Acad. Sci. U.S.A. 60,* 300-304.

Holley, R.W., and Kiernan, J.A. (1974) *Proc. Nat. Acad. Sci. U.S.A. 71,* 2908-2911.

Jones, K., and Allison, J. (1975) *Endocrinology 97,* 359-365.

Logan, J.E., Rossiter, R.J., and Barr, M.L. (1953) *J. Anat. 87,* 419.

Morgan, J., Gospodarowicz, D., and Owashi, N. (1976) *J. Clin. Endocrinology Metab.* in press, April.

Ross, R., Glomset, J., Kariya, B., and Harker, L. (1974) *Proc. Nat. Acad. Sci. U.S.A. 71,* 1207-1210.

Rudland, P.A., Seifert, W., and Gospodarowicz, D. (1974a) *Proc. Nat. Acad. Sci. U.S.A. 71,* 2600-2604.

Rudland, P.S., Eckhart, W., Gospodarowicz, D., and Seifert, W. (1974b) *Nature 250,* 337-339.

Singer, M. (1974) *Ann. N.Y. Acad. Sci. 228,* 308-322.

Stirling, C.A. (1975) *Brain Res.* 87, 130.

Todaro, G.J., and Green, H. (1963) *J. Cell Biol. 17,* 229-313.

Thomas, P.K. (1970) *J. Anat. 106,* 463.

Wood, P.M. and Bunge, R.P. (1975) *Nature 256,* 662-664.

HUMAN EPIDERMAL GROWTH FACTOR

GRAHAM CARPENTER
STANLEY COHEN

Department of Biochemistry
Vanderbilt University
Nashville, Tennessee

I. INTRODUCTION

Epidermal growth factor isolated from mouse submaxillary glands (mEGF) is a single chain polypeptide (MW 6045) that contains 53 amino acid residues and three disulfide bonds. The amino acid sequence and location of the disulfide bonds are shown in Fig. 1. *In vivo* and in organ culture systems, mEGF stimulates the proliferation and keratinization of epidermal tissue. Recent reports have shown that mEGF also is a potent mitogen when added to fibroblast cell cultures (Armelin, 1973; Hollenberg and Cuatrecasas, 1973; Cohen *et al.*, 1975). The biological and chemical properties of mEGF have been reviewed (Cohen and Taylor, 1974; Cohen and Savage, 1974; Cohen *et al.*, 1975).

Although mEGF promotes the proliferation of cells derived from a number of species (mouse, human, rabbit, chick) the presence of EGF-like molecules in species other than the rodent has

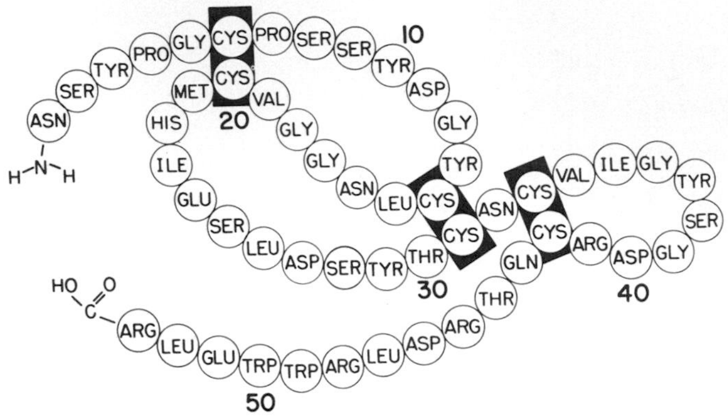

FIGURE 1 *Amino acid sequence and location of disulfide linkages of mouse-derived EGF. (from Savage et al., 1973).*

not, until very recently, been demonstrated. In this report we summarize our data regarding the isolation and biological properties of human EGF.

II. ISOLATION OF EGF FROM HUMAN MATERIAL

Two approaches were employed in our attempts to detect and isolate EGF-like molecules from human urine. The first method consisted of passing urine through an affinity column containing rabbit antibodies to mouse EGF; the second approach was a direct isolation from a protein concentrate of urine using an assay based on the competition of human EGF (hEGF) with ^{125}I-labeled mouse EGF for receptors present in human fibroblasts.

A. Affinity Chromatography

Starkey *et al.* (1975) were able to detect EGF-like molecules in an extract of human urine. Pooled human urine was passed through a column containing rabbit antiserum to mouse EGF covalently linked to cyanogen bromide-activated agarose beads. Adsorbed material was eluted from the column with formic acid and concentrated by lyophilization. The material was tested both by radioimmunoassay and by growth promoting activity in organ cultures. The competitive binding curves for mouse EGF and the human urine extract to rabbit antibodies to mouse EGF were very similar, suggesting that the substances are immunologically related. The biological activity of the preparation was assayed in organ cultures of the chick embryo cornea (Cohen and Savage, 1973). Mouse EGF produces a proliferation of the corneal epithelium; the urine extract produced the same, histologically evident, biological response. The biological effects of both mouse EGF and the human urine extract were completely blocked by the addition of antiserum to mouse EGF to the cultures.

These data suggested that the material in human urine eluted from the affinity column was biologically and immunologically similar to mouse EGF. The observation that the amount of human EGF detected by radioimmunoassay was orders of magnitude less than the amount detected by the bioassay suggested also that human EGF is less reactive to antibodies to mouse EGF than is mouse EGF itself (Starkey *et al.*, 1975).

B. Purification from Urine Powder

The isolation of microgram quantities of epidermal growth factor from a protein concentrate of human urine by conventional purification methods was dependent on the development of a quick, sensitive and specific assay. The lack of such an assay has hindered the attempts of many workers to isolate growth factors from biological material. However, the fibroblast receptor assay described below provides an excellent assay for the isolation of human epidermal growth factor.

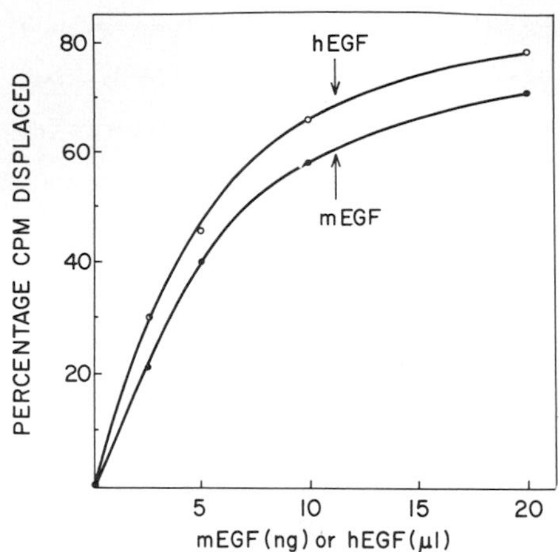

FIGURE 2 *Competitive fibroblast binding assay for mEGF and hEGF in the presence of ^{125}I-labeled mEGF. The binding medium consisted of 1.5 ml of an albumin-containing modified Dulbecco medium. Increasing concentrations of mEGF or hEGF were added simultaneously with ^{125}I-labeled mEGF (3.4 ng, 1.2 × 10^5 cpm) to a monolayer culture of human foreskin fibroblasts. Controls contained only the labeled polypeptide and bound approximately 6000 cpm/10^6 cells during the 1-hr incubation period. "Nonspecific" binding, determined by measuring the cell-bound radioactivity in the presence of excess mEGF (15 μg/ml), amounted to less than 3% of the total (from Cohen and Carpenter, 1975).*

1. *Fibroblast Receptor Assay for EGF*

The assay is based on the ability of both mouse and human EGF to compete with ^{125}I-labeled mEGF for binding sites on human foreskin fibroblasts. A typical standard curve obtained by ascertaining the effects of increasing quantities of mEGF (or hEGF) on the binding of a standard amount of ^{125}I-labeled mEGF is shown in Fig. 2. Under these conditions, 2-20 ng of mEGF are readily measurable. During the isolation procedures, the quantities of hEGF present in the fractions isolated from human urine are expressed as equivalents (by weight) of mouse EGF determined by the competitive binding assay. The absolute quantities of hEGF, as determined by amino acid analyses, were 33% to 50% of the values obtained as equivalents of mEGF by the competitive binding assay.

As previously reported (Carpenter *et al.*, 1975), no competition with ^{125}I-labeled mEGF could be detected with a wide variety of known peptide hormones.

2. Preparation of Human Epidermal Growth Factor

Human epidermal growth factor was isolated from urine protein concentrates as described by Cohen and Carpenter (1975). Ten grams of benzoic acid/acetone powder of urinary proteins were extracted with water at pH 9 and the insoluble material discarded. This extract contained approximately 800 mg of protein (Lowry *et al.*, 1951) and 600 μg of hEGF.

a. Gel filtration on Bio-Gel P-10 The material was applied to a Bio-Gel P-10 column (Fig. 3A). The fractions containing binding activity (between the arrows) were combined and lyophilized. Approximately 80 mg of protein and 550 μg of hEGF were recovered. (It should be noted that the A_{280} values shown in the figures result from both protein and pigment content.)

b. Gel filtration on Sephadex G-50 The lyophilized material derived from the Bio-Gel column was dissolved in 3 ml of water and applied to a Sephadex G-50 column (Fig. 3B). The fractions between the arrows were combined and lyophilized. Approximately 45 mg of protein and 400 μg of hEGF were recovered.

c. Passage through DE-52 cellulose, pH3.0 The lyophilized fraction after Sephadex G-50 chromatography was dissolved in 0.03 *M* formic acid (final pH 3.0) and applied to a column of DE-52 cellulose equilibrated with 0.03 *M* ammonium formate buffer, pH 3.0. The column was washed with 40 ml of the same buffer, and the eluate, containing approximately 28 mg of protein and 380 μg of hEGF, was lyophilized.

d. Ion-exchange chromatography on CM-52 cellulose At this stage of the purification, the lyophilized powders from two of the DE-52 preparations were combined. The sample was dissolved

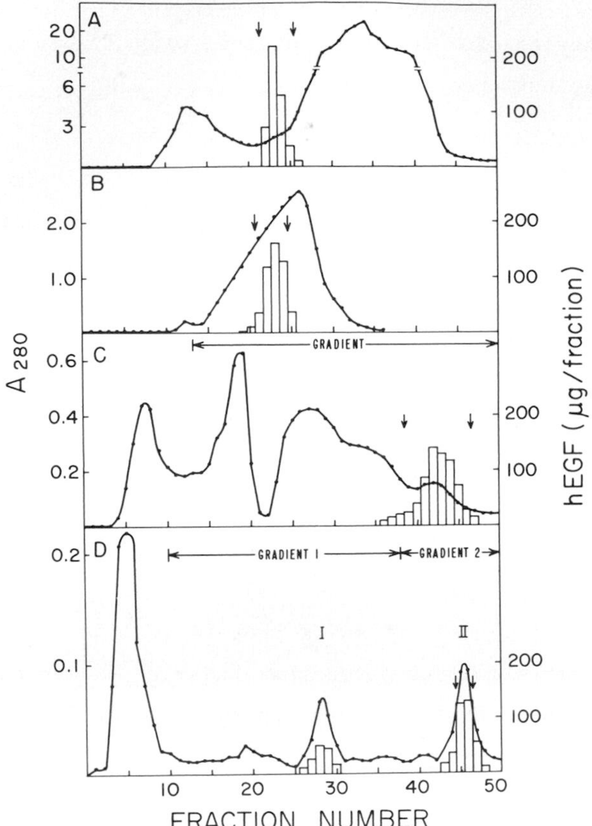

FIGURE 3 *Gel filtration and ion-exchange chromatography of hEGF. In each figure, the bar graph indicates the competitive binding potency of each fraction, expressed as equivalents (by weight) of mEGF. (A) The extract from 10 g of acetone powder, containing 800 mg of protein, was applied to a Bio-Gel P-10 column. (B) The Bio-Gel fraction, containing 80 mg of protein, was applied to a Sephadex G-50 column. (C) The DE-52, pH 3.0, fraction (two preparations), containing 56 mg of protein, was applied to a column of CM-52 cellulose. The protein was eluted with an ammonium acetate gradient. (D) The CM-52 fraction, containing 6 mg of protein, was applied to a column of DE-52 cellulose. The protein was eluted with two successive gradients of ammonium acetate buffer (from Cohen and Carpenter, 1975).*

in 0.04 *M* acetic acid (final pH 3.8 - 4.0) and applied to a column of CM-52 cellulose equilibrated with 0.04 *M* ammonium acetate, pH 3.8. After the column was washed with 50 ml of the buffer, a gradient (0.04 *M* to 2.0 *M* ammonium acetate, pH 3.8) was applied. A typical elution pattern is illustrated in Fig. 3C. The fractions between the arrows were combined and lyophilized. Approximately 6 mg of protein and 600 μg of hEGF were recovered.

e. Ion exchange chromatography on DE-52 cellulose, pH 5.6 The lyophilized powder from the CM-52 chromatographic separation was dissolved in 0.02 *M* ammonium acetate buffer, final pH 5.6, and applied to a column of DE-52 cellulose equilibrated with the same buffer. The column was washed with 20 ml of the buffer, and then a gradient buffer (0.02 - 0.2 *M* ammonium acetate, pH 5.6) was applied. Seventy milliliters of this gradient sufficed to elute a peak containing binding activity (peak I, Fig. 3D). A second gradient was applied (0.2 - 1.0 *M* ammonium acetate, pH 5.6). A second peak of active material (peak II, Fig. 3D) was thus obtained. All of the subsequent data were obtained with the pooled fractions of peak II, which contained approximately 250 μg of hEGF. The absolute amount of protein present in such small quantities is difficult to ascertain. As noted previously, the amount of protein is one-third to one-half of the value indicated by the competitive binding assay. The absolute amount of hEGF obtained in the pooled fractions of peak II, therefore, was approximately 80 - 125 μg of hEGF.

III. CHEMICAL PROPERTIES OF hEGF

A. Electrophoresis

The purity of the final preparation was examined by polyacrylamide disc gel electrophoresis under acid and alkaline conditions (Fig. 4). It may be seen that in each instance, hEGF migrates as a single band. Human EGF and mouse EGF migrate at approximately the same rate at pH 2.3. However, under alkaline conditions, hEGF migrates much more rapidly, suggesting that the

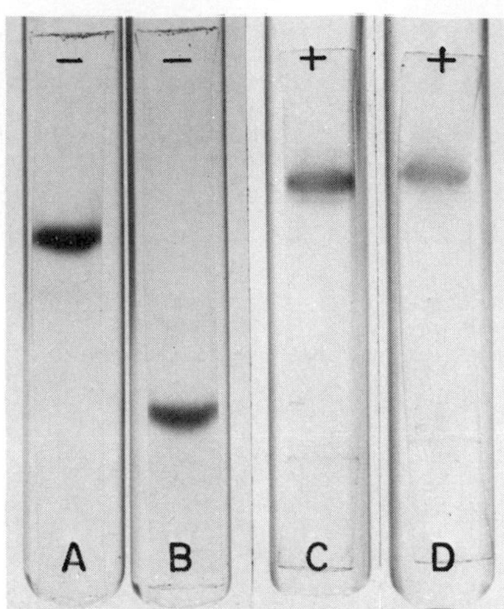

FIGURE 4 *Disc gel electrophoresis of hEGF and mEGF. Tubes A and C contain mEGF; tubes B and D contain hEGF. The pH of the gels in tubes A and B was 9.5 and in tubes C and D, 2.3. Samples of 10-20 μg of protein were applied (from Cohen and Carpenter, 1975).*

net charge of hEGF at pH 9.5 is more negative than its mouse counterpart.

To establish whether the competitive binding activity of hEGF was associated with the stained band observed in the gel, we performed the following experiment. The alkaline gel (gel B, Fig. 4) was sliced into 1 mm sections, and each segment was fragmented in 400 μl of 10% $NaHCO_3$ containing albumin and incubated overnight at 5°. The extract from each slice was assayed by competition with ^{125}I-labeled mEGF for binding to fibroblasts. Competitive binding activity was associated only with those fractions corresponding to the stained area of the gel.

B. Amino Acid Analysis

The results of amino acid analyses of 20 - 43 μg samples of hEGF are shown in Table I, together with a comparison with mEGF.

TABLE I Amino Acid Composition of Epidermal Growth Factors[a]

Amino acid	Residues per mole of protein: Probable composition of hEGF	Composition of mEGF
Lys	3	0
His	2	1
Arg	2	4
Asp	7	7
Thr	0	2
Ser	3	6
Glu	5	3
Pro	2	2
Gly	5	6
Ala	2	0
Half-Cys	6	6
Val	2	2
Met	1	1
Ile	2	2
Leu	4	4
Tyr	2	5
Phe	0	0
Trp	1	2
Total residues	49	53
Molecular weight	5458	6045

[a]From Cohen and Carpenter, 1975.

In view of the very small quantities of hEGF available for analysis, these data must be considered as preliminary. It is clear, however, that the two molecules exhibit both differences and similarities with respect to their amino acid compositions. The minimal molecular weight of hEGF was estimated to be approximately 5500, assuming five glutamic acid residues per mole of protein.

C. Molecular Weight

The molecular weight of hEGF, as estimated by gel filtration, was approximately 5700. Gel filtrations were carried out on a

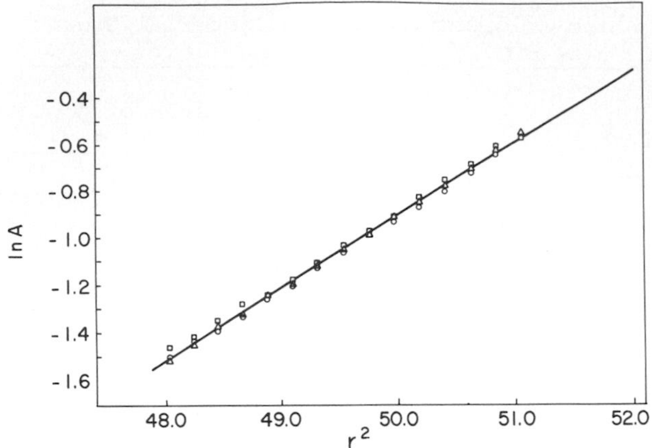

FIGURE 5 *Sedimentation equilibrium plot of ln A against the square of the radial distance (r) for hEGF. The three sets of symbols represent three separate scans. The least squares line is shown (from Cohen and Carpenter, 1975).*

column of Bio-Gel P-10, with cytochrome c, pancreatic trypsin inhibitor, and bacitracin as standard molecular weight markers in a solvent consisting of 0.1 *M* ammonium acetate. The elution volume of hEGF on the calibrated column was determined by the competitive binding assay.

The sedimentation equilibrium of hEGF was examined by Dr. Leslie Holladay at Vanderbilt University. Linear plots of ln concentration against r^2 (Fig. 5) revealed no significant heterogeneity, and a weight average molecular weight of 5291 was calculated.

IV. BIOLOGICAL PROPERTIES OF hEGF

The biological effects of mEGF include (i) stimulation of the growth of human foreskin fibroblasts, (ii) hypertrophy and hyperplasia of corneal epithelial cells in organ cultures, and (iii) induction of precocious eyelid opening when injected into newborn mice. All three effects have been duplicated, at least qualitatively, with pure hEGF.

A. Biological Activity of hEGF in Newborn Mice

Human EGF was assayed for precocious eyelid-opening activity by the daily subcutaneous injection into a newborn mouse (Cohen, 1962). Control mice opened their eyes at 14 days; mEGF (1 μg/g per day or 0.25 μg/g per day) resulted in eyelid opening on day 9 or 11, respectively; hEGF (0.4 μg/g per day) resulted in eyelid opening on day 11. Human EGF thus appeared to be active in this *in vivo* mouse assay.

B. Biological Activity of hEGF in Organ Culture

Mouse EGF and hEGF, when assayed with organ cultures of the chick embryo cornea (Savage and Cohen, 1973), were equally effective in causing the proliferation of the corneal epithelium. Histological results (not shown) identical to those previously described for mEGF (Cohen and Savage, 1974) and recently described for hEGF (Starkey *et al.*, 1975), purified partially by affinity chromatography, were observed. The biological effects of both mouse and human EGF on the corneal epithelium were completely and specifically inhibited by the addition of excess quantities of the gamma globulin fraction prepared from a rabbit antiserum against mEGF.

C. Biological Activity of hEGF on Human Fibroblasts

The effect of human epidermal growth factor (hEGF) on the growth of human foreskin fibroblasts (HF cells) *in vitro* was studied by measuring cell numbers, incorporation of labeled thymidine, and autoradiography.

1. Effect of hEGF on the Proliferation of Human Fibroblasts.

The data in Fig. 6 illustrate the effect of the addition of picomolar quantities (4 ng per ml) of hEGF on the growth of human fibroblasts plated at a low density in Dulbecco's Modified Eagle Medium (DM) + 10% calf serum. Cells grown in the presence of hEGF grew to a saturation density of 2×10^5 cells per cm^2, approximately four-fold higher than the final density of 5×10^4 cells per cm^2 reached by cells grown in the absence of hEGF. The saturation density reached by the control cultures was not limited by the depletion of nutrients from the medium since no net increase in cell number was observed when new growth medium was added. These results also indicate that the cells grown in the presence or absence of hEGF had average generation times of 37 hrs or 45 hrs, respectively.

Antibodies to hEGF prepared from rabbit serum were able to bind hEGF with a sufficiently high affinity to completely inhibit the growth promoting effect of hEGF.

The morphological appearance of cultures grown to their saturation densities in the presence or absence of hEGF, is shown in Fig. 7. The cells grown in the absence of hEGF formed a tightly packed cell monolayer with typical fibroblastic morphology and orientation. These cultures showed only occasional areas of nuclear overlap. In contrast, the cells cultured in the presence of hEGF, while, in general, retaining their fibroblastic appearance and orientation, grew in multiple layers and exhibited extensive areas of nuclear overlap.

The human fibroblasts used in this study did not grow well in medium containing 10% gamma globulin-free calf serum. In the presence of hEGF, however, HF cells grew well in this medium and reached a final cell density equivalent to that reached in a medium containing 10% calf serum (Table II). These results indicate that hEGF may be able to substitute for the undefined growth factor(s) present in the gamma globulin fraction of serum.

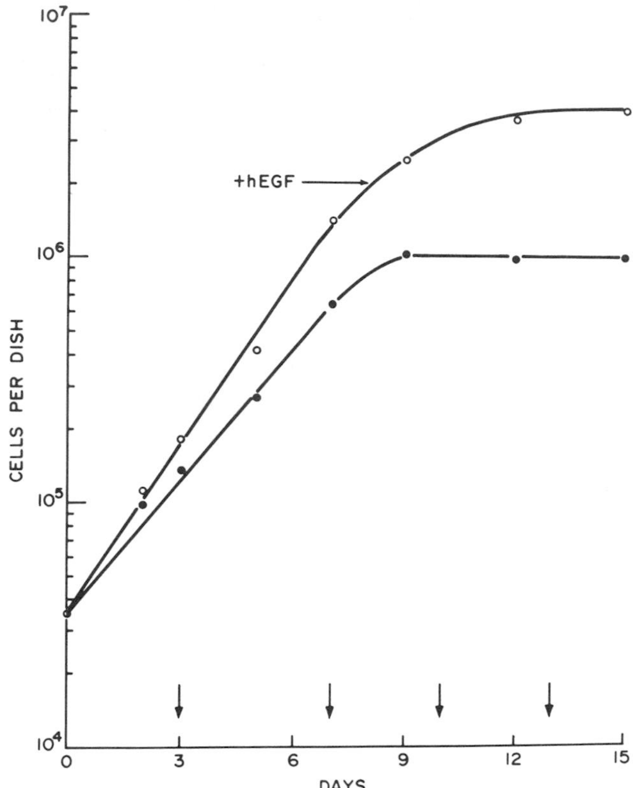

FIGURE 6 *Effect of hEGF on the growth of HF cells in medium containing 10% calf serum. HF cells were plated at approximately 3 × 10^4 cells per dish into 60 mm Falcon culture dishes containing DM plus 10% calf serum. The cells were incubated overnight, and hEGF (4 ng/ml) was added to one-half the dishes (day 0). At indicated times thereafter duplicate dishes from cultures growing in the presence (o) or absence (●) of hEGF were removed and the cell numbers determined. At the times indicated by the arrows, the medium in each set of cultures was removed, and fresh DM plus 10% calf serum was added. Human EGF was also added to the appropriate dishes at these times.*

Diploid human fibroblasts do not grow in a medium containing 10% gamma globulin-free calf serum or 1% calf serum. These cells do grow well in the presence of 10% calf serum but are subject to density dependent inhibition of growth (DDIG) when a confluent

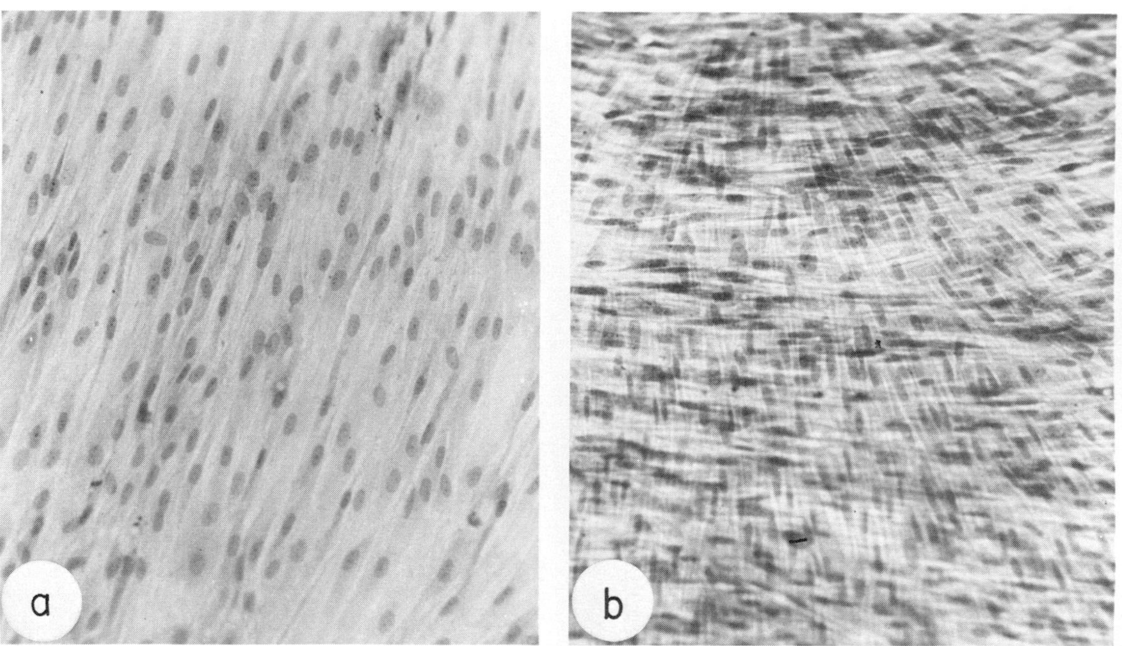

FIGURE 7 *Effect of hEGF on the morphological appearance of HF cells to saturation density in medium containing 10% calf serum. HF cells were grown to maximal saturation densities in DM containing 10% calf serum with (B) or without (A) hEGF.*

TABLE II Effect of hEGF on Growth of HF Cells in Presence of Gamma Globulin-free Calf Serum[a]

Days after plating	Cells per dish	
	-hEGF	+hEGF
6	184,600	622,400
11	288,467	1,667,000

[a]Trypsinized HF cells were washed twice with serum-free medium, and approximately 10^5 cells were added to 60 mm Falcon dishes containing DM plus 10% gamma globulin-free calf serum. Human EGF was added to one-half the dishes to a final concentration of 4 ng/ml. Duplicate dishes from each set of cultures were removed at indicated times and the cell numbers determined. The medium in each set of cultures was replaced with fresh medium on day 6.

monolayer of cells is formed. HF cells grown in a medium containing 10% calf serum and hEGF, however, do not stop dividing when a confluent monolayer is formed but grow to significantly higher saturation density with the formation of multiple cell layers. Also, HF cells grown in medium containing 10% gamma globulin-free calf serum plus hEGF or 1% calf serum plus hEGF proliferate and reach a saturation density typical of cells grown in the presence of 10% calf serum. The growth of HF cells in media containing hEGF, therefore, is not characterized by several growth parameters which would otherwise limit cell proliferation.

2. *Stimulation of DNA synthesis by hEGF*

Since, in the presence of hEGF, confluent monolayer cultures of HF cells are not as subject to DDIG as are cultures grown in the absence of this mitogen, the following experiments were performed to characterize some of the parameters which may be involved in the hEGF-mediated stimulation of proliferation of cells subject to DDIG.

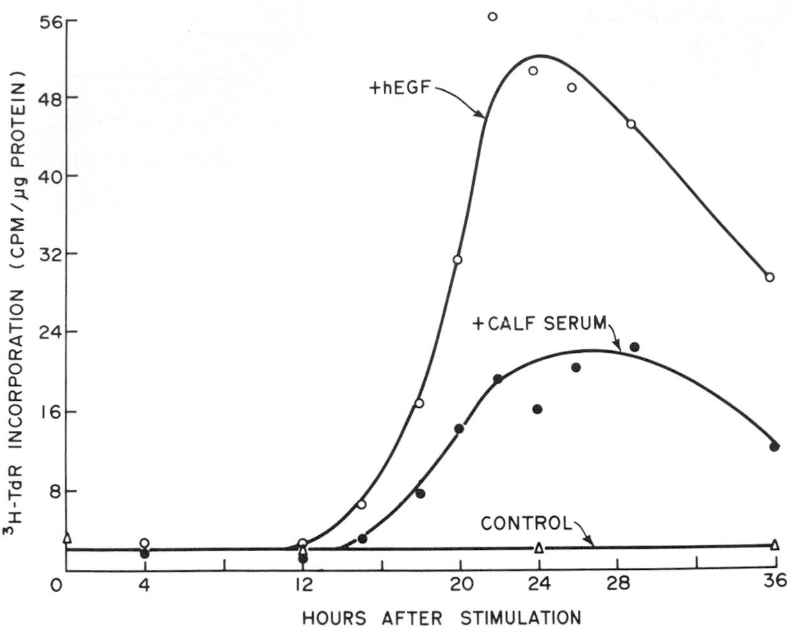

FIGURE 8 *Time course of ^{3}H-thymidine incorporation following stimulation of HF cells. Confluent, quiescent cultures of HF cells were stimulated by the addition of hEGF, 4 ng/ml (o) or fresh calf serum, 10% (●). Control cultures received no additions (Δ). At indicated times duplicate cultures were selected and labeled for one hour, 1 µC per ml and 0.15 µM with ^{3}H-thymidine.*

The time course of DNA synthesis, as judged by the incorporation of labeled thymidine, following the addition of hEGF or fresh calf serum to confluent, quiescent HF cells is represented in Fig. 8. Under these conditions, an increased rate of DNA synthesis was detectable after 12 hrs of incubation in the presence of hEGF or fresh serum; the maximal stimulation occurred at approximately 24 hrs. The effect of increasing concentrations of hEGF on the stimulation of DNA synthesis is presented in Fig. 9. Maximal stimulation of labeled thymidine incorporation occurred in the presence of 2 ng/ml (3.7×10^{-10} *M*) hEGF; half-maximal stimulation was observed at a concentration of approximately 0.25

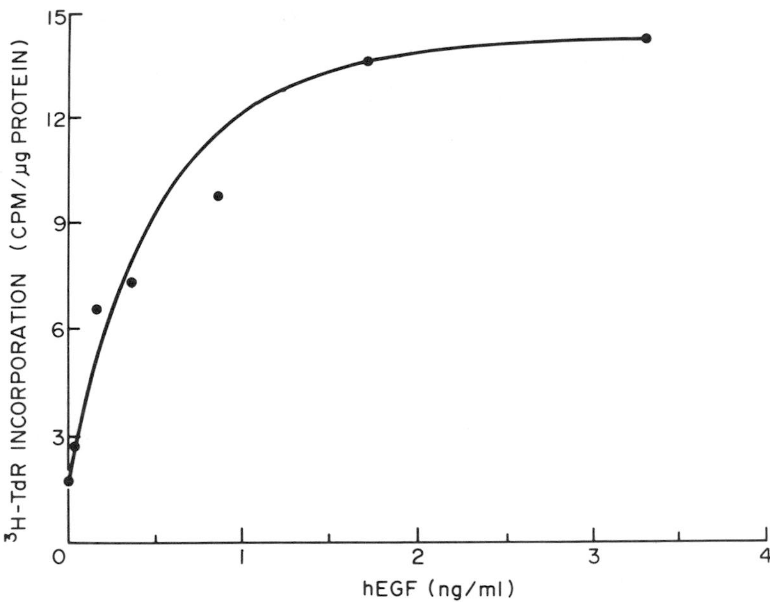

FIGURE 9 *Effect of hEGF concentration on the stimulation of ^{3}H-thymidine incorporation in HF cells. Varying concentrations of hEGF were added to confluent, quiescent cultures of HF cells. Twenty hours later ^{3}H-thymidine (1 μC per ml and 2.5 μM thymidine) was added, and the cells were labeled for 4 hrs.*

ng/ml (4.6×10^{-11} *M*) hEGF. The concentration of hEGF required for maximal stimulation is similar to the concentrations of mouse EGF, 3×10^{-10} *M* (Hollenberg and Cuatrecasas, 1973) and 8×10^{-10} *M* (Cohen *et al.*, 1975), required for the maximal stimulation of ^{3}H-thymidine incorporation into human fibroblasts under similar experimental conditions. These are concentrations comparable to the concentrations of mEGF in mouse plasma, 2.5×10^{-10} *M* (Byyny *et al.*, 1974), and hEGF in human urine, 3.3×10^{-8} *M* (unpublished results). The growth factor activity in human urine described by Holley and Kiernan (1968) may be due to hEGF.

The stimulation of DNA synthesis by hEGF in quiescent HF cells was significantly affected by the amount of serum in the medium and by the presence of added ascorbic acid. These data indicate that the response of HF cells to hEGF is greatest at serum concentrations above 2% and at approximately 2.5 µg/ml of ascorbic acid. A similar serum requirement has been found for the biological activities of mEGF (Cohen *et al.*, 1975) and fibroblast growth factor (Gospodarowicz and Moran, 1975) on human fibroblasts. Since HF cells incubated in serum-free medium remain fully responsive to the addition of fresh serum, it does not appear that serum is required simply to maintain cell viability under these experimental conditions. The results suggest that hEGF acts in a synergistic manner with undefined serum components. Although ascorbic acid is known to enhance the hydroxylation of collagen by fibroblasts, it is not known whether this is the mechanism whereby it promotes the mitogenic activity of hEGF.

To determine the percentage of cells which were stimulated to synthesize DNA following the addition of hEGF to quiescent HF cells, autoradiography of cells grown in the presence of labeled thymidine was performed. The results of these experiments (Table III) showed that following the addition of hEGF to confluent, quiescent cultures maintained in medium containing 1% or 5% calf serum, 21% or 41% of the nuclei, respectively, contained radioactivity. The addition of ascorbic acid plus hEGF to the medium increased the number of labeled nuclei to 34% or 56% in cultures maintained in 1% or 5% calf serum, respectively. In the absence of hEGF under the conditions employed, not more than 3.7% of the nuclei were labeled. The data in Figs. 6 and 7 suggest that confluent monolayers of HF cells which have reached their final saturation density do not respond to the addition of fresh serum; this was confirmed by the autoradiographic experiments which indicated that only 11-14% of the nuclei were labeled following the addition of fresh calf serum (10%).

To examine the question of how long a period of time a culture of HF cells must be exposed to hEGF before the stimulation of ^{3}H-thymidine incorporation is detected, antibody prepared against hEGF was added to parallel cultures of HF cells at vari-

TABLE III Autoradiographic Examination of HF Cells Stimulated by hEGF[a]

	Percent labeled nuclei	
Additions	Cells maintained in 1% serum	Cells maintained in 5% serum
None	0.9	3.0
hEGF	21.2	40.8
hEGF plus ascorbic acid	33.5	55.9
Ascorbic acid	1.3	3.7
Calf serum	14.6	11.2

[a]The medium in confluent cultures of HF cells, grown in the presence of 10% calf serum, was replaced by fresh medium containing either 1% or 5% calf serum. After incubation for 48 hrs in these media, the following additions were made; hEGF (4 ng/ml), ascorbic acid (25 μg/ml), fresh calf serum (10%). The cells were labeled with ^{3}H-thymidine and autoradiography performed.

ous times after the addition of hEGF. The results of this experiment (Fig. 10) indicate that the addition of antibody, at any time during the first 3 hrs of exposure of HF cells to hEGF, blocked any significant stimulation of thymidine incorporation. The addition of antibody at 7.5 hrs after incubation of cells with hEGF resulted in a stimulation that was 50% of that observed in the absence of antibody. These results suggest that hEGF must be present in the medium for several hours before the cells are committed to DNA synthesis. Similar data for the stimulation of human fibroblast cells by serum (Ellem and Mironescu, 1972), chick fibroblasts by serum (Rubin and Steiner, 1975) or multiplication-stimulating factors (Smith and Temin, 1974) and lymphocytes by Con A (Gunther *et al.*, 1974) have been reported.

Since the experiments described above were performed with confluent cultures of HF cells, the ability of hEGF to stimulate the incorporation of ^{3}H-thymidine in quiescent cells at different

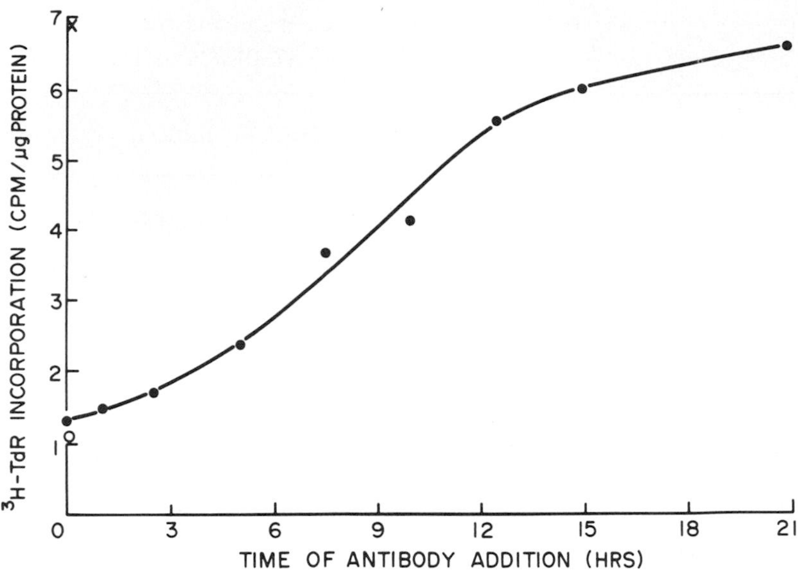

FIGURE 10 *Effect of the addition of antibody to hEGF on the stimulation of ^{3}H-thymidine incorporation into HF cells by hEGF. Human EGF (4 ng/ml) was added to quiescent, confluent cultures of HF cells and at various times thereafter 150 μg of DEAE purified antibody to hEGF were added to duplicate cultures. Twenty-one hours after the addition of hEGF to the cultures, ^{3}H-thymidine was added (1 μC per ml and 2.5 μM thymidine) and the cells labeled for 4 hrs. Control cultures received hEGF but no antibody (X) or no additions (O).*

cell densities was investigated. Human EGF was effective at all cell densities tested (5.6×10^3 to 6×10^4 cells/cm^2). Thus, hEGF is an effective mitogen in both confluent and sparse cell cultures.

V. DISCUSSION

Human EGF was isolated by utilizing its ability to compete with ^{125}I-labeled mouse EGF in binding to human foreskin fibroblasts in culture. Ten grams of the starting material (an acetone powder obtained from approximately 15 liters of urine from pregnant women) contained approximately 700 μg equivalents of hEGF. Approximately 100-150 μg of pure hEGF were isolated by the procedures outlined in this paper. Although we have made use of urine from pregnant women, hEGF is not unique to pregnancy in that it is also present in urine from adult males.

The biological effects of hEGF, at least qualitatively, are similar to those previously described for mouse EGF. These include the stimulation of the growth of cultured human foreskin fibroblasts and corneal epithelial cells of the chick embryo in organ culture, and the *in vivo* induction of precocious eyelid opening in newborn mice.

Although hEGF appears to be biologically similar to mEGF, the physical and chemical properties of the two molecules are not identical. Human EGF appears to have a slightly lower molecular weight and a greater net negative charge at an alkaline pH than the mouse EGF. The amino acid compositions of the two polypeptides indicate certain similarities, such as the absence of phenylalanine and the presence of one methionyl and six half-cystinyl residues per mole.

The radioreceptor assay proved far more sensitive than a radioimmune assay based on cross reaction with antibodies to the mouse polypeptide, suggesting a closer relationship between the receptor binding sites than between the antigenic sites on the human and mouse polypeptides.

There is at present no direct evidence for the role of these growth factors in normal development and cell control. However, the presence of EGF-like molecules in both man and mouse suggests that an important function does exist.

ACKNOWLEDGMENTS

The technical assistance of Mrs. Martha Reich and Mrs. Brenda Crews is greatly appreciated. These studies were supported by U.

S. Public Health Service Grant HD-00700. G. C. is a postdoctoral fellow of the USPHS, 5 F22 AM01176-01.

REFERENCES

Armelin, H. A. (1973). *Proc. Nat. Acad. Sci, U.S.A. 70:* 2702-2706.

Byyny, R. L., Orth, D. N., Cohen, S. and Doyne, E. S. (1974). *Endocrinology 95:* 776-782.

Carpenter, G., Lembach, K. J., Morrison, M. M. and Cohen, S. (1975). *J. Biol. Chem. 250;* 4297-4304.

Cohen, S. (1962). *J. Biol. Chem. 237:* 1555-1562.

Cohen, S. and Carpenter, G. (1975). *Proc. Nat. Acad. Sci., U.S.A. 72:* 1317-1321.

Cohen, S. and Savage, C. R., Jr. (1974). *In* "Recent Progress in Hormone Research" (R. O. Greep, ed.), Vol. 30, pp. 551-574. Academic Press, New York.

Cohen, S. and Taylor, J. M. (1974). *In* "Recent Progress in Hormone Research" (R. O. Greep, ed.), Vol. 30, pp. 533-550. Academic Press, New York.

Cohen, S., Carpenter, G. and Lembach, K. J. (1975). *Advances Metab. Dis.* (in press).

Ellem, K. A. O. and Mironescu, S. (1972). *J. Cell. Physiol. 79:* 389-406.

Gospodarowicz, D. and Moran, J. S. (1975). *J. Cell Biol. 66:* 451-457.

Gunther, G. R., Wang, J. L. and Edelman, G. M. (1974). *J. Cell Biol. 62:* 366-377.

Hollenberg, M. D. and Cuatrecasas, P. (1973). *Proc. Nat. Acad. Sci., U.S.A. 70:* 2964-2968.

Holley, R. W. and Kiernan, J. A. (1968). *Proc. Nat. Acad. Sci., U.S.A. 60:* 300-304.

Lowry, O. H., Rosebrough, N. J., Farr, A. L. and Randall, R. J.

(1951). *J. Biol. Chem. 193*, 265-275.

Rubin, H. and Steiner, R. (1975). *J. Cell. Physiol. 85:* 261-270.

Savage, C. R., Jr. and Cohen, S. (1973). *Exp. Eye Res. 15:* 361-366.

Savage, C. R., Jr., Hash, J. H. and Cohen, S. (1973). *J. Biol. Chem. 248:* 7669-7672.

Smith, G. L. and Temin, H. M. (1974). *J. Cell. Physiol. 84:* 181-192.

Starkey, R. H., Cohen, S. and Orth, D. N. (1975). *Science 189:* 800-802.

SUMMARY REMARKS

GORDON SATO

In this session, we have the signs of a revolution in progress. Its groundwork was laid with the discovery of tissue culture and its important elements have been slowly constructed over the past seventy years. With the accelerating accumulation of information, we can see that tissue culture will profoundly change the course of physiological research and our understanding of integrated physiology.

Let me shift our perspective slightly by describing some of the activities of my laboratory. We have been working for some time on growth factors and the development of hormone dependent cultures. Our experience has caused us to wonder why we always add serum to culture medium, and the hypothesis we have developed is that serum is mainly a source of hormones. Media are not made with mixtures of classical hormones replacing serum. Obviously there are hormones yet to be discovered and many will have to be discovered by *in vitro* techniques. *In vitro* techniques have played an important role in the elucidation of the factors described in this session. However, a better example to illustrate my point is the somatomedins which were discovered by *in vitro* techniques. They had to be discovered in this way because there is no organ of storage and the organ of synthesis (presumably the liver) cannot be conveniently extirpated to observe inhibition of growth. The classical route of discovery (gland extirpation or injection of tissue extracts) could not be used.

Today, the somatomedins are respectable hormones and should be viewed as the first genuine physiological effectors which were discovered in culture. The reports of our session are ample evidence that this is a continuing process.

We were inspired by a conversation with the late Gordon Tomkins to frame our hypothesis and to put it to a direct test. Ms. Izumi Hayashi in my laboratory has shown that medium in which serum is replaced by four hormones, and the iron transport protein, transferrin, will support the sustained growth of GH_3 cells. The hormones are somatomedin, tri-iodothyronine, parathormone and TRH. We know that similar media will support the growth of a wide variety of, and perhaps all, cells established in culture. We are also confident that new hormones discovered by culture techniques will play an important role. The implication of this observation is that every cell in body requires a specific constellation of hormones for maintenance and these will be worked out in culture.

Since its beginning, tissue culture has offered the hope that this technique would provide radical advances in our understanding of cell biology and physiology. Today, we have seen hints that this hope may at long last be fulfilled.

Symposium III

LARGE-SCALE PRODUCTION

PROBLEMS ENCOUNTERED IN LARGE SCALE CELL PRODUCTION PLANTS

H. C. GIRARD

Foot & Mouth Disease Institute
Ankara, Turkey

I. INTRODUCTION

Prophylaxis against Foot-and-Mouth Disease (FMD) in countries where the disease is enzootic is mainly based on systematic and periodic vaccination of the entire livestock.

Production of Foot-and-Mouth Disease vaccine requires a large quantity of FMD virus. This quantity is all the more huge since FMD vaccines are generally polyvalent to face the numerous types or sub-types of the virus.

A new era in this production opened in 1962 when successively;

- Mac Pherson and Stoker at the Glasgow Institute of Virology isolated the Baby Hamster Kidney (BHK) cell line.
- Mowat and Chapman at Pirbright virus disease laboratory showed the receptivity of BHK cell to FMD virus.
- And still during the same year when Capstick, Telling, Chapman, Stewart, also at Pirbright laboratory, adapted the BHK cell to suspension cultures.

However, it was only in 1967 that Brescia FMD Institute in Italy with Ubertini, Nardelli, Dal Prato, Panina, Barei started BHK suspended cell cultures for large scale production.

During this lapse of time 1962-1967, various studies were carried out. They were mainly undertaken at Pirbright. They dealt with all factors affecting the production of FMD virus such as the culture vessel, pH control, cell concentration, the culture medium, the oxygen consumption and so forth. These studies provided the basic elements for the extension of large scale plants concerning BHK cell, and FMD virus production.

Indeed, since 1967 large scale plants have developed throughout the World, either under private organizations or under Governmental Institutions. Today England, Italy, Germany, Spain, Denmark, France as well as various countries in South America can exhibit industrial plants.

Personally I started in 1969 to assist under United Nations support, some developing countries in applying these techniques. I continued in 1972, establishing a pilot plant able to produce 10 million doses of vaccine and today I am in charge of organizing a plant able to grow 500,000 litres of cell cultures per year.

Last but not least, in the University of Alabama, Lynn and Acton established a suspension culture pilot plant to process Mammalian cells and, on a broader angle, people are now turning to the large scale production of biomass through the production of a single cell protein such as yeast and bacteria which emphasize the fermentation technology.

II. INDUSTRIAL PLANT

There are no doubt many solutions for establishing a large scale plant for cell production. They depend on the purpose and the scale of the production, on architectural, geographical, industrial, and biological conditions, not to mention hazard and personal factors. Thus, a plant installed in any of the developing countries must not be too computerized, because maintenance of automatic machines is sometimes difficult!

In any case, the equipment at my disposal can undertake suspension cell culture from the laboratory to the industrial scale, from 1/2 litre to 3000 litres. After various trials we now prefer the use of vibrofermenters[1] which deserve a short description.

A *vibrofermenter* is a container of differing capacities built in glass or in stainless steel according to the volume; either spherical or with a hemispheric bottom and equipped with a vibromixer.[1]

The *vibromixer* is composed of a stirrer and a vibrator.

The *stirrer* is a round stainless steel plate or disc fixed along a shaft vertically positioned. This disc is perforated with conical holes. The orientation of these cones determines the direction of the agitation and the flow runs towards the top of the cone in the same way as lava is ejected from a volcano. The stirrer can therefore be installed with the narrow ends of the holes pointing upwards to direct the main flow towards the surface of the liquid. In the reverse, the top of the cones pointing downward, directs the main flow towards the bottom of the fermenter. Naturally, stirrer discs are of different sizes which within certain limits must be adapted to the diameter of the container. In special cases, two stirrers in reverse position can be installed on the same shaft. For sterile work, on a laboratory scale, the shaft is fitted through an absolutely airtight sealing membrane assembly, inserted in a rubber stopper in the neck of the glass vessel. On a large scale, in a closed fermenter, a flanged membrane fitted to the shaft is inserted at the top of the vessel.

[1]Vibromixer & vibrofermenter are trademarks of CHEMAP AG, Maennedorf, Switzerland. Exclusive distribution in the United States of this equipment by Chemapec, Inc., 1 Newark Street, Hoboken, New Jersey 07030.

The *vibrator* is an electromagnetic power-unit causing vertical vibration. These vibrations are generated at the frequency of the alternating current supply: 60 cycles in the U. S., generally 50 throughout Europe. Moving the stirrer plate up and down in a liquid creates pressure differences in the tapered holes causing the liquid to stream through continuously. Naturally the amplitude of the stirrer is adjustable from nearly nothing to a maximum of about 3 mm thanks to a variable transformer or an electric "tyristor."

We illustrate these brief descriptions in Figs. 1-11, which show the scaling up of some installations.

FIGURE 1 *One of the first laboratory "VIBROFERMENTERS" as it was described in 1965. The Fermenter consists of a glass balloon with jacket for temperature control.*

FIGURE 2 *A "VIBROMIXER" in action. The even distribution of particles with a smooth turbulence on the surface is noticeable.*

FIGURE 3 *The laboratory "VIBROFERMENTER" in sizes 0.5, 2 and 6 Liter made from reinforced glass are connected to a panel where pH, temperature and oxygen supply is controlled.*

FIGURE 4 *A 30 Liter "VIBROFERMENTER" with all its accessories, built for sealing up of the production.*

FIGURE 5 *Two 100 Liter "VIBROFERMENTERS" with control panel and water heater.*

FIGURE 6 *Our Cell Centrifuge with a flowmeter of 600 L/hour.*

FIGURE 7 *A 400 Liter "VIBROFERMENTER" installed at the FMD Institute in Padova, Italy.*

FIGURE 8 *Perhaps an underdeveloped installation with its rubber connections - but producing 400 Liter of cells 3 times a week with Mixer for medium preparation, Filter, "VIBROFERMENTER," continuous cell Centrifuge and control panel.*

FIGURE 9 *This 400 Liter "VIBROFERMENTER" is connected to a "SEITZ" Filter for medium supply.*

FIGURE 10 *A 3000 Liter Fermenter with air/CO_2, pH and temperature control.*

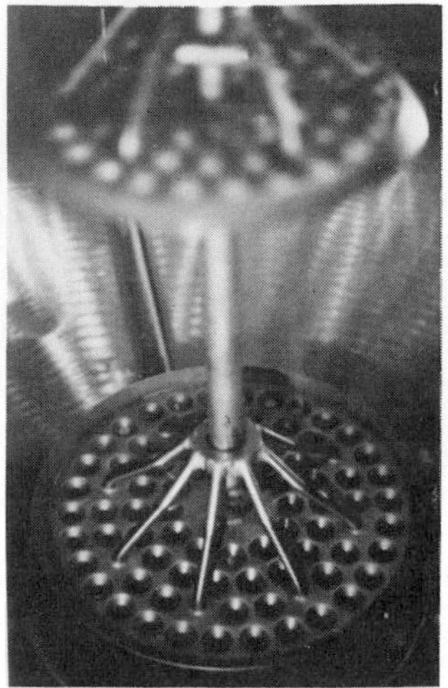

FIGURE 11 *A view of the "VIBROMIXER" stirrer installed in the 3000 Liter "VIBROFERMENTER."*

III. PROBLEMS RAISED BY LARGE SCALE PRODUCTION

Figures 1-11 show that fermenters to ensure continuous cell cultures can be equipped as required with steam sterilizable pH electrodes, dissolved oxygen electrode, temperature probe, various sterilizable filters, reflux cooler, air incinerator, light and view ports, steam sterilizable sampler, solenoid valves for the regulation of various fluids, electronic devices to measure the medium needed, and additional openings for medium distribution, cell collecting, etc. Since engineers, bioengineers as well as microbiologists have given all their attention to the design of large scale suspension cultures facilities, many problems of cell growth have been elucidated and it is not my intention to recall them. I shall elaborate only on certain problems with which I have had to deal as a user.

A. Mechanical Problems

First of all I shall give the reasons for prefering the vibrating form of agitation. I have used, and occasionally still do at the laboratory scale a magnetic stirrer, both Belco type and mainly Pirbright type, at the intermediary scale the impeller and the slow up and down movement of a large plain disc as in cattle tongue epithelium culture and at the large scale the Archimedean screw and the turbine wheels. With most of these means of agitation BHK cell growth was obtained. On the other hand my very first trials with a laboratory vibrofermenter made me suspect that this type of agitator produced mild shock waves in a similar way to an ultrasonic machine. But very quickly further trials demonstrated and regularly confirmed that the vibromixer, like any stirrer, either magnetically coupled or mechanically driven, can be regulated for suitable agitation. And as the vibromixer is capable of many variations in mixing any substance from which its universal use stems, several factors must be taken into consideration for regulating it properly. Among these, the diameter, the position and the location of the plate, and above all the amplitude adjustment of the stirrer give a solution to any problem of agitation. Once this agitation was regulated, we experienced at any level of the production, all other conditions being similar, that the cell production either in yield or in morphology was always better than those obtained with any other types of stirrers. Aggregates and cell clumps as well as low yield were a common lot at one period of our work when we were unfortunately experiencing water or serum of low quality. During this difficult phase we invariably noted the advantage of the vibrating system of agitation from the standpoint of cell growth, morphology and the dispersion of the cells produced. This fact is very likely due to a better air exchange since the suspension constantly crosses the air medium interphase. Moreover in FMD virus replication it was noted that a stronger agitation promoted the dislocation of the cells and released the virus, thus enabling a shortening of the process of producing high parameter virus. Our experiments have been ongoing for several years.

In Alabama, Lynn and Acton have found that the vibrating disc could be equally as good as the spinner flask for cell growth, but that the expression of particular cell surface components was reduced. Hence, their results at least confirm ours on the importance of the agitation system.

Two other items still deserve some comments; the supply of oxygen and the pH regulation. Concerning the supply of oxygen, we entirely adhere to the conclusion of Radlett, Telling, Whitside and Maskell: "if a steam sterilizable dissolved oxygen electrode is useful, it is not essential for routine cultures once the oxygen solution rates obtainable in the vessel and the oxygen demand of the cells are known." About the pH, experience has shown that BHK can easily grow in an acid environment. Basic pH beyond 7.6 is more detrimental to cells than a slightly acid, pH 6.6. However, pH 7.2 is the most suitable for correct growth. To some extent, surface air can compensate the development of acidity consecutive to cell metabolism and noticeable with high cell concentration. In this way the agitation system plays a major role. This fact is particularly apparent at the beginning of the virus replication. Nevertheless, sparged air must be used to satisfy oxygen demand. As it was shown that excessive sparging could be detrimental to cell growth, air distribution is ensured not continuously, but alternatively.

B. Biological Problems

Many factors influencing the growth of BHK suspended cells and the replication of FMD virus on these cells have already been described. But it is also known thatmost of the Institutes dealing with the same production were paralyzed for months at a time for various reasons and it is to these reasons for these momentous and sometimes repeated failures that I now turn.

I shall deal with water, steam, and serum. I must also emphasize the problems of BHK cell behavior and above all of FMD virus replication.

1. Water

It may seem trivial to tackle such a problem, but the conditions encountered in developing countries are very different from those found in developed countries, although pollution is now a general problem with which the world is concerned.

For years the water used in Ankara FMD Institute was extracted from a well. This raw water was distilled at the beginning. Later on, with the development of cell production, this distilled water was insufficient in quantity and a plant for water demineralization was installed. For a long time distilled or demineralized water seemed suitable for cell development. A preliminary warning appeared sometime in 1972: after a heavy storm, the raw water was noted to contain fine brown particles which were not retained by the resins of the columns of the demineralizer. These particles, due to an insoluble organic substance were considered detrimental to cell cultures even though they do not really affect the resistivity of the treated water. The solution, advocated by a specialist, was to pre-treat the water and flocculate these organic substances before filtration through gravel and active carbon. But such a transformation could naturally not be carried out instantaneously and as a temporary measure after washing the resins several times, the organic substance was no longer discernible.

One and a half years later, after 24 hours of culture, BHK cells in five production fermenters were found dead for no apparent reason. However, at this time it was noted that the resistivity of the demineralized water had dropped entirely. The reason was due to a substitution of the well water by raw water from a stagnant lake in spite of so-called previous treatment. The effect of this substitution was similar to a tidal wave, everything was polluted: softener, demineralizer, boiler, piping, washer. Even the distilled water was not satisfactory for cell multiplication. Two engineers were needed to regenerate the demineralizer. Although demineralizer and distillator were again connected to the well water, the lake water continued for some time to supply the tap water and the boiler and we experienced some phenomenon of subtoxicity, characterized either by the non development of cells or a progressive extinction of the cell through various passages.

I shall close the description of these events in spite of the fact that they stopped our production for months by summarizing the solutions retained. These were the use of demineralized water after pretreatment of the raw water for general washing, and supplying a distillator able to produce bi-distilled water for medium preparation.

2. *Steam*

A demonstration of the harmful action of steam was shown in the first fermenters we were given. They were built to be sterilized by direct expansion of sprayed steam. They had to be thoroughly washed with sterilized treated water after normal steam sterilization if cell development was to succeed. Steam sterilization of fermenters through a double jacket is the only advisable process and the use of steam originating from treated water is required to sterilize filters and supply autoclaves.

3. *Serum*

The question of the serum is another chapter. While so much care is given to the composition of any growth medium it is still paradoxical to be obliged to add X% serum as if serum was a standard product of well-defined composition.

When serum is used at the dose of 10% of the growth medium, it immediately raises problems of supply either in quantity or quality. The quantity required obliges most Institutes to harvest blood at the abattoir during slaughtering. Collecting blood from slaughtered animals creates the problem of quality. Fortunately our own experience, already based on several years, shows that FMD antibodies found in the blood do not prevent BHK cell growth and do not modify the receptivity of BHK cells to FMD virus replication, although cells must be removed from medium containing serum during the virus replication. However, the quality of serum is seriously compromised in cattle fed with artificial food and in Italy they now use only the serum from dairy cattle since these animals generally receive natural fodder to avoid modifications in milk quality and flavor.

Serum, as it is harvested in the abattoir on any slaughtered animal raises the possibility of bacterial or viral contamination and it is a very serious limiting factor. The constant threat of contamination by Mycoplasma, so difficult to detect in cell cultures and so tremendously heterogeneous in their metabolic activities, is another concern.

Sterilization of serum by inactivation at 56°C was recommended but BHK cells did not proliferate with this treated serum. Bartling, from Lelystad Institute in Holland recommends the use of polyethylene glycol to precipitate virus, microbes and antibodies since BHK cell growth with PEG-serum appears to be normal.

Last but not least, the era when BHK cell cultures will be carried out by chemicals instead of serum is now not very far away. The first hope in this area came from Tomei and Issel of Plum Island Animal Disease Center who developed BHK cells in a serum-free chemically defined medium and from Keay at Washington University who is pursuing similar studies. This ultimate solution will greatly help the problems of large scale cell culture.

4. BHK Cells

I am not far from believing that each Institute finally has its own BHK cells in spite of their common origin: BHK-21 clone 13 of Mac Pherson and Stoker isolated in 1962. Indeed, according to their sources, BHK cells vary in size, shape and behavior.

When with my colleagues Okay and Kivilcim, we began our work on BHK in suspension our cells were provided by Brescia Institute Italy. Later on we received some BHK cells from Razi Institute, Teheran, Iran. It did not take long to notice different characteristics and properties between them. Most of the Brescia cells, more adapted to suspension cultures, were round and multiplied extensively. Razi BHK cells were bigger in size, more elongated or polygonal with a rounded outlined shape. Several FMD virus replicate well and nearly always the first time on Razi BHK, whereas the same viruses generally require several passages to be adapted to Brescia BHK.

At one time, we had to begin again with a new master seed and we received a new batch of Brescia cells. This strain had only received 9 passages in Brescia Institute. It had come from Pirbright Laboratory where it had been adapted to suspension. Curiously enough there were many similarities between these Brescia No. 2 BHK cells and the Razi cells, even in their behavior in front of FMD virus.

Another time Lelystad FMD Institute from Holland sent us their BHK cells, one of them after 74 passages in suspension, the other after 362 passages, each differing by its characteristics and properties already mentioned for Razi and Brescia No. 1, even the BHK 362 cannot grow in monolayer.

Perhaps it should also be mentioned as a reminder that the original BHK cells, the Glasgow cells, not adapted to suspension culture are of the fibroblastic type in regular spindle shape.

Thus it seems that the progressive adaptation of BHK cells to suspension culture through passages modifies their morphological character from the spindle shape to a regular round form with a long intermediary step of polygonal rounded angle cells. It is worthwhile noting that these variations of forms are most obvious when these cells are cultivated in monolayer culture. At the last stage of adaptation to suspension, monolayer culture becomes rather difficult.

Experience shows us that at the large scale production it is better to use BHK cells recently adapted to suspension.

5. *BHK Cells and FMD Virus Inter-reaction*

In the course of performing routine plaque assays on type Asia 1 FMD virus, Cowan from Plum Island Center with Erol and Whiteland, all working in Ankara FMD Institute found that virus grown in our BHK monolayer section (roller bottles) produced mainly large plaques whereas virus from our BHK suspended cell unit produced predominantly small plaques. This casual observation led to the demonstration of different antigenic characteristics which in turn resulted in a difference of immunological properties. Vaccine prepared from the Asia 1 large plaque was found at least 3 times more effective in inducing immunity in cattle against the field virus than was a vaccine prepared from the small plaque Asia 1 virus.

Moreover Razi BHK cells grown in monolayer proved highly susceptible to Asia 1 field virus but passaging of this virus resulted in a rather rapid change from a predominantly large plaque population to a mainly small plaque. Consequently, to produce large plaque virus it is necessary to remain at a low passage level. On the other hand Brescia No. 1 BHK cells were not susceptible to Asia 1 virus. Adaptation of this virus required a series of preliminary passages in Razi BHK during which the virus shifted toward a predominantly small plaque population. This explains why the virus produced in the monolayer section using Razi BHK cells was predominantly large plaque whereas virus adapted to submerged Brescia cells and produced in BHK suspended cell unit were small plaques. Therefore, to produce an Asia 1 virus that is mainly of the large plaque type it is necessary to remain at a low passage level of virus and used BHK cells susceptible to virus infection such as Razi or Brescia No. 2 BHK cells.

It is worthwhile noting that our last findings, show that the preliminary passages of FMD field virus, mainly Asia 1, on Glasgow BHK cells, *stabilize* the predominant virus population, as if they were cloned. These virus can then be replicated on BHK suspended cells with no other transformation at least during the first stages.

Recent studies also confirm the progressive loss of susceptibility to FMD virus of BHK cells following consecutive cell passages. Between the 30th and 80th passage large plaque Asia 1 virus infectivity titer was reduced by 2.5 log. Concerning 0 virus its behavior is entirely different. Replication on suspended Brescia or Razi BHK does not really alter the type of plaque, however it becomes smaller through passages.

In conclusion, the production of FMD virus antigenically similar to the field virus required the use of BHK cells that were not specifically adapted to grow in suspension and also the use of seed virus very close to the field virus.

Thus we can see that the problem of large scale virus production on BHK is not only a problem of quantity but also a problem of quality.

REFERENCES

1. Barteling, S. J., *(1974)*. Personnel communication. Consultantship to Ankara FMD Institute and *(1974)* Bull. off. Int. Epizo., *11-12*, 1243-1254.
2. Capstick, P. B., Telling R. G., Chapman, W. G., and Stewart, D. L., (1962). *Nature, 195*, 1163-1164.

3. Cowan, K. M., Erol, N., Whiteland, A. P., (1975). XIVth Conference of the OIE Commission on FMD Paris 11-14 March, (1974) Bull. off. Int. Epiz., *11-12*, 1271-1298.

4. Keay, L., (1975), Biotech. Bioeng., 14, 745-764.

5. MacPherson, I., and Stoker, M., (1962). *Virology*, 16, 147-151.

6. Mowat, G. N., and Chapman, W. G., (1962). *Nature*, *194*, 253-255.

7. Lynn, D. J., Acton, R. T., (1975). *Biotech. Bioeng.*, *17*, 659-673.

8. Radlett, P. S., Telling, R. C., Whitside, J. P. and Maskell M. A. (1972), Biotech. Bioeng., *14*, 437-445.

9. Tomei, L. D., and Issel, C. J., (1975), Plum Island Animal Disease Center Greenport. Biotech. Bioeng., *12*, 265-178.

10. Ubertini, B., Nardelli, L., Del Prato, A., Panina, Gl, and Barei, S., (1967). Zentr Veterinarmed., *14*, 432-441.

LARGE-SCALE PRODUCTION OF MURINE LYMPHOBLASTOID CELL LINES EXPRESSING DIFFERENTIATION ALLOANTIGENS

RONALD T. ACTON[a]
PAUL A. BARSTAD[b]
R. MICHAEL COX
ROBERT K. ZWERNER[c]
KIM S. WISE[c]
J. DANIEL LYNN

Department of Microbiology and
the Diabetes Research and Training Center
University of Alabama in Birmingham
University Station
Birmingham, Alabama

I. INTRODUCTION

The mammalian cell has become the "New Frontier" of present day biology. Although discovered over a century ago scant few of the mysteries concerning its ability to communicate with and react

[a]This work was done during tenure of an Established Investigatorship of the American Heart Association.
[b]Supported by a fellowship from the Alabama Heart Association.
[c]Supported by U.S. Public Health Service Institutional Research Fellowship No. GM-00130 from the National Institute of General Medical Science.

to environmental stimuli, including other cells, have been unraveled. Consequently, our knowledge of neoplastic and degenerative diseases, developmental abnormalities and immunologic failures is still quite primitive. We do know that certain molecules on the cell surface react to environmental stimuli and somehow notify the interior of the cell of this interaction. The nature of the cell surface receptor and the mechanism by which it effects communication with the internal workings of the cell is still poorly understood. For this reason much biological research effort has been directed toward the isolation and structural-functional characterization of cell surface components.

Our laboratory is presently interested in several alloantigens expressed on normal murine lymphocytes as well as lymphoblastoid cell lines. In particular, three cell surface antigens found on the murine T-lymphocyte, Thy-1 (1), H-2 (2), and gp70 (3-6), have been chosen for study. The Thy-1 molecule consists of a single glycoprotein (approximately 25,000 daltons) found in large quantities on central nervous system tissue and on thymocytes of the mouse and rat where it reflects the state of T-lymphocyte differentiation (7-15). This antigen exists as one of two allelic forms, Thy-1.1 and Thy-1.2, in the various strains of inbred mice, although only the Thy-1.1 form has been found in the rat. While the function of the Thy-1 antigen is still unknown, the tissue distribution and developmental appearance of this molecule suggests it plays a major role in cell-cell interaction. Presently, the H-2 molecule is one of the most extensively studied cell surface antigens (2). It consists of two polypeptide chains of 45,000 and 12,000 daltons and is expressed in varying amounts on nearly all tissues of the body (2, 16, 17). Recent evidence suggests that this molecule also plays an important role in intercellular interaction, especially in the immune system (18). Finally, gp70 is an evelope glycoprotein of oncoviruses, which often is expressed on normal as well as transformed murine cells (4-6). It has been suggested that endogenous viral antigens ex-

TABLE I Expression of Selected Membrane Components by Several Cell Types

Membrane component	Cell type	No. of molecules/cell[a]	References
Thy-1	Mouse thymocytes	440,000	25
	Mouse lymphocytes	61,000	25
	Rat thymocytes	500,000	11
H-2	Mouse thymocytes	87,000	25
	Mouse lymphocytes	570,000	25
HLA	Human lymphocytes	140,000	26
TL	Mouse thymocytes	45,000	25
Ig	Rat lymphocytes	150,000	27
C3b	Human Raji cell line	400,000	28
β_2-microglobulin	Human lymphocytes	520,000	26
Glycophorin	Human erythrocytes	500,000	29
PHA receptors[b]	Human erythrocytes	450,000	30
Insulin receptors	Rat liver cells	250,000	31
	Human monocytes	16,000	31
	Rat adipocytes	50,000	31
	Turkey erythrocytes	3,000	31
Growth hormone receptor	Human lymphocytes	4,000	31
Glucagon receptor	Rat liver cells	110,000	31
β-adrenergic receptor	Frog erythrocytes	1,800	31
	Turkey erythrocytes	600	31
Acetylcholine receptor	Electric tissue of eel	>1,000,000	32
Cytochalasin receptor	Human lymphocytes	1,000,000	33
	Human erythrocytes	300,000	33

[a]In every case data has been selected from the literature that demonstrated the maximal number of each component.
[b]PHA = phytohemaglutinin.

pressed on the cell surface may have a functional role in cellular differentiation (19, 20). In addition to these particular cell surface components there are obviously several other lymphocyte alloantigens of similar interest including Ia, TL, and Ly (21-24).

The largest obstacle confronting the total characterization of cell surface molecules is the paucity of each component available. An estimate of the quantities per cell of several cell surface molecules is shown in Table I. It should be mentioned at this point that "lectin receptors" are probably not a single species of molecule but rather represent a large number of cell surface components containing a particular carbohydrate moiety; usually glucose, galactose, or mannose. Indeed the lentil lectin receptors on our cell lines are Thy-1, H-2, and gp70 as well as several other molecules. The other membrane components listed in Table I are present in amounts ranging from >1,000,000 to less than 600 molecules per cell. Thy-1 is present at approximately 500,000 molecules per cell, thus 100 nanomoles, which in our opinion is the minimum amount required to begin extensive structural analysis, would represent approximately 10^{11} cells. Since the recovery of small amounts of material during a several thousand fold purification is likely to be as low as 10% (34), 100 nanomoles of Thy-1 would require 10^{12} cells.

One major consideration in our approach was that a homogenous cell line grown *in vitro* would eliminate any question of contamination by other cell types. Thus, any antigen found on these cells, even in low quantities, is made by the cell and not by another cell type present in low amounts. Furthermore, we desired to study systems like established murine lymphoblastoid cell lines, which are available to other investigators so that we could correlate our findings and observations with theirs.

Cell culture is usually conducted in one of two forms; monolayer or suspension. In monolayer culture cells attach to an inert surface for growth and are usually detached by EDTA or trypsin treatment. In suspension culture, on the other hand, cells are harvested without treatment simply by pouring off the desired amount of cell-containing culture fluid. In both cases cells are separated from the culture fluid by centrifugation. In order to eliminate the possibility of modification of cell surface antigens by such agents as trypsin or EDTA we selected suspension culture as the method to produce large quantities of established murine lymphoblastoid cell lines.

Murine lymphoblastoid cell lines often reach densities of 2×10^6 cells/ml or more in suspension culture (35, 36), thus, 10^{12} cells would represent 500 liters of culture fluid. Since these cell lines double approximately every 24 hr, only one half the volume of each vessel can be harvested each day under semi-continuous culture conditions. Thus, a total working volume of 150 liters or more for culture vessels is required to produce 10^{12} cells in a 4 day period. With these requirements in mind, in addition to the absolute need for sterility, we have designed, constructed and are operating a large scale, mammalian cell culture system. Our experience to date with the system will be the subject of this review.

II. DESIGN AND OPERATION

The design of the facility at the University of Alabama in Birmingham (UAB) was influenced by our own experience as well as by consultation with others in academic institutions and industrial concerns.

A. Architectural Considerations

The primary concern in housing the facility was centered around potential biohazards associated with growing and processing large quantities of cells. The renovation was effected such that the area conforms to the guidelines described by the National Cancer Institute for work with moderate risk oncogenic viruses. The facility also conforms to the P3 requirements of the Recombinant DNA Research Guidelines of the National Institutes of Health. The laboratory is separated from areas that are open to the general public. The surfaces of walls, floors, bench tops,

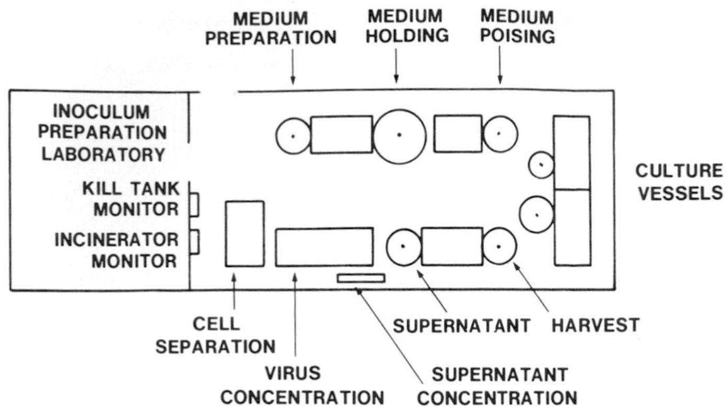

FIGURE 1 *Primary equipment configuration. Cell culture center.*

FIGURE 2 *Photograph of primary equipment.*

and ceilings are easily cleanable. The ventilation system is balanced to provide a supply of inflow air from the access corridor into the facility. Exhaust air is not recirculated.

As depicted in Fig. 1 the area is divided into two laboratories. The smaller laboratory is for the preparation of inocula for the large culture vessels and contains a biological safety hood, incubator for spinner flasks, freezer for serum storage and a New Brunswick 14-liter bacterial fermentor modified for mammalian cell culture as previously described (35, 36). The larger laboratory houses the large-scale equipment consisting of stainless steel vessels, control cabinets and product separation equipment (Fig. 2).

B. Serum Preparation

It is economically impractical to purchase sterile serum that had been screened for adventitious agents to use in the large-scale culture facility. We, therefore, purchase high quality serum collected under sterile conditions and stored frozen, which is processed within the facility. This "raw" serum is processed by slow thawing at 5°C after which it is introduced into the medium poising vessel (to be described presently in section C), where the temperature is raised to 56°C for 30 min under constant agitation at 100 rpm. This treatment inactivates most adventitious agents, such as virus and mycoplasma, which are difficult to remove by filtration. The temperature of the serum is then depressed to 5°C and is filter sterilized with equipment designed by the Millipore Corporation. This unit consists of a 31" AP-20 Lifegard depth filter cartirdge, a 31" AP-15 Lifegard depth filter cartridge, a 10" CW-06 Milligard prefilter cartridge, a 10" CW-03 Milligard prefilter cartridge, a 31" MF Millitude final filter (0.22 micrometer porosity) and a 293 millimeter diameter membrane final disc filter (0.22 micrometer porosity). With this apparatus raw serum can be sterilized in 60-100 liter batches depending upon the species in about two hr. The serum is filtered directly into sterile 20 liter stainless steel vessels (37) and stored at 2°C until use.

C. Medium Preparation, Holding, and Poising

Medium is prepared by adding commercially available powdered concentrate and processed serum to high purity water in a 650 liter open vessel. A built in agitation system insures rapid and total solution of all the components of the medium. This system is composed of a magnetically driven, offset, marine impeller located in the bottom of the vessel (38). This configuration obviates the need for seals or diaphrams in that there is no penetration of the vessel. Also, by virtue of its eccentricity, there is no need of baffles or other turbulence causing devices. The impeller is magnetically coupled to a gearhead motor outside the vessel. The speed of the agitator is monitored and controlled on a panel adjacent to the vessel. The pH of the medium may be adjusted by the addition of acid or base by a pH controller. The final adjustment of pH is usually made in the poising vessel just prior to addition to the culture.

After the medium has been prepared, it is then filter sterilized. The medium is pumped through a quarter-turn, stainless steel ball valve with Teflon packing by virtue of a centrifugal pump through a pair of presterilized (in place) Pall Trinity cartridge filters into the medium holding vessel. The two series cartridge filters (AB2AA8P pre-filter, and AB2AR8P bacterial of 0.22 micrometer porosity) are in parallel with a similar pair of filters. Each of these filters has 10 square feet of surface area. In the event one set becomes blocked, the alternate set of filters is utilized. The medium holding vessel is a 625 liter stainless steel pressure vessel with a jacket around its circumference to allow cooling fluid to circulate and keep the temperature of the vessel depressed. The vessel is also insulated on the outside to avoid condensation of water vapor. This vessel, along with the remaining vessels of the system, is sterilized in place along with being jacketed and of the pressure vessel type. Medium can be stored in this vessel for extended periods of time and removed as needed. Medium in this vessel is normally kept at $2^{o}C$ with the agitator operating to eliminate temperature gradients and inhibit precipitation. On a control panel adjacent to the medium holding vessel are valves for regulation of air pressure in the vessel, circulation of coolant and steam sterilization. The panel also contains temperature controlling, monitoring and recording instrumentation along with that for agitation.

The next step in the process is to adjust the pH and temperature of the medium prior to addition to the culture. This procedure is called "medium poising." During the poising procedure, the medium is heated to $37^{o}C$. The pH is also lowered to its appropriate value by sparging CO_2 through the medium. The temperature, pH and E_h are monitored, controlled and recorded by instruments located in a relay rack adjacent to the vessel. This rack contains a set of valves (similar for all vessels) for temperature control, gas regulation, and steam sterilization. Transfer of medium from the holding vessel to the poising vessel is effected by a flexible teflon lined stainless steel hose and quick-connects. After attachment, steam is circulated through the hose for sterilization. After this cycle is completed, valves at both ends of the line are opened and a pressure differential established between the two vessels causes fluid to flow from the holding to the poising vessel. A liquid level indicator located on both vessels indicates the amount of fluid transferred. All transfers between vessels are made in this manner.

D. Cell Culture and Harvest

To begin large-scale culture, 20 liters of poised medium is transferred to the 70 liter culture vessel and "seeded" with a 12 liter culture of cells produced in the 14 liter culture vessel (35).

Several parameters are controlled, monitored, and recorded during cell growth. The control panel adjacent to the culture vessels contains instruments for monitoring pH, dissolved oxygen, temperature, agitation rate, turbidity and CO_2 output. Since we may be growing cells that are producing oncogenic viruses, from this point on each vessel is connected to an incineration system for all exhaust gases and to a kill tank system for liquids disposed. These two biohazard control systems will be discussed later.

After 24 hr in the 70 liter culture vessel cells have usually doubled and the vessel can be filled by the addition of poised medium. After an additional 24 hr the contents of the 70 liter culture vessel is used to inoculate the 200 liter culture vessel. The mixture of inoculum and the poised medium should result in a density in accordance with the lower log phase density of that cell line (39). After 48 hr, the 200 liter vessel is ready for harvest, and cells are directed to the harvest vessel. The harvest vessel is similar to the holding vessel except for the biohazard consideration. The harvest vessel is used as a temporary holding vessel to prevent any possible contamination from the centrifugation system to the culture vessels and is emptied through the flexible line to a Sharples Laboratory centrifuge enclosed in an isolator to contain hazardous aerosols. The operator works through rubber gloves in the isolator and after the process is over, sprays the interior with formaldehyde to guard against viral contamination. The isolator is attached to the aforementioned incinerator system.

The supernatant from the Sharples is directed either into a supernatant holding vessel for continued processing or to the kill tanks for disposal. If one is interested in products other than the cells themselves, such as viruses or macromolecules in the supernatant, additional processes are required. One may concentrate the supernatant on a system developed by the Amicon Corporation. With 100,000 molecular weight cutoff hollow fiber cartridges in this system viruses can be concentrated 30 to 60 fold and then further concentrated and partially purified by use of a Beckman L5-50 preparative ultracentrifuge equipped with a CF-32 continuous-flow rotor. This centrifuge is contained in a laminar flow hood fitted with rubber gloves. By using fiber cartridges of lower molecular weight cutoff one may separate additional macromolecular fractions found in the supernatant.

Adjacent to the cell separation device are monitors for the kill tanks and incinerator located in other areas.

E. Ancillary Equipment

The primary equipment described in the previous section requires a number of services for its operation. These ancillary service items are located on the floor below the primary facility and also on the roof above. Figure 3 is a schematic representation of the ancillary equipment located on the floor below. An electric steam generator produces all the steam required to sterilize the vessels, filter system and transfer lines on the floor above. Behind the steam generator is a refrigeration plant that provides chilled water for the jacketed vessels on the floor above and also is used to regulate the temperature of the culture and poising vessels. Figure 4 is a schematic representation of all components of the water system. Normal tap water is fed through two prefilters to remove particulate matter and then through the reverse osmosis (RO) unit. From this unit the water goes into the large 250 gal polypropylene reservoir (Fig. 3). A valve and pump is located at the bottom of the reservoir to circulate the water within the loop. The water in the RO loop goes to a water heater, which in turn directs the water to the steam generator or upstairs to hot water taps. Also, water in the loop goes to cold water taps upstairs. This hot and cold water is used for washing down the vessels and to produce high purity water that is used in the preparation of medium. High purity water is produced by pumping water from the RO loop through an organic absorber, two de-

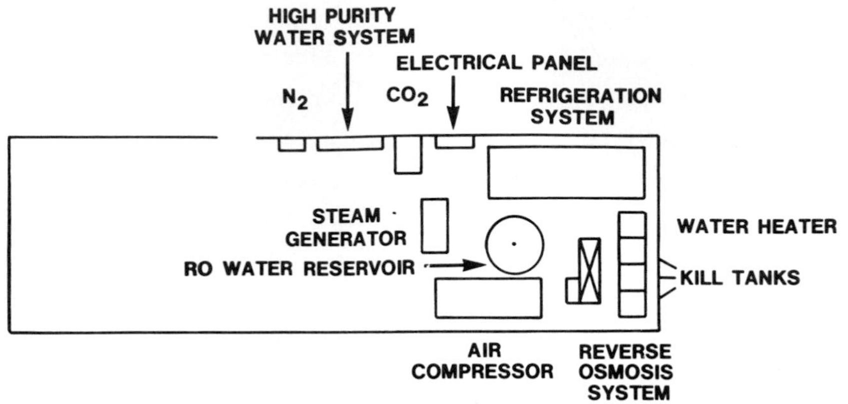

FIGURE 3 *Ancillary equipment configuration. Cell culture center.*

ionizer cartridges and a micro-filter. When the water leaves these four cartridges it is very pure, (18 megohms resistance) and sterile: the properties desired for growth medium. Not shown on the schematic are a liquid level controller in the reservoir and also a vacuum switch to shut off the pump if the reservoir goes dry. A number of check valves are required in the system that are not shown for the sake of clarity. On the other side of the high purity water system are located gas cylinder manifolds. On the left side is nitrogen, which can be used to regulate the reduction-oxidation potential of the medium during poising and on the right is the CO_2 manifold system, which is used for pH control. An air compressor provides air for aeration during culturing of cells, transfer of liquid from one vessel to another, and operation of pneumatic control valves. Located to the right of the compressor are four industrial water heaters. One is used to provide hot water to the steam generator and the facility while the other three have been converted to be used as the kill tanks as described later.

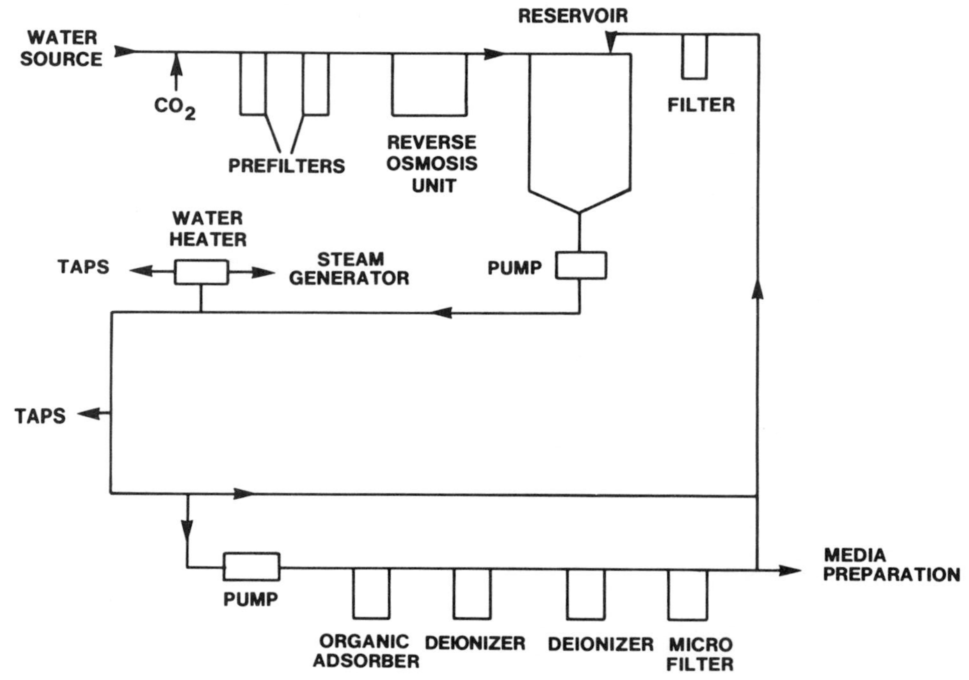

FIGURE 4 *Schematic of water system. (Check valves & minor components omitted.)*

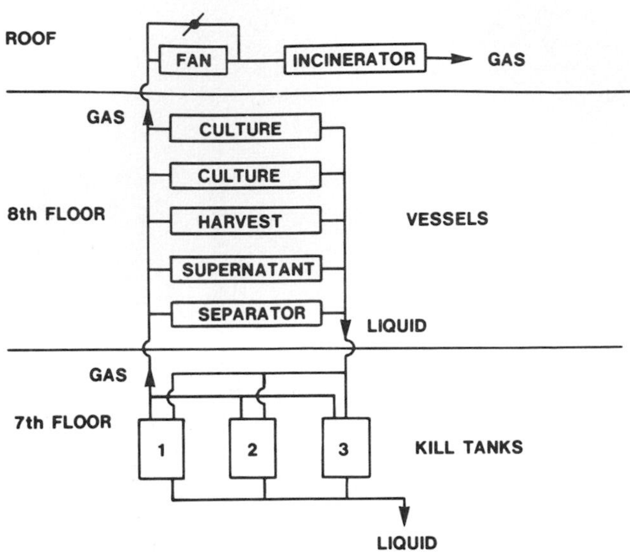

FIGURE 5 *Schematic of biohazard containment system. (Check valves & minor components omitted.)*

The roof above the cell culture facility houses other items of equipment required for the operation of the system. A large exhuast fan is used to exhaust air from the two biological safety cabinets located in the facility. A HEPA-filtered air handler serves exclusively the cell culture area. Also located in this area is a fan and an incinerator, which is part of the biohazard containment system (Fig. 5).

The containment system is separated into components on the three floors that house the primary and ancillary equipment. This system is capable of handling all fluids and gases emitted from the culture system. Prior to disposal, liquids from the two culture vessels, the harvest and supernatant vessels and the cell separator are directed to industrial water heaters converted to be used as kill tanks located on the floor below. Fluids can be transferred from one kill tank to another but they may not be released to the drain automatically, this must be done with a hand valve to avoid dumping contaminated fluid due to an electrical malfunction. A monitor in the cell culture area indicates the status of each of the tanks, i.e. fluid contained and length of residence. All gases in the system, that is the gases from the kill tanks as well as cell culture, harvest and supernatant vessels pass through the incinerator by virtue of a fan located on the roof prior to release. This incinerator has three heating elements whose status can be monitored in the cell culture area. This system is designed such that if one element fails, the incinerator will still be effective in killing pathogens. Above the fan is located a balancing loop such that a constant flow of air passes through the incinerator to protect the elements and insure proper residence time of all potential pathogens in the incinerator. There are a number of check valves and other components in this system which have been omitted for the sake of clarity.

III. PRODUCT GENERATION

A. Selection of Cell Lines

We are particularly interested in cell surface components of thymus-derived lymphocytes (T-cells) termed "differentiation alloantigens." These include Thy-1, TL, H-2, and G_{IX} (gp70) whose qualitative and quantitative expression reflects the state of differentiation of T-lymphocytes (1-20). Since most mouse lymphocytic leukemias are of T-cell origin, one would expect cell lines derived from these tumors to continue expressing the normal complement of T-lymphocyte cell surface molecules (40-43). We have analyzed a number of established murine lymphoblastoid cell lines (35, 36). Approximately 20 cell lines were obtained from various laboratories, adapted to suspension culture and analyzed for growth properties as well as expression of alloantigens.

TABLE II Properties of BW5147 Murine Lymphoblastoid Cell Line

Property	Data	References
Mouse strain of origin	AKR/J	44
Method of tumor induction	Spontaneous	44
Cell surface phenotype		
Determined	Thy-1.1, H-2^k, GCSA, GP>0	36, 47, 52
Expected	Ly-1.2, Ly-2.1, Ly-3.1, Ly-5.1, $G_{IX}2$, TL 1, 2, 4	
Type of virus produced	Gross	44
Population doubling time	10-16 hr	35, 36
Other observations	Cortisol and PHA sensitive	42, 46

Most of these cell lines grew reproducibly in semi-continuous suspension culture to densities of 2-4 × 10^6 cells/ml if the proper vessel dimensions, medium and pH were maintained (35, 36). From these early studies several cell lines expressing various specificities of Thy-1, TL, and H-2 were selected for further analysis. Considerable information has been obtained on the cell surface components of the BW5147 line (42) and will be reviewed in this section. Table II summarizes some of the properties of this line. In addition to maintaining a cell surface phenotype analogous to a murine thymocyte the cell line is also sensitive to cortisol and phytohemaglutinin (PHA). The line is a relatively good producer of Gross virus and doubling times often equal that calculated for a thymocyte *in vivo*.

B. Cell Growth

The generation of "seed" culture in 50 ml to 4 liter spinner flasks for inoculation into the 14 liter fermentor has previously been described (35). We have established certain parameters that must be maintained to maximize the generation of cells (39). In all vessels, regardless of size, it is important to maintain cells in a logarithmic stage of growth which often require that they be cut every 12 hr. If cells are allowed to reach the stationary phase, they may either die rapidly or fail to divide as fast when fresh medium is added (35, 36, 39). Moreover, the expression of certain cell surface components decreases when cells enter the stationary phase of growth (35, 36, 47). In Fig. 6 the growth of cells in a typical scale-up from 14 liters to the 70 liter and 200 liter vessels is shown. Although environmental parameters can be monitored and controlled in the 14 liter vessel, more control capability is available in the large scale system. In the 70 liter and 200 liter culture vessels, pH, dissolved oxygen (DO), oxidation-reduction potential (ORP) and CO_2 output can all be continuously recorded by use of strip charts. In the 14 liter culture vessel the pH, temperature, agitation rate, gas flow rate, and vessel pressure can be controlled to various set-point levels but with the exception of pH are not continuously recorded.

In a typical run utilizing the BW5147 cell line pH is normally 6.95 following the introduction of seed culture into freshly poised medium. Air is introduced through the sparge and overlay lines at a continuous rate of 1-2 lpm and 4-5 lpm for the 70 and 200 liter culture vessels, respectively. The pH is maintained at 6.95 ± 0.05 by the introduction of CO_2 at a rate of 0.3 lpm and 1.2 lpm for the 70 and 200 liter vessels, respectively, or by the addition of 1 molar sodium carbonate. During a typical run with BW5147 in the 200 liter vessel 200 ml of sodium carbonate is utilized in a 24 hr period for maintaining pH at the set point.

The BW5147 cell line grows well in 2% horse or fetal calf serum (51). As can be seen in Fig. 6, this line can be reproducibily grown to densities of 4×10^6 cells/ml with population doubling times averaging 14 hr. We have continuously grown BW5147 cells for 38 consecutive days in our facility producing greater than 10^{13} cells. A number of other cell lines have been grown for shorter periods with similar success.

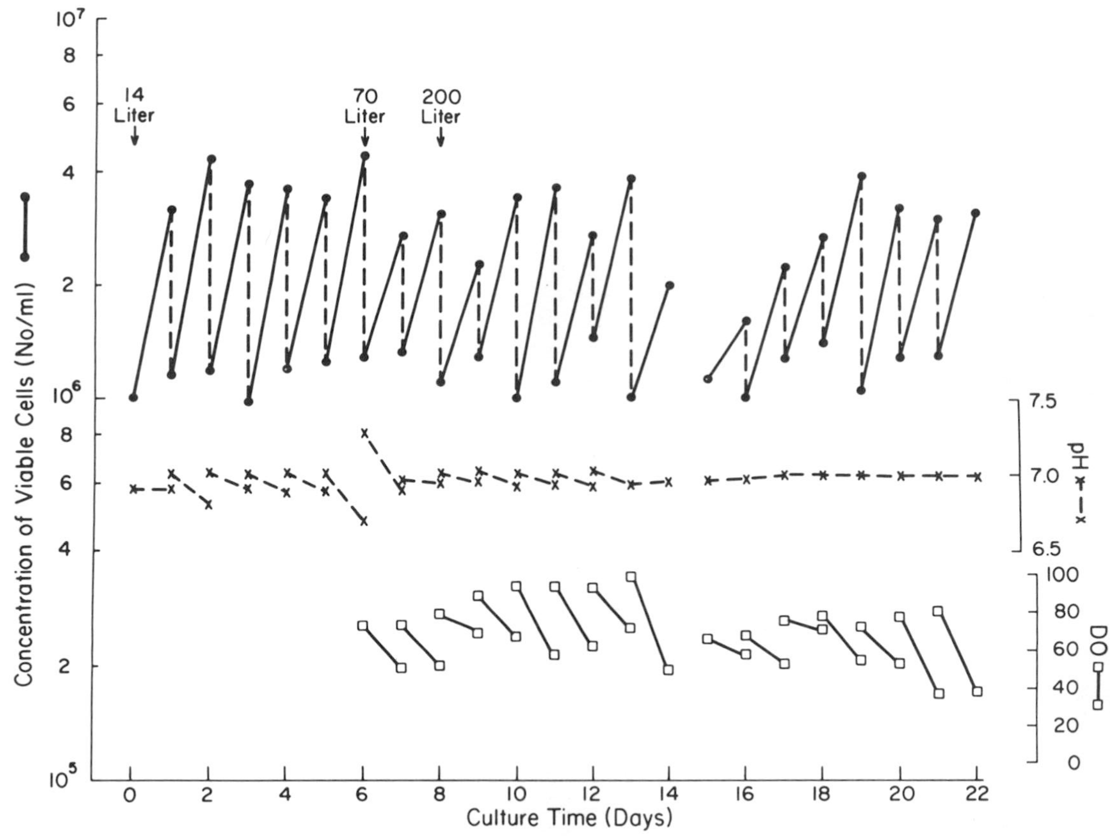

Concentration of Viable Cells (No/ml)
10^7
8
6
4
2
10^6
8
6
4
2
10^5
14 Liter
70 Liter
200 Liter
7.5
7.0
6.5
pH
100
80
60
40
20
0
DO
0
2
4
6
8
10
12
14
16
18
20
22
Culture Time (Days)

FIGURE 6 *The growth of BW5147 under semi-continuous culture conditions in 14, 70, and 200 liter culture vessels utilizing RPMI-1640 supplemented with 0.3 grams/liter asparagine, 2.5 grams/liter glucose, and 2% fetal calf serum as the growth medium. On day 14 the entire fluid volume of the 200 liter culture vessel was harvested in order to make a repair. On day 15 the vessel was inoculated again from the 70 liter vessel that had been maintained in an active state from day 6. Each day an aliquot of the vessel fluid was removed for inoculation into a large vessel or for harvest and replaced with fresh culture medium.*

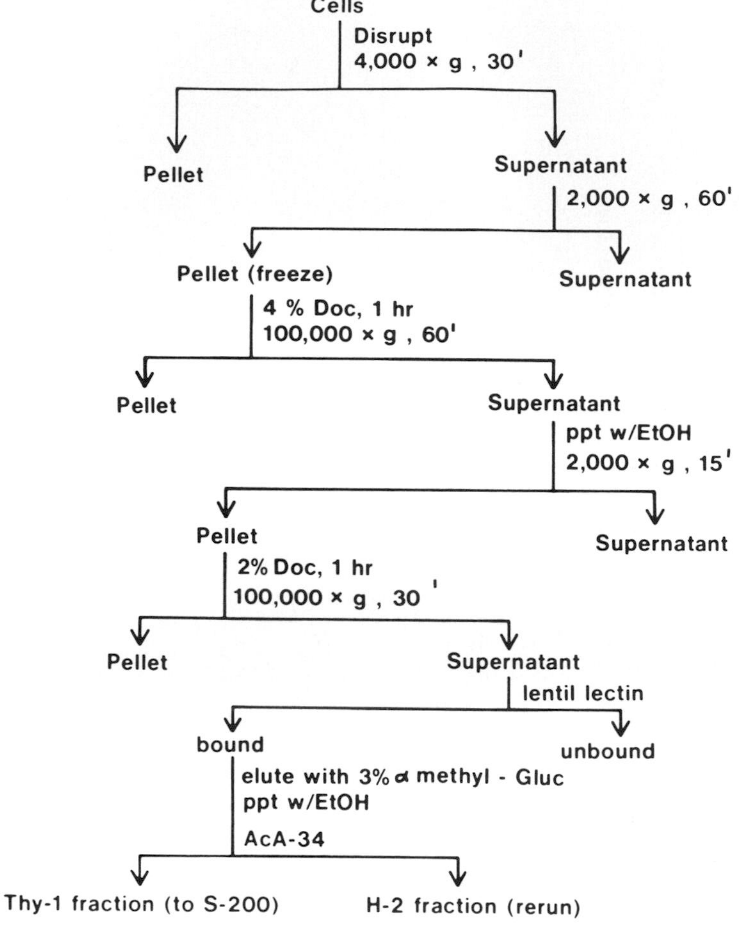

FIGURE 7 *Membrane component isolation scheme. Purification scheme for membrane components. Crude membranes are isolated by differential centrifugation.*

C. Isolation of Membrane Components

A generalized scheme for the purification of several membrane components is shown in Fig. 7. It involves disruption of the harvested cells, isolation of the crude membrane fraction, detergent solubilization and the isolation of membrane components by use of affinity chromatography and gel filtration.

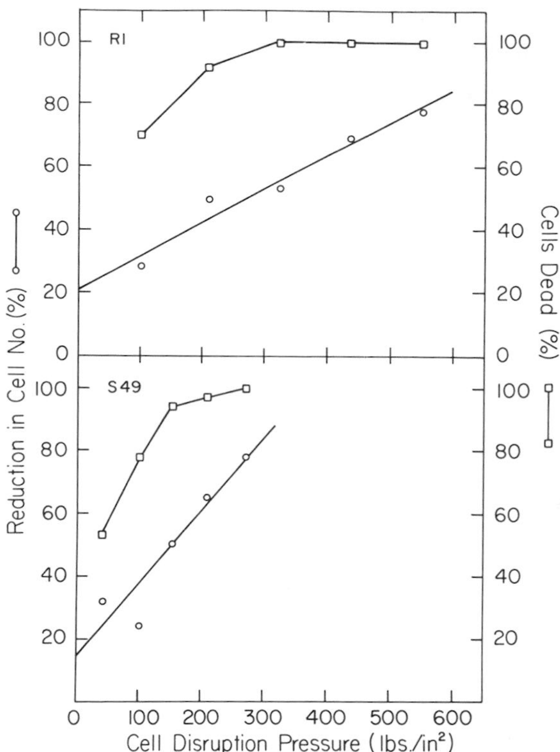

FIGURE 8 *The controlled disruption of R1 and S49.1 cell lines at a concentration of 1 × 10⁸ cells/ml over a range of disrupting pressures. The cell suspension was pumped at a rate of approximately 3 liters/hour.*

During an average run our facility generates 3×10^{11} cells per day. This represents a packed cell volume of approximately 300 mls, which is somewhat of a problem to handle in an efficient manner for the subsequent generation of membrane components. Cells are disrupted by pumping a cell suspension under pressure through a controlled orifice. We presently are utilizing a Stansted Model A0612 cell disruption pump equipped with a model 716 disruption valve. As can be seen in Fig. 8 controlled disruption measured by the severity of total cellular disintegration and cell viability as determined by trypan blue exclusion was examined by pumping the cell suspension through the orifice at pressures varying between 150 and 325 psi. Optimal membrane yields are achieved under conditions that result in a 50% reduction in cell number with the remaining cells being less than 10% viable (48). We have observed that the cell disruption pressure required to achieve the optimal yield of membrane varies from one cell line to another. As shown the disruption profiles for the murine lymphoblastoid cell lines Rl and S49 are distinctly different. Moreover, we have observed that a given cell line will also vary with regard to the degree of pressure necessary for disruption if harvested during different stages of the growth curve. The temperature of the cell suspension and the rate of pumping also influence the degree of disruption at a given pressure. These conditions for disruption must be determined for each cell line utilized.

Membranes produced during a production run are usually frozen at -20°C until use. Approximately 5×10^{11} membranes are thawed, homogenized and solubilized in 500 ml 4% deoxycholate (DOC), 40 m*M* Tris, 10 m*M* iodoacetamide at pH 8.3. The mixture is incubated for one hour at 4°C and centrifuged at 100,000 × g for one hour. The supernatant is precipitated with ethanol as described by Letarte-Muirhead *et al.* and Barclay *et al.* (13, 14). The protein pellet is redissolved in 500 ml of 2% DOC, 20 m*M* Tris, 10 m*M* iodoacetamide at pH 8.3 and centrifuged again at 100,000 × g for one hour. This supernatant is then applied to a 200 ml *Lens culinaris* lentil lectin column equilibrated in .5% DOC, 10 m*M* Tris at pH 8.2 prepared and utilized according to Allan *et al.* (49). The glycoprotein fraction is eluted from the lectin column with 3% alpha-methyl-D-glucoside, precipitated with ethanol and redissolved in .5% DOC, 10 m*M* Tris at pH 8.2 and applied to a 2.5 × 100 cm AcA 34 column in the same buffer. Antigenic activity is monitored by inhibition of cytotoxic activity as previously described (34, 47).

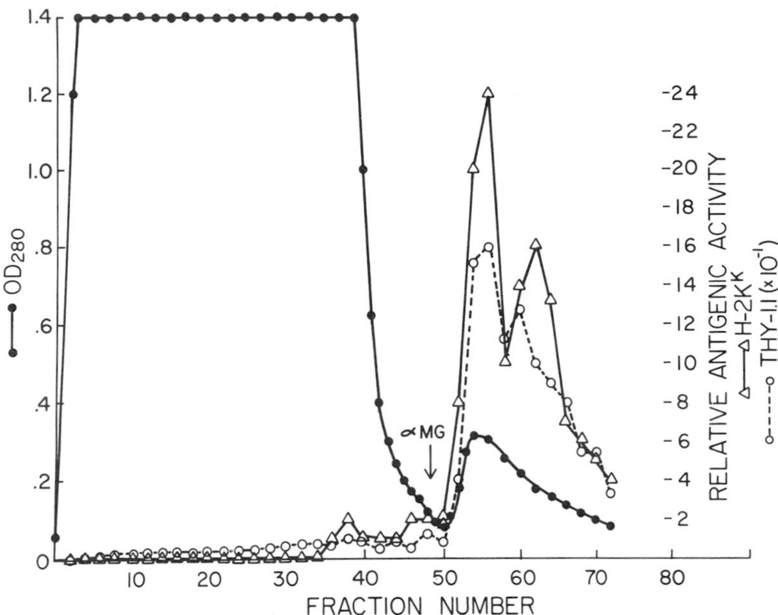

FIGURE 9 *Lens culanaris lectin chromatography of BW5147 proteins. The DOC solubilized 100,000 xg supernatant was applied to the column and washed with 10 mM Tris-HCl, pH 8.2 containing 0.5% DOC. Elution of the glycoprotein fraction was accomplished by 3% alpha-methyl-D-glucoside, which was applied at fraction 48 in the above buffer. H-2K^k and Thy-1.1 antigenic activities were monitored.*

The elution profile of DOC solubilized BW5147 cell membrane components from a lentil-lectin affinity column is shown in Fig. 9. All detectable H-2K^k and Thy-1.1 alloantigens bind to the lentil column and elute with 3% alpha-methyl-D-glucoside. Cell surface gp70 also elutes in the same position (not shown). As shown in Fig. 10 the Thy-1.1 and H-2K^k activities are separated on an AcA 34 column in agreement with their proposed molecular weights of approximately 57,000 daltons for H-2 (including β2-microglobulin) and 25,000 daltons for Thy-1.1. As shown in Fig. 11 gp70 antigenic activity elutes in a position identical to that for H-2K^k. Polyacrylamide gel electrophoretic analysis of these fractions revealed that significant purification had been

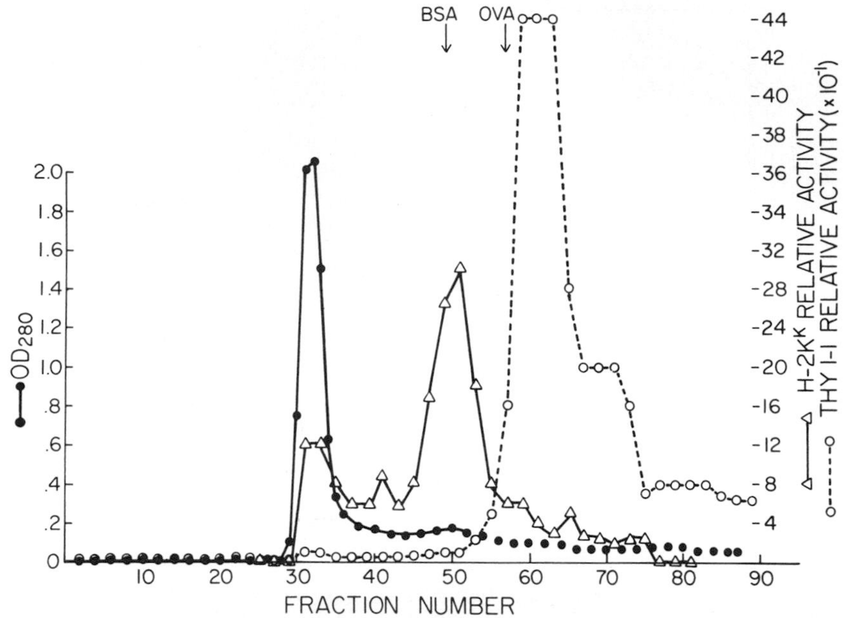

FIGURE 10 *Gel filtration on an AcA-34 column of lectin-purified BW5147 membrane glycoproteins. Column buffer was 10 mM Tris-HCL, pH 8.2 containing 0.5 ml DOC. H-2K^k and Thy-1.1 antigenic activities were monitored.*

achieved at this step. The Thy-1 fraction from the AcA 34 column can be rechromatographed on a Sephracyl S-200 column for final purification. At this stage the molecule is pure as judged by SDS polyacrylamide gel electrophoresis (34, 50). Table III tabulates the data for the purification of Thy-1.1. The Thy-1.1 molecule can be isolated in yields of approximately 11% and represents greater than a 2000 fold purification. Although H-2K^k and gp70 have not been totally purified we have been able to obtain a 1000 fold purification of the H-2K^k and gp70 components over intact cells with an overall yield of antigenic activity of 5 to 10% (51, 52). These results demonstrate the feasibility of isolating sufficient quantities of several cell surface components from the same cell line for structural and functional investigations.

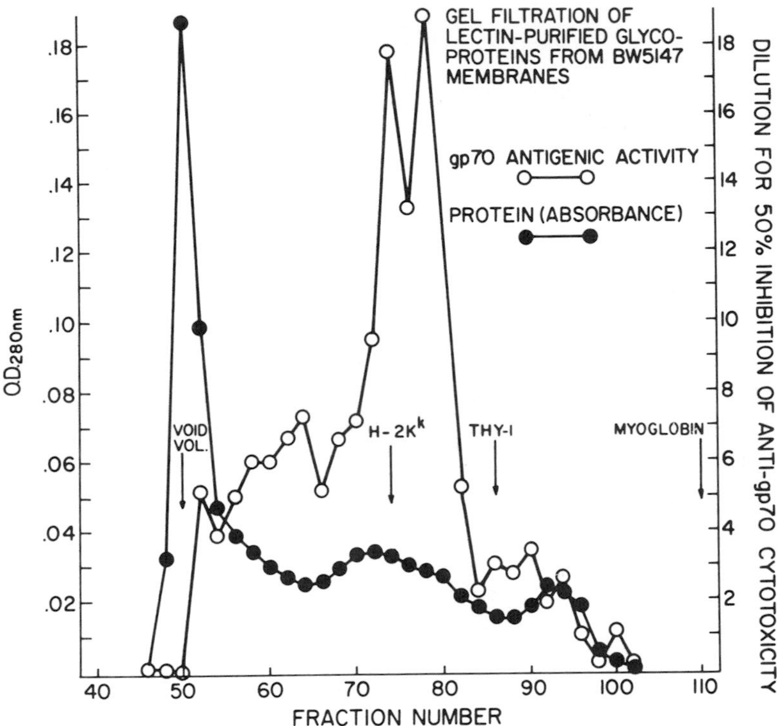

FIGURE 11 *Elution profile of H-2K^k, Thy-1. and gp70 from an AcA-34 column under identical conditions described in Fig. 10.*

D. Isolation of Murine Leukemia Virus

Five hundred liters of supernatant generated from 3 days of cell harvests of BW5147 cells is subjected to the isolation scheme shown in Fig. 12 for the purification of murine leukemia virus (MuLV). It is concentrated by use of Amicon Diafiber cartridges in conjunction with an air operated dual diaphragm pump. The viruses are separated from the fluid by use of five H10P100 cartridges whose fibers will retain molecular species larger than 100,000 daltons. The 30-60 fold concentrate is clarified by centrifugation at 6000 × g for 30 minutes. This pellet contains small membrane fragments that can be solubilized with DOC for the extraction of membrane components. The super-

TABLE III Purification of Thy-1.1 from BW5147 Lymphoblastoid Cells

Fraction	Total protein (mg)	Total activity (units/ fraction)	Protein %	Purification	Yield %
Cells	20,160	471.2	100	1	100
400 xg supernatant	11,495	351.6	57.0	1.3	74.6
400 xg pellet	4,960	141.0	24.6	1.2	30.0
4,000 xg supernatant	7,550	364.2	37.4	2.1	77.3
4,000 xg pellet	768	75.0	3.8	4.2	15.9
20,000 xg supernatant	8,250	152.6	40.9	0.8	32.4
20,000 xg pellet	1,150	125.1	5.7	4.7	26.8
Acetone pellet in deoxycholate	758	170.1	3.8	9.6	36.1
100,000 xg supernatant	289	150.0	1.4	22.2	31.8
100,000 xg pellet	437	3.6	2.2	0.4	0.8
Lectin column concentrate	11	94.3	0.06	356.5	20.0
AcA-34 column concentrate	1	50.0	0.005	2139.0	10.6

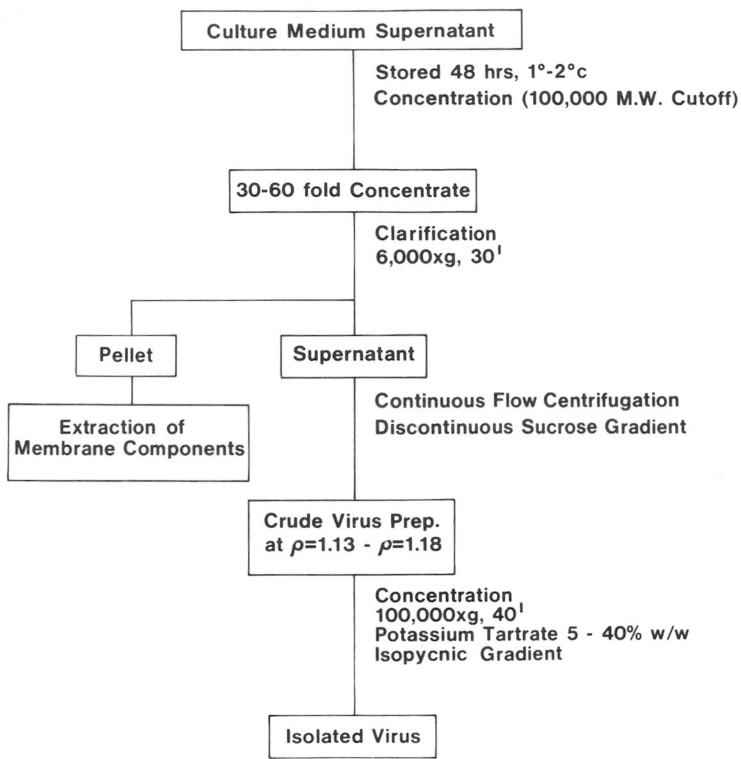

FIGURE 12 *Isolation scheme for murine leukemia virus from the media used to grow BW5147 cells.*

natant containing the virus is partially purified by continuous-flow, zonal centrifugation using a Beckman CF-32 rotor and a discontinuous sucrose gradient. The virus containing fraction from this run is slightly diluted, centrifuged at 100,000 × g for 45 min and the resuspended pellet subjected to isopycnic centrifugation on potassium tartrate gradients to obtain purified, concentrated virus. Although the BW5147 cell line is not a prolific producer of Gross virus we have been able to generate approximately 100 mg of pure virus from a 7 day production run. This material should be of great value as we attempt to correlate the chemical nature of normal differentiation alloantigens with envelope glycoproteins of oncoviruses found on the surface of cells that are indicative of malignant transformation (4-6).

IV. FUTURE POSSIBILITIES

Hopefully the data presented have provided insight into the capabilities of our Cell Culture Center. Obviously there is still room for several areas of improvement. We are particularly concerned with the yield of membrane presently obtained. Efforts are now underway to improve this process. Improvement of product acquisition coupled with the capabilities for producing large numbers of cells will allow the pursuit of a variety of scientific investigations concerning the mammalian cell at a relatively low cost. We are particularly excited about the possibility of totally deciphering the structural and functional properties of the major membrane glycoproteins from a murine lymphoblastoid cell line such as BW5147. We feel much greater knowledge of the mammalian cell will be gained as a consequence of these endeavors.

V. ACKNOWLEDGMENTS

These investigations were supported by grant No. GB-43575X from the Human Cell Biology Section of the National Science Foundation, the Diabetes Trust Fund, Birmingham, Alabama, Grant No. IM-33A from the American Cancer Society, Public Health Service Grant No. CA-15338 and No. CA-18609, and Cancer Center Core Support Grant No. CA-13148 from the National Cancer Institute.

The expert technical assistance of Beth Porter, Susan Snead, Barbara Patterson, and Susanne Henley as well as the advice of Dr. J. Claude Bennett is gratefully acknowledged.

REFERENCES

1. Reif, A. E., and Allen, J. M. V. (1964). *J. Exp. Med. 120,* 413-433.
2. Klein, J. (1975). "Biology of the Mouse Histocompatibility-2 Complex." Springer-Verlag, New York.
3. Lilly F., and Steeves, R. (1974). *Biochem. et Biophys. Acta 355,* 105-118.
4. Del Villano, B. C., Nave, B., Croker, B. P., Lerner, R. A., and Dixon F. J. (1975). *J. Exp. Med. 141,* 172-187.
5. Obata, Y., Ikeda, H., Stockert, E., and Boyse, E. A. (1975). *J. Exp. Med. 141,* 188-197.
6. Tung, J.-S., Vitetta, E. S., Fleissner, E., and Boyse, E. A. (1975). *J. Exp. Med. 141,* 198-205.
7. Reif, A. E., and Allen, J. M. (1966). *Nature 209,* 521-523.
8. Reif, A. E., and Allen, J. M. (1966). *Nature 209,* 523.
9. Raff, M. C. (1969). *Nature 224,* 378-379.
10. Douglas, T. C. (1972). *J. Exp. Med. 136,* 1054-1062.
11. Acton, R. T., Morris, R. J., and Williams, A. F. (1974). *Eur. J. Immunol. 4,* 598-602.
12. Letarte-Muirhead, M., Acton, R. T., and Williams, A. F. (1974). *Biochem. J. 143,* 51-61.
13. Letarte-Muirhead, M., Barclay, A. N., and Williams, A. F. (1975). *Biochem. J. 151,* 685-698.
14. Barclay, A. N., Letarte-Muirhead, M., and Williams, A. F. (1975). *Biochem. J. 151,* 699-706.
15. McClain, L. D., Acton, R. T., and Bridgers, W. F. (1976). *American Society for Neurochem. 7,* 65.
16. Schwartz, B. D., Kato, K., Cullen, S. E., and Nathenson, S. G. (1973). *Biochem. 12,* 2157-2164.
17. Silver, J., and Hood, L. (1974). *Nature 249,* 764-765.
18. Munro, A., and Bright, S. (1976). *Nature 264,* 145-152.
19. Kurth, R. (1975). *Nature 256,* 613-614.

20. Lerner, R. A., Wilson, C. B., Del Villano, B. C., McConahey, P. J., and Dixon, F. J. (1976). *J. Exp. Med. 143,* 151-166.

21. David, C. S. (1976). *Transplant. Rev. 30,* 299-322.

22. Boyse, E. A., and Bennett, D. (1974). *In* "Cellular Selection and Regulation in the Immune Response." (G. M. Edelman, ed.), pp. 155-176. Raven Press, New York.

23. Cantor, H., and Boyse, E. A. (1975). *J. Exp. Med. 141,* 1376-1389.

24. Cantor, H., and Boyse, E. A. (1975). *J. Exp. Med. 141,* 1390-1399.

25. Hämmerling, U., and Eggers, H. J. (1970). *Eur. J. Biochem. 17,* 95-99.

26. Plesner, T. (1976). *Scand. J. Immunol. 5,* 1097-1102.

27. Jensenuis, J. C., and Williams, A. F. (1974). *Eur. J. Immunol. 4,* 91-97.

28. Theofilopoulos, A. N., Bokisch, V. A., and Dixon, F. J. (1974). *J. Exp. Med. 139,* 696-711.

29. Marchesi, V. T., Tillack, T. W., Jackson, R. L., Segrest, J. P., and Scott, R. E. (1972). *Proc. Nat. Acad. Sci. U.S.A. 69,* 1445-1449.

30. Economidou, J., Hughes-Jones, N. C., and Gardner, B. (1967). *Vox Sang. 12,* 321-328.

31. Kahn, C. R. (1976). *J. Cell Biol. 70,* 261-286.

32. Cohen, J. B., and Changeux, J. P. (1975). *Annu. Rev. Pharmacol. 15,* 83-103.

33. Parker, C. W., Greene, W. C., and MacDonald, H. H. (1976). *Exp. Cell Res. 103,* 99-108.

34. Zwerner, R. K., Barstad, P. A., and Acton, R. T. (1977). *J. Exp. Med.* (in press).

35. Zwerner, R. K., Runyan, C., Cox, R. M., Lynn, J. D., and Acton, R. T. (1975). *Biotechnol. Bioeng. 17,* 629-657.

36. Zwerner, R. K., and Acton, R. T. (1975). *J. Exp. Med. 142,* 378-390.

37. Lynn, J. D., and Acton, R. T. (1975). *Biotechnol. Bioeng. 17,* 659-673.
38. Cameron, J., and Godfrey, E. I. (1968). *J. Appl. Bact. 31,* 405-410.
39. Acton, R. R., and Lynn, J. D. (1977). *Adv. Biochem. Eng.* (in press).
40. Gross, L. (1959). *Proc. Soc. Exptl. Biol. Med. 100,* 325-328.
41. Chazan, R., and Haran-Ghera, N. (1976). *Cell. Immunol. 23,* 356-375.
42. Ralph, P. (1973). *J. Immunol. 110,* 1470-1475.
43. Hyman, R., Ralph, P., and Sarkar, S. (1972). *J. Natl. Cancer Inst. 46,* 173-184.
44. Green, E. L. (1968). *In* "Handbook on Genetically Standardized Jax Mice." pp. 51-59. The Jackson Laboratory, Bar Harbor, Maine.
45. Acton, R. T., Blakenhorn, E. P., Douglas, T. C., Owen, R. D., Hilgers, J., Hoffman, H. A., and Boyse, E. A. (1973). *Nature New Biol. 245,* 8-10.
46. Ralph, P., and Nakoinz, I. (1973). *J. Natl. Cancer Inst. 51,* 883-890.
47. Barstad, P. A., Henley, S. L., Cox, R. M., Lynn, J. D., and Acton, R. T. (1977). *Proc. Soc. Exp. Biol. Med. 155,* 296-300.
48. Runyan, C. C., and Acton, R. T. (1975). *Fed. Proc. 34,* 1013.
49. Allan, D., Auger, J., and Crumpton, M. J. (1972). *Nature New Biol. 236,* 23-25.
50. Zwerner, R., Barstad, P., and Acton, R. (1977). *In* "Protides of the Biological Fluids" (H. Peeters, ed.). Vol. 25. Pergamon Press, Oxford (in press).
51. Barstad, P., Zwerner, R., and Acton, R. (1977). *In* "Protides of the Biological Fluids" (H. Peeters, ed.). Vol. 25. Pergamon Press, Oxford (in press).

52. Wise, K., and Acton, R. (1977). *In* "Protides of the Biological Fluids" (H. Peeters, ed.). Vol. 25. Pergamon Press, Oxford (in press).

COMPUTER CONTROL
OF MAMMALIAN CELL
SUSPENSION CULTURES

LASZLO K. NYIRI

Fermentation Design, Inc.
Division of New Brunswick Scientific Co., Inc.
Bethlehem, Pennsylvania

I. INTRODUCTION

Attempts to produce eucaryotic cells of animal origin *in vitro* resulted in the development of various methods and types of equipment (Telling and Radlett, 1970; Litwin, 1971; VanHemert, 1972; Knazek *et al.*, 1972; Kruse and Patterson, 1973) with different performance characteristics.

The known sensitivity of animal cells to environmental conditions requires strict control of the environmental variables (T, P, pO_2, pCO_2, pH, nutrient feed, etc.) (Nyiri, 1974). This fact limits the applicability of many of the cell culture apparatus and techniques for large scale cell mass production. The most adaptable technique is the cell suspension culture (CSC). This offers the following advantages:

(1) Cell growth takes place in three dimensions, therefore, the effect of contact inhibition is reduced.

(2) The basic culture technique is similar to cultivation of microbial cells about which much applicable technical literature exists.

(3) Animal cells with "anchorage dependency" can also be cultured in suspension using microcarriers, and

(4) Scale-up can be easily achieved with respect to space requirements, and unit operation.

Information available on cell suspension cultures to date indicate that this technique will be given wider applicability in the future.

The direction of this development raises the question: "What sort of control equipment and techniques are needed to satisfy the handling of such complex processes like animal cell suspension cultures?"

This paper is an attempt to review the problems of control of such complex biological processes. The most recent achievements in cell suspension culture apparatus design and construction (Nyiri, 1974; Acton, 1975) make this review timely.

This paper addresses four major areas of interest, namely:

(1) The interactive nature of the biological processes,

(2) Sensors with the capability to measure the status of the biological process,

(3) Interactive control strategy,

(4) Overall control of a mammalian cell mass-metabolite production line.

Although the present discussion stems from observations made while working with microbial cells, certain fundamental laws can be applied to animal cells too, carefully pondering, however, the major differences between the two cellular systems (Table I).

TABLE I Some Features of Microbial and Animal Cells

	Microbial cell	Animal cell
Size	about 1 μ	10-100μ
Metabolism	Individual regulation	Individual + hormonal regulation
	Flexible bioconversion	Less flexible metabolic pathways
Doubling time	Minutes	Hours
Environment	Self-response to the changes	Decreased capability to adapt to the changes
	Competitiveness	Sensitivity to predators (virus)
Genetics	Mutation $\alpha \rightleftharpoons \alpha$	Diploid → haploid conversion

II. INTERACTIONS IN CELL CULTURE SYSTEMS

A. The Term Interaction

Since this word is extensively used in the biological-engineering-computer control literature for a variety of functions, it is necessary to define our understanding with regard to the interactions with specific respect to multicomponent systems.

In our understanding, interaction is a process in which a minimum of two constituents of a system mutually influence each other resulting in reversible or irreversible changes in each constituents' quantity and/or quality. The interaction may take place between any constituents of the whole system.

B. Process of Interaction

Take three constituents of a multicomponent system (Figure 1), with the ability to selectively exercise an effect on the other constituents. Constituents A_i and C_i perform reciprocal actions on each other causing changes in the quantity and/or quality of the affected constituent. This is called *direct interaction* between constituents. At the same time constituent B_i also in-

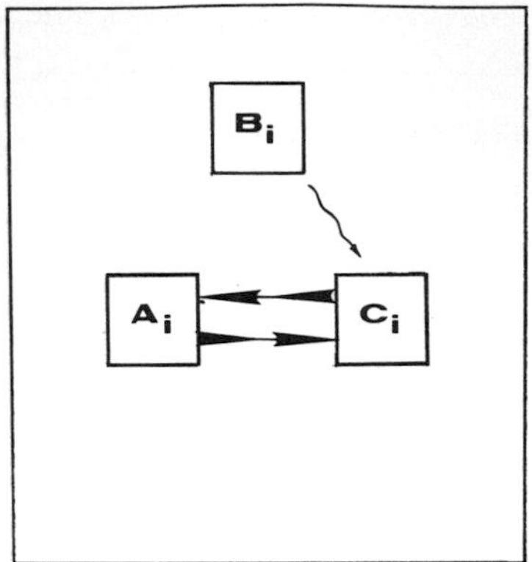

FIGURE 1 *Principle of interaction between system constituents.*

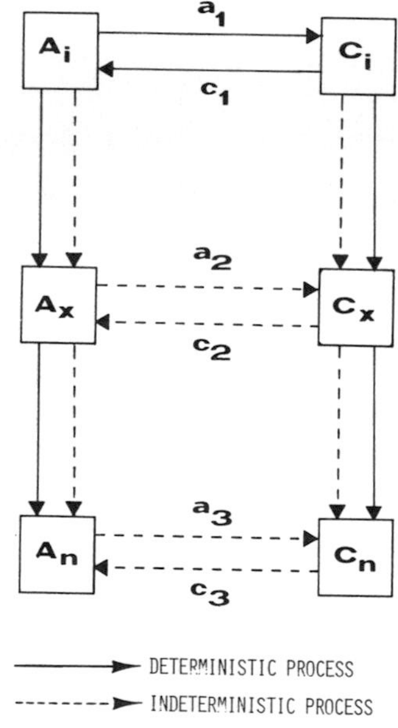

FIGURE 2 *Basic mechanism of interaction between system constituents.*

fluences C_i, however, C_i does not interact with B_i. The effect of B_i on C_i can amplify or diminish the effect of A_i on C_i. This is called *indirect interaction* between constituents A_i and B_i, which can be *synergistic* or *antagonistic*, respectively. As a result of the effect of the "third party" B_i on C_i, the latter's direct interactive response toward A_i can be greater or less than that if B_i is not present in the system.

During the interaction, constituent C (being in state C_i) receives input (a_1) from constituent A (while it is in state A_i) (Figure 2). This input, after a unit delay, offsets the original state of C_i and brings it into a transition (C_x), then into another more stable state (C_n). Conversely, C_i while in its original state influences A_i bringing it through transition A_x to a final A_n state. It is also assumed that the intermediate and final states also interact with each other. Based on experiments in enzyme catalized reactions, in biological systems the interaction between two system constituents lasts throughout several step changes in the states (Cleland, 1970). If the sequence of process is known, the interaction can be called deterministic. However, because of the existence of indirect interaction (third party's effect) (Nyiri and Toth, 1975) the biological system can be considered indeterministic, at least probabilistic. In other words, the biological system is considered an indeterministic one with a probability distribution over the initial and the possible next states during the interactions.

C. Interactions in Biological Systems

In a biological system (in our case: *in vitro* culture of eucaryotic cells) the main interacting system constituents are (Figure 3): (1) Cell(s) and (2) environmental factors (including compounds with positively (Y^+) or negatively (X^-) charged functional groups, electrons (e^-), free radicals (Z), and energy bearing physical constituent(s), e.g. temperature (T)). A minimum of two major types of interactions are taking place, one at the *cell-environment* and an other at the *cell-cell* level. (We

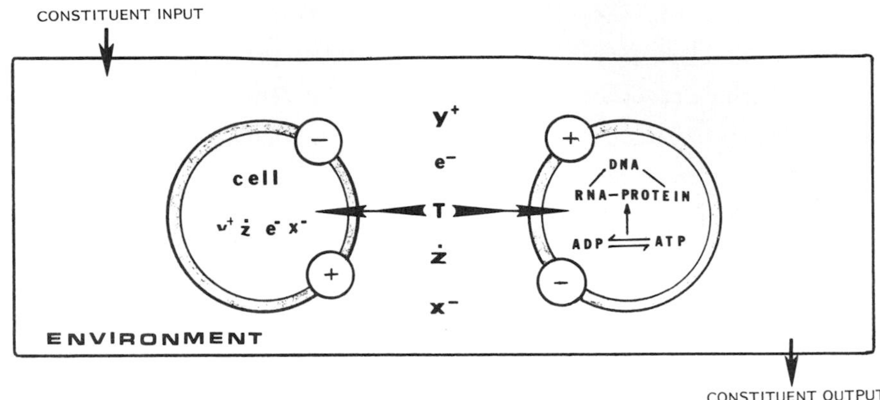

FIGURE 3 *Interactions between system constituents in a biological system.*

the cell-environment and another at the *cell-cell* level. (We omit, knowingly, the intracellular interactions between the cell constituents).

Cell-environment interaction occurs via the cell membrane primarily by means of altering the charge conditions of the reactants. Only the heat energy is considered to penetrate the cell membrane without directly interacting with cell membrane constituents.

Cell-cell interaction takes place either by direct contact between the cells or through mediators released into or taken up via the medium's main constituent (H_2O). This process also involves charge condition changes in the constituents. Some evidences has been reported indicating that the cell-cell interactions are influencial to the cells nuclear replication, too (Mueller, 1971; Mastro *et al.*, 1974).

Whatever the level of interaction, ultimately the interactive effect of one constituent on another takes place at the electron distribution level in the affected compounds as it is exemplified in case of acetylcholine-acetylcholinesterase reaction (Figure 4). The change in the electron distribution primarily depends on the

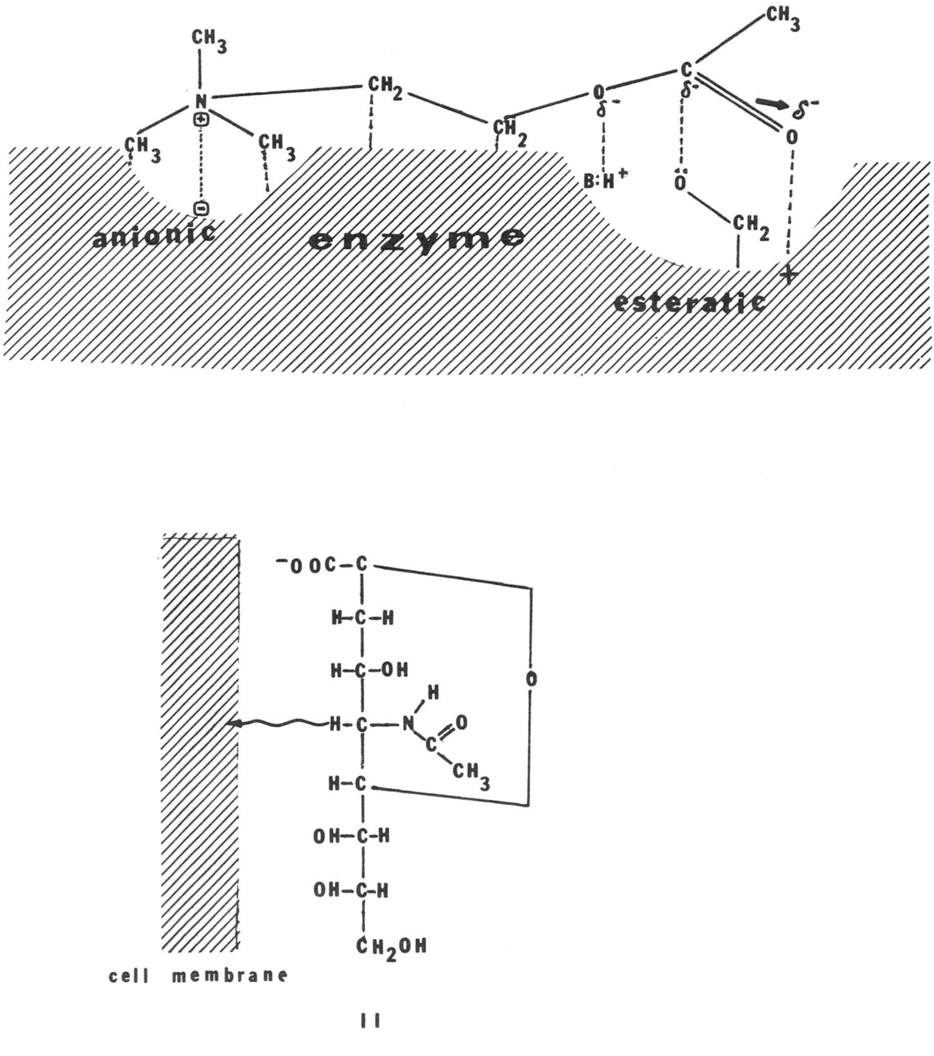

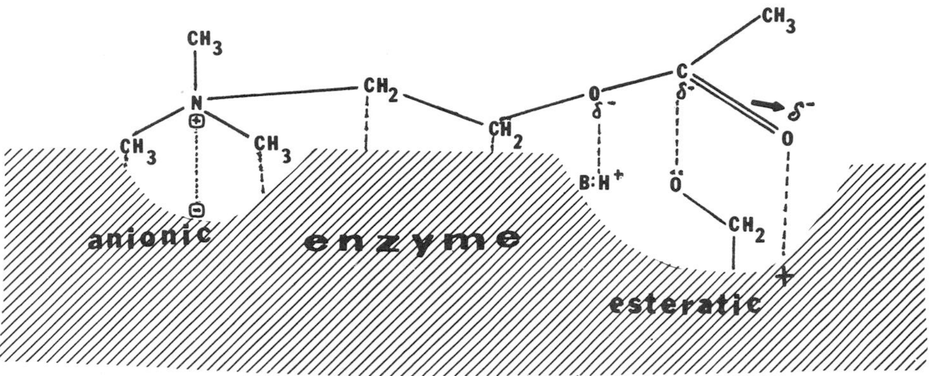

cell membrane

II

FIGURE 4 *Ultimate level of interaction: electron proton exchange. I. Interaction between acetylcholinesterase and acetylcholine. II. Sialic acid as potential electron donor and acceptor system.*

available free energy for the process which, thusly, is considered as one of the most important factors in *interactions*. Compounds with the capability to dislocate electrons and transfer protons are considered to be essential during these processes. In animal cell membranes, the sialic acid (II in Figure 4) and the receptor molecules (Marchesi, 1975) are regarded to be intimately involved in cell-environment and cell-cell interactions.

Because of the importance of charge condition changes in the molecules of cell membrane and in the intercellular medium, sensors monitoring changes in the proton or electron intensity and capacity of the culture are extremely useful from process control point of view (Toth, 1977).

D. Interactions between the Biological and Physical System Constituents in an In Vitro Culture System

The *in vitro* cultured cells are grown in a properly designed culture vessel with its auxiliary equipment (pipes, valves, pumps, motors, instruments) the ultimate purpose of which is the maintenance of optimum environmental conditions.

Because of the interactions between the cells and their environment and the changing characteristics of the cells in the culture the environment can not be considered neutral.

Taking the simplest substrate (S) related growth (X) which is the function of temperature, and pH a complex interactive condition is obtained (Figure 5) (Young, 1970). Any change in S will involve changes in X and pH which ultimately influence the uptake of S hence, again, X. The interaction becomes more complex with the introduction of new system constituents and with the requirement to control the process variables. This constant modification of the environment involves the process status identification sensing the process variables and performing control action by actuation of valves, motors, pumps of the physical system.

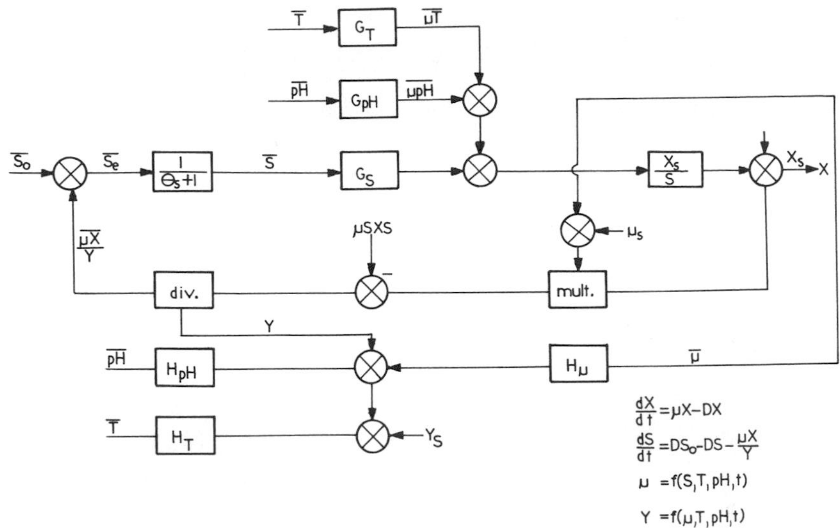

FIGURE 5 *Interaction between the system's biological constituents (S,X) and mechanical-electronic elements.*

III. IDENTIFICATION OF STATUS OF BIOLOGICAL PROCESSES

A. Analysis and Determination of Interactions

Means for on-line, real-time detection and quantitative measurement of the state of constituents of the system is inevitable. In the practice of culturing cells *in vitro* the means are *sensors* which generally detect and measure the status of constituents. It is noted that, generally, the measurable status (e.g. pH) is the *end result* of several interactions and not the interaction itself. Practices in biochemical engineering, however, indicate that this type of sensing seems to be adequate for process control if the proper correlation between the sensed status of the constituent (state variable) and the process is defined (Nyiri, 1972; Aiba *et al.*, 1973; Nyiri *et al.*, 1976).

B. Sensors in Eucaryotic Cell Cultures

Instruments and sensors for cell cultures may be divided into two broad categories (1) DIRECT MEASUREMENT types and (2) NET EFFECT SENSOR types.

Most instruments currently in use in the cell culture systems belong to the first category and are used for the purpose of monitoring as well as control of process variables (Table 2). Devices of this type can provide information only on the physical and physico-chemical conditions of the culture (e.g. T, pH, DO), whereas the physiological conditions are usually of greater interest. There are, of course, a small number of instruments that do offer some information on the cellular metabolism (e.g. exit O_2 and CO_2 analyzers, most recently automatically reporting gas chromatographs) but they have not found wide application in control of the cell cultures' environment.

From a process analysis point of view, therefore, a more promising approach is the application of so-called *Net Effect Sensors* (Nyiri and Jefferis, 1973). The net effect sensor is a composite of various individual sensor elements and measures a process variable related to a (preferably main) metabolic pathway. The outputs of individual sensors are coupled electronically in order to produce a new signal relative to a otherwise non-measurable process variable (e.g. O_2 uptake, CO_2 release rates, or respiratory quotient). An implementation of the net effect sensor concept is the on-line, real-time application of computers with a Data Analysis oriented program to obtain information on the process status. Here with multivariation of the directly available signals otherwise inaccessible state variables (process indicators) can be obtained. These process indicators reflect the physico-chemical, physiological and biochemical conditions of the culture (Humphrey, 1971; Nyiri, 1972; Nyiri and Krishnaswami, 1974; Nyiri and Toth, 1975; Nyiri *et al.*, 1975).

TABLE II Direct Sensing Devices Applied to Cell Culture Vessels

Sensor	Measured variable	Manipulated variable	Intermediate	Controlled variable	Disturbance
Thermometer	Temperature	Coolant/steam flow valve position		Broth temperature	Metabolic heat mixing heat
Tachometer	Motor speed	Motor speed		Tip speed	Broth viscosity
Gasflow meter	Gasflow rate	Gasflow valve position		Gas input into broth	Vessel head pressure
Pressure	Vessel head pressure	Exhaust gas valve position		Vessel head pressure	
pH probe	H + concentration	Reagent flow gasflow	Gas transfer into the liquid	H^+/OH^-	pH shifting metabolic products
DO probe	DO level	Motor speed gasflow valve pressure regulator	Gas transfer into the liquid	pO_2	Oxygen uptake detergents temperature broth density
eH probe	ORP	Gasflow valves	N_2-CO_2 transfer	eH	Metabolism
Light cell	Turbidity	Valves	Nutrient flow	Cell mass	Oscillations in the system
Exit O_2 analyzer		O_2% in exit gas stream			
Exit CO_2 analyzer		CO_2% in exit gas stream			
Gas chromatograph		Various volatile metabolites in exit gas stream			

C. Process Status Identification by Computer

Differences between the eucaryotic microbial cells (e.g., yeast) and cells of metazoan origin (Table 1) made it necessary to define some specific pivotal process status indicators for animal cell cultures. Figure 6 is a tentative flowchart of a computer program constructed specifically for analysis of conditions in animal cell cultures. The ultimate goal of such an operation is the proper status identification on the basis of which complex (interactive) control of the environment can be performed. Among the process indicators we list the Reynolds number (NRE) which is the indicator of shear effects and may be correlated with population-density regulation of growth (Holley, 1974). The on-line, real-time computed respiratory quotient (RQ) was found in case of eucaryotic microbial cells as a pivotal indicator of the cells' physiological conditions (Fiechter and VonMeyenburg, 1968; Nyiri *et al.*, 1975) and it is expected to serve the same purpose in the case of animal cell cultures. The oxidation-reduction potential (ORP) is considered probably the most important indicator of changes in charge conditions in animal cell cultures. On the basis of literature evidence ORP seems to be a key factor in the proliferation of normal and malignant cells as well as an indicator (forecaster) in changes of specific growth rate and indicator of cell-environment interactions (including virus induced transformations) (Wiles and Smith, 1969; Daniels *et al.*, 1970; Taylor *et al.*, 1971; Klein *et al.*, 1971; Pardee *et al.*, 1974; Toth, 1977). Correlation between pO_2 and ORP (Taylor *et al.*, 1971), on the other hand, makes it necessary to compute k_La, dpO_2/dt, dpH/dt, dCO_2/dt, deH/dt on the basis of which interactive control of pO_2, pCO_2, pH and eH can be achieved.

As an example of process status identification real-time computed, process indicators are shown in Figure 7. In this case a PDP-11/20 computer was interfaced with a highly instrumented cul-

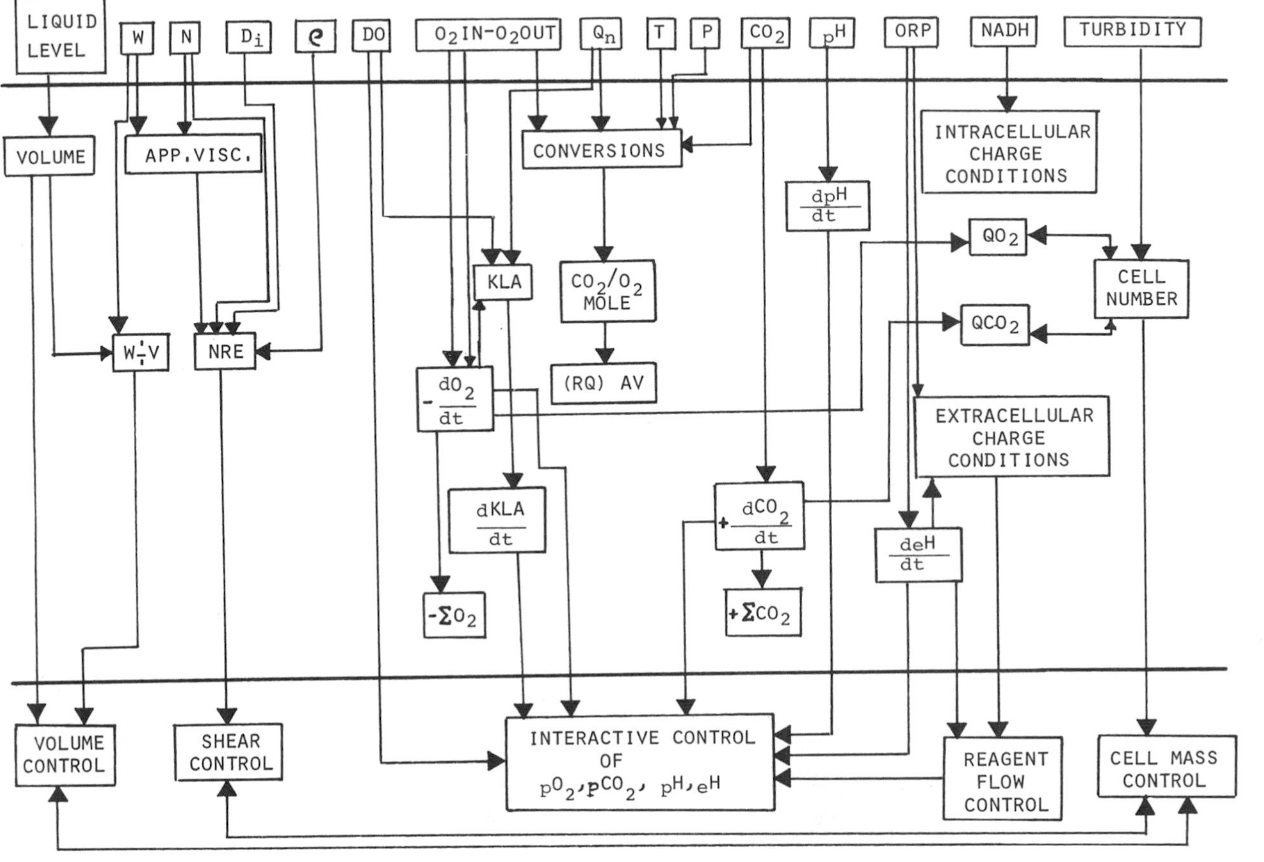

FIGURE 6 *Non detailed flowchart for computer based process status identification and interactive control of animal cell suspension culture.*

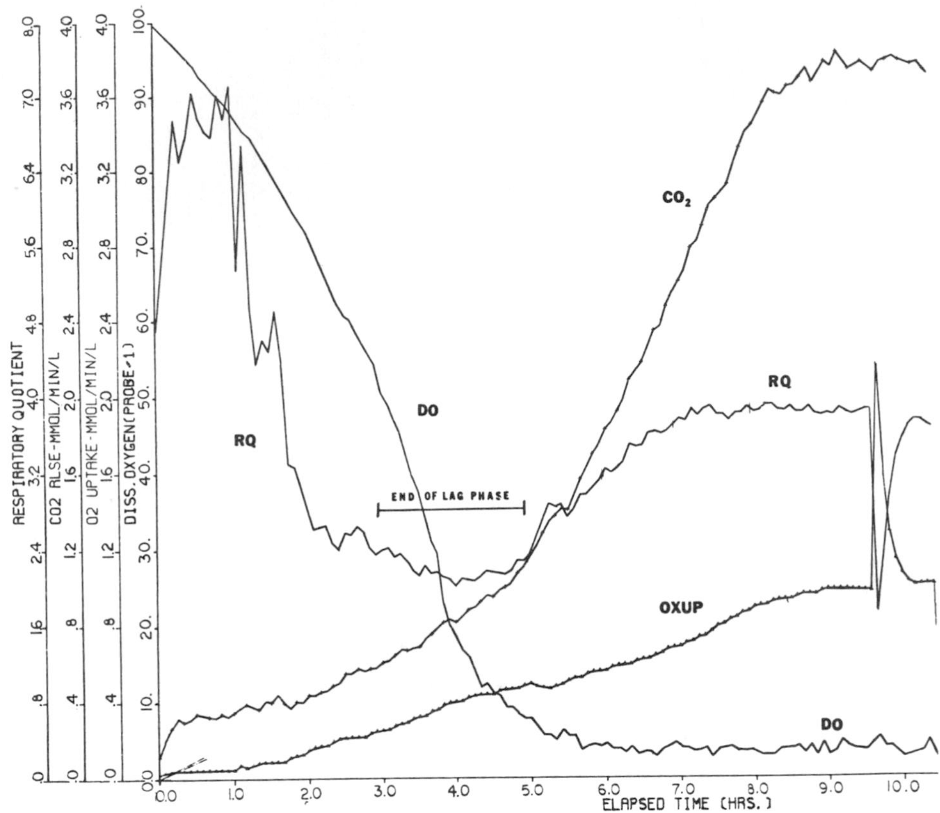

FIGURE 7 *Identification of process status by on-line, real-time operating computer (C. utilis culture).*

ture vessel (Harmes, 1972) in which *C. utilis* cells were grown under specified conditions. (Nyiri and Krishnaswami, 1974; Nyiri *et al.*, 1976).

The ability to define transitions from lag (resting) to logarithmic (proliferative) phases of growth is of particular interest. Ross attempted to provide such capability by means of numberical analysis of available fermentation data (1974).

However, such an approach requires constant wet chemical analysis and off-line calculations. The presented picture shows the computer analysis on the gas exchange conditions in the early phase of growth.

DO level exhibited a gradual decrease to a minimum of 25% of saturation over a period of 6 hours. Similar "smooth" transitions were also found by both the computed oxygen uptake (OXUP) and CO_2 production (CO_2) rates. The smooth transitions exhibited by these process indicators made them particularly inappropriate for the purpose of defining the transition between lag and logarithmic phases. On the other hand, the RQ went through drastic changes during this early period of cell growth. Cells transferred from inoculum responded to the altered environmental conditions with a rapid increase in RQ (→7) which after the first hour dropped to 2.0 and then increased again to a steady value of 3.8 (EFT = 7.0 hrs.). RQ values higher than 1.0 at the start of fermentations and during ethylalcohol production of *S. cerevisiae* cultures have been reported by other investigators (Gorts, 1967; Fiechter and VonMeyenburg, 1968).

Chemical analysis of the culture revealed significant correlations between RQ and some biochemical events including: (1) nucleic acid, (2) protein, and (3) ethylalchol synthesis (Figure 8).

The drop in RQ value during the elapsed time period of 1-3.0 hours coincided with the increase in specific nucleic acid content of the culture, while the minimum RQ (EFT = 3-4 hrs.) coincided with the start of increased specific protein concentration. The increase in RQ value in the fourth hour coincided with the start of ethylalcohol formation (shaded area). Correlation obtained between the culture's metabolic activity detected by wet chemical methods and the respiratory quotient obtained via computer operation demonstrate that the latter, after the definition of correlations, can be used to determine certain transition conditions in eucaryotic cell cultures.

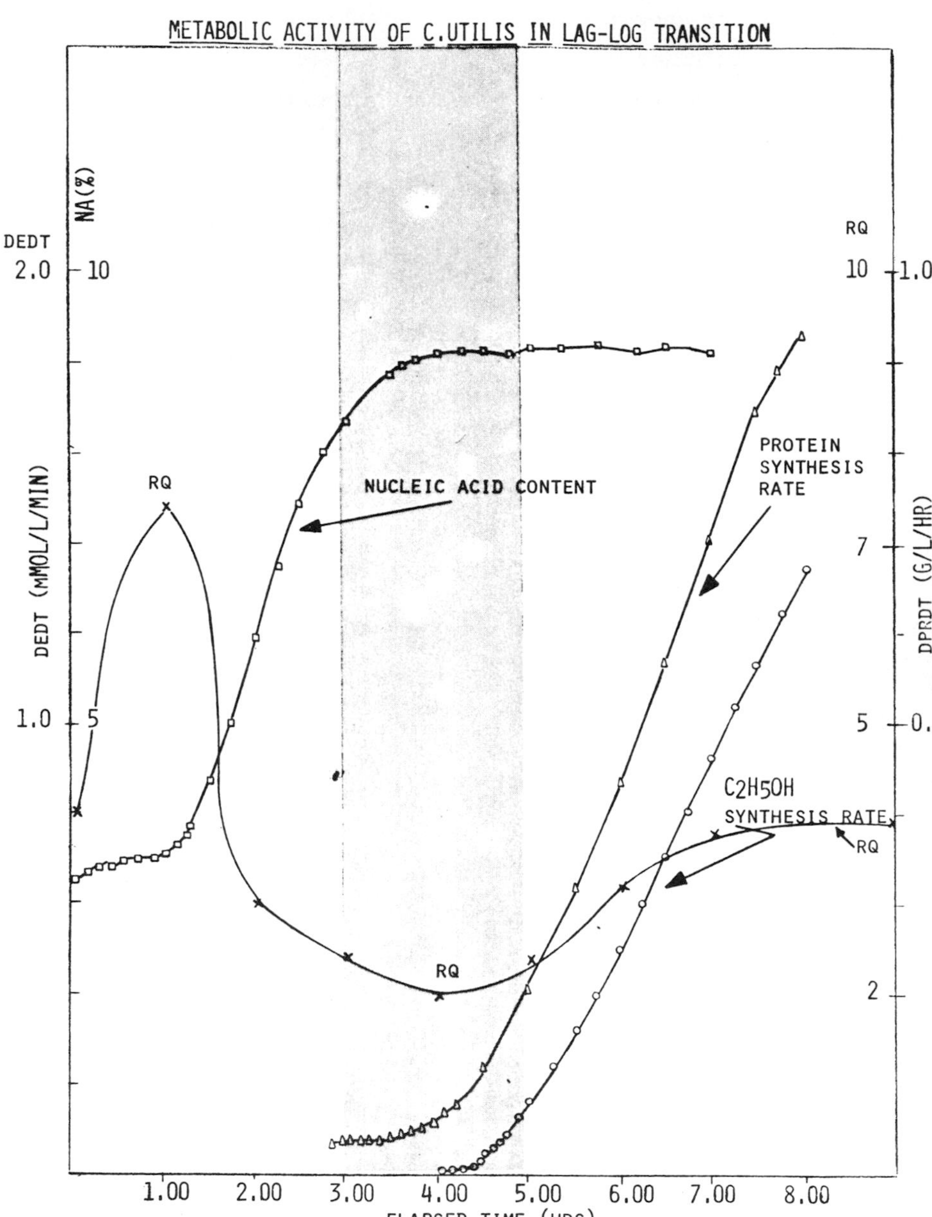

FIGURE 8 *Correlation between the changes in RQ specific nucleic acid content (NA%) protein synthesis rate (DPRDT) and ethylalcohol synthesis rate (DEDT) in C. utilis culture.*

IV. INTERACTIVE PROCESS CONTROL

There are two major objectives to be achieved by means of interactive control of environmental conditions during cell culture. These are

1. Control the environment to make it suitable to the genetically altered conditions, and
2. Control the environment in order to shift the metabolism toward a desired direction.

Accordingly, the need for corrective actions to the environmental conditions assumes extensive knowledge of the process status. This, as it was demonstrated previously, can be substantially improved using computer for process status identification.

In addition to the cell-environment interactions, there are interactions between the effect of individual control elements. The effect of agitation (tip) speed and airflow rate, as well as temperature and gas pressure on the culture's gas exchange conditions is among the most convincing arguments.

The question is how a control strategy works which takes the above mentioned interactions into consideration.

Figure 9 presents two different control concepts. In the first case, environmental variables V_1, V_2 and V_3 are maintained at their target values according to the PID control law or one of its variants. In this case, the control actions are taking place independently from each other. This technique was characteristically followed for more than a quarter century in the antibiotic fermentation industry.

In the second case (B), the target values of the V_1, V_2, V_3 environmental variables are maintained by separate control means (as in the first case), however, the setpoints are changed according to the (formerly detected) interactions between V_1, V_2

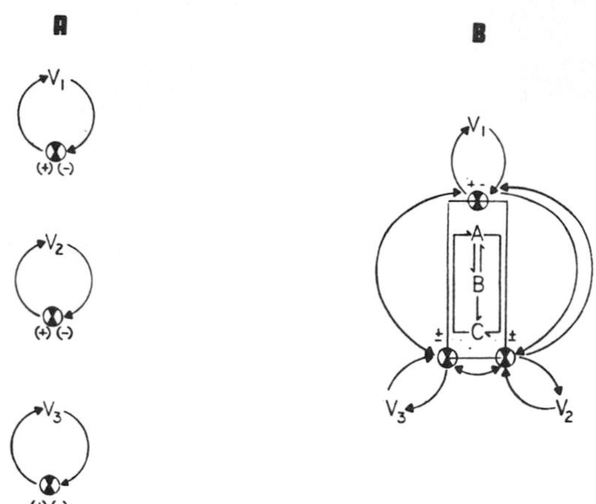

FIGURE 9 *Control concepts in cell culture technology: A. Independent control loop approach, B. Interactive control.*

and V_3 as well as the interactions between A, B and C biological variables as influenced by environmental variables V_1, V_2, V_3.

This control operation may take place on several levels (in several loops) (Figure 10). In this case, for instance, the target values of environmental variables (T, N, Q, P, DO, S, pH, etc.) are maintained by otherwise known techniques (e.g., with control elements using supervisory control procedure, or DDC) and the setpoint for each environmental variable is altered according to the biological status of the culture and taking into consideration the formerly defined interactive effects of the constituents.

As an example, (Figure 11) shows the performance of a computer during interactive control of environmental conditions. In this case, the previously defined optimum carbon to nitrogen ratio (condition for RQ = 1) was maintained by a computer during the logarithmic phase of growth of *C. utilis* cells. Maintaining other environmental variables (T, P, pH) constant at their predefined optima and altering others (e.g., DO on the basis of k_La) the computer used the RQ value to control the feed rate of carbohydrate and nitrogen sources in order to maintain the RQ around

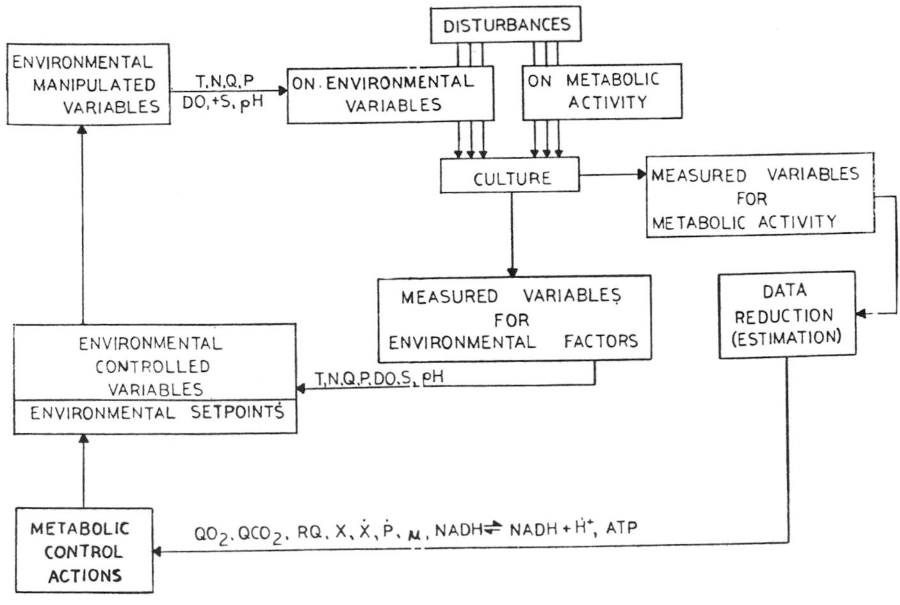

FIGURE 10 *Implementation of interactive control (cascaded control of environment).*

1.0. In this case, according to the scheme, the control-loop was closed via a computed process indicator (RQ) which reflected the pivotal physiological condition of *C. utilis*. At Ø.ØØ EFT (which is the starting time of the experiment in a fed-batch culture where molasses was added at a rate of 24 $G.L^{-1}.HR^{-1}$) the growth rate was 5.0 $G.L^{-1}.HR^{-1}$ and RQ was below the SP (RQSP = 0.5 min-1.2 max). As a consequence, feed of $(NH_4)_2HPO_4$ started in order to reach a proportion of 2:1, being previously defined as optimum for RQ=1 by pulse-response experiments. As a result, the RQ started to increase and reached the region of RQ=0.8-0.9. At 4:20 EFT RQ started to decrease and reached the region of RQ=0.8-0.9. Pump started to feed $(NH_4)_2HPO_4$ again. A subsequent increase in RQ coincided with increases in oxygen uptake and CO_2 release rates indicating the enhanced metabolic activity and

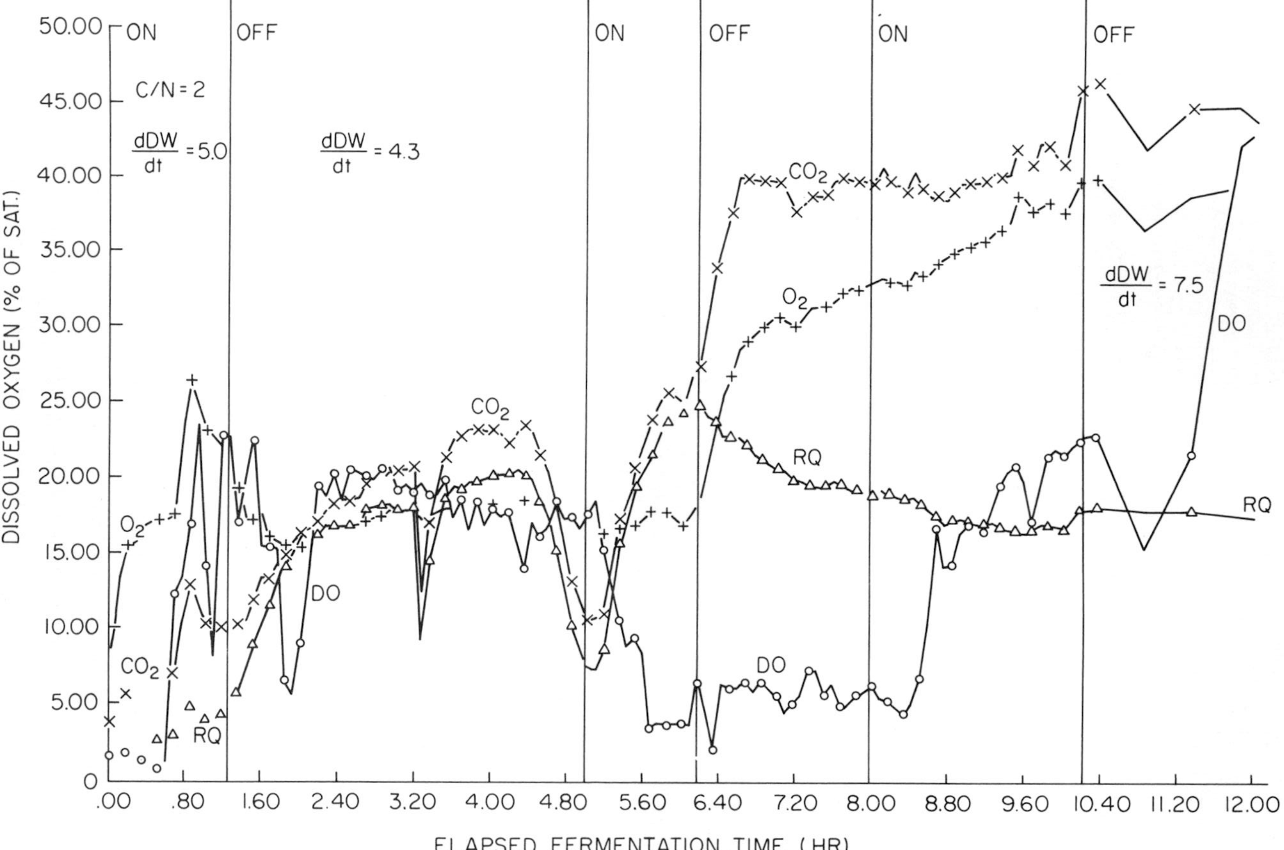

FIGURE 11 *Interactive control of nutrient feed on the basis of RQ in C. utilis culture.*

growth. Here RQSP was narrowed to 0.9 min-1.2 max. Reaching RQ=1.2 the pump was turned OFF resulting in decrease in the RQ below 1.0 but not below 0.9. Here the SP was adjusted to RQSP=1.0$\pm$0.05. The N feed pump was turned ON and OFF keeping the RQ in this domain. During this operation samples were taken and the OD was determined. There was a gradual increase in growth rate from the 5:ØØ EFT peaking at 7.5G.L^{-1}.HR^{-1} value at 1Ø:2Ø EFT. This growth rate is corroborated by the oxygen uptake rate of the culture which peaked at 3.9 mMO_2.L^{-1}.Min^{-1} (=234 mMO_2.L^{-1}.HR^{-1}).

Oscillations in process variables were observed mostly in the first five hours of the experiment and are mainly due to "third party" effects, where new pulses reached the process before the effect of the previous pulse was not completed (Nyiri and Toth, 1975).

Although this experiment is considered to be extremely crude, to our best knowledge, this is the first one where automatic control of environmental condition was implemented on the basis of a pivotal physiological variable (RQ).

V. OVERALL CONTROL OF MAMMALIAN CELL CULTURES

A. Animal Cell Culture Plant

Recent developments and prospects in molecular biology, cancer and virus research indicate an increased need for animal cell production in mass quantities. This can be implemented in culture systems of various size and complexity. Such a system is depicted in (Figure 12).

The flow of operation incorporates medium preparation including sterilization and medium transfer (A→B), storage of medium (B), medium poising (C), development of seed culture (D), inoculation of main culture vessel (E), semi-continuous operation withdrawing the cell containing culture liquid into a harvest

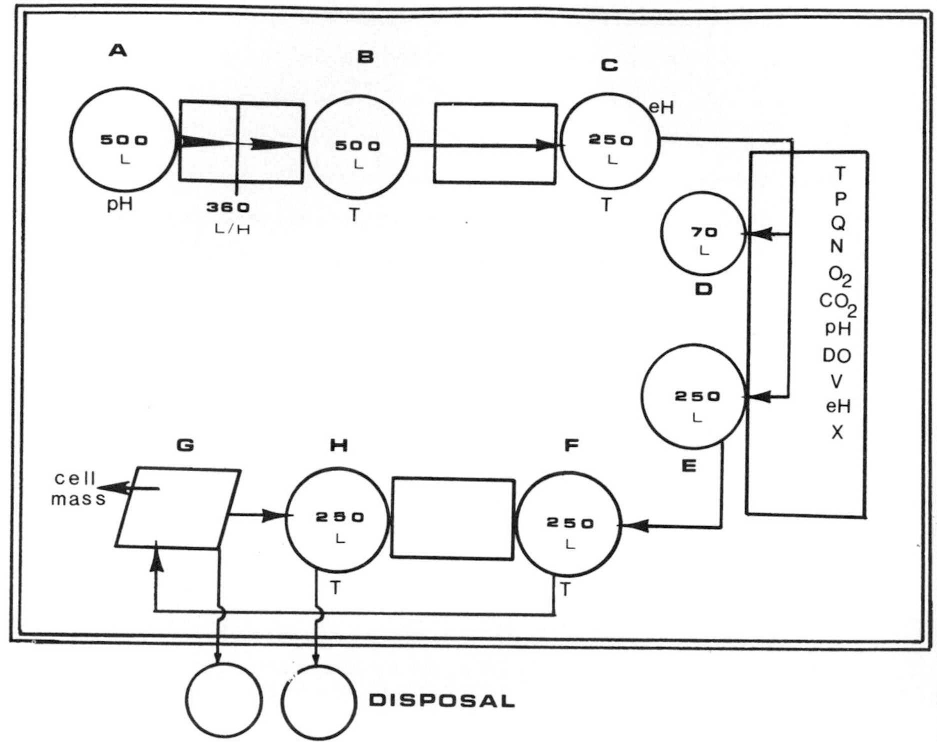

FIGURE 12 *Flow diagram of mammalian cell suspension culture system.*

tank (F) with a subsequent replenishment of the culture with fresh medium from the poising vessel (C); removal of cells from the culture liquid by centrifugation (G) followed by a storage of the spent medium in a supernatant vessel (H). The system also contains auxiliary, support equipment (air compressor, air incinerator, spent medium and effluent disposal system).

B. Computer Control of Plant Operation

This type plant is considered an ideal system for computer control. Three major computer functions can be implemented, viz., (1) sequential control of flow of operations, (2) interactive control of the culture, and (3) process management (Figure 13).

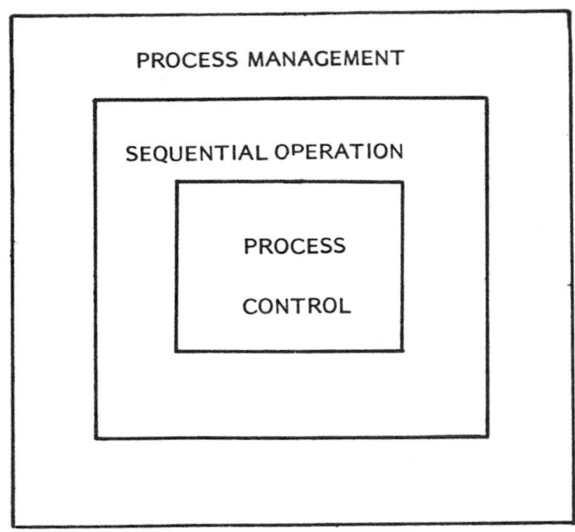

FIGURE 13 *Computer control hierarchy in large scale mammalian cell mass production system.*

The core of the computer operation is identification of the process status (Figure 14). This is performed using the suitable data analysis technique which incorporates frequency spectrum analysis of the acquired signal (Nyiri *et al.*, 1975), noise filtering or rejection and data analysis on the basis of multivariation of the individual signals (c.f. Figure 6).

After the identification of the process status which includes the conditions and position of valves, motors, etc. and the condition of the culture a predictor technique is used. The predictor uses the current values of the process output and determines the necessary step in the operation sequence. A typical example is the determination of the optimum transfer time of seed culture into the main vessel. (Figure 15(a)) shows the changes in DO concentration during a dynamic assessment of QO_2X and K_La values in a murine lymphoblastoid (L251A) cell culture (cell count at the time of temporary stop of airflow was $8.5x10^5$ cells/ml.). The slope (dC/dt) gives information not only on the QO_2X but from the correlation between the specific O_2 uptake (QO_2) and specific growth rate (Q_x) the proper seed transfer time can be determined.

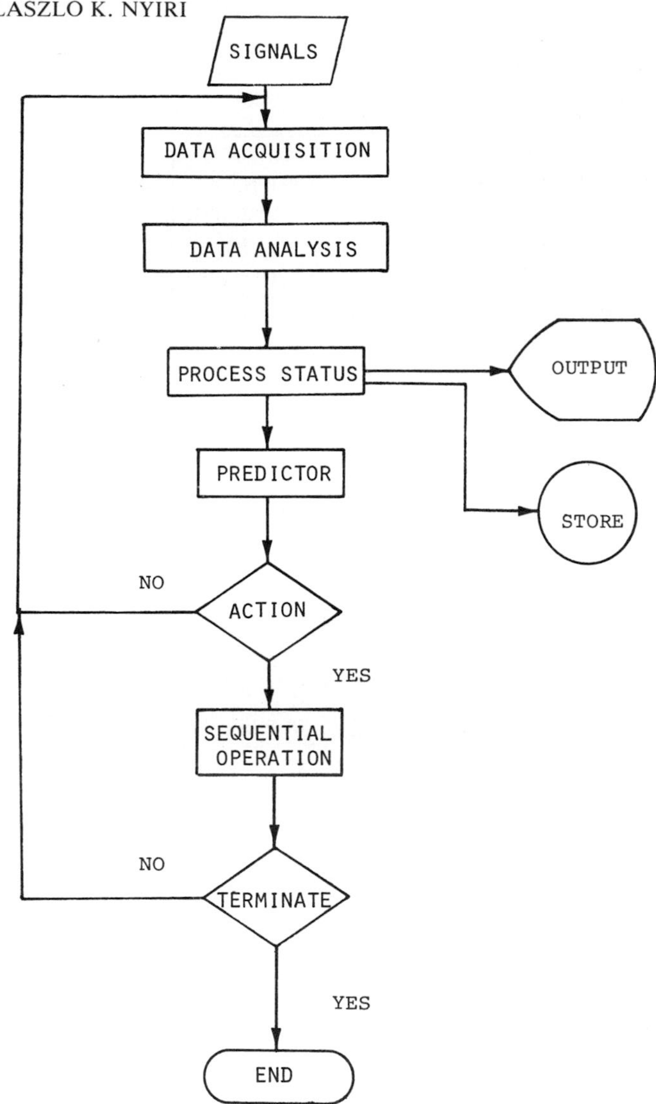

FIGURE 14 *Basic flowchart of overall computer control of large scale mammalian cell mass production system.*

On this basis a sequential operation is implemented to sterilize and fill the main culture vessel in time so it will be ready to accept the seed at the proper time.

Difficulties arising from stopping and restarting airflow can be avoided having an exit O_2 gas analyzer by which QO_2X can be simply computed (Figure 15(b)). In this picture (observed with

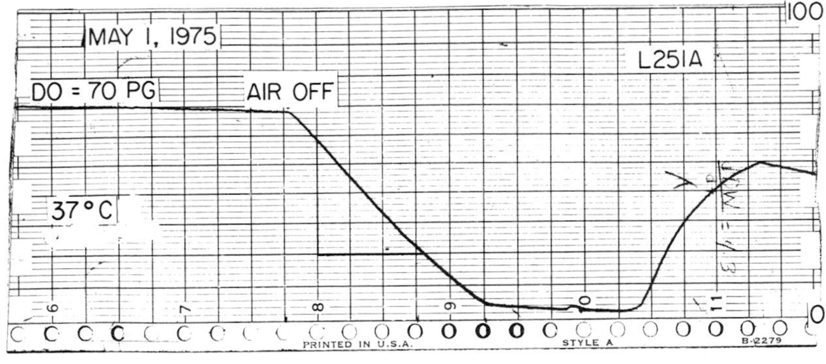

a

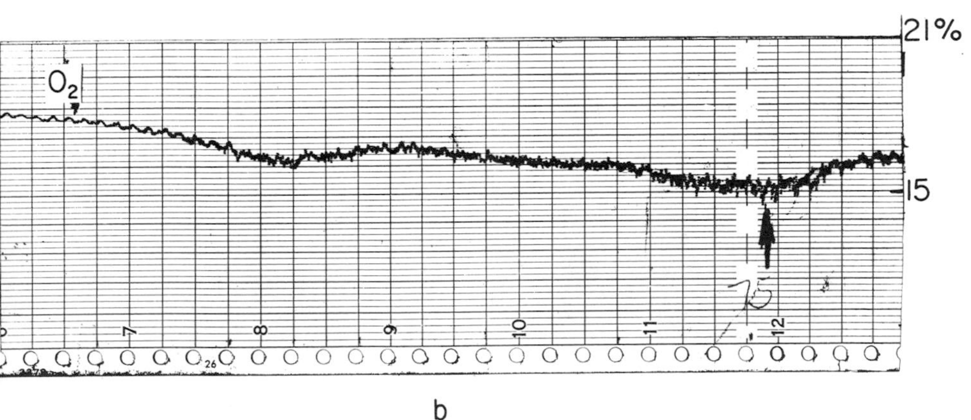

b

FIGURE 15 *Predictors in computer based overall control of biological process. Definition of physiological status on the basis of (a) rate of decrease in DO concentration, (b) changes in oscillation of exit* O_2 *gas analysis signal.*

C. utilis culture) the maximum respiratory activity concurred with maximum oscillation in the signal. In the case of a high scanning rate, the computer can define the (in)stability of the signal thus avoiding the time consuming computations in search of maximum (or minimum) of the signal.

Such an operation can be used in defining the time of withdrawal of part of the culture during semi-continuous operation or the time of fresh nutrient feed in case of fed-batch cultures. The flow rates in between the vessels are, in this case, defined according to the culture conditions which therefore governs the entire operation.

The control of individual process variables (DO, S feed rate, pH) is implemented by other *interactive control technique* some examples of which were already mentioned.

An equally important part of the computer operation is the management of the process. This includes tasks of logging the acquired and analyzed data as well as the control operations, alarm signal production and inventory of the supplies.

VI. CONCLUDING REMARKS

The significance of pondering the installation of a computer controlled mammalian cell plant can be summerized in the following:

(1) Development of such a system requires extensive analysis of correlations between the available sensor signals and the events of cell metabolism. This work enhances our knowledge about the cell-environment and cell-cell interactions.

(2) The work recognizes the need for development of new sensors particularly in the area of cell-cycle analysis and direct determination of various metabolic functions related to alteration in cell membrane and cell differentiation.

(3) If implemented, interactive control will result in environmental conditions which will more approximate those conditions from which the cells are originated.

(4) In the case of hazardous operations (cancerogenic virus research, plasmids, etc.) computer control of the plant is an inevitable part of a Class III containment system.

Developments and achievements in computer applications to microbial fermentation processes demonstrates not only their economic advantages but even in its short period of existence enormously enhanced our understanding of the living processes. On the basis of these findings, it is anticipated that works now facing intriguing puzzles in mammalian cell culture will be faster and more satisfactorily performed.

NOTATION

CO_2	Carbon dioxide analyzer; output signal of CO_2 analyzer (%).
D_i	Diameter of impeller (cm).
DO	Dissolved oxygen analyzer and controller; output signal from DO analyzer (%).
eH	Redox potential analyzer; output signal of ORP analyzer (mV).
K_La	Volumetric oxygen gas transfer coefficient (1/hr).
N	Agitation speed (RPM).
O_2	Oxygen analyzer; output signal from O_2 analyzer (%).
ORP	Oxidation-Reduction Potential.
OXUP	Volumetric oxygen uptake rate (mMO_2/min/L).
Q, Qn	Air-gas flow rate (SLPM).
QCO_2	Specific carbon dioxide release rate ($mMCO_2$/G cell/time).
QO_2	Specific oxygen uptake rate (mMO_2/G cell/time).
P	Vessel headpressure (atm.).
T	Temperature of medium or culture liquid (^oC).
V	Liquid volume (Liter).
W	Power requirement for fluid mixing (watt.hr).
X	Cell number (cell-mass) (G/L; cell count/ml).
ρ	Liquid density (g/cm^3)

REFERENCES

Aiba, S., Humphrey, A.E. and Millis, N.F. (1973). Biochemical Engineering. University of Tokyo Press.

Acton, R.T. (1975). Proceedings of Cell Culture Congress.

Cleland, W.W. (1970). In "The Enzymes" (P.P. Boyer, Ed.), Third Edition, Vol. II. pp. 1-65.

Daniels, W.F., Garcia, L.H. and Rosensteel, J.F. (1970). Biotechnol. Bioeng. *12,* 419-428.

Fiechter, A. and VonMeyenburg, K. (1968). Biotechnol. Bioeng. *10,* 535-549.

Gorts, C.P.M. (1967). Antonie VanLeeuwenhock, J. Microbiol. Serol., *33,* 451-453.

Harmes, C.S. III. (1972). In "Developments in Industrial Microbiology." Vol. *13,* pp. 146-150. A Publication of the Society for Industrial Microbiology. American Inst. of Biol. Sciences. Washington, D.C.

VanHemert, P.A. (1972). Paper presented at the first Int'l. Symp. on Adv. in Microbiol. Engineering. Marianske Lazne. Czechoslovakia.

Holley, R.W. (1974). In "Control of Proliferation on Animal Cells," (B. Clarkson and R. Baserga, Eds.). Cold Spring Harbor Conferences on Cell Proliferation. Vol. 1., pp. 13-18. Cold Spring Harbor Laboratory Publ.

Humphrey, A.E. (1971). Proc. of LABEX Symposion on Computer Control of Fermentation Processes. London.

Klein, F., Mahlandt, B.G. and Lincoln, R.E. (1971). Appl. Microbiol. *22,* 145-146.

Knazek, R.A., Gulino, P.M., Kohler, P.O. and Kedrk, R. (1972). Science *178,* 64.

Kruse, P.F., Jr. and Patterson, M.K., Jr. (eds.) (1973). Tissue Culture Methods and Applications. Academic Press, New York.

Litwin, J. (1971). Process Biochemistry. *6,* 15-18.

Marchesi, V.T. (1975). "Miles International Symposium on Cell

Membrane Receptors," John Hopkins Medical Institute.

Mastro, A.M., Beer, C.T. and Mueller, G.C. (1974). Biochem. Biophys. Acta *352*, 38-51.

Mueller, G.C. (1971). In "The Cell Cycle and Cancer," (R. Basergal, Ed.) pp. 269-307. Marcell Dekker, Inc., New York.

Nyiri, L. and Jefferis, R.P. III. (1973). In "Single Cell Protein," (S.R. Tannenbaum and D.I.C. Wang, Eds.). Vol. *II*, pp. 105-126. MIT Press. Cambridge, Massachusetts.

Nyiri, L.K. (1974). In "Enzyme Engineering," (E.K. Pye and L.B. Wingard, Jr., Eds.). Vol. 2, pp. 31-40. Plenum Press.

Nyiri, L.K. and Krishnaswami, C.S. (1974). 75th National Meeting of ASM, Chicago, Illinois.

Nyiri, L.K. and Toth, G.M. (1975). Paper presented at the 1975 Annual Meeting of Society of Industrial Microbiologists. Kingston, Rhode Island.

Nyiri, L.K., Toth, G.M. and Charles, M. (1975). Biotechnol. Bioeng., *17* 1663-1678.

Pardee, A.B., deAsua, L.J. and Rozengurt, E. (1974). In "Control of Proliferation in Animal Cells," (B. Clarkson and R. Baserga, Eds.). Cold Spring Harbor Conferences on Cell Proliferation. Vol. 1., pp. 547-561.

Ross, L.W. (1974). Biotechnol. Bioeng. *16*, 555-556.

Taylor, G.W., Kondig, J.P., Nagle, S.C. and Higuchi, K. (1971). Appl. Microbiol. *21*, 928-933.

Telling, R.C. and Radlett, P.J. (1970). Vol. 13, pp. 91-119. In "Advances in Applied Microbiology (Perlman, D., Ed.).

Toth, G.M. (1977). In "Advances in Biochemical Engineering," (A. Fiechter, Ed.). (in press) Springer Verlag. Berlin.

Wiles, C.C. and Smith, V.C. (1969). Am. Inst. Chem. Engr. Symp., (Bioeng. Biotechnol.), pp. 85-93.

Young, T.B. (1970). ACS National Meeting, Div. Microbiol. Chem. Technol. No. 39, Chicago, Ill.

OPTIMIZING PARAMETERS FOR GROWTH OF ANCHORAGE-DEPENDENT MAMMALIAN CELLS IN MICROCARRIER CULTURE

D. W. LEVINE
D. I. C. WANG
W. G. THILLY

The Cell Culture Center
Department of Nutrition and Food Science
Massachusetts Institute of Technology
Cambridge, Massachusetts

ABSTRACT

We examined parameters affecting growth of diverse cell types in microcarrier culture. We found nutrient absorption by the DEAE-Sephadex A50 carriers to be a major problem in the growth of the strain of normal human fibroblasts, HEL 299, and an obstacle to increased vessel productivity. We discovered that this effect can be offset by the use of carboxymethylcellulose in the growth medium. Using this technique in conjunction with medium perfusion, we have obtained rapid exponential phase growth to a level of 2×10^6 cells/ml. Other cell types which have been successfully cultivated are secondary chick fibroblasts, mouse bone marrow epithelial cells and a transformed mouse fibroblast. The transformed mouse fibroblast has been grown to maximum cell density of 6×10^6 cells/ml. Current research efforts include the synthesis of improved microcarriers, application of non-trypsinized cell transfer among beads as a means of continuous culture and development of a microcarrier fermentor unit for mass production of cells synchronized in mitosis.

I. INTRODUCTION

Progress in large scale mammalian cell propagation has been limited to cell lines capable of growth in suspension culture. Such cells lend themselves to the techniques of microbial fermentation. There has been no comparable progress for the propagation of anchorage-dependent animal cells.

Current operational techniques for large scale propagation of anchorage-dependent cells are based on linear expansion. Cell culture plants utilize a large number of low yield, batch reactors, in the forms of dishes, prescription bottles, roller tubes and roller bottles. Each bottle is a discrete unit, an isolated batch reactor; individual environmental controls must, of economic necessity, be of the most primitive type. Variation in nutrients is corrected by a medium change, an operation requiring two steps, medium removal and medium addition. Since it is not uncommon for a moderately sized facility to operate hundreds of these batch reactors at a time, even a single change of medium requires hundreds of operations, all of which must be performed accurately, and under exacting sterile conditions. Any multiple step operation, such as cell transfer or harvest, compounds the problem accordingly. Costs of equipment, space and manpower are great for this type of facility.

Recently, experimentors in the Netherlands, (van Wezel, 1967; van Hemert, *et al.*, 1969; van Wezel, 1973) reported the growth of anchorage-dependent cells on microcarriers in a stirred tank.

They used DEAE-Sephadex beads as carriers. Since the ultimate limit to productivity of any bottle is the surface area available for cell attachment, the most obvious advantage to a microcarrier system is the great increase in the attainable ratio of growth surface to vessel volume (S/V). A one liter fermentor with an S/V equal to 48 cm^2/cm^3, has a surface area equivalent to that of 50, 2 liter roller bottles. This increase in productivity, in turn, allows the construction of a single-unit homogeneous batch (or semi-batch) propagator of high volumetric productivity. A unit of this type has two distinct advantages over a multiple unit batch system. First, a single, stirred-tank vessel with simple feedback control of pH and pO_2 presents a homogeneous environment for a large number of cells, thus eliminating the necessity for expensive and space consuming, controlled environment incubators. Secondly, the use of a single, high-productivity reactor drastically reduces the total number of operations required per unit of cells produced. Microcarriers, then, offer economies of capital, space and manpower in the production of anchorage-dependent cells, relative to current porduction methods.

Another advantage gained by use of microcarriers is the environmental continuity afforded cells growing in a controlled environment. Although little data are available to document the effects of constant versus variable environment on cell growth, Ceccarini and Eagle (1971) have shown that the final attainable cell concentration is very much dependent of the pH history of the culture. Certainly, best cell growth will be obtained if all cells being cultured are exposed to constant, optimal conditions and not forced to contend with exposure to suboptimal conditions.

Microcarriers are not, however, the only alternate method to linear scale up of small batch cultures. Six promising techniques have been reported as being in the development stage: plastic bags, stacked plates, spiral films, glass bead propagators, artificial capillaries and microcarriers. These techniques are presented in schematic form in Fig. 1.

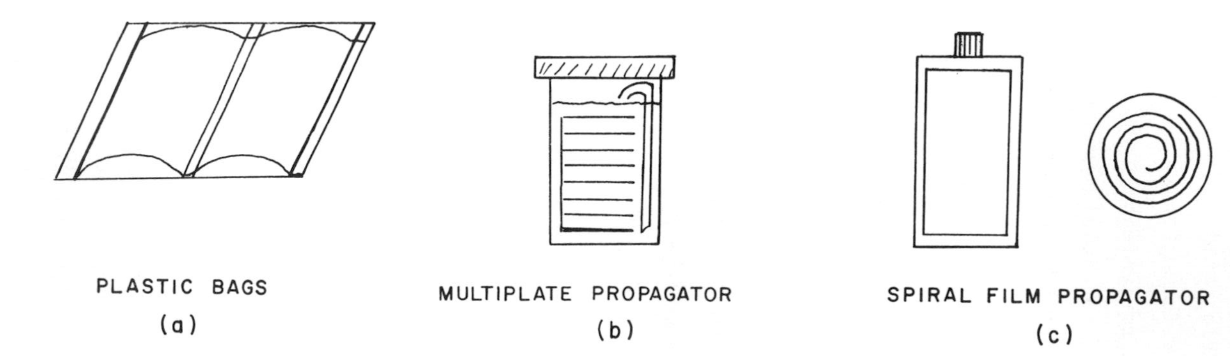
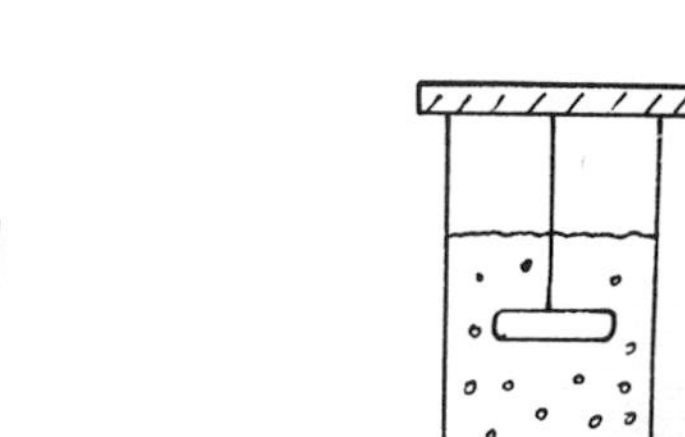
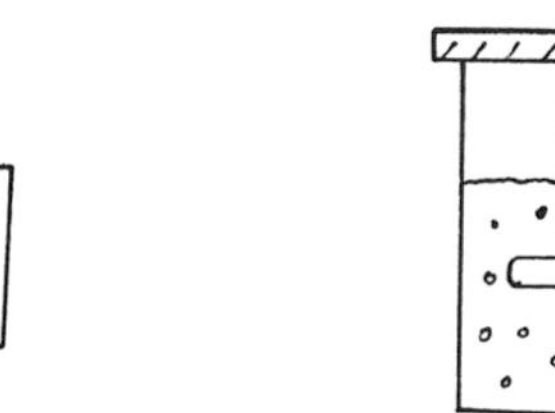

FIGURE 1 *New techniques of mammalian cell propagation.*

TABLE I A Comparison of S/V Value Estimates for Cell Propagators

System	S/V (cm^2/cm^3)	Reference
Roller bottles	.2-.7	
Plastic bags	5.0	Munder, *et al.*, 1971
Multiplate propagator	1.7	Weiss and Schleicher, 1968
Spiral film	4.0	House, *et al.*, 1972
Glass Bead propagator	10.0	Wöhler, *et al.*, 1972
Artificial capillary	31	Knazek, *et al.*, 1972
Microcarriers	36	van Wezel, 1973
	Potential S/V increase	Potential Physical scale-up
Roller bottles	-	-
Plastic bags	± (≤2 fold)	-
Multiplate propagator	-	+
Spiral film	± (≤2 fold)	+
Glass bead propagator	-	-
Artificial capillary	-	±
Microcarriers	+ (≤10 fold)	+

A comparison of these six methods must account for two major criteria: potential volumetric cell productivity (cell/reactor volume) and system potential for large scale operation, that is, physical scale up. Table I presents a comparison of these techniques for currently attained S/V, potential for S/V increases and potential for physical scale-up. These data show the microcarrier system to be approached in current S/V only by the glass bead propagator and artificial capillaries. Neither of these systems has practical potential for physical scale-up, and neither of these systems, nor any of the other techniques have the potential for virtually order of magnitude increases in S/V which the microcarrier system affords.

The plastic bag culture vessels developed by Munder, *et al.* (1971), are non-rigid polymer sheet bags, containing growth medium and cells (Fig. 1a). Their main advantage over roller bottles is their increased S/V (∿5) effected by the elimination of dead volume (airspace) from the interior of the culture vessel. This is possible because the films employed are permeable to O_2 and CO_2. While these bags are an improvement over conventional techniques, this method is still a multiple unit batch system of better, but limited, surface to volume ratio, without increase in environmental control potential.

The multiplate propagator (Fig. 1b), (Weiss and Schleicher, 1968) is ideally suited to single unit batch operations, being essentially a homogeneous reactor with discs as surfaces for cell growth. The use of discs however, tends to limit the maximum obtainable S/V. As more plates are included per volume, static medium is trapped between the plates. The consequences of this entrapment are nutrient depletion in localized areas, and a general loss of culture homogeneity, leading to a loss of cell productivity.

The spiral film propagator (Fig. 1c) consists of a 2 liter roller bottle which contains a spiralled film of plastic suitable for cell attachment. The system described by House, *et al.* (1972), has an S/V of 4. This system is a single-unit batch and so has potential for environmental control similar to microcarrier culture. However, the scale-up potential of this system is limited. As with the multiplate propagator, as surface area is increased, medium flow between two stationary surfaces in close proximity decreases. Maintenance of uniform environment throughout the vessel would, therefore, be difficult.

The glass bead propagator (Fig. 1d) of Wöhler, *et al.* (1972), is actually a limiting case of the microcarrier system. In this case, the beads are stationary and medium is constantly recirculated. By almost all of the previous criteria, this system seems to be an outstandingly good one. A major drawback, however, is that the glass bead bed must remain stationary. Any movement results in the death of cultured cells by grinding (Wöhler, *et al.*, 1972). This means that the propagator must be very carefully packed, and fluid flow through the vessel must also be carefully planned to prevent any shifting of the packing. Secondly, packing must eliminate dead spaces which would create non-homogeneous growth conditions. Thirdly, as volume increases in scale-up, so must volumetric flow rate. This means that superficial velocity of the medium will also increase. This increase will further compound problems of channelling, packing shifts and shear exposure of the cells. Thus, as scale increases, environmental continuity, and, hence, productivity must suffer.

The artificial capillary of Knazek, *et al.*, (1972), is also a serious alternate propagation system. In this case, cells are grown on the surfaces of hollow fibers, set in a dialysis shell. Medium is circulated through an external pumping loop, where pH control and oxygenation take place. This system's major advantages are a high S/V (31) and ease of synthesized product recovery by dialysis through the growth surfaces. Knazek, *et al.*, have demonstrated the ability to recover dialyzable components (human chorionic gonadotrotrophin) throughout the course of cellular growth, with no threat of contamination. As a mode of cell production, however, there are some disadvantages. As with other non-homogeneous type reactors, constant environment is maintained only by adequate medium perfusion, a factor which becomes increasingly difficult to maintain with increasing scale. Furthermore, direct examination of culture growth is difficult.

A comparison of the microcarrier system (Fig. 1f) outlined above to alternative systems of propagation shows it to be already the equal in S/V ratio of any other system, while maintaining a greater potential for further increase. Van Hemert, *et al.* (1969), have achieved an X/V of 36. However, there is no *a priori* reason why this value could not be increased as much as tenfold. The system provides all the advantages of a homogeneous propagator, particularly the potential for physical scale-up. Furthermore, the system has shown itself to be well suited to procedures involved in separating cells from medium by simple settling of the beads. This property is extremely useful in virus production, for example, or for growth systems requiring medium renewal.

Van Wezel and his colleagues have introduced a major new concept with this system, the concept of combining multiple surfaces with movable surfaces. We feel that this innovation carries the potential for innovative cellular manipulations, advantages in scale-up and advantages in environmental controls. Curiously, few laboratories have exploited van Wezel's technique, probably because of the few remaining difficultues in its application to cell production. These problems were described by van Wezel as a "toxic effect" of the DEAE-Sephadex, which prevents cell attachment and growth. We confirmed van Wezel's observations, extended them to other cell strains and lines of general interest and found a simple means of overcoming the "toxicity" of the anion exchange resins.

II. MATERIALS AND METHODS

A. Cell Types

Human embryo lung cells (ATCC HEL 299) were used at passage 7 and 8. These cells were obtained frozen from ATCC at passage 4 and maintained in dishes for two passages (at 1:10) before being refrozen and stored under liquid nitrogen. A continuous strain of mouse bone marrow epithelial cell and a murine leukemia virus producing strain of mouse fibroblast were obtained from Dr. R. Weinberg of the M. I. T. Cancer Research Center. Secondary chick fibroblasts were obtained from the Cell Culture Center at M. I. T.

B. Media

For all experiments, DMEM medium was used. For the growth of HEL 299 cells, and the epithelial mouse cells, the medium was supplemented with 10% fetal calf serum. For the growth of chick fibroblasts, the medium was supplemented with 1% calf serum, 1% chick serum and 2% tryptose phosphate broth. For the growth of the mouse fibroblasts, the medium was supplemented with 10% calf serum. The pH was controlled by a 5% CO_2:Air overlay and a bicarbonate buffering system. Carboxymethylcellulose used in spinners was Hercules CMC-Gum 7H4F.

C. Culture Vessels

Cells were grown in spinner cultures of 100 ml working volume with approximately 150 ml head space. The glass bottles were siliconized to prevent adhesion of both cells and microcarriers to the sides. Cultures were agitated by suspended magnetic spinners. The ratio of bottle diameter to spinner length is 1.4. Agitation speed was fixed at 80-90 rpm.

D. Microcarriers

For all spinner cultures DEAE-Sephadex A50 was used. The dry Sephadex was sieved to obtain a uniform bead size $>90\mu<105\mu$. These were then suspended in phosphate buffered saline and autoclaved, excess PBS removed, and the sequence repeated five times. This washing procedure was found to serve as well as the acid/base washes employed by van Wezel. The beads were stored in PBS at room temperature. Beads were prepared in stock preparations of 10 mg/ml. Before use, the beads were settled and excess PBS was removed. Beads were then suspended in the appropriate amount of medium. At 1 mg/ml, the spinners have an estimated surface area of 4.5 cm^2/ml of culture volume.

E. Cell Count

Growth of cells in microcarrier culture was followed by taking a well mixed sample (3 ml), centrifuging, removing medium, and adding 3 ml of a 0.1% crystal violet solution in aqueous 0.1 *M* citrate. After ∿1 hour incubation at 37°C, the samples were sheared by repeated pipetting. The cell nuclei, which were stained and liberated by this procedure, could then be counted in a hemocytometer.

F. Microcarrier Culture

Excess PBS is removed from stock microcarriers; the appropriate amount of medium is added, and the suspension transferred to the growth vessel. Plated cells are harvested with trypsin according to standard procedures and added directly to the agitated growth vessel at a concentration of $\sim 10^5$/ml. Aliquots of growth medium are renewed as required by culture growth. This operation is simplified by the settling rate of the carriers, which is on the order of a few minutes. All cell growth is carried out at 37.5°C with a 5% CO_2:Air overlay.

III. RESULTS AND DISCUSSION

A. Optimization of Microcarrier Culture

The main objective of this study is to develop a convenient and simple procedure of microcarrier culture for the growth of anchorage-dependent mammalian cells. Our criteria of optimization is based on comparison of microcarrier culture with cell growth in plastic dishes. Growth kinetics for strain HEL 299, a human embryonic fibroblast, in dishes are shown in Fig. 2. Three parts of the curve are of concern: that portion of the inoculum which is available for culture growth; the overall growth rate of the culture; and the final cell density obtained. In most growth

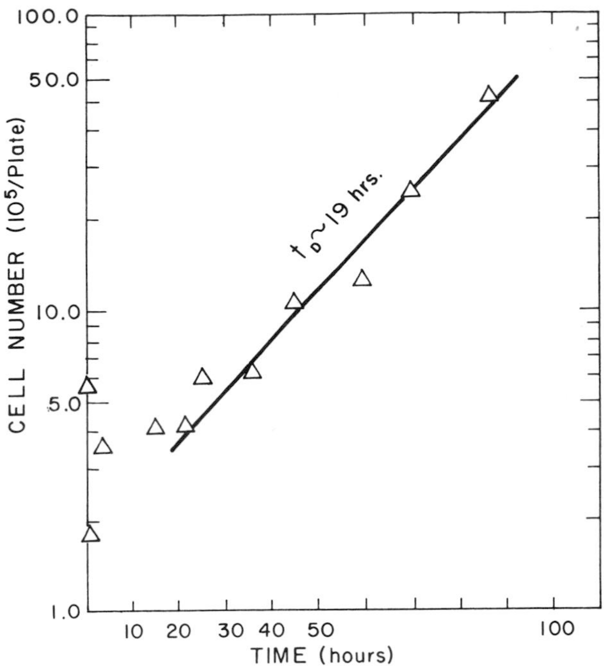

FIGURE 2 *Growth of HEL 299 in dishes.*

systems, the saturation cell density is limited by surface area available for growth. Increase in this saturation density is a direct reflection of increased carrier concentration, and a principal advantage of microcarrier culture. In the case of HEL 299 in plastic dishes, approximately 60% of the inoculum attaches to the growth surface, and is, therefore, potentially part of the growing population. The culture doubling time is 18-20 hours. Comparing this growth curve to Fig. 3, which is a curve for cell growth in DEAE-Sephadex A50 microcarrier culture, we see that only 30% of the inoculum remains in the culture after 24 hours. We also see that once growth begins, the culture doubling time is greater than 30 hours. An increase in microcarrier concentration decreases the potential growing population to as little as 10% of the inoculum in 24 hours, and increases culture doubling time.

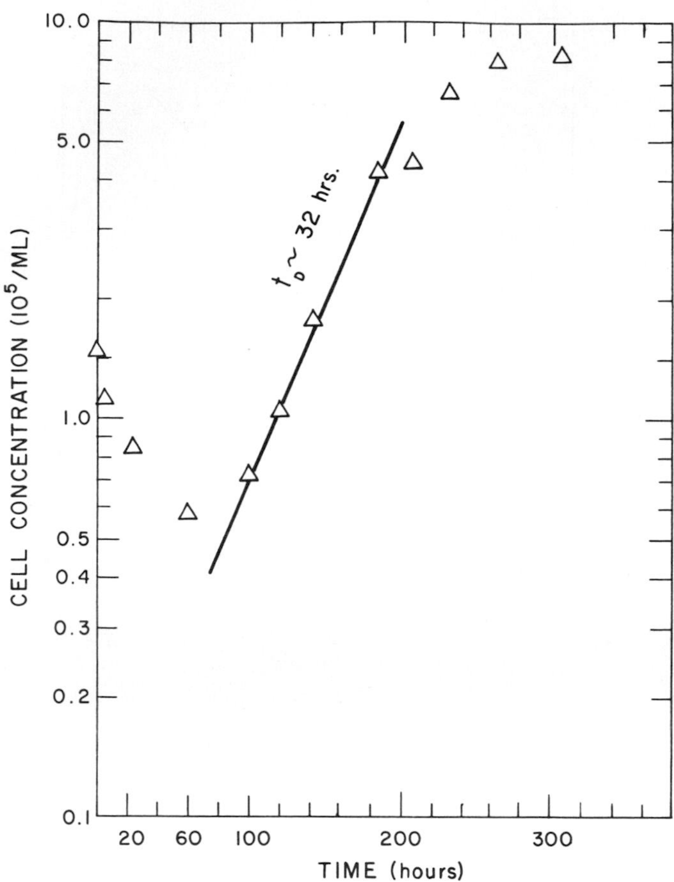

FIGURE 3 *Microcarrier culture of HEL 299, at 1 mg/ml DEAE-Sephadex A50.*

These results are in agreement with van Wezel's observations, and represent a major obstacle to development of microcarrier technique.

We postulated either the presence of a soluble toxic factor released by the beads or absorption of medium nutrients by the carriers to explain these adverse effects (Levine, 1974). To test these hypotheses, we first established a "toxic" limit for carriers. In this case, we suspended varying amounts of DEAE-

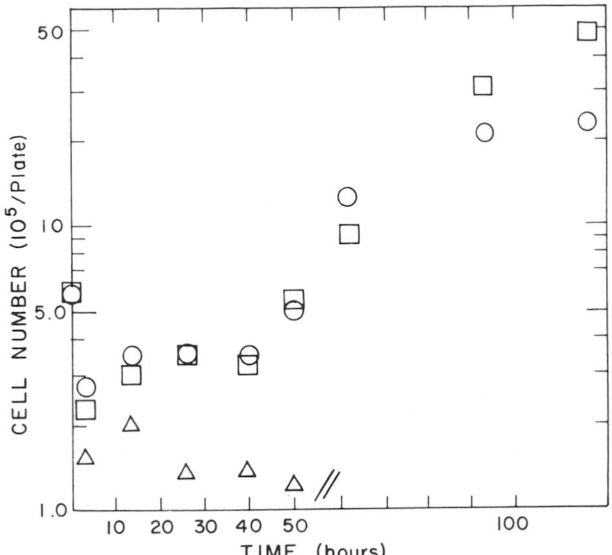

FIGURE 4 *Plate growth of HEL 299 in medium contacted with varying amounts of DEAE-Sephadex A50.* ○ *0.0 mg/ml,* □ *0.1 mg/ml,* △ *0.25 mg/ml.*

Sephadex A50 in growth medium at 37°C for 24 hours. After removing the beads, we used this "conditioned" medium for plating cells. These results are presented in Fig. 4. At a DEAE-Sephadex concentration greater than 0.1 mg/ml, no cell growth was observed. We then suspended 5 mg/ml DEAE-Sephadex in medium and equilibrated it as above. After removing the carriers, we mixed, at varying ratios, the "conditioned" medium with normal growth medium. If a soluble toxic compound were present, then we would expect to see growth inhibition even at relatively dilute concentrations of the "conditioned" medium, corresponding roughly to the ratio of the toxic limit of 0.1 mg/ml to 5 mg/ml. If nutrient absorption were the effect, we would expect the varying ratio of "conditioned" medium in the plating medium to have little effect on growth. Figure 5 gives the results of this experiment. In fact- there is no evidence of a soluble toxic factor. This

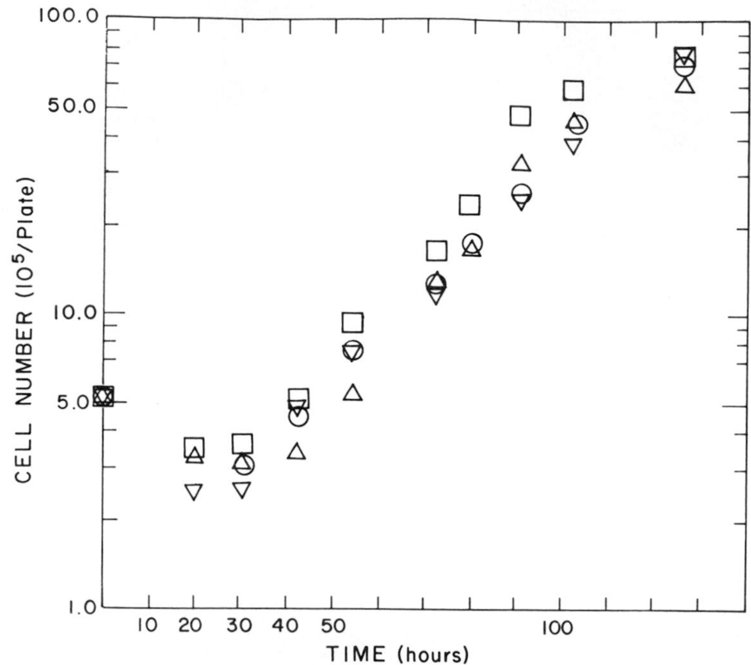

FIGURE 5 *Plate growth of HEL 299 in varying levels of medium contacted with 5 mg/ml DEAE-Sephadex. Contacted medium was diluted with fresh medium in the following ratios (conditioned medium:final volume):* □ *1.2,* △ *1:5,* ▽ *1:10,* ○ *0:1.*

conclusion is supported by the observation that there are significant amounts of protein associated with DEAE-Sephadex beads which have been exposed to growth medium. We feel, therefore, that there is significant uptake of nutrients by the beads. This conclusion was also reached by Horng and McLimans (1975) in their work with microcarriers.

Our solution to the postulated problem of nutrient leaching is to add to the medium a negatively charged non-nutritive component to compete with the positively charged exchange sites on the beads. Use of carboxymethylcellulose (CMC), a polyanion, has given excellent results. That, in fact, the CMC is competing for adsorptive sites on the DEAE-Sephadex is shown in Fig. 6. Here, as

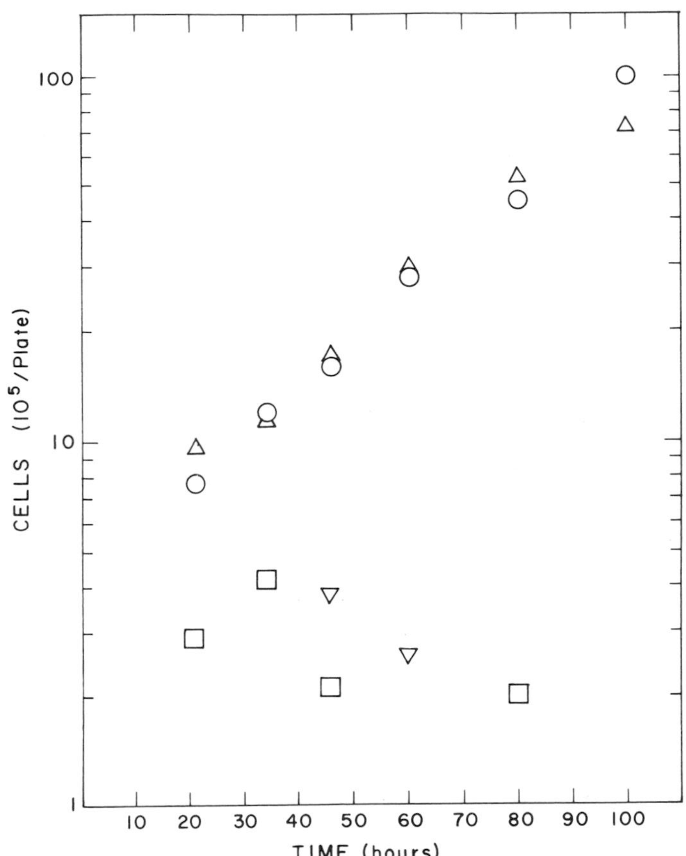

FIGURE 6 *Plate growth in medium equilibrated with DEAE-Sephadex A50, in the presence and absence of CMC.* o *Control,* CMC^-, Δ CMC^+, □ CMC^-, ∇ *Control,* CMC^+.

before, medium is equilibrated with low levels of the microcarriers; the carriers are removed and cells are plated in dishes. When CMC is not present during contact time, no growth occurs. This indicates that the CMC either completes with medium components for adsorptive sites on the beads, or otherwise renders the sites inaccessible to the critical nutrients.

Using CMC as an additive to growth medium, we have grown cells on the carriers with excellent results. At low microcarri-

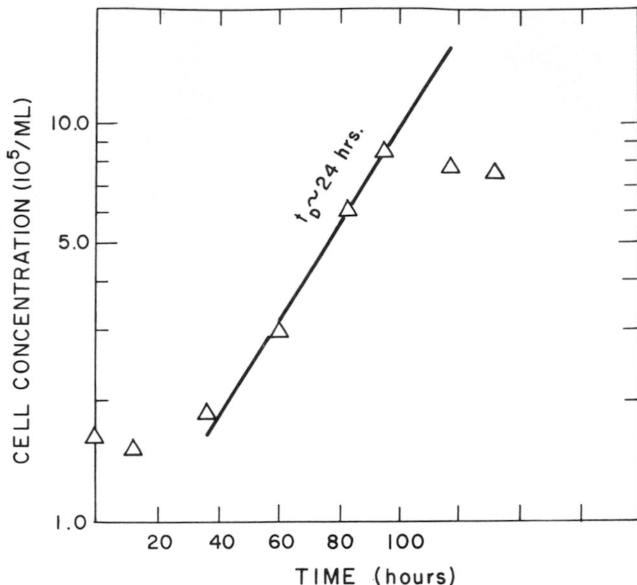

FIGURE 7 *Microcarrier culture of HEL 299 at 1 mg/ml DEAE-Sephadex A50, in presence of CMC.*

er concentrations (<2 mg/ml), 0.01 gm/l CMC is added to growth medium, and the culture is inoculated as above. At higher microcarrier concentrations (>2 mg/ml), microcarriers are first suspended in growth medium and CMC (at a ratio of .01 mg of CMC per mg of carrier), and then after 12 hours, resuspended in growth medium with .01 gm/l CMC. Figure 7 shows the growth of normal human fibroblasts (HEL 299) in microcarrier culture at a concentration of 1 mg/ml of DEAE-Sephadex A50. In this case, very little of the initial inoculum is lost. Poor culture growth was seen with these conditions, however, unless additional media was provided by periodic replacement of a portion of growth medium as was done in Fig. 7. In this case, 25% of the culture volume was renewed daily, and culture growth rate is comparable to growth in dishes. Similarly, in Fig. 8, with a carrier concentration of 4 mg/ml, we see little decrease in cell inoculum and strong exponential phase growth up to a final density of 2×10^6 cells/ml.

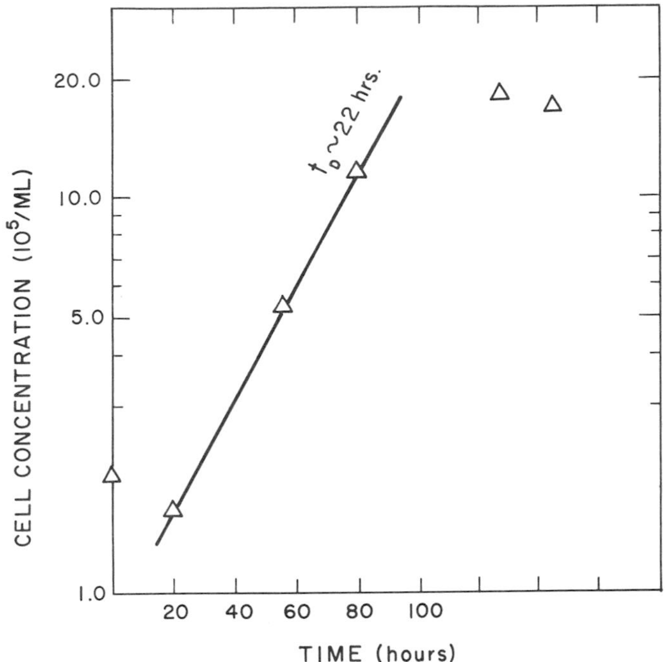

FIGURE 8 *Microcarrier culture of HEL 299 at 4 mg/ml DEAE-Sephadex A50 in presence of CMC.*

Here, daily medium renewal was unnecessary. Good growth was obtained with only two 25% media renewals, one at 24 hours, and one at 72 hours. This final density compares favorably with densities obtained in suspension cultures of established human cell lines such as HeLa. Typically, less than 2% of the total cell population is in free suspension; cell yield on a medium volume basis is comparable to that obtained with dishes, and cells are easily recovered from the carriers by standard trypsinization procedure. Of course, the facility of recovery is much greater for the microcarrier culture, since only one vessel is involved in cell recovery.

A second approach to the problem of nullifying nutrient absorption is also being pursued. In this case, we are attempting to synthesize a carrier which is optimal for cell adhesion and

TABLE II Screening of Support Materials

Material	Adsorption	Growth
DEAE-Sephadex A50	+	+
DEAE-Sephadex A25	+	+
Sephadex G-10	–	–
Sephadex C-25	–	–
Sephadex LH-20	–	–
Amberlite IR-45	+	+
Amberlite XAD-2	±	±
BioRad AG-21K	+	+
Biosil BH	–	–
Florisil	–	+
Silica gel (1)	–	±
Silica gel (2)	–	±

growth. A survey of possible carrier materials is shown in Table II. The survey experiments consisted of bringing together cells, medium and test materials in plates. Two criteria for successful spinner operation were applied: adhesion of cells to the carrier surface and cell spreading with accompanying overgrowth. Our results confirmed the desireability of a positively charged surface. Therefore, we are in the process of establishing minimum workable charge densities for cell adhesion and growth and are attempting to place charged groups on both impervious bead supports (such as polyethylene and polystyrene) and on uncharged Sephadex G50. In the case of our studies with modified Sephadex G50, the key concepts of our efforts are to either concentrate all charges at the surface, leaving the center of the bead uncharged, or simply to reduce the total milliequivalents per gram of carrier, and thus reduce the total nutrient uptake. Adding impetus to this effort have been apparent variations in cell growth behavior in DEAE-Sephadex microcarrier culture, which seem to correlate with changes in commercial DEAE-Sephadex lots. We feel that this approach, alone or in conjunction with CMC addition, could yield an order of magnitude increase in workable S/V.

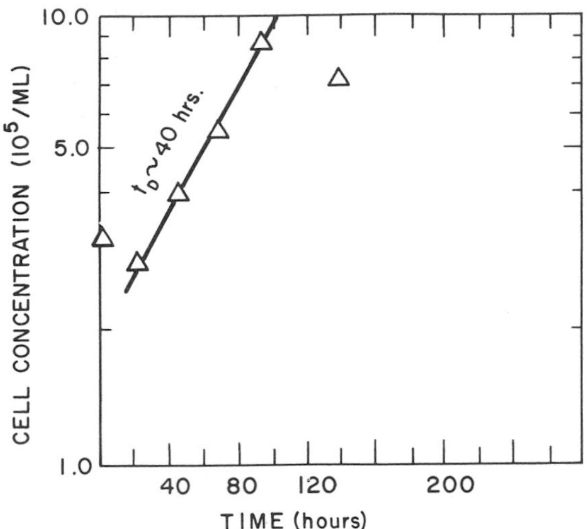

FIGURE 9 *Microcarrier culture of secondary chick fibroblasts at 1 mg/ml DEAE-Sephadex A50.*

B. Application of Microcarrier Culture to Diverse Cell Types

In addition to our studies with normal human fibroblasts, we have examined the applicability of microcarrier culture to growth of other cell types. Figure 9 shows preliminary observations of secondary chick fibroblasts. As can be seen, a relatively high inoculum yields good results. Figure 10 shows the growth of an established line of mouse bone marrow epithelial cells. As shown, after a slight loss of cells, growth is rapid to 1.2×10^6 cells/ml. Therefore, the use of positively charged microcarriers seems to be generally adaptable to both fibroblastic and epithelial-like cells and to cells of diverse species of origin. To our knowledge, no failures in adapting anchorage-dependent cells have been encountered in other laboratories. Certainly we have observed no difficulties once the nutrient leaching effect is overcome. In all cases, at maximum cell densities, less than 2% of the cell population is free in suspension.

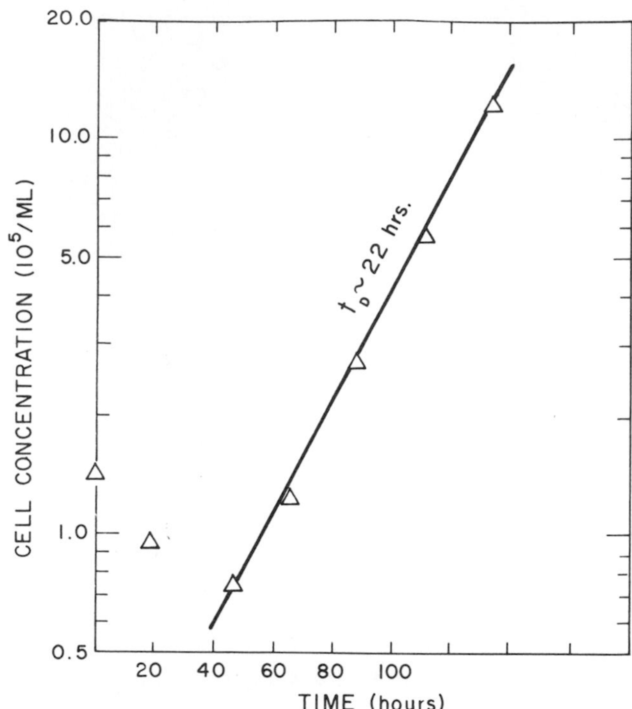

FIGURE 10 *Microcarrier culture of epithelial mouse bone marrow cells at 1.5 mg/ml DEAE-Sephadex A50.*

C. Applications of Microcarrier Culture

A pressing need of virologists and cancer researchers interested in oncogenic viruses, is a constant and sufficiently large supply of virus for experimentation. We have examined in microcarrier culture the growth and virus producing characteristics of a Moloney murine leukemia virus-producing strain of mouse fibroblasts. Figure 11 shows growth of this line at a microcarrier concentration of 4 mg/ml. With medium renewal, a simple matter with the microcarrier system, we were able to obtain the remarkably high cell density of 6×10^6 cells/ml. Growth was exponential up to approximately 2×10^6 cells/ml. From there on to 6×10^6 cells/ml, growth was linear, as shown in Fig. 12, a linear replotting of the previous data. The linear nature of growth in-

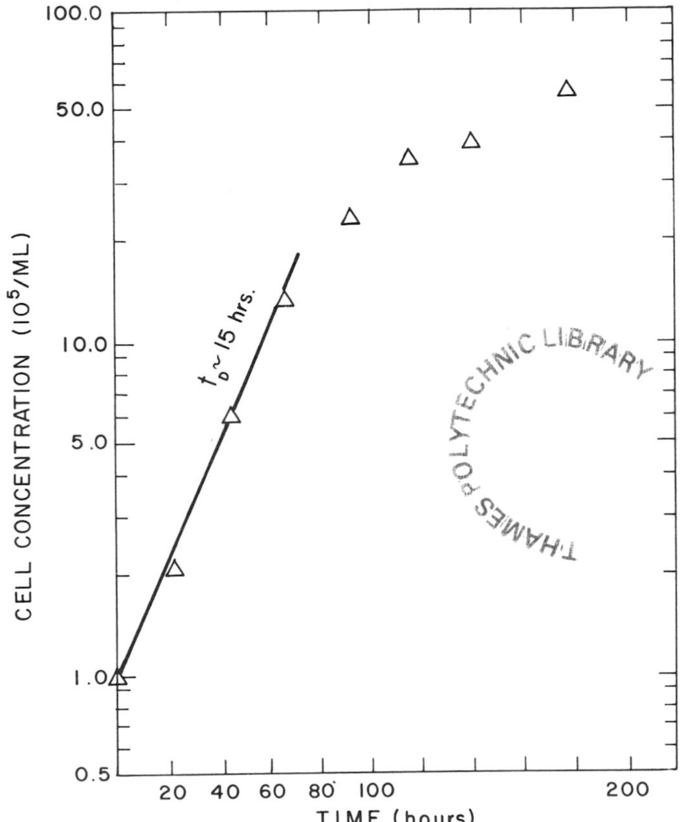

FIGURE 11 *Microcarrier culture of mouse fibroblast at 4 mg/ml of DEAE-Sephadex A50.*

dicates that gas exchange may be involved. In addition to good cell growth, we have also found that murine leukemia virus production for these cells continues at a rate comparable to controls grown in monolayer cultures.

Finally, we have begun to investigate cellular manipulations made possible by the microcarrier concept. We have demonstrated that cell transfer can occur between a confluent bead and an unseeded bead by direct transfer, as shown in Fig. 13. This suggests a means to eliminate the trypsinization process involved in cell passaging. The implications of Fig. 13 are that cells in

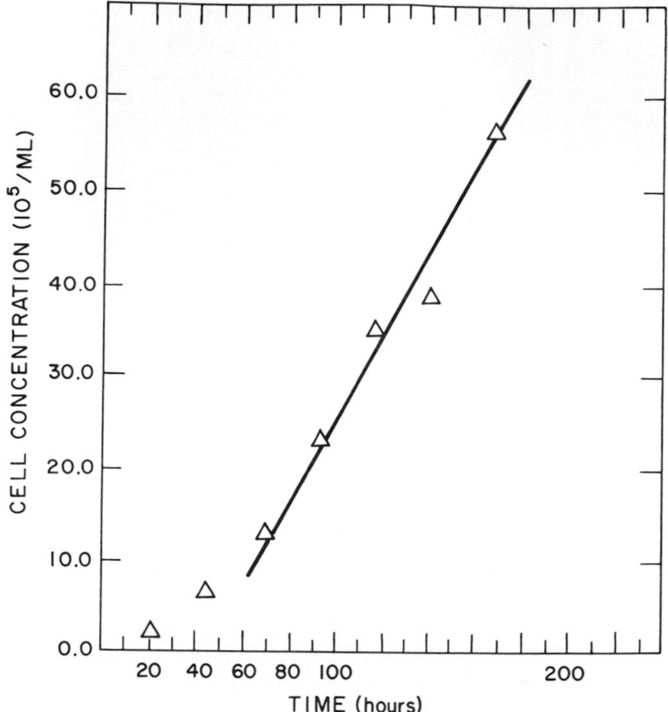

FIGURE 12 *Microcarrier culture of mouse fibroblast at 4 mg/ml of DEAE-Sephadex A50.*

microcarrier culture may be maintained, from initial isolation to senesence, without the necessity of the stress of the physiologically disruptive process of trypsinization. Furthermore, the multi-step trypsinization process in cell passaging is a major drain of time and manpower for cell production facilities. Replacement of this process with single-step passaging, as with suspension cultures, would be a most welcome development, both psychologically for workers and economically for management.

Another potentially useful application of microcarriers is the facile production of large quantities of cells in mitosis. Our own studies of synchronization procedures (Thilly, *et al.*, 1974; Thilly, *et al.*, 1975a, 1975b) have concentrated on anchorage-independent cell cultures, but the frequent appearance of mi-

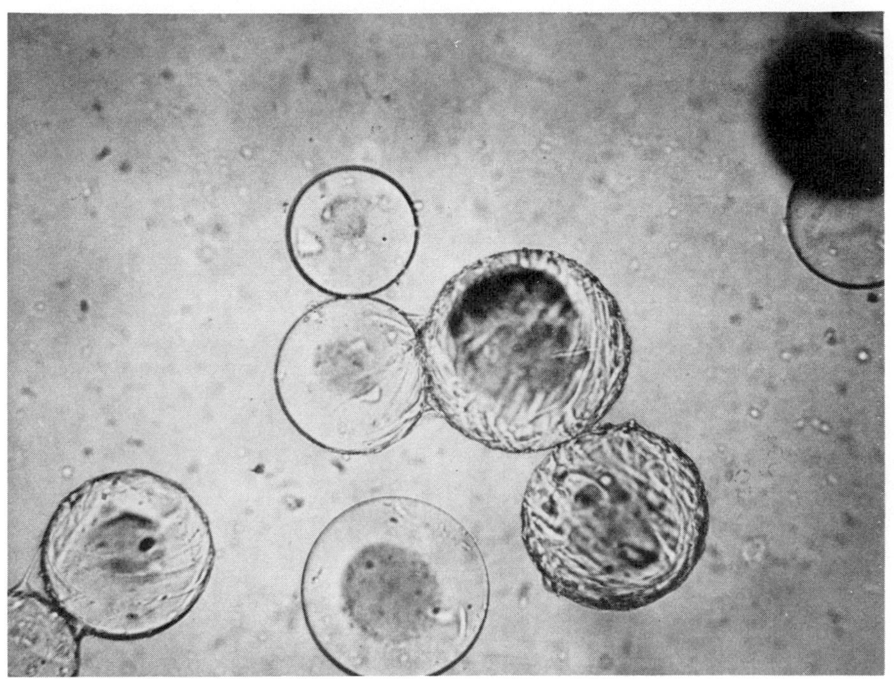

FIGURE 13 *Non-trypsinized cell transfer between a confluent carrier and an unseeded carrier.*

totic cells unattached to microcarriers (Fig. 14) leads us to believe that comparable facility can be obtained for anchorage-dependent cells. The procedure under development is based on the observations of Terasima and Tolmach (1963) that mitotic cells can be detached from monolayers by shear forces insufficient to detach non-mitotic cells. The microcarrier fermentor operating at 10^6 cells/ml and 24 hour doubling time has 2.93×10^4 cells/ml (assuming 1 hour for mitosis) in mitosis. Thus, a one liter culture contains 2.9×10^7 mitotic cells at any particular moment. These should be easily isolated by applying appropriate minimal shear conditions, separating beads from freely suspended cells by simple settling of the culture and pelleting mitotic cells from the beadless supernatant. The cells could be used for cell cycle studies, and the medium returned to the microcarrier culture.

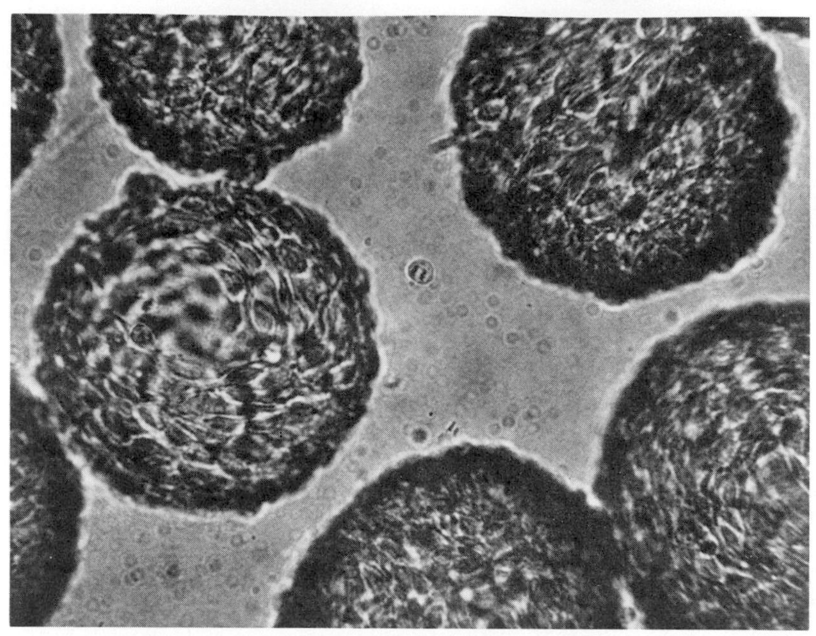

FIGURE 14 *Microcarrier culture of mouse fibroblast showing a mitotic cell in free suspension.*

This scheme is particularly attractive, since the cells would be selected under normal culture conditions and available in sufficient numbers for biochemical analysis. Our preliminary observations indicate that cell lines vary with regard to strength of attachment to microspheres at mitosis: some lines detach under normal agitation shear forces, while others are considerably more sticky.

ACKNOWLEDGMENTS

Technical assistance was provided by Thomas Skopek of the M. I.T. Undergraduate Research Opportunities Program. We are grateful to Dr. Robert Weinberg for discussions of oncogenic virus production and for performing plaque assays for murine leukemia virus.

We gratefully acknowledge the financial support of the National Science Foundation, Grant No. BMS 740 5676A01. David Levine is the recipient of an M.I.T. Health Science Fellowship.

REFERENCES

Ceccarini, C. and Eagle, H. (1971). "pH as a Determinant of Cellular Growth and Contact Inhibition." *Proc. Nat. Acad. Sci. 68,* 229.

Horng, C. and McLimans, W. (1975). "Primary Suspension Culture of Calf Anterior Pituitary Cells." *Biotechnolo. Bioeng. 17,* 713.

House, W., Shearer, M. and Maroudas, N. G. (1972). "Method for Bulk Culture of Animal Cells on Plastic Film." *Exp. Cell Res. 71,* 293.

Knazek, R. A., Gullino, P. M., Kohler, P. O. and Dedrick, R. L. (1972). "Cell Culture on Artificial Capillaries." *Science 178,* 65.

Levine, D. W. (1975). *Ph.D. Thesis Progress Report.* Department of Nutrition and Food Science, Massachusetts Institute of Technology, Cambridge, Massachusetts.

Munder, P. G., Modolell, M. and Wallach, D. F. H. (1971). "Cell Propagation on Films of Polymeric Fluorocarbons as a Means to Regulate Pericellular pH and pCO_2 in Cultured Monolayers." *FEBS Letters 15,* 191.

Terasima, J. and Tolmach, J. (1963). "Growth and Nucleic Acid Synthesis in Synchronously Dividing Populations of HeLa S3 Cells." *Exp. Cell Res. 30,* 344.

Thilly, W. G., Arkin, D. I., Nowak, Jr., T. S. and Wogan, G. N. (1975a). "Maintenance of Perpetual Synchrony in HeLa Cell Suspension Cultures: Question of Unbalanced Growth." *Biotechnolo. Bioeng. 17,* 703.

Thilly, W. G., Arkin, D. I., Nowak, Jr., T. S. and Wogan, G. N. (1975b). "Behavior of Subcellular Marker Enzymes During the HeLa Cell Cycle." *Biotechnolo. Bioeng. 17*, 695.

van Hemert, D., Kilburn, D. G. and van Wezel, A. L. (1969). "Homogeneous Cultivation of Animal Cells for the Production of Virus and Virus Products." *Biotechnolo. Bioeng. 11*, 875.

van Wezel, A. L. (1967). "Growth of Cell Strains and Primary Cells on Microcarriers in Homogeneous Culture." *Nature 216*, 64.

van Wezel, A. L. (1973). "Microcarrier Cultures of Animal Cells." *Tissue Culture, Methods and Applications*. Kruse, P. F. and Patterson, M. K., eds. Academic Press, New York. p. 372.

Weiss, R. E. and Schleicher, J. B. (1968). "A Multisurface Tissue Propagator for the Mass-scale Growth of Cell Monolayers." *Biotechnolo. Bioeng. 10*, 601.

Wöhler, W., Rüdiger, H. W. and Passarge, E. (1972). "Large Scale Culturing of Normal Diploid Cells on Glass Beads Using a Novel Type of Culture Vessel." *Exp. Cell Res. 74*, 571.

SUMMARY REMARKS

J. J. CALLIS

Plum Island Animal Disease Center
NER, ARS, USDA
Greenport, New York

In Symposium I, the theme was "Cell Culture for the Study of Differentiation and Cell Cycle." In Symposium II, the theme was "In Vitro Stimulators of Growth." In the symposium this morning, the theme is "Large-Scale Production" of tissue cultures when the cells are to be used for production of viruses for cell antigens.

One of the advantages of this Congress, where we all have reflected on tissue cultures, is the fact that we have an opportunity to talk with people who use tissue cultures for such a wide variety of purposes. It is indeed interesting to contemplate the developments that have already taken place in this field and, even more important, those that will surely take place in the future. Some of us have been concerned with production of cells for producing viruses, and we therefore are really fascinated by the studies on use of cells in relation to growth factors and for the production of enzymes. We can even contemplate the production of a variety of biological products in large-scale tissue culture systems or even perhaps tissue for repair in cases of trauma.

Carrel was among the first to show that tissue cells might be grown almost like microorganisms. Enders and his associates showed that cells cultivated in vitro could be used as a substrate for the multiplication of poliomyelitis virus; and a few years later, Earle and his associates succeeded in growing single cells as monolayers on glass. Since that time the in vitro cultivation of tissue cells has become a general tool for the production of virus vaccines and viral products. Generally, established cell lines are used for this production, and they can be subcultured indefinitely in vitro and in many instances can be grown in suspension. BHK cell lines are used extensively for the production of foot-and-mouth disease virus. Established cell lines are rarely diploid; therefore, they are not allowed for use in production of human virus vaccines. Rather, the primary cell cultures are used for this purpose. Hayflick and Moorhead developed diploid cell lines, starting with human embryonic tissue. These cells may be propagated many times but still retain the diploid karyotype and can be tested for extraneous viruses and oncogenic activity; however, the fact that the cells grow only on monolayer culture presents a problem of space in large-scale cultivation. Because the limiting factor in growing cells in monolayer culture is the surface, many workers have developed culture systems in which larger culture areas are available for the cells. Examples are roller bottles, perfusion systems, multi-surface propagators and microcarrier. This morning we will hear about cultivation of cells in suspension and in microsphere culture.

I. SUMMARY REMARKS

We have heard this morning from Dr. Girard (FAO-Ankara, Turkey) about problems encountered in large-scale production of cells for virus production. These problems may be manifold, especially in the developing countries. One of these problems is the quality of water and steam. The availability of serum in quantity and quality is a special problem in the large-scale cultivation of cells. Contamination is the scourge of serum production, especially when large quantities are required. Another problem in large-scale tissue culture production is the cell that is being used. For example, each institute may now have its own BHK cell, originally described by MacPherson and Stoker. Samples of such cells from various institutes look different, have different multiplicities and produce differing plaque types of FMDV. Therefore, Dr. Girard has suggested that cells should not be used when repeatedly passaged in suspension. Finally, one must also be mindful of the quality and quantity of virus from the cultures.

Dr. Acton, one of our hosts, discussed the operation of a large-scale tissue culture unit in which murine lymphoblastoid cell lines are propagated in suspension. Dr. Acton described the new cell culture laboratory at the University of Alabama. He stressed the importance of the control of the many factors affecting cell growth, such as pH and dissolved oxygen. The purposes of the cell center are to produce lymphoblastoid cells for the differentiation of alloantigens, T. L. and Thy-1 components of the cell membrane. The qualitative and quantiative measure of these alloantigens reflects the differentiation state of the lymphoid precursor cell that normally matures in the thymus. The Thy-1 antigen was also identified on nerve tissue and epidermal cells.

Dr. Nyiri (Fermentation Design, Inc., Pennsylvania) discussed the fact that the technology is now available to measure and regulate by computer, the different physical and biochemical parameters in a biological system such as a cell culture. In his discussions, he gave examples of how, with information on oxygen utilization and carbon dioxide production, these data could be used to calculate the respiratory quotient. This quotient could, in turn, be adjusted for the optimum for cell growth by regulation of the components of the media.

Dr. Levine (M.I.T., Cambridge, Massachusetts) described application of Van Wazel's 1973 report concerning DEAE sephadex beads as a mobile support for growing cells in suspension. The beads are pretreated with carboxymethyl cellulose to saturate absorption sites that might otherwise remove nutrients from the medium. Cells grow on these microspheres at concentrations ranging up to 2×10^6 cells per ml. The cells are trypsinized off the beads, and there is 95% viability. The method promises to be applicable to many cell types and to have great potential for expansion if synthetic beads with better physical characteristics can be obtained.

Symposium IV

CELL CULTURE FOR THE STUDY OF DISEASE

REGULATION OF PROLIFERATIVE CAPACITY DURING AGING IN CELL CULTURE

V. J. CRISTOFALO
J. M. RYAN
G. L. GROVE

The Wistar Institute of Anatomy and Biology
Philadelphia, Pennsylvania

I. INTRODUCTION

Biological aging can be characterized as a process which (1) occurs in all members of a population; (2) is progressive and eventually harmful to the organism; (3) is irreversible under usual conditions; (4) is characterized by changes in functional capacity which result in the loss of the organism's ability to respond to environmental change. Loss of capacity for adaptive response appears to occur across a wide range of biological organization within the animal organism. In approaching complex biological problems, model systems have proven useful in elucidating fundamental mechanisms by which cells carry out specific processes or functions. In this discussion, we will be considering an approach to the study of aging using diploid human cells in culture as a model system.

It is now reasonably well established that populations of normal human fibroblast-like cells can proliferate in culture for only limited periods of time. Typically, after explantation, there is a period of rapid proliferation during which the cultures can be subcultivated relatively often. This gives way to a period of declining proliferative capacity during which the cells become granular, debris accumulates, and ultimately the culture is lost. Detailed descriptions of the natural history of these cultures which establish the generality of the phenomenon are contained in Swim and Parker (1957), Hayflick and Moorhead (1961), Hayflick (1965), Hay (1967), and Cristofalo (1972).

Initially, the inability of human cell cultures to proliferate indefinitely was ascribed to various technical difficulties. However, Hayflick and Moorhead (1961) showed that when mixtures of young and old populations (male and female), distinguishable by karyotypic markers (Barr Bodies), were co-cultivated in the same culture vessel, the older population was lost after it had undergone a total of approximately 50 population doublings. The younger population continued to proliferate until the 50 or so population doublings had been completed. These results would seem to rule out any simple direct effect of nutritional deficiency, microcontaminants or the presence of toxic substances in the culture.

This limited lifespan phenomenon in terms of proliferative capacity has been well documented for human (Hayflick, 1965; Hayakawa, 1969), chick (Haff and Swim, 1956), bovine (Stenkvist, 1966), and tortoise (Goldstein, 1974) cells in culture. It is significant that the *in vitro* lifespan (proliferative capacity) of a given tissue is species-specific and reproducible. These observations suggest that the limited lifespan phenomenon is intrinsic to the cell and, in part at least, is genetically programmed. In addition, recently Dell'Orco *et al.* (1973; 1975) have maintained human cells in a quiescent state for as long as six months in medium in which the serum supplement is too low to support significant proliferation. When these cells were returned to growth medium, all cultures resumed proliferation and completed the same number of population doublings as control cultures that were continually subcultured. Thus, the cellular mechanism for regulating proliferative lifespan seems to involve the counting of division events.

To one interested in the study of senescence, however, other important questions are: (1) Does aging *in vitro* bear any relationship to aging in the whole organism? and (2) Do studies carried out in cell culture have any relevance to aging in the intact animal?

These questions have been discussed in detail elsewhere (Cristofalo, 1972; Hayflick, 1972). Clearly, there is no evidence to support the notion that animals die because they exhaust the supply of proliferating cells; nor has this been seriously suggested. In addition, there is not, at present, sufficient evidence to relate the time course of this loss of proliferative capacity *in vitro* to the maximum lifespan of the donor animal. On the other hand, there is evidence in the literature indicating that in some tissues *in vivo* there are declines in proliferative capacity with age. Similarly in some serial transplantation studies, a decline in proliferative capacity has been shown (for review see Cristofalo, 1972).

Perhaps the strongest evidence supporting a relationship between aging *in vivo* and *in vitro* derives from a variety of work that suggests a cumulative effect of *in situ* plus *in vitro* aging by showing a relationship between the age of the cell donor and the proliferative capacity of the cells derived from that donor. For example, Hayflick (1965) showed a striking difference in proliferative lifespan between cultures derived from human embryonic and human adult lung. Goldstein and his co-workers (1969) found that an inverse correlation existed between the age of the donor and the proliferative lifespan for a series of human skin cultures.

Martin *et al.* (1970) studied over 100 mass cultures of fibroblast-like human skin cells from a variety of donors ranging from newborn to 90 years. They found a significant negative regression of growth potential as a function of age of the donor. More recently, these findings have been confirmed by Le Guilly *et al.* (1973). These workers used 100 cell lines derived from human liver and found a statistically significant negative correlation between the age of the donor and the proliferative lifespan of the culture and the number of doublings in phase II. Goldstein *et al.* (1969) and Martin *et al.* (1970) have shown that cell cultures derived from patients with progeroid syndromes or with Werner's syndrome (diseases associated with premature aging) have a reduced proliferative lifespan as compared with cultures derived from normal donors of the same ages.

Correlative metabolic findings show that cell cultures derived from adult lung have lysosomal enzyme characteristics after only 12-14 passages similar to those of fetal cells after 35-45 passages (Cristofalo *et al.*, 1967). In addition, adult skin cells show cell cycle characteristics after only a few passages similar to those of degenerating fetal cultures (Macieira-Coelho and Ponten, 1969; Macieira-Coelho, 1970).

In any case, as we mentioned previously, aging *in vivo* is characterized, in part, by a decline in functional capacity. In our model system in cell culture, there is also a decline in functional capacity viz. proliferative capacity. In the studies to be described below, we are concerned with understanding how the regulation of cell proliferation changes during *in vitro* aging.

Some years ago (Cristofalo, 1970), we reported that hydrocortisone (cortisol), at a concentration of 5 μg/ml, increased the proliferative lifespan of WI-38 cells. Macieira-Coelho (1966) had reported a similar finding for human cells grown with cortisone (5 μg/ml).

As a result of our initial studies, the principal features of the hydrocortisone effect can be summarized as follows: (1) Replicate experiments based on cell counts have shown that the lifespan in terms of actual population doublings was extended 30-40% by the continuous inclusion of 5 μg/ml hydrocortisone in the medium; (2) This effect seemed to be maximal with 5 μg/ml hydrocortisone; (3) There is no rejuvenation with the hormone, i.e. once a culture could no longer achieve confluency, hydrocortisone would not reverse this condition; (4) If hydrocortisone was added at different periods in the lifespan, the magnitude of the extension of lifespan was in direct proportion to the amount of time the culture was grown in the presence of the hormone; (5) The saturation density of the culture was increased in the presence of the hormone; (6) The overall effect on proliferative capacity was not due to improved attachment of the cells to the glass or plastic growth surface.

This hydrocortisone effect on cell lifespan represented the action of a chemically defined modulator of cell division and population lifespan. We have further documented these observations in the hope of using hydrocortisone as a probe for modulating the regulation of cell division and the limitation of proliferating capacity of these diploid cell cultures.

II. MATERIALS AND METHODS

Except where noted, all studies were done with human diploid cell lines WI-38 and WI-26 (Hayflick, 1965; Hayflick and Moorhead, 1961). Starter cultures were obtained either from frozen stock maintained here at The Wistar Institute or from Dr. Leonard Hayflick of Stanford University.

The cells were grown as previously described (Cristofalo and Sharf, 1973). Cell population doublings were calculated by comparing the cell counts/vessel at seeding and when the cultures reached confluency. All cell counts were done electronically using a Coulter Counter.

Autoradiography was carried out by our standard procedures (Cristofalo and Sharf, 1973). The autoradiographs were analyzed microscopically by scoring the percentage of cells with labeled nuclei (5 silver grains or more) in random fields throughout the coverslips. At least 400 cells were counted on each coverslip and all coverslips were prepared in duplicate.

Cell cycle analysis was carried out essentially according to the method of Quastler and Sherman (1959) and analysis of the data was done according to the method of Mendelsohn and Takahashi (1971).

For evaluation of the effect of various steroids on DNA synthesis, the hormones were purchased from commercial suppliers at the best grade of purity available. Active hormones were also checked for the presence of impurities by standard thin-layer chromatography. For autoradiographic analysis of their biological activity, the procedures of Cristofalo and Sharf (1973) were used. All hormones were used at a concentration of 5 μg/ml. Where solubility was limiting, the hormones were dissolved in 100% ethanol. Subsequent dilution was carried out in medium to give a final concentration of ethanol of no more than 0.5%. Paired controls were always run with an identical concentration of ethanol in the medium. For all active hormones, the change in the labeling index was correlated with direct cell counts.

To determine the effect of hydrocortisone on transcriptional activity, cells between the 29th and 39th population doubling were used. Transcriptional activity of exponentially growing cells or confluent cells stimulated to divide was determined using three different methods: (1) Incubation of isolated chromatin from hydrocortisone-treated and control cells with *E coli* RNA polymerase, CTP, GTP, ATP and ^{3}H-UMP incorporated into RNA (Ryan and Cristofalo, 1975); (2) Formation of nuclear monolayers by treating cells with 0.5% NP40 and comparing the rate of ^{3}H-UMP incorporated into RNA (Tsai and Green, 1973; Ryan and Cristofalo, 1975); (3) Isolation of nuclei from hormone-treated and control cells by Dounce homogenization and differential centrifugation followed by incubation of nuclear suspension (3-5 μg DNA) with ^{3}H-UTP and the remaining nucleotide triphosphates (Marzluff *et al.*, 1973).

III. RESULTS

We have shown previously that aging in human diploid populations is characterized by a gradual decline in the number of cells in the rapidly proliferating pool (Cristofalo and Sharf, 1973). It was of interest to determine the effect of hydrocortisone on the kinetics of this decline. Figure 1 shows the percent labeled nuclei as a function of time in the presence and absence of exogenously added hydrocortisone over a portion of the lifespan of the culture. Initially, nearly 100% of the cells were labeled in both cases. As the lifespan progressed, the decline in the fraction of labeled cells was more rapid in the control cultures and they phased out well before the hydrocortisone-treated cultures; however, the pattern of decline in labeling was the same in both cases. Hydrocortisone seemed to retain the cells in the actively proliferating pool for longer periods.

Finally, it is important to note that the differences between hydrocortisone and control were greater as the culture aged. However, the percentage of cells responding to the stimulus for division declined with age. The rate of decline was slower in the hydrocortisone-treated culture.

In order to better understand the action of hydrocortisone in extending lifespan, it was of interest to determine the effect during a single growth cycle.

Table I shows the effect of nuclear labeling of previously untreated cultures of older cells in which hydrocortisone was present for either 24 or 48 hours while the ^{3}H-dT was present for 24 hours. There was a statistically significant increase in the percentage of cells in both young and old cultures incorporating

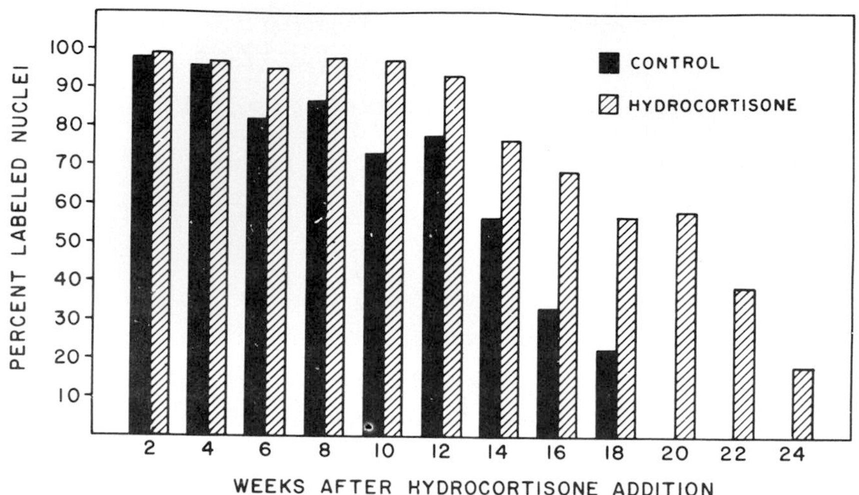

FIGURE 1 *The effect of hydrocortisone on the fraction of cells incorporating ^{3}H-dT during a 30 hour pulse.*

^{3}H-dT in the presence of hydrocortisone after 48 hours. As shown above, parallel cultures showed an increase in cell number in the presence of the hormone and the results are consistent with stimulation of both DNA synthesis and cell division.

In other experiments reported elsewhere (Cristofalo, 1975), we have shown by chemical determination that the rate of DNA synthesis is higher in hydrocortisone-treated cultures and that ^{3}H-dT incorporation parallels DNA synthesis in both treated and control cultures. Thus, the hydrocortisone-mediated increase in DNA synthesis seems to be due, in part at least, to an increase in the fraction of cells in the rapidly proliferating pool. Hydrocortisone appears to amplify the stimulus for proliferation.

To evaluate the effect of hydrocortisone on DNA synthesis, further experiments were designed to determine if the biological activity was specific for this molecular structure or whether it was simply a non-specific effect. To test these alternatives, various steroids were assayed for their biological activity with

TABLE I The Effect of Hydrocortisone for 24 or 30 Hours on the Fraction of Cells Incorporating ^{3}H-dT During a 24 Hour Pulse

Treatment	Exposure to hydrocortisone (hours)	Percent labeled nuclei[a] Young	Old
None	24	41.5 ± 6.6 (5)[b]	38.6 ± 10.3 (3)[b]
	48	79.8 ± 5.7 (6)[c]	45.9 ± 3.4(15)[c]
Hydrocortisone	24	60.0 ± 6.5 (5)[b]	50.0 ± 10.2 (3)[b]
	48	92.3 ± 1.9 (6)[c]	64.6 ± 4.1(14)[c]

[a]Mean ± Standard error of mean (number of samples).
[b]$p < .005$
[c]$p < .001$

WI-38 cells as determined by the increase in the percentage of labeled nuclei in the culture.

Three groups of compounds could be distinguished: those that were very active, hydrocortisone being the most active in stimulating DNA synthesis; those that were clearly inhibitory to DNA synthesis, testosterone being the most inhibitory; and a third group in which there may have been moderate borderline effects but which were not significantly different from the control.

Figure 2 shows a comparison of the molecular structures of hydrocortisone and the other five stimulatory steroids. Note that all have the 3 keto, Δ 4 configuration in Ring A, and 11β-hydroxy group and a keto group at C-20. Other functional groups such as the 17α and the 21-hydroxy groups, present in hydrocortisone but not in some of the others, or the C-18 methyl group present in all but aldosterone and 19-hydroxycorticosterone seemed of minor and variable significance to the action of the hormone. In general, the specificity observed here is very similar to that reported for stimulation of cell division in confluent 3T3 mouse fibroblasts by Thrash *et al.* (1974).

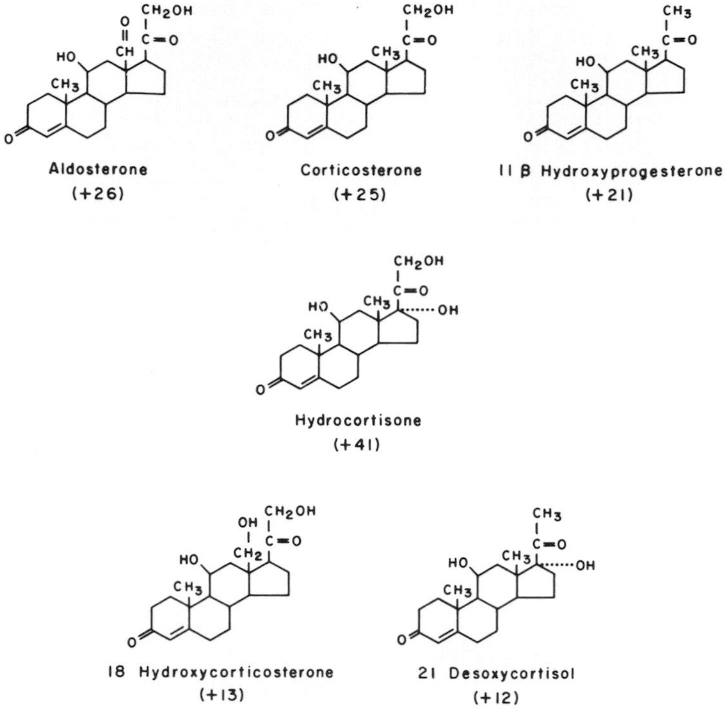

Aldosterone
(+26)

Corticosterone
(+25)

11 β Hydroxyprogesterone
(+21)

Hydrocortisone
(+41)

18 Hydroxycorticosterone
(+13)

21 Desoxycortisol
(+12)

FIGURE 2 *Molecular structure of steroids which stimulate cell division in WI-38 cells.*

Initially, the increase in proliferative capacity seemed enigmatic. Typically, hydrocortisone has been reported to inhibit cell proliferation in tissue culture (Ruhmann and Berliner, 1965) and we were interested in determining whether our observations were specific for this cell type and whether under identical conditions we could duplicate the inhibitory effects reported by others for other cell types.

Table II shows a partial listing of the results of experiments which are representative of a larger number of cell cultures that we have screened for their response to hydrocortisone. In the upper portion of the table are seven human fibroblast-like populations. These include human fetal lung cell cultures from four

TABLE II EFFECT OF HYDROCORTISONE ON DNA SYNTHESIS AND CELL DIVISION IN VARIOUS CELL LINES

Origin				Percent change with hydrocortisone	
Species	Tissue	Designation	Passage	Cell count	Percent labeled nuclei
Human	Fetal lung	WI-38	40	+40	+38
Human	Fetal lung	WI-26	38	--	+33
Human	Fetal lung	MRC-5	39	--	+51
Human	Fetal lung	HEL-31	50	+31	+20
Human	Fetal foreskin	HF	9	+24	+20
Human	Infant skin	#2	36	--	+27
Human	Adult skin (Werner's syndrome)	GM-167	5	+60	+47
Human	Fetal kidney (epith)	HEK-31	5	0	0
Human	SV40 transformed WI-38	WI-38-VA13A	-	-25	0
Human	SV40 transformed WI-26	WI-26-VA4	-	-30	0
Human	SV40 transformed mucosa	W-18-VA2	-	-31	0
Human	Cervical carcinoma	HeLa	-	-31	-20
Monkey	Kidney	CV-1	40	-54	-18
Hamster	Whole embryo	---	2	-61	-15
Mouse	Areolar tissue	L	-	-34	-24
Chick	Whole embryo	---	11	-9	-16
Lizard	Whole embryo	GE-1	33	-20	-35
Frog	Whole embryo	RP	45	-20	-40
Fish	Adult minnow tail	FHM	17	-67	-50

individuals and at comparable passage levels, and cultures of fetal, infant and adult human skin. The adult skin was received from Dr. George M. Martin of the University of Washington and was obtained at autopsy from a patient with Werner's syndrome, a disease of premature aging. For all of these human cells, both DNA synthesis and cell division were stimulated by hydrocortisone.

In more recent studies, we have assayed a series of human skin cell lines from various aged individuals including embryonic, neonatal, and adult cell cultures and have found that, based on the percent labeled nuclei assay, the response to hydrocortisone was quite variable. Effects ranged from stimulation to inhibition of DNA synthesis. We are currently extending these studies to other diploid human skin cell cultures to determine the relationship, if any, between hydrocortisone response and age or tissue of origin.

Human fetal kidney cultures of epithelial-like cells, obtained from Dr. Warren Nichols of the Institute for Medical Research, Camden, New Jersey, and studied at passage 5 were unresponsive to hydrocortisone.

In contrast to their normal progenitor strains, the 3 SV40-transformed cultures (WI-38-VA13A, WI-26-VA4, W-18-VA2) showed an inhibition of cell division of about 30%, but no reduction in the percentage of cells synthesizing DNA. HeLa cells, on the other hand, showed an inhibition of both DNA synthesis and cell division, thus suggesting the possibility of separate effects of the hormone on mitosis and DNA synthesis.

Finally, in the lower portion of the table, we show the effect of cortisol on cell lines derived from species representing all the vertebrate classes: three mammalian lines and representatives of lines derived from birds, reptiles, amphibians, and fish. For all of these, both DNA synthesis and cell division were inhibited by hydrocortisone.

By employing the percent labeled mitosis method (Quastler and Sherman, 1959) we have been able to characterize the cell cycle of

TABLE III The Effect of Hydrocortisone on Cell Cycle Parameters During Aging in WI-38 Cells

	Control					Hydrocortisone				
Weeks after HC addition	t_1	t_s	t_2	t_c	CV	t_1	t_s	t_2	t_c	CV
3	4.1	10.4	4.5	19.0	29.6	3.5	10.5	4.4	18.4	22.7
9	4.2	10.6	4.6	19.4	32.0	4.8	10.3	4.6	19.7	29.6
15	14.8	10.5	5.4	30.7	37.0	7.5	10.7	4.6	22.8	28.9
21						18.0	10.9	5.8	34.7	48.9

these cultures at several different times during their finite lifespan. Table III shows the results of this study. Our findings show the intercellular variation in cell cycle times is a characteristic of these cultures which becomes even more pronounced in late passage populations. This increased heterogeneity in senescent cultures has also been observed by Absher *et al.* (1974) in their time lapse cinemicrophotographic studies of this phenomenon of *in vitro* cellular aging. The duration of mitosis has previously been shown not to appreciably change with culture age by Macieira-Coelho *et al.* (1966), thus changes in t_1 and t_2 which include a portion of mitosis are an indication of changes in G_1 or G_2 respectively. The data obtained shows that the cell cycle increases with advancing age while the duration of DNA synthesis remains relatively constant. In those late passage cultures where proliferative activity is extremely low, the duration of G_2 is also prolonged. However, it is the changes in the duration of G_1 which account for almost all the differences in cell cycle times. These data are in agreement with the previous studies of Macieira-Coelho *et al.* (1966) and Macieira-Coelho (1973). Hydrocortisone-treated cultures show the same increases. However, we have found that in the presence of the hormone these changes are significantly delayed.

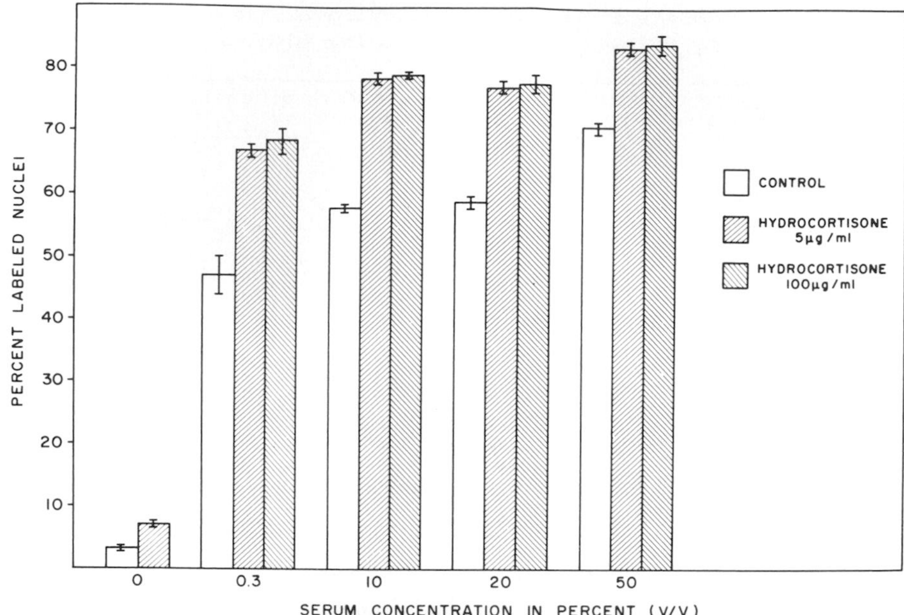

FIGURE 3 *The effect of serum and hydrocortisone on the fraction of cells incorporating* ^{3}*H-dT during a 30 hour pulse.*

The changes in proliferative patterns obtained in our study are consistent with the notion that cellular senescence is due to a transition from a rapidly cycling state to one or a series of more slowly cycling states and finally to a state in which the cells are arrested or cycling so slowly as to be unable to repopulate the culture vessel (Cristofalo, 1975; Macieira-Coelho *et al.*, 1975). One simple interpretation of the hydrocortisone extension of culture lifespan is that the steroid delays these transitions.

In considering this effect, a question of major interest was whether hydrocortisone stimulated division directly or functioned as an amplifier of the serum mitogen. Figure 3 shows that in the absence of serum there was essentially no division; the primary signal for division was clearly serum. With graded increases in serum there was a graded response in the fraction of cells incorporating labeled precursor. Note that in the presence of hy-

TABLE IV The Effect of the Time of Exposure to Hydrocortisone on Cell Proliferation

HC exposure	DNA synthesis (CPM/filter)
Time (hours)	Ratio HC/control
0	1.00
3	1.23
6	1.49
9	1.44
12	1.34
15	1.39
18	1.25
21	1.32
48	1.38

drocortisone, 0.3% serum gave a response higher than 10% serum without hydrocortisone, and 10% serum plus hydrocortisone was equivalent to or greater than 50% serum. Thus, the hormone functions as an amplifier of the serum signal.

Confluent monolayers of cells can be stimulated to divide by the addition of fresh serum or complete medium. Upon stimulation, there is a parasynchronous wave of DNA synthesis and cell division and thus the effects of various agents and their time course of action can be resolved. When various combinations of serum and hydrocortisone were added to confluent monolayers of WI-38 cells, amplifying effects of hydrocortisone similar to those described above were observed.

In order to approach the mechanism of action of the hormone, we first designed experiments to determine the time course of the hormone effect. For these experiments, confluent monolayers were stimulated by the addition of serum or serum plus hydrocortisone. At the times indicated in Table IV, the hydrocortisone-containing media was removed and replaced with similarly conditioned media

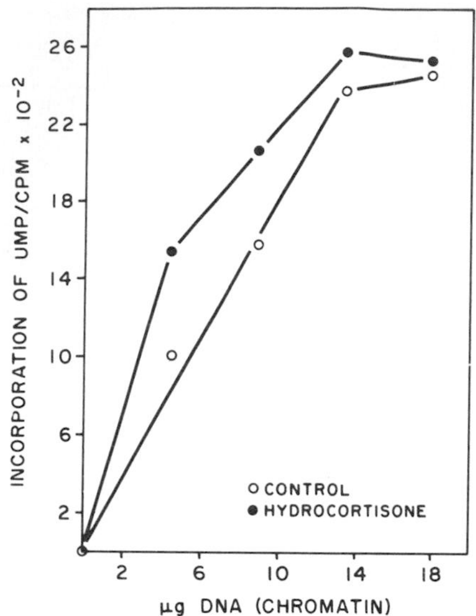

FIGURE 4 *The effect of hydrocortisone on transcriptional activity in WI-38 cells.*

from dummy flasks and then the cells were fixed after a total incubation of 48 hours. At each time point, controls were handled in exactly the same way. The results when the hormone is present for possibly three but certainly the first six hours after stimulation, are as if the hormone had been present continually for the entire 48 hour period. Furthermore, we have found that if we add the hormone after six hours no enhancement of proliferation is seen indicating that the hormonal effect occurs somewhere during these first six hours of the prereplicative phase.

The effects of Hydrocortisone can also be resolved at the molecular level. Figure 4 shows that exponentially growing WI-38 cultures in the presence of 14 μM hydrocortisone for 48 hr have an increased transcriptional activity over control cells. To further evaluate the kinetics of hormone-induced alterations in chromatin activity and to identify when in the cell cycle increased tran-

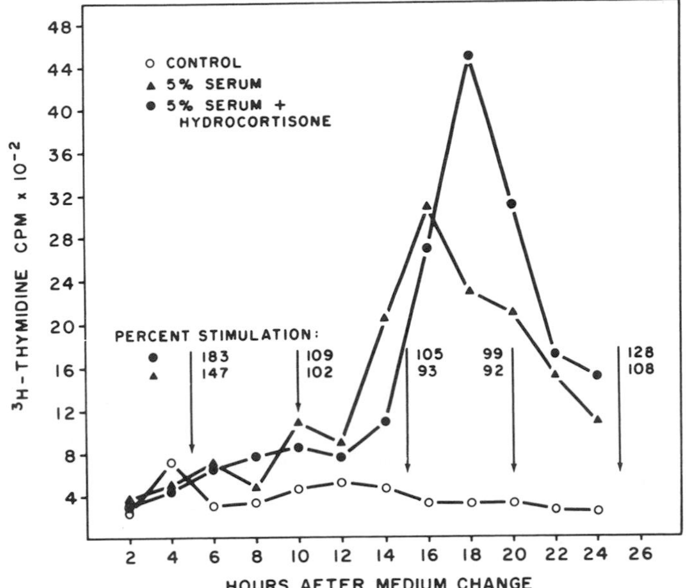

FIGURE 5 *Time course of DNA synthesis and transcriptional activity in cultures stimulated to divide in the presence and absence of exogenous hydrocortisone.*

scriptional activity occurred, chromatin was isolated and assayed for transcriptional activity at five hr intervals following stimulation of confluent cultures as described above. Results, based on ^{3}H-thymidine incorporation, (Fig. 5) showed that the pre-DNA synthesis period of control and hormone-treated cultures was about 12 hours and the period of active DNA synthesis lasted 10-12 hours in both cultures, consistent with the cell cycle data described above. There was essentially no difference in the time required for hormone-treated and control cells to enter DNA synthesis, or in the time required to complete DNA synthesis, although the magnitude of ^{3}H-dT incorporation was greater in hormone-treated cultures. The transcriptional activity (arrows) of serum-stimulated and serum-hydrocortisone-stimulated cultures increased early in the pre-DNA synthesis period, however, serum-hydrocortisone-treated cells had an increased activity over serum-treated cells

(183 vs 147%). As cells from both groups approached and entered the S phase of the cell cycle, the increased transcriptional activity of stimulated cultures was abolished (99 and 92%) but reappeared to a slight extent as cells entered the post-DNA synthesis period (128 and 108%).

It is important to note here that the increase in transcriptional activity in the presence of hydrocortisone occurs during the first five hour period following stimulation and this time course is consistent with the time course shown above for the period when hydrocortisone is active in increasing cell proliferation.

One possible mechanism by which hydrocortisone increases transcriptional activity is by permitting a greater quantity of cells to initiate preparations for DNA synthesis and cell division. This would result in a greater fraction of active chromatin which would be reflected in "increased" chromatin transcriptional activity. In addition, there may be possible quantitative and/or qualitative differences in gene products of the hormone-treated cells.

IV. DISCUSSION

The data presented above show that (1) hydrocortisone increases the fraction of cells in the rapidly proliferating pool; (2) There is a delay in the lengthening of the G_1 phase of the cell cycle that occurs during aging; (3) Hydrocortisone is not a primary mitogen but acts by amplifying the serum signal for division; (4) Certain critical events take place within six hours following stimulation for division; (5) Chromatin template activity is increased in the presence of hydrocortisone; (6) The most significant increase in template activity occurs within the first five hours.

In the past, hydrocortisone, cortisone and other corticosteroids have been reported to prolong the *in vitro* survival time of several cell types (Arpels *et al.*, 1964; Yuan and Chang, 1969; Yuan *et al.*, 1967). These studies have been concerned with post-mitotic maintenance of the cultures. It was Macieira-Coelho (1966) who first reported the increase in proliferative lifespan of diploid, fibroblast-like cells in culture with glucocorticoids. Division stimulating effects with hydrocortisone have also been reported by Castor and Prince (1964) for cartilage cells and by Smith *et al.* (1973) for human fetal lung cells. Recently, Thrash and Cunningham (1973) have shown the stimulation of division by hydrocortisone in density-inhibited 3T3 cells and Armelin (1973) and Gospodarowicz (1974) have shown that hydrocortisone amplifies the activity of the pituitary and brain-derived polypeptides that stimulate cell division. Thus, division-stimulating effects for hydrocortisone seem now to be well documented in the literature.

Perhaps the best documented effects of glucocorticoids involves hormone-dependent alterations in transcriptional activity. The data described above indicate that there is an increase in transcriptional activity following stimulation to divide and that the magnitude of increase is enhanced when exogenous hydrocortisone is present in the medium.

On the other hand, we must consider the possibility that, although changes in transcriptional activity represent the best defined action of glucocorticoids, there is a wide range of other, less well understood actions of steriods which are termed permissive effects (Thompson and Lipmann, 1974). The increased transcriptional activity, described above, need not reflect a direct action of the hormone on transcription but could reflect a permissive effect allowing more cells to respond to the serum mitogen and enter the rapidly proliferating pool.

If we assume that aging of the population is reflected at the cellular level by a transition from a rapidly cycling state to one or a series of more slowly cycling states, and finally, to a sterile state, i.e., a state in which the cells are arrested or are cycling so slowly as to be unable to repopulate the culture vessel (Cristofalo, 1975; Macieira-Coelho, 1975), then, somehow, the steroid appears to delay these transitions.

Several explanations can be offered for this effect. One interpretation is that the cells change responsiveness to the mitogenic stimuli. Possibly young cells are responsive to 10% serum initially, but eventually lose their responsiveness to this concentration of serum and undergo a transition to a second state where 10% serum plus hydrocortisone are required to elicit a division response. There may be a second population in the culture which succeeds the first and which will only proliferate in the presence of serum plus hydrocortisone. We might also postulate that slowly cycling or arrested cells inhibit the growth of those cells still capable of division, either simply by the space they occupy or by eliciting an inhibitor or chalone into the environment. In these rapidly dividing cells hydrocortisone has no effect. In the slowly cycling cells, however, hydrocortisone could work either by some inactivation of the hypothetical inhibitor or by killing the cells responsible for its secretion. Thus, the young cells in the population are not inhibited to the same extent. Our current work is designed to clarify these possibilities.

ACKNOWLEDGMENTS

The expert technical assistance of Barbara Sharf, Joan Kabakjian, E. Donald Kress, and Joanne Cochran in various parts of these studies is gratefully acknowledged.

The work reported above was supported by grants HD02721 and HD06323 from the National Institute of Child Health and Human Development and a Young Investigator Pulmonary Research Award, HD17224, to J. M. Ryan.

REFERENCES

Absher, P. M., Absher, R. G., and Barnes, W. D. (1974). *Exptl. Cell Res. 88,* 95-104.

Armelin, H. A. (1973). *Proc. Nat. Acad. Sci. U.S.A. 70,* 2702-2706.

Arpels, C., Babcock, V. I. and Southam, C. M. (1964). *Proc. Soc. Exptl. Biol. Med. 115,* 102-106.

Castor, C. W. and Prince, R. K. (1964). *Biochim. Biophys. Acta 83,* 165-177.

Cristofalo, V. J. (1970). *In* "Aging in Cell and Tissue Culture" (E. Holečková and V. J. Cristofalo, eds.), pp. 83-119. Plenum Press, New York.

Cristofalo, V. J. (1972). *Adv. Gerontol. Res. 4,* 45-79.

Cristofalo, V. J. (1975). *In* "Cell Impairment in Aging and Development" (V. J. Cristofalo and E. Holečková, eds.), pp. 7-22. Plenum Press, New York.

Cristofalo, V. J. and Sharf, B. B. (1973). *Exptl. Cell Res. 76,* 419-427.

Cristofalo, V. J., Parris, N., and Kritchevsky, D. (1967). *J. Cell Physiol. 69,* 263-272.

Dell'Orco, R. T. (1975). *In* "Cell Impairment in Aging and Development" (V. J. Cristofalo and E. Holečková, eds.), pp. 41-49. Plenum Press, New York.

Dell'Orco, R. T., Mertens, J. G. and Kruse, P. F., Jr. (1973). *Exptl. Cell Res.* 77, 356-360.

Goldstein, S. (1974). *Exptl. Cell Res. 83,* 297-302.

Goldstein, S., Littlefield, J. W. and Soeldner, J. S. (1969). *Proc. Nat. Acad. Sci. U.S.A. 64,* 155-160.

Gospodarowicz, D. (1974). *Nature 249,* 123-127.

Haff, R. F. and Swim, H. E. (1956). *Proc. Soc. Exptl.*

Biol. Med. 93, 200-204.

Hay, R. J. (1967). *Adv. Gerontol. Res. 2,* 121-158.

Hayakawa, M. (1969). *J. Exp. Med. 48,* 171-179.

Hayflick, L. (1965). *Exptl. Cell Res. 37,* 614-636.

Hayflick, L. (1972). *In* "Aging and Development" (H. Bredt and J. W. Rohen, eds.), pp. 1-15. Mainz Academy of Science and Literature, Mainz, Germany.

Hayflick, L. and Moorhead, P. S. (1961). *Exptl. Cell Res. 25,* 585-621.

Le Guilly, Y., Simon, M., Lenoir, P., and Bourel, M. (1973). *Gerontologia 19,* 303-313.

Macieira-Coelho, A. (1966). *Experientia 22,* 390-391.

Macieira-Coelho, A. (1970). *In* "Aging in Cell and Tissue Culture" (E. Holečková and V. J. Cristofalo, eds.), pp. 121-132. Plenum Press, New York.

Macieira-Coelho, A. (1973). *In* "Frontiers of Matrix Biology" (L. Robert and B. Robert, eds.), Vol. II, pp. 46-77. S. Karger, Basel.

Macieira-Coelho, A. and Ponten, J. (1969). *J. Cell Biol. 43,* 374-377.

Macieira-Coelho, A., Loria, E. and Berumen, L. (1975). *In* "Cell Impairment in Aging and Development" (V. J. Cristofalo and E. Holečková, eds.), pp. 51-65. Plenum Press, New York.

Martin, G. M., Sprague, C. A., and Epstein, C. J. (1970). *Lab. Invest. 23,* 86-92.

Marzluff, W. F., Murphy, E. C. and Huang, R. C. C. (1973). *Biochemistry 12,* 3440-3446.

Mendelsohn, M. L. and Takahashi, M. (1971). *In* "The Cell Cycle and Cancer" (R. Baserga, ed.), pp. 58-95. Marcel Dekker, Inc., New York.

Quastler, H. and Sherman, F. G. (1959). *Exptl. Cell Res. 17,* 420-438.

Ruhmann, A. G. and Berliner, D. L. (1965). *Endocrinology 76,* 916-927.

Ryan, J. M. and Cristofalo, V. J. (1975). *Exptl. Cell Res. 90,* 456-458.

Smith, B. T., Torday, J. S. and Giroud, C. J. P. (1973). *Steroids 22* 515-524.

Stenkvist, B. (1966). *Acta Path. et Microbiol. Scand. 67,* 180-190.

Swim, H. E. and Parker, R. F. (1957). *Am. J. Hyg. 66,* 235-243.

Thompson, E. B. and Lipmann, M. E. (1974). *Metabolism 23* 159-202.

Thrash, C. R. and Cunningham, D. D. (1973). *Nature 242,* 399-401.

Thrash, C. R., Ho, T., and Cunningham, D. D. (1974). *J. Biol. Chem. 249* 6099-6103.

Tsai, R. L. and Green H. (1973). *Nature New Biol. 243,* 168-170.

Yuan, G. C. and Chang, R. S. (1969). *Proc. Soc. Exptl. Biol. Med. 130,* 934-936.

Yuan, G. C., Chang, R. S., Little, J. B., and Cornil, G. (1967). *J. Gerontol. 22,* 174-179.

ISOLATION AND CHARACTERISATION OF HLA ANTIGENS FROM CULTURED CELLS AND THEIR CONTRIBUTION TO THE STUDY OF DISEASE*

DAVID SNARY
MICHAEL J. CRUMPTON

National Institute for Medical Research
London, United Kingdom

PETER GOODFELLOW
WALTER F. BODMER

Genetics Laboratory
Department of Biochemistry
University of Oxford
Oxford, United Kingdom

I. INTRODUCTION

The major human histocompatibility system HLA was originally defined as a series of antigenic specificities identified by serological techniques using peripheral blood lymphocytes. These

*The abbreviation SDS is used for sodium dodecylsulphate.

specificities were assigned to two separate series corresponding to two closely linked loci called A (formerly LA) and B (formerly 4) each with multiple alleles. Recent studies, especially on the H-2 system, the mouse analogue of HLA, as well as in man have shown that the genetic region identified by these two loci also contains many other genes controlling cell surface determinants, immune response differences, some components of the complement system and perhaps other related functions connected in general with cell-cell recognition (Shreffler and David, 1975; Kissmeyer Nielsen, 1975). The original impetus behind the analysis of the HLA system was its use for matching for transplantation; this has turned out to be more complex than was at one time hoped. The demonstration of genetic linkage between the murine major histocompatibility complex H-2 and resistance to virus induced leukemogenesis (Lilly *et al.*, 1964) and specific immune responses (Benacerraf and McDevitt, 1972) has however stimulated an extensive search for association between antigens of the HLA system and specific diseases. Though early data showed only weak associations with Hodgkins and some other diseases, subsequent studies have shown very striking associations between antigens especially of the HLA-B locus and a number of diseases having a presumptive or suspected autoimmune etiology including for example, ankylosing spondylitis, coeliac disease, myaesthemia gravis and multiple sclerosis. In the case of ankylosing spondylitis the association is so strong that it is already of some possible diagnostic value (McDevitt and Bodmer, 1974; Transplantation Reviews, 1975).

These associations have been interpreted in terms of the effects of linked immune response genes in the HLA region rather than direct effects of the antigens of the HLA-B or A loci themselves. It has thus already been established that mixed lymphocyte culture determinants controlled by the HLA-D locus show more significant association with multiple sclerosis than do antigens of the HLA-A or B loci. This emphasizes the importance of developing direct methods for identifying genes of the HLA region which may be more directly involved in disease association.

The complete characterisation of gene products and functions depends ultimately on a detailed analysis of their chemical structure. Much progress has recently been made in the elucidation of the chemical structure of the HLA and H-2 serological determinants and the aim of this chapter is to review some of our own studies in this direction. Among the gene products recently identified in the HLA and H-2 region are a new set of serological specificities called Ia for immune associated. Unlike the H-2D and H-2K, or HLA-A or B antigens the Ia antigens have a tissue distribution that is relatively specific, including especially B but excluding T lymphocytes. (Shreffler and David 1975; Jones *et al.*, 1975). These antigens seem to be closely associated with mixed lymphocyte culture determinants of the HLA-D locus and promise to become a powerful tool in disease association and related studies (Kissmeyer Nielsen, 1975). There is in addition the definite possibility that the Ia specificities will be of more direct importance for the problem of clinical transplantation than the specificities of the original defined HLA-A and B loci. The chemical characterisation of these products which is an obvious concomitant of the work on the HLA-A and B locus antigens, may therefore turn out to be of considerable interest. Apart from fundamental advances which follow from a further understanding of the chemical nature of these products, it seems most probably that valuable antisera for HLA typing may be produced by immunisation with the purified antigens.

II. SOURCE OF HLA

It is preferable that the tissue used for the isolation of HLA antigens should satisfy the following requirements: (1) it should be readily available; (2) it should contain large numbers of cells with a high content of HLA antigens; (3) it should express selected HLA specificities; and (4) it should be fresh (i.e., not autopsy material) in order to reduce possible degradation. The readily available tissues such as red blood cells (Doughty *et al.*, 1973) and placenta (Goodfellow *et al.*, 1977) contain low levels of HLA antigens only. Tissues such as liver, spleen, white cells and leukemic cells have higher levels (Amos and Ward, 1975) but, because of the polymorphism of HLA antigens, are basically unsuitable except for the isolation of small amounts of antigen. In contrast, lymphoblastoid cells grown in culture express high levels of HLA antigens on their surface, and represent a reproducible antigen source. They can be grown in suspension to relatively high densities (2 to 4 × 10^6 cells/ml) and, given mass culture facilities, in unlimited numbers. Lymphoblastoid cell-lines also offer the opportunity of choosing particular HLA specificities and of developing homozygous cell-lines to promote the separation of allogeneic activities as well as doubling the antigen yield. The studies on HLA antigens described here were carried out using the lymphoblastoid cell-line BRI8 (HLA-A1, A2, B8, B13, 4a and 4b).

III. ISOLATION OF HLA

HLA antigens are intimately associated with the cell surface (plasma) membrane (Wilson and Amos 1972; Snary *et al.*, 1974) and are insoluble in conventional aqueous solvents. Two approaches to solubilisation have been used namely, partial degradation and detergent extraction. The former method using proteolysis with papain (Sanderson 1969; Cresswell *et al.*, 1973) or 3M KCl extraction

(Reisfeld *et al.*, 1971) yields antigenically active but degraded products. Although such materials are suitable for examination of the molecular basis of HLA specificity, they provide no information on the antigen's mode of integration into the cell membrane. Fragments of molecules are also generally less immunogenic than the whole molecules and may induce a different spectrum of immune responses (for example, humoral *versus* cellular immunity). In contrast, detergents solubilise the complete molecule.

Two approaches to detergent solubilization are available. Firstly, direct solubilisation of the surface membrane of whole cells by non-ionic detergents such as Nonidet P-40 and Triton X-100 which fail to solubilise the nuclear membrane (Schwartz & Nathenson 1971) and secondly, detergent solubilization of the isolated cell surface membrane (Snary *et al.*, 1974; Springer *et al.*, 1974). The direct approach gives good solubilisation of HLA antigens without significant losses in HLA activity although the concomitant release of proteolytic enzymes from the cell (Grayzel *et al.*, 1975) may lead to degradation. Detergent solubilization of a crude cell membrane preparation has the same associated problems, whereas a purified plasma membrane preparation greatly reduces the possibility of enzymic degradation, and has the added advantage that it gives a significant degree of purification.

HLA antigens are known to be glycoproteins (Sanderson *et al.*, 1971; Snary *et al.*, 1974) and advantage can be taken of this fact during fractionation. Thus, lectins which bind carbohydrate residues can be used to separate detergent solubilised membrane glycoproteins from the non-glycosylated proteins (Allan *et al.*, 1972; Hayman and Crumpton, 1972).

A. Plasma Membrane Preparation

BR18 cells were broken in a mechanical pump by shear (Stansted Fluid Power Ltd., Stansted, Essex, CM24 8HT, U.K.). The cells were pumped past a ball bearing held in a small orifice by a spring the tension on which was adjusted so that the plasma membrane was stripped from the cell (Crumpton and Snary, 1974). The plasma membrane was separated from the cell homogenate by differential centrifugation and purified by discontinuous sucrose density gradient centrifugation. The sucrose densities were selected by analysis of a continuous sucrose density gradient (Crumpton and Snary, 1974). The plasma membrane preparation was essentially free from other subcellular organelles as judged from assays of various marker enzymes. Evidence for the surface membrane localization of HLA antigens was obtained by comparing the HLA antigen distribution amongst the subcellular fractions with the distribution of marker enzymes for the plasma membrane (5'-nucleotidase and Na^+/K^+ ATPase) (Crumpton and Snary, 1974). The preferential location of HLA antigens on the cell surface is also supported by the similar distribution of HLA antigens and 5'-nucleotidase on continuous sucrose density gradient centrifugation of the microsomal fraction (Snary *et al.*, 1974). A representative example of the purification and recovery of the plasma membrane and its associated HLA antigens is shown in Table I. A relatively high recovery of plasma membrane was achieved (average of 10 experiments about 35°/o), together with a 45-fold purification of the HLA antigens.

B. Fractionation of Solubilized Membrane

The BRI8 plasma membrane (4mg of protein/ml) was solubilised in 2°/o (w/v) sodium deoxycholate-10mM tris buffer, pH 8.2. After centrifuging for 3×10^6g min greater than 86°/o of the HLA activity and 90°/o of the membrane protein were present in the supernatant fraction. Non-ionic detergents such as Triton X-100, Lubrol-PX and Nonidet P-40 were found to be much less efficient solubilising agents (less than 60°/o of the HLA activity being solubilised even at 1mg of membrane protein/ml).

The sodium deoxycholate solubilised BRI8 plasma membrane was fractionated by gel-filtration on a column of Ultragel AcA 34 (LKB Instruments Ltd.). A single peak of HLA activity was recovered

TABLE I Distribution of HLA-A2 Antigenic Activity and 5'-Nucleotidase on Fractionation of Disrupted BRI8 Cells[a]

Fraction	Protein (°/o of homogenate)	5'-Nucleotidase	HLA-A2
		(specific activity relative to homogenate)	
Homogenate	100	1.0	1.0
Nuclei and mitochondria	61.8	0.5	0.7
Cytosol	39.7	0.2	0.3
Endoplasmic reticulum	0.98	9.1	11.1
Plasma membrane	0.6	51	45

[a]Experimental details are described in Crumpton and Snary (1974) and Snary *et al.* (1974). BRI8 cells were grown by Searle Diagnostics, High Wycombe, Bucks., U.K.

which accounted for 93°/o of the HLA activity loaded onto the column (Fig. 1). The degree of purification of HLA antigens in the fraction recovered from gel-filtration was 135-fold relative to whole cells (Table II).

The recovered material was further fractionated by affinity chromatography in 0.5°/o sodium deoxycholate on a column of *Lens culinaris* lectin attached to Sepharose 4B. This lectin binds the majority of lymphocyte plasma membrane glycoproteins (Hayman and Crumpton, 1972) including HLA antigens (Snary *et al.*, 1974). The bound glycoproteins were eluted with methyl-α-D-mannopyranoside (Fig. 2). The glycoprotein fraction accounted for 19°/o of the HLA activity of the original cells. The degree of purification was 1240-fold relative to the whole cells (Table II).

Since HLA antigens contain β_2-microglobulin (Grey *et al.*, 1973; Solheim and Thorsby, 1974) they can be selectively separated from other components by using immunoadsorbents against β_2-microglobulin. The above glycoprotein fraction was passed down a column containing the immunoglobulin fraction of a rabbit anti-

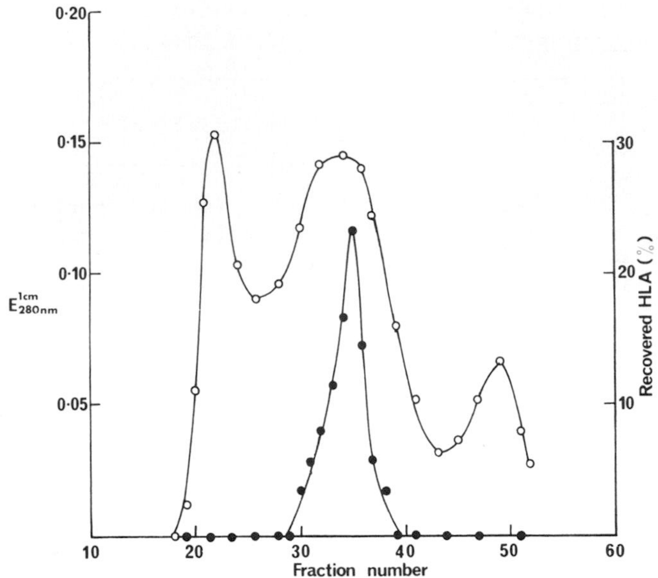

FIGURE 1 *Fractionation of BRI8 plasma membrane solubilised in sodium deoxycholate by gel-filtration on Ultragel AcA 34. The column (91 × 1.6cm) was eluted with 0.5°/o sodium deoxycholate at a flow-rate of 4.9ml/h. Fractions (2.45ml) were assayed for protein (○; E^{1cm}_{280nm}) and HLA-A1 activity (●).*

serum to human β_2-microglobulin attached to Sepharose 4B. The bound HLA antigens were subsequently eluted with 3M potassium thiocyanate in 0.5°/o sodium deoxycholate.

C. Characterisation of HLA Antigen Fractions

Figure 3 shows the polypeptide compositions of the above HLA antigen fractions revealed by polyacrylamide gel electrophoresis in SDS after reduction with dithiothreitol. The material which was adsorbed by and eluted from the anti-β_2-microglobulin column, comprised primarily two polypeptide chains of molecular weights 43 000 and 12 000. The larger chain most probably corresponds to the glycosylated polypeptide of HLA antigens that carries the polymorphic allogeneic specificities, whereas the smaller chain

TABLE II Purification of HLA Antigens from BRI8 Cells[a]

	Protein (°/o)	HLA (°/o)	Degree of purification of HLA
BRI8 cells	100	100	1
Plasma membrane	0.6	27	45
Na deoxycholate solubilised membrane	0.56	22.8	41
Gel-filtration[b] fraction	0.156	21.2	136
Glycoprotein fraction[c]	0.0153	19.1	1240

[a]Results are expressed relative to the initial BRI8 cells.
[b]See Fig. 1.
[c]See Fig. 2.

represents β_2-microglobulin (Tanigaki and Pressman, 1974). The minor bands are probably due to elution of some immunoglobulin from the column. The glycoprotein fraction possessed both of these polypeptide chains together with polypeptides of molecular weights 39 000, 33 000 and 28 000. The latter polypeptides are of unknown origin and function, although the Ia antigens expressed on the surfaces of B-lymphocytes and lymphoblastoid cell-lines (Jones *et al.*, 1975) are known to co-fractionate with the HLA antigens on gel-filtration and by affinity chromatography on *Lens culinaris* lectin-Sepharose (Snary and Goodfellow, unpublished observations). Furthermore, mouse Ia antigens have been shown to comprise two or three polypeptide chains with slightly different molecular weights of about 30 000 (Sachs *et al.*, 1975). These results suggest that the polypeptide chains of molecular weight 28 000 and 33 000, and possibly 39 000, (Fig. 3) could be the human counterparts of the mouse Ia antigens.

The above preparations of HLA antigens isolated from BRI8 cells expressed four private specificities (A1, A2, B8 and B13).

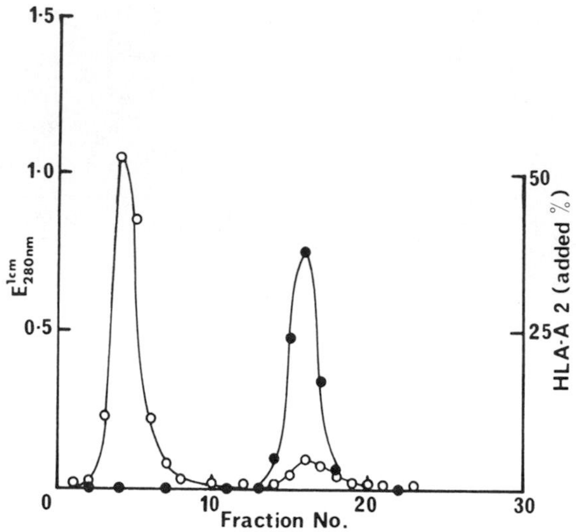

FIGURE 2 *Separation of glycoprotein by affinity chromatography on Lens culinaris lectin-Sepharose. The HLA antigen fraction from gel-filtration (Fig. 1) was added to a column (6 × 10cm) of Lens culinaris lectin attached to Sepharose 4B (1mg of lectin/ml of Sepharose) in 0.5^{o}/o sodium deoxycholate. The adsorbed glycoproteins were eluted with 2°/o (w/v) methyl-α-D-mannopyranoside in 0.5^{o}/o sodium deoxycholate (commencing at fraction no. 12). The fractions were assayed for protein (○; E^{1cm}_{280nm}) and HLA-A2 activity (●).*

Some separation of these specificities has been achieved by isoelectric focusing in 1°/o Triton X-100. Figure 4 shows that the HLA-A2 antigen (isoelectric point, pH 6.2) separates partly from the other three HLA types (isoelectric point, pH 5.5). However, no separation of the HLA-A1, -B8 and -B13 antigens or of the Ia antigens was detected. The problem of the separation of different HLA specificities may be in part circumvented by careful selection of homozygous cell-lines with HLA antigens of different isoelectric points.

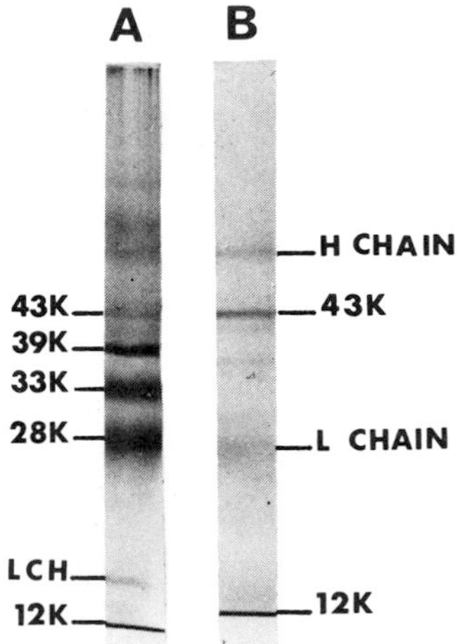

FIGURE 3 *Polyacrylamide gel electrophoresis patterns of HLA antigen fractions in SDS. Samples were reduced with dithiothreitol and run on 10°/o (w/v) gels using the tris-glycine buffered system of Laemmli (1970). The gels were stained with Coomassie blue. Sample A, the glycoprotein fraction recovered in Fig. 2. Sample B, the glycoprotein fraction was added to a column (10 × 1.2cm) of rabbit anti-human β_2-microglobulin - Sepharose 4B (15mg of immunoglobulin/ml) and the adsorbed protein was eluted with 3M potassium thiocyanate in 0.5°/o sodium deoxycholate. Molecular weights were calculated from the mobilities of transferrin, bovine serum albumin, ovalbumin, immunoglobulin L chain, soybean trypsin inhibitor and cytochrome c.*

Abbreviations: LcH, Lens culinaris lectin; L- and H-chain, immunoglobulin G L- and H-chains.

IV. THE MOLECULAR STRUCTURE OF HLA ANTIGENS

Recently, it has been proposed that HLA antigens have an immunoglobulin-like structure in which two basic units of one each of the polypeptide chains of molecular weight 43 000 and 12 000 are linked either noncovalently (Rask *et al.*, 1974) or via a disulfide bridge (Strominger *et al.*, 1974; Cresswell & Dawson, 1975). A considerable amount of information on the molecular structure of HLA antigens can be obtained directly even though they are not available in a pure form. Thus, biological activity can be used to measure their location on gel-filtration and su-

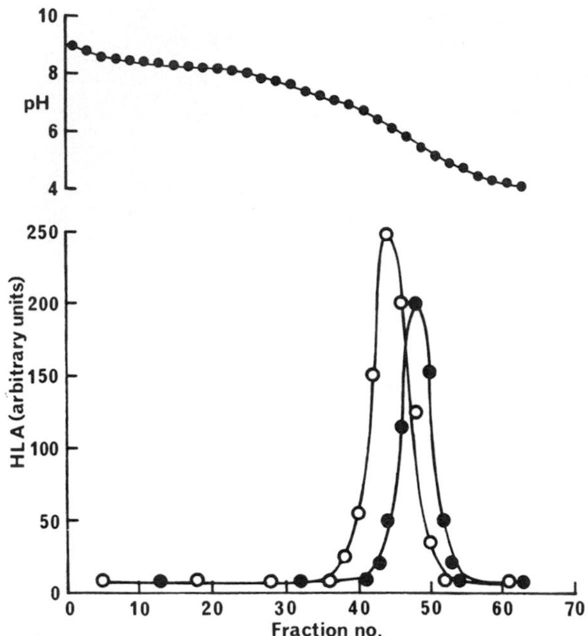

FIGURE 4 *Distribution of HLA antigenic activities on isoelectric focusing. BRI8 plasma membrane solubilised in 1°/o Triton X-100 was fractionated on a pH 3 to 10 gradient in a column (55 × 1cm) containing a sucrose gradient from 5 to 60°/o (w/v). The electrode buffers were 1°/o ascorbic acid in 60°/o sucrose, and 1°/o ethanolamine. Fractions were assayed for HLA-A1 (●) and HLA-A2 (○) activities.*

crose density gradients of deoxycholate-solubilised plasma membrane, and their size and shape can be calculated from these measurements (Siegel and Monty, 1966).

Gel-filtration of sodium deoxycholate-solubilised BRI8 plasma membrane showed (Fig. 1) a single peak of HLA-A1 activity with no evidence of aggregates or gross polydispersity. HLA activities of the A and the B series and the HLA antigens of the plasma membrane fraction from a human spleen occupied identical positions with that shown for HLA-A1. The Stokes radius of the detergen-solubilised HLA molecule (Table III) was calculated (Siegel and Monty, 1966) using the calibration curve shown in Fig. 5. The

TABLE III Molecular Nature of HLA Antigens in Sodium Deoxycholate

HLA antigens	Antigen series	$s_{20,w}$	Stokes radius (Å)	Molecular weight	f/fo
Sodium deoxycholate, solubilised	A	5.15	44.0	88 000	1.49
	B	4.55	44.0	78 000	1.55
Papain, solubilised	A } B	3.68	29.8	46 000	1.26

sedimentation coefficients of the HLA antigens were obtained by analysis of deoxycholate-solubilised BRI8 membrane on sucrose density gradients (Fig. 6). The A and B series antigens possessed slightly different $S_{20,w}$ values (Table III). Molecular weights calculated for the A and B series antigens from the Stokes radii and $S_{20,w}$ values were 88 000 and 78 000 respectively. Provided dissociation had not occurred, these molecular weights are too low to accommodate a four chain, immunoglobulin-like dimer of molecular weight 110 000 (based on the β_2-microglobulin and HLA polypeptide chains having sizes of 12 000 and 43 000 respectively as determined by polyacrylamide gel electrophoresis in SDS). Dissociation was, however, considered unlikely for two reasons. Firstly, so far as can be determined the HLA antigens had not been subject to reducing conditions during the isolation, solubilisation and fractionation of the BRI8 plasma membrane. Secondly, sodium deoxycholate failed to promote the dissociation of non-polar interactions between subunits in multimeric proteins such as hemoglobin (Snary *et al.*, 1975). The most likely explanation for the difference between the calculated values for the molecular weights of the deoxycholate-solubilised antigens and those estimated by polyacrylamide gel electrophoresis in SDS (minimum molecular weight 55 000) is that HLA antigens resemble other inte-

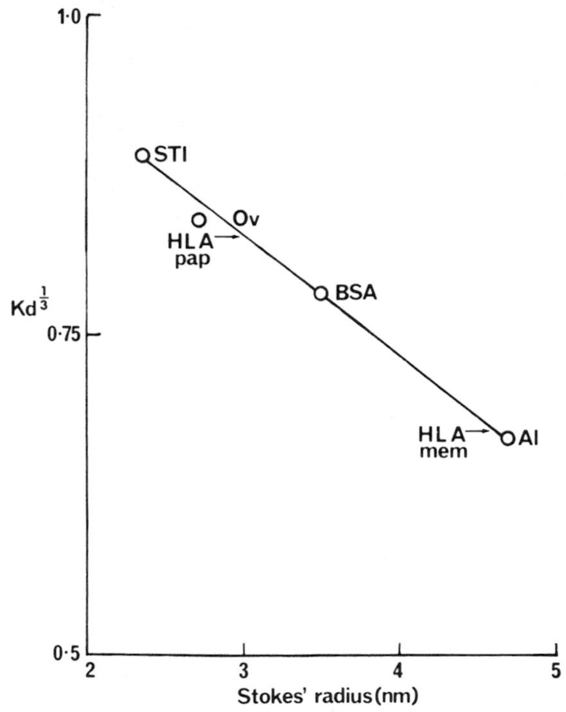

FIGURE 5 *Relationship between elution positions ($K_d^{1/3}$) of standard proteins from Ultragel AcA 34 in 0.5°/o sodium deoxycholate and Stokes radii. $K_d^{1/3}$ was calculated as described by Siegel and Monty (1966). The Stokes radii were those reported by Andrews (1970) except for soybean trypsin inhibitor which was calculated from the data of Birk et al. (1963). Al, aldolase; BSA, bovine serum albumin; Ov, ovalbumin; STI, soybean trypsin inhibitor. The arrows indicate the measured $K_d^{1/3}$ for the whole HLA antigens (mem) and for the papain solubilised HLA antigens (pap).*

gral membrane proteins, such as cytochrome b_5 (Spatz and Strittmatter, 1973), in possessing a hydrophobic domain which binds detergent (Helenius and Simons, 1975). As a result the deoxycholate-solubilised HLA antigens will possess anomalously high molecular weights when assessed relative to water-soluble globular

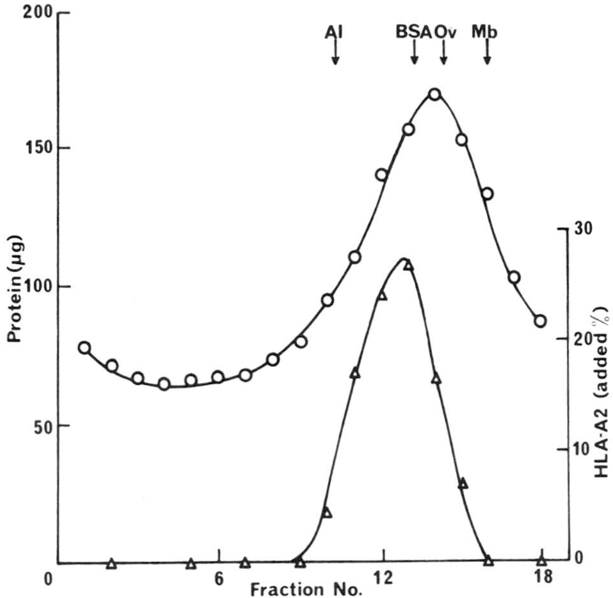

FIGURE 6 *Sucrose density gradient centrifugation of BRI8 plasma membrane solubilised in 1°/o sodium deoxycholate. Solubilised membrane (1ml) was layered onto a linear sucrose gradient (9 to 28°/o sucrose) containing 1°/o sodium deoxycholate. Marker proteins: aldolase (Al) 7.5S; bovine serum albumin (BSA) 4.41S; ovalbumin (Ov) 3.55S; sperm whale myoglobin (Mb) 2.08S were added to identical gradients. The gradients were centrifuged at 4°C in a Beckman SW 41 rotor for 48h at 38 000 r.p.m. Fractions (0.67 ml) were pumped from the bottom of the gradients and were assayed for protein (0) and HLA-A2 activity (Δ).*

proteins. In contrast, examination of papain-solubilised HLA antigens by gel-filtration and sucrose density gradient centrifugation in sodium deoxycholate gave a molecular weight of 46 000 (Table III) which is in close agreement with the value found by polyacrylamide gel electrophoresis in SDS (Turner et al., 1975). This result implies that the hydrophobic domain which is inserted in the lipid bilayer and which binds detergent is cleaved by papain (Fig. 7), whereas the antigenically-active region binds little or no detergent. An analogous situation has been previously reported for cytochrome b_5 (Spatz and Strittmatter, 1973). De-

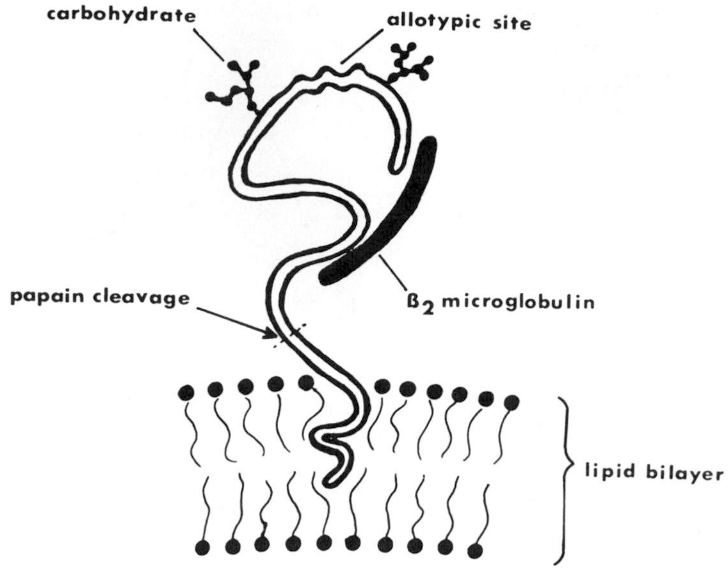

FIGURE 7 *Model of membrane-bound HLA antigen.*

tergent binding is also consistent with the higher molecular weights reported for HLA and H2 antigens in detergents of higher micelle size than sodium deoxycholate (460 000 and 380 000 for Brij 99 and Nonidet P40 respectively; Springer *et al.*, 1974 and Schwartz *et al.*, 1973), since these probably express the different micelle sizes of the bound detergent. Estimates of the amount of deoxycholate bound by the HLA antigens gave 0.57 and 0.39g of deoxycholate/g of protein for the A and B series antigens respectively. These values are in accord with those determined for membrane proteins (within the range 0.3-0.7g of deoxycholate/g of protein; Helenius and Simons, 1975 and Snary *et al.*, 1975). The difference between the amounts of deoxycholate bound by the A and B series antigens probably arises from a difference in hydrophobicity, in which case it may indicate that the A series antigens are inserted deeper into the lipid bilayer than B series antigens.

In contrast to previous work (Strominger *et al.*, 1974; Cresswell and Dawson, 1975), no evidence was obtained for the presence of dimers. The reason for this difference is not known. However, one possible explanation is that the disulfide-bonded dimers found by the other workers arose by disulfide-interchange. Thus, the gel-filtration pattern of dimers and larger aggregates reported by Cresswell and Dawson (1975) for deoxycholate-solubilised HLA antigens resembles that given by an aged sample of bovine serum albumin which had undergone aggregation through disulfide-interchange (Hellsing, 1969). The crude membrane preparations used by Strominger *et al.* (1974) and Cresswell and Dawson (1975) most probably contain relatively large amounts of endoplasmic reticulum which is rich in an enzyme catalysing disulfide-interchange (de Lorenzo *et al.*, 1966). If the above interpretations are valid, then the 4-chain, dimeric, immunoglobulin-like model that has been proposed for membrane-bound HLA antigens is unlikely to be correct.

V. SUMMARY

Cultured lymphoblastoid BRI8 cells provide a very convenient source of HLA antigens. The antigens were isolated from a purified preparation of the cell surface membrane by solubilisation in sodium deoxycholate, gel-filtration and affinity chromatography on *Lens culinaris* lectin attached to Sepharose 4B. The product contained primarily HLA and IA antigens which were separated by using an immunoadsorbent against β_2- microglobulin.

HLA antigens are composed of a glycosylated polypeptide chain of molecular weight 43 000 that is non-covalently bonded to β_2-microglobulin (mol. wt. 12 000). Estimates of the molecular weights of deoxycholate-solubilised HLA antigens by sucrose density gradient centrifugation and gel filtration gave values of 88 000 and 78 000 for the A and B series antigens respectively. These values most probably reflect the binding of deoxycholate, presumably to a hydrophobic domain which is normally inserted into the lipid bilayer of the membrane and which is removed by papain (Fig. 7). No evidence for the presence of dimers was obtained. It was concluded that the present results cast doubt upon the validity of the dimeric, immunoglobulin-like, model for HLA antigens. On the other hand, the absence of a dimeric, immunoglobulin-like structure does not rule out the possibility of a common evolutionary origin for HLA antigens and immunoglobulin.

ACKNOWLEDGMENTS

We are grateful to Dr. A. R. Sanderson for a gift of papain solubilised HLA. The work was suported in part by a grant from the Medical Research Council to The Genetics Laboratory, Oxford. P. Goodfellow is the recipient of a Research Scholarship from the Medical Research Council.

REFERENCES

Allan, D., Auger, J., Crumpton, M. J., (1972), *Nature New Biol.* *236*, 23-25.

Amos, D. B., Ward, F. E., (1975), *Physiological Rev.* *55*, 206-246.

Andrews, P., (1970), *Meth. Biochem. Anal.* *18*, 1-53.

Benacerraf, B., McDevitt, H. O., (1972), *Science* *175*, 273-279.

Birk, Y., Gertler, A., Khalef, S., (1963), *Biochem. J.* *87*, 281-284.

Cresswell, P., Turner, M. J., Strominger, J. L., (1973), *Proc. Nat. Acad. Sci. U.S.A.* *70*, 1603-1607.

Cresswell, P., Dawson, J. R., (1975), *J. Immunol.* *114*, 523-525.

Crumpton, M. J., Snary, D., (1974), *in* "Contemporary Topics in Molecular Immunology," (Ada, G. L. ed.), Vol. III, pp. 27-56, Plenum Press, New York.

de Lorenzo, F., Goldberg, R. F., Steers, E., Givol, D., Anfinsen, C. B., (1966), *J. Biol. Chem. 241,* 1562-1567.

Doughty, R. W., Goodier, S. R., Gelsthorpe, K., (1973), *Tissue Antigens 3,* 189-194.

Goodfellow, P. N., Barnstable, C. J., Bodmer, W. F., Snary, D., Crumpton, M. J., (1977), *Transplantation,* in press.

Grayzel, A. I., Hatcher, V. B., Lazarus, G. S., (1975), *Cell. Immunol. 18,* 210-219.

Grey, H. M., Kubo, R. T., Colon, S. M., Poulik, M. D., Cresswell, P., Springer, T., Turner, M., Strominger, J. L., (1973), *J. Exptl. Med. 138,* 1608-1612.

Hayman, M. J., Crumpton, M. J., (1972), *Biochem. Biophys. Res. Comm. 47,* 923-930.

Helenius, A., Simons, K., (1975), *Biochem. Biophys. Acta 415,* 29-80.

Hellsing, K., (1969), *Biochem. J. 114,* 145-149.

Jersild, C., Dupont, B., Fog, T., Hansen, G. S., Nielsen, L. S., Thomsen, M., Svejgaard, A., (1973), *Transplant. Proc. 5,* 1791-1798.

Jones, E. A., Goodfellow, P., Bodmer, J. G., Bodmer, W. F., (1975), *Nature 256,* 650-652.

Kissmeyer Nielsen, F., (ed), (1975), "Histocompatibility Testing 1975," Munksgaard, Copenhagen.

Laemmli, U. K., (1970), *Nature 227,* 680-685.

Lilly, F., Boyse, E. A., Old, L. J., (1964), *Lancet ii,* 1207-1209.

McDevitt, H. O., Bodmer, W. F., (1974), *Lancet i,* 1269-1275.

Rask, L., Ostberg, L., Lindblom, B., Fernstedt, Y., Peterson, P. A., (1974), *Transplant. Rev. 21,* 85-105.

Reisfeld, R. A., Pellegrino, M. A., Kahan, B. D., (1971), *Science 172,* 1134-1136.

Sachs, D. H., Cullen, S. E., David, C. S., (1975), *Transplant, 19,* 388-393.

Sanderson, A. R., (1969), *Nature 220,* 192-194.

Sanderson, A. R., Cresswell, P., Welsh, K. I., (1971), *Nature New Biol. 230,* 8-12.

Schwartz, B. D., Nathenson, S. G., (1971), *J. Immunol. 107,* 1363-1367.

Schwartz, B. D., Kato, K., Cullen, S. E., Nathenson, S. G., (1973), *Biochemistry 12,* 2157-2164.

Shreffler, D. C., David, C. S., (1975), *in* "Adv. in Immunol," (Dixon, F. J., Kunkel, H. G., eds.), Vol. XX, pp. 125-195, Academic Press, New York.

Siegel, L. M., Monty, K. J., (1966), *Biochem. Biophys. Acta 112,* 346-362.

Snary, D., Goodfellow, P., Hayman, M. J., Bodmer, W. F., Crumpton, M. J., (1974), *Nature 247,* 457-461.

Snary, D., Goodfellow, P., Bodmer, W. F., Crumpton, M. J., (1975), *Nature 258,* 240-242.

Solheim, B. G., Thorsby, E., (1974), *Tissue Antigens 4,* 83-94.

Spatz, L, Strittmatter, P., (1973), *J. Biol. Chem. 248,* 793-799.

Springer, T. A., Strominger, J. L., Mann, D. L., (1974), *Proc. Nat. Acad. Sci. U.S.A. 71,* 1539-1543.

Strominger, J. L., Cresswell, P., Grey, H., Humphreys, R. E., Mann, D., McCune, J., Parham, P., Robb, R., Sanderson, A. R., Springer, T. A., Terhorst, C., Turner, M. J., (1974), *Transplant. Rev. 21,* 126-143.

Tanigaki, N., Pressman, D., (1974), *Transplant. Rev. 21,* 15-34.

Transplantation Reviews (1975), (Moller, G., ed.), Vol. XXII, Munksgaard, Copenhagen.

Turner, M. J., Cresswell, P., Parham, P., Strominger, J. L., Mann, D. L., Sanderson, A. R., (1975), *J. Biol. Chem. 250,* 4512-4519.

Wilson, L. A., Amos, D. B., (1972), *Tissue Antigens 2,* 105-111.

NEW APPROACHES TO HUMAN LYSOSOMAL STORAGE DISEASES USING ANIMAL CELLS AND ANIMAL VIRUSES

WILLIAM S. SLY

The Edward Mallinckrodt Department of Pediatrics
Washington University School of Medicine
The Division of Medical Genetics
St. Louis Children's Hospital
St. Louis, Missouri

I. INTRODUCTION

In the early 1960s, the dramatic successes of molecular genetics and microbiology made it fashionable to prophesy that what was true for *E. coli* would be true for elephants, and for everything in between. The implication was that there was little need to bother with research on more complicated organisms for, when the *E. coli* workers were finished, they would have said it all. It hasn't turned out that way. To everyone's delight, both the human fibroblast and the human lymphocyte have proved extraordinarily useful experimental tools for the study of human disease. I would like to cite a few examples which we now take largely for granted. The short term lymphocyte culture was a relatively simple cell culture development. Yet, this simple cell culture development made human chromosome analysis possible. This in turn has had a profound effect on diagnosis and counseling in patients with suspected chromosome imbalance. It also paved the way for the elegant developments in human cytogenetics that have proved so important to human chromosome mapping by somatic cell hybridization. Its extension to the mixed lymphocyte culture has provided a major tool in immunogenetic studies which has also had practical implications for human organ transplantation. Likewise, the diploid human fibroblast system has become an important tool with profound effects on the delineation of the basic defects in a steadily growing list of enzyme deficiency diseases. That there is still great potential for this simple experimental system was recently demonstrated by Brown and Goldstein (1974) in their elegant studies defining an altered membrane receptor for low density lipoprotein in familial hypercholesterolemia. The fibroblast culture system has contributed to understanding of DNA repair defects (Setlow *et al.*, 1969). It has also allowed the demonstration of the validity of the single active X hypothesis in a number of X-linked hereditary diseases (Krooth *et al.*, 1968). It provided the foundation for prenatal diagnosis of chromosome imbalances and biochemical defects in high-risk pregnancies (Nadler, 1972). It contributed substantially to somatic cell hybridization studies for complementation analysis, linkage studies, and studies mapping genes on the human chromosomes (Ruddle and Kucherlapati, 1974). Finally, it has provided an *in vitro* model system for experimental treatments aimed at correction of a number of metabolic diseases.

Clearly, cell culture has had a tremendous role in many exciting developments in medicine through the past 15 years. I would like to focus on one small area of this excitement, the area of lysosomal storage diseases, which I have found tremendously stimulating. I would like to briefly review the considerable progress that has been made in this area and then mention two departures from the traditional approaches with cell culture which have proved so powerful in approaching human diseases.

II. LYSOSOMAL STORAGE DISEASES

The lysosome is a mammalian cellular organelle which serves a disposal function. It does so through some 30 to 50 enzymes which participate in the sequential degradation of macromolecules which the cell either makes or ingests, but which must be degraded to small molecules by the lysosomal apparatus to be eliminated from the cell. Deficiency of one of these enzymes leads to intracellular accumulation of undigested material (Neufeld *et al.*, 1975).

The concept of a lysosomal storage disease was elaborated by Hers (1965) on the basis of one carefully studied example. Since then, a large number of diseases have been added to the list, each characterized by intracellular storage and, with one class of exceptions, each due to a deficiency of a single degradative enzyme. In crude terms, the lysosomal apparatus is the cell's garbage disposal. An enzyme deficiency disrupts the catabolic process involved in the disposal of macromolecules leading to progressive constipation of lysosomes with undigestible material. The chemistry of the macromolecular garbage depends on which enzyme is missing. The cell types affected in the organism and the clinical results depend both on the degree of enzyme deficiency, and on the importance of that particular degradative enzyme in a given differentiated cell type. The general classes of lysosomal storage diseases so far described include the mucopolysaccharidoses, the sphingolipidoses, disorders of glycoprotein degradation, and also single enzyme deficiencies involved in the degradation of glycogen, cholesterol esters, and phosphate esters (Neufeld *et al.*, 1975).

A. Mucopolysaccharide Storage Diseases

I will focus on the mucopolysaccharide storage diseases because they illustrate the dramatic role that cell culture can play in the elucidation of mechanisms in human biochemical diseases. Only 7 years ago, it was known that there were a group of 6 mucopolysaccharide storage disorders. Although they could be distinguished on the basis of clinical manifestations, mode of inheritance, or the nature of the urinary excretion products, nothing was known of the biochemical basis for this group of diseases. Neufeld and her co-workers (Fratantoni *et al.*, 1968a) blew this field wide open with the demonstration that cultured fibroblasts from these patients accumulate exaggerated amounts of ^{35}-SO_4 into intracellular mucopolysaccharides. Even more dramatic was the observation that this abnormal accumulation could be "corrected" by adding to the medium secretions of cells or concentrates of urine, provided that the donor of the cells or the urine was not afflicted with the same disorder as the cells to be corrected (Fratantoni *et al.*, 1968b; Fratantoni *et al.*, 1969). These observations provided definitive evidence that this group of diseases resulted from a degradative defect in mucopolysaccharide metabolism, ending the life of the alternate hypothesis that exaggerated biosynthesis of mucopolysaccharides was responsible for these disorders. These studies proved as practical as they were important.

Immediately, this *in vitro* cross-correction complementation assay was used to classify the mucopolysaccharide storage diseases into complementation classes. The Hurler and Scheie syndrome cells proved not to complement one another, suggesting that they were deficiencies for the same enzyme (Barton and Neufeld, 1971). The Sanfilippo syndrome was found to contain two complementary types, suggesting two different enzyme deficiencies which could not be distinguished clinically (Kresse *et al.*, 1971). Even before any information was available on the enzyme defects, the *in vitro* complementation assays became clinically relevant by allowing one to distinguish Hurler syndrome patients from Hunter syndrome patients. Thus, one could give precise genetic counseling to patients with these two disorders which differ in their mode of inheritance, but often cannot be distinguished clinically. However, the major use of the cell culture assay for corrective factors was its use to guide the purification of individual corrective factors, each later to be identified as an enzyme involved in mucopolysaccharide degradation (Neufeld, 1974).

Thus, the corrective factor for the Hurler and the Scheie syndromes was identified as α-iduronidase. The Hunter corrective factor proved to be iduronate sulfatase. The Sanfilippo A corrective factor is heparin-N-sulfatase. The Sanfilippo B corrective factor is N-acetyl-α-glucosaminidase. Clearly, the unfolding of the biochemical mechanism underlying this group of disorders represents a cell culture success story. In many cases, the corrective factors were highly purified initially using the cell culture assay for correction, and their enzymatic activity established later by analysis of the hydrolytic activity of the purified factor with either natural or artifical substrates. Once the hydrolytic activities for all of these factors had been identified, a rational explanation for the particular accumulated products in a given mucopolysaccharidosis became apparent. For example, iduronidase and iduronate sulfatase were both implicated in the breakdown of dermatan sulfate and heparin sulfate and a deficiency of either enzyme leads to accumulation of both storage products. On the other hand, the two different Sanfilippo factors are involved in the degradation of heparin sulfate, but not in the degradation of dermatan sulfate, and one finds as one might expect the predominant accumulation product to be heparin sulfate (Neufeld, 1974; Neufeld *et al.*, 1975).

B. β-Glucuronidase Deficiency Mucopolysaccharidosis

Although we contributed in a small way to this exciting story, and a number of other laboratories also made significant contributions, most of these contributions were logical extensions of the highly original work of Neufeld and her co-workers. Our contribution involved the discovery of the clinical entity β-glucuronidase deficiency mucopolysaccharidosis (Sly *et al.*, 1973) and subsequent studies with skin fibroblasts from this patient (Sly *et al.*, 1974; Brot *et al.*, 1974; Glaser *et al.*, 1975). The finding by Hall *et al.* (1973) that fibroblasts from this patient were corrected by added bovine β-glucuronidase stimulated us and others to consider the possibility that deficient patients could be corrected by administered enzyme. Using this *in vitro* model for enzyme correction, we and others began to study the requirements for correction of β-glucuronidase deficient fibroblasts *in vitro*. We naturally hoped that these studies would provide information that would be helpful to guide us in clinical trials with purified enzyme in the deficient patient.

The studies on fibroblasts already published provided evidence for the following general conclusions:

1. Purified human β-glucuronidase is corrective (Hall *et al.*, 1973; Lagunoff *et al.*, 1973; Sly *et al.*, 1974; Brot *et al.*, 1974).

2. Human β-glucuronidase preparations vary tremendously in corrective potency depending on the organ source from which the enzyme is purified (Brot *et al.*, 1974; Nicol *et al.*, 1974). Although enzyme from urine or plasma is corrective, enzyme from placenta was nearly 10 times as active in uptake and correction as enzyme from urine and plasma. Enzyme from blood platelets was 100 times as corrective as the same amount of catalytic activity from urine or plasma.

3. Human β-glucuronidase with high corrective potency represents a specific isoelectric fraction of the β-glucuronidase preparation, regardless of the source of origin (Glaser *et al.*, 1975).

4. Enzyme taken up by fibroblasts is quite long-lived. In fact, a half disappearance time in excess of two weeks was found (Sly *et al.*, 1973b).

C. Use of Cell Culture to Select an Animal Model for Enzyme Replacement Studies

Although a few trials with infusions of fresh frozen plasma and one with purified platelet enzyme produced limited and transient evidence of biochemical and clinical improvement (Sly *et al.*, 1973b) we became convinced that we would need an animal model to answer questions about the uptake, organ distribution, and fate of infused human β-glucuronidase which we could not answer from human studies. Among the questions we would like to answer are (1) does the enzyme defined as high-uptake and highly corrective enzyme in fibroblasts have a different uptake, fate, and distribution in the whole animal? (2) Can we cross the blood-brain barrier with infused enzyme? (3) Is the long half-life of human enzymes seen in fibroblasts also seen *in vivo* following infusion?

An ideal animal model would be an enzyme deficient animal with evidence of a storage disease, in which one could test the entire concept of enzyme replacement, including the response to therapy. We had no such model for β-glucuronidase deficiency. Yet, human β-glucuronidase is so stable, so readily purified, and so well characterized that we did not want to abandon it. Because we knew that bovine enzyme was corrective for human fibroblasts, we suspected that other animals might share the recognition system for the uptake and localization of enzymes in fibroblasts. We therefore asked whether animal fibroblasts would recognize the human enzyme and take it up like human fibroblasts. If they did, we reasoned that we might be able to use an animal to study the uptake, organ distribution and fate of infused human enzyme. However, since the animal would have its own β-glucuronidase, we would have to have a means of distinguishing the human enzyme from the endogenous animal enzyme. To approach this question, we first established fibroblasts from a number of animal species and attempted to define conditions for differential inactivation of animal β-glucuronidase (Frankel *et al.*, 1975). A simple method was devised which took advantage of the remarkable heat stability of human β-glucuronidase. We discovered that human β-glucuronidase could withstand heating up to 65°C for up to 120 minutes without significant loss of activity. By contrast, animal enzymes from a variety of species were considerably more heat labile. Using deoxycholate to solubilize the fibroblasts from the tissue culture plates, we found a single heat inactivation condition which would inactivate nearly every animal β-glucuronidase we examined without inactivating human enzyme. This then allowed us to expose the fibroblasts from all of these animal species to high-uptake human β-glucuronidase from blood platelets and to compare the animal fibroblasts to human fibroblasts for the ability to take up the human enzyme (defined as the heat-stable enzyme in the animal fibroblasts after exposure to human enzyme). These experiments demonstrated that bovine, rat and hamster fibroblasts took up β-glucuronidase 80%, 60%, and 50% as well as deficient human fibroblasts. Every animal species that we examined showed some specific uptake of the high-uptake form of human β-glucuronidase. However, bovine, hamster and rat fibroblasts were similar enough to human fibroblasts in their uptake of β-glucuronidase to lead us to study these fibroblasts further. These fibroblasts also distinguished between "high-uptake" and "low-uptake" human enzyme. The rat was chosen over the possibly slighly superior bovine model largely for reasons of space. Studies with rat fibroblasts showed that the rat fibroblasts had the same kinetics of uptake as human fibroblasts for the high-uptake enzyme, and that enzyme had the same long half-life in rat fibroblasts once it was taken up (Frankel *et al.*, 1975). For these reasons, the rat was considered a favorable model for *in vivo* studies.

Only minor modifications of the heat inactivation conditions were required to identify human enzyme in rat organs. A large number of infusions have been carried out using this animal system (Achord *et al.*, 1975). To date, we have found that the highly purified human placental enzyme (predominantly low-uptake enzyme in the human fibroblast assay system) is rapidly cleared from the rat plasma following infusion and is taken up chiefly by liver and spleen. No enzyme from this source was identified in skeletal muscle, heart muscle, brain, or kidney. The studies of the subcellular distribution of infused enzyme in liver 24 hours after infusion indicated preferential localization of enzyme in lysosomes. The half-life of infused human enzyme in rat liver was 2.6 days and in spleen was 5.8 days.

We do not yet know whether high-uptake enzyme will have a different localization in the whole animal, or a different survival time in any given organ. However, this model should give us clear-cut answers to these important questions. It should also provide a model which would allow us to ask whether any modification of the host or a modification of the enzyme, such as packaging in liposomes, will modify the uptake, the organ distribution, and the rate of disappearance of the infused β-glucuronidase in the whole animal.

Since a large family of lysosomal enzymes share the same mechanism for lysosomal enzyme recognition and uptake by fibroblasts, the observations with this single enzyme replacement model should have some general significance to the problem of lysosomal enzyme replacement. I relate this story here, even thougl it is an unfinished story, because it represents a somewhat novel departure from the traditional use of cell culture for the study of a human disease. Here we have used cell culture to select what we think is a favorable model for animal experimentation, an approach which clearly could have other applications.

III. MEMBRANE MUTANT OF MAMMALIAN CELL AFFECTS LYSOSOMAL ENZYME GLYCOSYLATION

Another general question about lysosomal enzymes which is of great interest to us is the nature of the recognition signal which accounts for their localization in lysosomes. There is a disorder called I-cell disease, or multiple lysosomal hydrolase deficiency, which is especially relevant to this question. I-cell disease is a Hurler-like storage disease associated with a striking deficiency of a whole family of lysosomal enzymes in cultured fibroblasts, and enormous excesses of these hydrolases in culture medium *in vitro,* and in body fluids of the affected patient (Wiesmann and Herschkowitz, 1974). Mucolipidosis III, or pseudo-Hurler polydystrophy, appears to be a mild form of this disease, at least in biochemical and cytological terms (Glaser *et al.*, 1974). Hickman and Neufeld (1972) demonstrated that I-cell fibroblasts take up normal enzyme normally and retain it normally once taken up. The enzyme secreted by I-cell fibroblasts, however, is not recognized and taken up by other fibroblasts. This led them to postulate that this whole family of hydrolases is normally modified by addition of a recognition component which accounts for their localization in fibroblasts. I-cell disease was thought to result from failure to assemble this recognition component. Since most of the enzymes are glycoproteins, and since gentle periodation preferentially inactivates uptake activity compared to catalytic activity, they suggested that the recognition component might be carbohydrate (Hickman *et al.*, 1974).

Because I-cell fibroblasts were unusually sensitive to freezing, we suspected that I-cell fibroblasts might have a membrane defect as well. One might have abnormal lysosomal enzymes and also abnormal membranes if both of these cellular components share one step in biosynthesis which is defective. One obvious step which lysosomal enzymes and membranes might share in their biosynthesis is a glycosylation step involved in the addition of sugars to the carbohydrate of glycoproteins. We approached this question in two steps. We first asked whether a mutation which affects glycosylation of cell plasma membranes might also affect glycosylation of lysosomal enzymes. Kornfeld and associates (Gottlieb *et al.*, 1974; Gottlieb *et al.*, 1975) had isolated a mutant Chinese hamster cell line defective in glycosylation of membrane glycoproteins as a ricin resistant mutant in cell culture. This mutant was found to be deficient in a glycoprotein N-acetylglucosaminyltransferase activity. This deficiency has a profound effect on host membrane glycoproteins. We found that this mutant also has a probable defect in lysosomal enzyme glycosylation. When lysosomal enzymes were extracted from the parent cell and passed over a ricin-Sepharose affinity column, at least three lysosomal enzymes were specifically bound to the column and eluted by 0.1 *M* lactose. In contrast, the lysosomal enzymes derived from the ricin resistant mutant showed no specific binding to the ricin-Sepharose affinity column (Sly, Gottlieb and Kornfeld-unpublished observations). These observations suggested to us that lysosomal enzymes in Chinese hamster cells share at least some glycosylation steps with host membrane glycoproteins.

IV. USE OF ANIMAL VIRUS PROBLE TO SHOW MEMBRANE DEFECT IN I-CELL DISEASE

If membrane glycoproteins and lysosomal enzymes share some glycosylation steps, we have at least one mechanism whereby I-cell disease, which is postulated to be a disorder of abnormal lysosomal enzyme glycosylation, might also be a disorder with a membrane abnormality in cultured fibroblasts. To test the hypothesis of a membrane abnormality in I-cell disease, we chose to use Sindbis virus as a membrane probe. Sindbis virus is a simple enveloped RNA virus which contains a lipid envelope derived from the host plasma membrane. It contains only three proteins. One of these proteins contains no carbohydrate and is associated with the viral nucleic acid. The two other proteins are both glycoprotein components of the virus envelope. Both are glycosylated by host enzymes (Keegstra *et al.*, 1975). Sindbis virus was grown in normal human fibroblasts and in fibroblasts derived from patients with I-cell disease and the properties of the viruses produced were compared. Although Sindbis virus grows well in normal human fibroblasts and in I-cell fibroblasts, the virus produced by

I-cell fibroblasts has two striking differences from that produced in normal fibroblasts (Sly *et al.*, 1975). These differences are: (1) A striking sensitivity to freeze-thaw. The virus from I-cell fibroblasts loses nearly a log of infectivity with each cycle of freeze-thaw. The virus from normal fibroblasts is quite stable to freezing and thawing. (2) Virus produced in I-cell fibroblasts shows a nearly hundred fold increased sensitivity to inactivation by detergents such as Triton X-100. Thus, the Sindbus virus produced in I-cell fibroblasts appears to have an envelope defect which it derives from the host plasma membrane. Passage of the abnormal Sindbis virus through mouse L cells produces normal virus in a single passage, as one would predict if the I-cell virus is only phenotypically altered by growth in the membrane-defective host (Sly *et al.*, 1976).

These observations clearly indicate that I-cell disease is a disease which involves the cell membrane as well as a disease involving the lysosomal enzymes. The freeze sensitivity and the sensitivity to detergent suggest a membrane lipid abnormality, although they could conceivably be related to a protein abnormality with secondary disturbances in lipid-protein interactions in the virus envelope.

We can think of at least two possible single genetic defects which might produce these abnormalities in lysosomal enzymes and in the I-cell membrane. One possible defect could be a defect in a host glycosylation step in I-cell fibroblasts which results in abnormal glycosylation of lysosomal enzymes and abnormal glycosylation of membrane glycoproteins and possibly glycolipids. An alternate hypothesis would be that the I-cells have a primary defect in lipid metabolism producing a disturbance in the membrane lipid bilayer which has secondary effects on certain membrane-associated enzymes. In other words, the primary defect could be a lipid membrane abnormality, of which freeze-thaw and detergent sensitivity are symptoms, and the abnormal glycoproteins (lysosomal enzymes and possibly membrane glycoproteins) might be secondary manifestations of this membrane abnormality which disturbs the function of membrane-associated glycosylation enzymes.

In either case, the viral glycoproteins and the viral lipids are much simpler to analyze than the more complicated host membrane components and the carbohydrate structure of the host lysosomal enzymes. Thus, we have hopes that analysis of the differences in the envelope lipids and glycoproteins between the viruses produced by I-cell fibroblasts and those produced by normal fibroblasts may shed light on the basic defect in I-cell disease. If this analysis does reveal the basic defect in I-cell disease, these studies will likely allow us to infer something about the normal recognition component for lysosomal enzyme that appears to be either absent or defective in I-cell disease.

Again, this represents a new cell culture approach to study of a human disese that might have other important applications.

ACKNOWLEDGMENTS

The author gratefully acknowledges the work of Drs. Daniel Achord, Frederick Brot, Janet Glaser, Kenneth Roozen and Mr. Harry Frankel in many of the studies cited in this paper, many of which have not yet appeared other than in abstract form. The collaboration of Drs. Charlene Gottlieb and Stuart Kornfeld in the Chinese hamster mutant studies is gratefully acknowledged. The collaboration of Dr. Elizabeth Lagwinska and Sondra Schlesinger in the animal virus work is gratefully acknowledged. The author is supported by grants from the U. S. Public Health Service (Training grant 5 T01 GM01511 and Research grant 1 P01 GM21096), and also by the Ranken Jordan Trust Fund for Crippling Diseases in Children.

REFERENCES

Brown, M. S. and Goldstein, J. L. (1974) *Proc. Nat. Acad. Sci. U. S.A. 71*, 788-792.

Setlow, R. B., Regan, J. B., German, J. and Carrier, W. L. (1969) *Proc. Nat. Acad. Sci. U.S.A. 64*, 1035-1041.

Krooth, R. S., Darlington, G. A. and Velazquez, A. A. (1968) The Genetics of Cultured Mammalian Cells *in* Ann. Rev. Genet. 2, 141-164.

Nadler, H. L. (1972) Prenatal Detection of Genetic Disorders *in* Advances in Human Genetics (H. Harris and K. Hirschhorn, eds.), Vol. III, pp. 1-37. Plenum Press, New York.

Ruddle, F. H. and Kucherlapati, R. S. (1974) *Scientific American 231*, 36-44.

Neufeld, E. F., Lim, T. W. and Shapiro, L. J. (1975) Ann. Rev. Biochem. *44*, 357-376.

Hers, H. G. (1965) *Gastroenterology 48*, 625-633.

Fratantoni, J. C., Hall, C. W. and Neufeld, E. F. (1968a) *Proc. Nat. Acad. Sci. U.S.A. 60*, 699-706.

Fratantoni, J. C., Hall, C. W. and Neufeld, E. F. (1968b) *Science 162*, 570-572.

Fratantoni, J. C., Hall, C. W. and Neufeld, E. F. (1969) *Proc. Nat. Acad. Sci. U.S.A. 64*, 360-366.

Barton, R. W. and Neufeld, E. F. (1971) *J. Biol. Chem. 246*, 7773-7779.

Kresse, H., Weismann, U., Cantz, M., Hall, C. W. and Neufeld, E. F. (1971) *Biochem. Biophys. Res. Commun. 42*, 892-898.

Neufeld, E. F. (1974) *in* Progress in Medical Genetics (A. G. Steinberg and A. G. Bearn, eds.), Vol. X, pp. 81-101. Grune and Stratton, New York.

Sly, W. S., Quinton, B. A., McAlister, W. H. and Rimoin, D. L. (1973) *J. Pediat. 82*, 249-257.

Sly, W. S., Brot, F. E., Glaser, J. H., Stahl, P. D., Quinton, B. A., Rimoin, D. L. and McAlister, W. H. (1974) *in* Skeletal

Dysplasias, (D. Bergsma, V. A. McKusick, J. Dorst and D. Siggers, eds.), pp. 239-245. The American Elsevier Publishing Co., New York.

Brot, F. E., Glaser, J. H., Roozen, K. J., Sly, W. S. and Stahl, P. D. (1974) *Biochem. Biophys. Res. Commun. 57,* 1-8.

Glaser, J. H., Roozen, K. J., Brot, F. E. and Sly, W. S. (1975) *Arch. Biochem. Biophys. 166,* 536-542.

Hall, C. W., Cantz, M., Neufeld, E. F. (1973) *Arch. Biochem. Biophys. 155,* 32-38.

Lagunoff, D., Nicol, D. M. and Pritzl, P. (1973) *Lab. Invest. 29,* 449-453.

Nicol, D. M., Lagunoff, D. and Pritzl, P. (1974) *Biochem. Biophys. Res. Commun. 59,* 941-946.

Sly, W. S., Brot, F. E., Chavalitdhamorong, P. and Stahl, P. D. (1973b) *Pediat. Res. 7,* 156.

Frankel, H. A., Glaser, J. H. and Sly, W. S. (1975) *Pediat. Res. 9,* 313.

Achord, D. T., Frankel, H. A., Glaser, J. H., Brot, F. E. and Sly, W. S. (1975) Enzyme Therapy: Rat Model for β-Glucuronidase Replacement Studies. Abstracts of the 27th annual meeting of the American Society of Human Genetics, Oct. 8-11, Baltimore, Maryland, p. 10a.

Weismann, U. N. and Herschkowitz, N. N. (1974) *Pediat. Res. 8,* 865-870.

Glaser, J. H., McAlister, W. H. and Sly, W. S. (1974) *J. Pediat. 85,* 192-198.

Hickman, S. and Neufeld, E. F. (1972) *Biochem. Res. Commun. 49,* 992-999.

Hickman, S., Shapiro, L. J. and Neufeld, E. F. (1974) *Biochem. Biophys. Res. Commun. 57,* 55-61.

Gottlieb, C., Skinner, A. M. and Kornfeld, S. (1974) *Proc. Nat. Acad. Sci. U.S.A. 71,* 1078-1082.

Gottlieb, C., Baenziger, J. and Kornfeld, S. (1975) *J. Biol. Chem. 250,* 3303-3309.

Keegstra, K., Sefton, B. and Burke, D. (1975) *J. Virol.* *16*, 613-620.

Sly, W. S., Lagwinska, E. and Schlesinger, S. (1975) *Clin. Res.* Vol. 23, No. 3, p. 399A, April, 1975.

Sly, W. S., Lagwinsba, E., and Schlesinger, S. (1976) *Proc. Nat. Acad. Sci.* *73*, 2443-2447.

SUMMARY REMARKS

ROLE OF CELL CULTURE IN THE STUDY OF DISEASE

JERE P. SEGREST

Departments of Pathology, Biochemistry and Microbiology
Comprehensive Cancer Center
Institute of Dental Research
University of Alabama in Birmingham Medical Center
Birmingham, Alabama

Development of techniques for the growth of mammalian cells in culture has opened up new vistas for investigations into the cause and cure of disease. Culture of an individual mammalian cell type under controlled conditions allows a more detailed comparison of normal versus diseased cells than is possible under *in vivo* conditions.

I. PROBLEMS

The development of cell culture as a technique has given rise to a number of associated "biotechnical" problems. We will briefly touch on a few of these problems here.

The problem of contamination is an ever-present worry for those involved with maintaining cells in culture. Contamination can be either microbial (viral, bacterial, mycoplasmal) or cellular (i. e. contamination with unwanted cell types, such as fibroblasts).

A second problem concerns difficulties encountered in producing sufficient cells for structural analyses. This hopefully is a purely technical problem capable of being solved by the new large capacity facilities at the University of Alabama in Birmingham and the Massachusetts Institute of Technology.

Culture of non-neoplastic mammalian cells has, in general, proven difficult. The apparent preprogrammed limit to the number of cell divisions that non-neoplastic mammalian cells in culture can undergo may be related to this problem. By contrast, most neoplastic cells act in culture as if they are immortal. Solutions to these problems may very well lead to a better understanding of basic molecular and cellular mechanisms involved in cancer and aging.

The most unique and possibly the most difficult problem associated with the use of mammalian cell cultures to study disease at the present time is the uncertainty as to whether any established cell line accurately represents the nature of the corresponding cell *in vivo*. This is an exceedingly difficult problem involving the apparent tendency for established cell lines to dedifferentiate in culture. One explanation is that standard culture conditions provide selective pressure which favors a dedifferentiated varient over the parent cell. However, this is an assumption. A solution to these difficulties requires new ideas and techniques including a better understanding of the effects of environment on cell growth. Meanwhile, use of cell culture to study disease must be tempered by the ever-present uncertainty of identity between the cell in culture and the corresponding cell *in vivo*.

II. USES

Isolation and maintenance of a specific cell type (often a cloned single cell) under controlled conditions is the key to the use of cell culture to study disease by a comparison of "normal" versus corresponding "diseased" cells.

1. Isolated cells can be examined for their intracellular and extracellular molecular products. These products, combined with structural parameters, constitute the phenotype of the cell.
2. The effects of environmental factors (chemical, physical as well as cellular) on cell growth and metabolism can be readily examined. This constitutes a study of external cell regulation.
3. The influence of internal factors on cell growth and metabolism (internal cell regulation) can also be studied. An example is the study of the cell cycle.
4. A cloned, established mammalian cell in culture provides a unique means of studying mammalian genetics.
5. The "normal" cell in culture provides a controllable means of studying the basis for cellular senescence.

In addition to these basic uses of mammalian cells in culture to study disease, cell culture is suitable for applied investigation of disease.

Attempts at cell modification to alleviate disease is one way of using applied cell culture research to study disease; modification would involve use of drugs or other substances. A recent and highly promising avenue for cellular modification is the stimulation of uptake of liposomes (phospholipid vesicles) by cultured cells. Pinocytotic uptake provides a means of replacement of defective or missing lysosomal enzymes. Uptake by liposome-membrane fusion provides a means of modifying the surface membrane of diseased cells. In addition, appropriate receptors attached to liposomes potentially can produce cell specific liposomes capable of "homing" to specific cell types (e. g. a malignant cell) thus providing a means of specifically directing drugs or other substances to cells *in vivo*.

Another applied use of mammalian cells in culture is their use as a culture media for micro-organisms. The most important application at present is the use of cell culture for viral replication for basic research or for production of viral products (e. g. vaccines) or production of viral-stimulated cell products (e. g. interferon).

III. EXAMPLES

The four papers presented in this session provide concrete examples of the use of cell culture to study disease. Dr. Benditt's paper describes one way that cell culture has been used to study atherosclerosis. Vascular and connective tissue cells in culture provide a means of isolating for study environmental factors, cellular products and genetic parameters that may play a role in atherogenesis and provide an experimental system in which to attempt cellular modification relevant to control of atherosclerotic plaque formation.

Dr. Cristofalo's paper makes use of the senescence properties of "normal" cells in culture to study the general process of aging. His system is suitable for detecting changes in product regulation with cellular senescence and for examining the genetics of aging and provides a means of studying the possibility of modification of the aging phenomenon as expressed in individual cells.

Dr. Snary's paper describes the use of cell culture to study a cell surface antigen. Cell culture provides a means of studying the genetics of this antigen, the effects of environment on its expression (and/or production), and methods for alteration of expression (such as use of liposomes).

Dr. Sly's concluding paper describes the use of cell culture to study lysosomal storage disease. For example, the genetics of the disease and the effects of product regulation are amenable to study. Further, the use of liposomes as cell modifiers provides a possible means of specific lysosomal enzyme replacement.

Plenary Sessions

MYCOPLASMA CONTAMINATION OF CELL CULTURES: A STATUS REPORT

MICHAEL F. BARILE

DHEW, Food and Drug Administration
Bureau of Biologics
Division of Bacterial Products
Mycoplasma Branch
Bethesda, Maryland

Robinson *et al.* (1956) reported the first isolation of a mycoplasma from a contaminated cell culture. Subsequently, mycoplasmas have been shown to be common contaminants capable of altering the activity of cells. Because many of the viral vaccines prepared for human use are produced in cells, we have maintained a continuing study for the past 18 years to examine the effect of mycoplasmas as contaminants in cell cultures. In this report, I will review our findings and present an updated status report on (1) the epidemiological aspects of cell culture contamination, (2) discuss the incidence, prevalence and sources, as well as the procedures recommended for prevention and elimination of mycoplasma contamination; (3) review the general aspects of mycoplasma-virus-cell interactions; (4) discuss some specific effects of mycoplasmas on cell function and virus propagation; and (5) summarize the advantages and disadvantages of procedures used for primary isolation and detection of mycoplasmas (Barile, 1973a, b; 1974; Barile *et al.*, 1973).

I. EPIDEMIOLOGICAL ASPECTS OF MYCOPLASMA CONTAMINATION

A. Incidence of Contamination

Primary cell cultures are rarely contaminated (zero to four percent), whereas, continuous, stable cell lines are frequently contaminated (57 to 92 percent, Table I). Therefore, the data shows that the original tissues used to prepare the primary cell culture are not major sources of contamination, and that most contamination occurs during cell propagation and originates from outside sources.

1. Effect of Cell Volume on Contamination

Because contamination comes from outside sources, the risk and/or frequency of contamination should be directly proportional to the volume of cells and the number of containers used. Because the laboratories that use cells for virus propagation and the commercial suppliers use or produce large volumes of cells, they should have the highest risk and, indeed, the highest incidence of contamination; and, in fact, this is precisely what our studies showed many years ago (Table II). The virologist had 76% and the suppliers had 91% contamination. In recent years, however, the suppliers have shown a marked improvement in their efforts to provide mycoplasma-free cultures to the scientific community. Nonetheless, these data clearly show that large scale production of cell cultures presents a greater risk of mycoplasma contamination, and that the large volume producer or user must be cognizant of these risks and establish vigorous quality control procedures to prevent contamination (see Prevention I.G.).

TABLE I Incidence of Mycoplasma Contamination: A Survey*

Cell cultures	Number positive/ Number tested	Percent	Investigators
PRIMARY CELLS	0/37	0.	Rothblatt and Morton, 1959
	12/>1500	0.8	Barile et al., 1962, 1963, 1968, 1972
	0/26	0	Herderschee et al., 1963
	?	3-4	Levashov and Tsilinskii, 1964
		(0-4%)	
CONTINUOUS CELLS	22/37	59	Rothblatt and Morton, 1959
	94/166	57	Pollock et al., 1960
	140/234	57	Barile, et al., 1962, 1963, 1968, 1972
	32/55	58	Herderschee et al., 1963
	45/49	92	Rakavskaya, 1965
	52/60	87	Ogata and Koshemizu, 1967
		(57-92%)	

*From: Barile, (1973)

TABLE II The Effect of Cell Usage on the Rate of Mycoplasma Contamination*

Nature of Cell Culture Studies	Number of Laboratories	No. Positive / No. Tested	Percent Positive
All Laboratories	15	48/92	52
Nutritional Studies	4	0/31	0
Virus Propagation	9	38/50	76
Commercial Cell Culture Suppliers	2	10/11	91

1. The larger the number of containers & volume of cells, the greater the risk
2. Investigators who examine cell daily for morphological & nutritional changes have lower rates of contamination because problem detected earlier

* From: Barile *et. al,* (1962)

B. Identification and Sources of Mycoplasma Contamination

Table III summarizes our findings on the identification and prevalence of mycoplasma species isolated from contaminated cell cultures. To date, of 11,000 primary and continuous cell culture specimens examined, 1374 mycoplasmas have been isolated and identified. About 8 percent of the cells had mixed contamination with two or more species. Ninety nine percent of the mycoplasma contaminants were either human, bovine, or swine species (Barile *et al.*, 1973).

1. Human Sources

The human oral and genital mycoplasma species represent the largest group of contaminants (Table III), and therefore, the laboratory personnel are a major source of contamination. Mouth pipetting is the major vehicle of human oral contamination. Saliva is introduced into the cell culture: the antibiotics present in the medium destroy the bacteria, but do not inhibit mycoplasma. In cells grown without antibiotics, the bacteria destroy the cells and the culture is discarded. Thus, antibiotics tend to mask mycoplasma contamination and should be avoided. Contamination by human genital strains is probably due to inadequate sterile procedure.

2. Bovine Sources

The bovine species of mycoplasmas are the second major group of contaminants. At least eight distinct bovine species have been isolated from contaminated cell cultures (Table III).

3. Contaminated Bovine Sera

Barile and Kern (1971) have shown that commercial bovine serum is frequently contaminated with mycoplasmas and that it is the major source of bovine mycoplasma contamination of cell cultures. In these studies (Barile *et al.*, 1973), 285 mycoplasmas

TABLE III Identification and Speciation of Mycoplasmas Isolated from Contaminated Cell Cultures*

Mycoplasma and *Acholeplasma* species	Natural habitat	Isolations	
		Number (%)	Totals (%)
M. orale	HUMAN, oral	513 (37.4)	
M. hominis	genital	85 (6.1)	605 (43.9)
M. fermentans	genital	5 (0.36)	
M. buccale	oral	2 (0.14)	
M. arginini	BOVINE oral & genital	295 (21.4)	
M. agalactiae subsp. *bovis*	nasopharyngeal	12 (0.87)	
M. bovoculi	conjunctivae	1 (0.07)	
Mycoplasma sp. 70-159	?	67 (4.8)	533 (38.8)
A. laidlawii	oral & genital	117 (8.5)	
A. axanthum	oral	2 (0.14)	
Acholeplasma sp.	oral & genital	39 (2.8)	
M. hyorhinis	SWINE, nasal	219 (15.9)	219 (15.9)
M. arthritidis	MURINE, nasopharyngeal	5 (0.36)	
M. pulmonis	nasopharyngeal	4 (0.29)	9 (0.65)
M. gallisepticum	AVIAN, nasopharyngeal	3 (0.21)	
M. gallinarum	nasopharyngeal	2 (0.14)	5 (0.36)
M. canis	CANINE, oral & genital	1 (0.07)	
Mycoplasma sp. 689	genital	2 (0.14)	3 (0.21)
Total		1374	

*From: Barile *et.al.*, 1973

TABLE IV Isolation of Mycoplasmas from Contaminated Commercial Bovine Sera[a]

Sera[b]		
Source	Origin	Isolations[c]
Supplier	Calf	22/55
Supplier	Fetal bovine	159/438
Manufacturer	Fetal bovine	104/395
Totals		285/888

[a]From: Barile *et al.* (1973).
[b]Supplier denotes unprocessed sublots of raw sera and manufacturer denotes final lots of sera produced for market.
[c]Number positive/number tested. Lots were not selected at random and data do not reflect incidence of contamination.

TABLE V Mycoplasma Contamination of Commercial Bovine Sera: Identification & Speciation†

Species	Natural Habitat	Numbers of Isolations	Percent
M.arginini	bovine: oral & genital	65	33.7
M.alkalescans	bovine: oral & genital	6	3.1
M. agalactiae subsp. *bovis*	bovine, genital	5	2.6
M.bovoculi	bovine, conjunctivae	3	1.6
Mycoplasma sp 70-159*	(cell cultures)	3	1.6
M.hyorhinis	swine, nasal	1	0.5
A.laidlawii	bovine, genital	87	45.1
A.axanthum	(cell cultures)	1	0.5
Acholeplasma sp*	?	22	11.4

* Distinct unspeciated serotypes unrelated to established species.

† From: Barile *et al,* (1973)

were isolated from 888 lots of raw, unprocessed sera obtained from suppliers or from final lots of sera produced for market (Table IV). Successful isolations were made with our most sensitive large volume broth culture procedure (Barile and Kern, 1971).

4. *Mycoplasma Species Isolated from Contaminated Bovine Sera*

The bovine mycoplasmas isolated from contaminated bovine sera were similar or identical to the bovine species isolated from contaminated cell cultures (Table V) and support the findings that contaminated bovine sera are the major source of bovine mycoplasma contamination. In addition, one mycoplasma was a swine strain of *Mycoplasma hyorhinis*, suggesting that bovine serum may also be a source of swine mycoplasma contamination (Barile *et al.*, 1973).

5. *Swine Sources*

The third major group of contaminants are the swine mycoplasmas (Table III). Although trypsin has been incriminated, all attempts to grow out mycoplasmas from commercial trypsin have failed. Because swine and cattle are frequently processed through the same abattoir, bovine sera may also become contaminated with swine mycoplasmas during collection and manufacture.

6. *Original Tissue Sources*

Occasionally primary cell cultures are contaminated with mycoplasmas derived from murine, avian, and canine origin (Table III). Because these mycoplasmas were only isolated from primary cells derived from murine, avian and canine tissues, and were host specific, the probable source of these contaminants were the original tissues used to produce the primary cell culture (Barile, 1973a; Barile *et al.*, 1973).

7. *Sources of Mycoplasma Contamination*

To summarize, the major sources of contamination are of human origin - the investigator, his colleagues and his environment; of bovine origin - primarily contaminated bovine sera; and of swine origin (Table VI). About one percent of the tissues used to produce primary cell cultures are also contaminated (Barile *et al.*, 1973). In addition, specimens collected for virus isolation studies may also contain mycoplasmas, and thereby contaminate the cell culture. Moreover, many of these mycoplasma contaminants have been mistaken for viruses because they produce cytopathic effects, are filterable, can hemadsorb, and are neutralized by antisera (Barile, 1973).

C. Prevalence of Contamination

The amount of contamination from human, bovine and swine sources was determined for the period 1960 through 1972 (Fig. 1).

TABLE VI Sources of Mycoplasma Contamination of Cell Cultures

1. Human sources: (42%)
   a. Oral: mouth pipetting (saliva); antibiotics destroy bacteria & permit mycoplasma to fluorish (35%)
   b. Genital: inadequate sterile procedures (7%)
2. Bovine sources: (38%)
   a. Sera: contaminated, commercial bovine sera
3. Swine sources: (19%)
   a. Trypsin?
   b. Contaminated commercial bovine sera
4. Original tissue source (1%)
   a. Tissue used to prepare primary cell culture; esp., kidney, tumor or infected tissues
   b. Embryonic avian fluids & tissues
   c. Human & animal specimens used for virus isolation studies may contain mycoplasmas

The major source of contamination in the early 1960's was the human oral strains; in the mid-1960's it was the swine strains, and in the early 1970's it was the bovine strains (Barile, 1973a). Today, the most prevalent contaminants (55%) are the human species, followed by the bovine and swine species which cause an equal amount of contamination.

D. Spread of Contamination

In addition to the original sin, mycoplasma contamination can spread from cell to cell by aerosols of contaminated cell culture material or by contaminated saliva or nasal secretions, genital and intestinal droppings of man and animals. Contamination can spread also by contaminated virus pools, antisera, and other reagents, or by contaminated equipment commonly used in cell culture studies (Barile, 1973a,b).

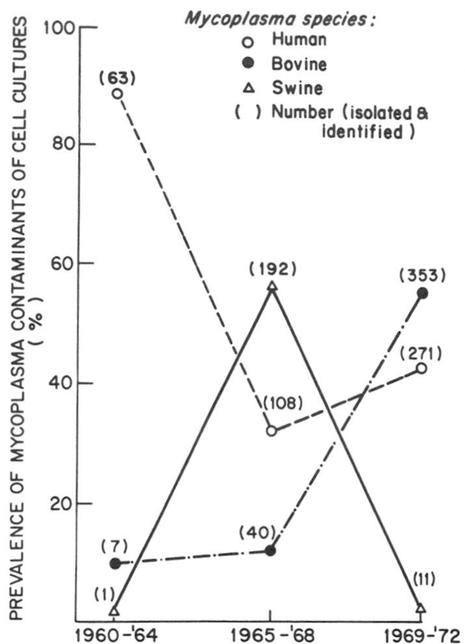

FIGURE 1 *The prevalence of mycoplasma contamination from human, bovine and swine sources during the period 1960-1972. The numbers in parentheses are the numbers of mycoplasmas isolated from contaminated cell cultures for each time period.*

E. Cell Susceptibility to Mycoplasma Contamination

Cell cultures derived from all tissues, organs and animals examined were found to be contaminated, including cells from mammalian, avian, reptilian, fish, insect and plant origin. All types of cells were found subject to contamination, including cells from normal, infected or neoplastic tissues, primary or continuous; diploid, heteroploid, fibroblastic and epithelial cells, and cells grown in monolayer or suspension. Most mycoplasma contaminants produce titers in most cell cultures which range from 10^5 to 10^9 logs per ml of medium fluid (Barile, 1973a).

TABLE VII Procedures Used to Eliminate Mycoplasmas from Contaminated Cell Cultures

1. Antibiotics:
   Oxytetracycline: Robinson *et al.*, 1956
   Chlorotetracycline: Hearn *et al.*, 1959
   Kanamycin: Pollock & Kenny, 1960
   Tetracycline: Carski & Sheppard, 1961
   Novobiocin: Balduzzi *et al.*, 1964
   Tylosin: Friend *et al.*, 1966
2. Prolonged heat:
   41°C for 18 hrs: Hayflick, 1960
3. Specific, neutralizing antisera:
   Barile, 1962, Pollock & Kenny, 1963
4. Aurothio malate:
   Sheddon & Cole, 1966
5. Triton X:
   Reynolds & Hetrick, 1969
6. Tricine:
   Spendlove *et al.*, 1971
7. Specific high titered neutralizing antisera plus Pretested effective antibiotic, Barile, 1973
   Last resort: very difficult to eliminate because resistant strains develop; prolonged R_x toxic; both cells & mycoplasmas enclosed by membrane & have similar toxicities; may induce cell selection; 10% cells have mixed, multiple mycoplasma contaminants

F. Eliminating Mycoplasma Contamination

Because mycoplasma contamination is a common event, we have maintained a continuing study to examine and to develop procedures for eliminating mycoplasmas from a contaminated cell. It has been our experience that there is no single, fail-safe procedure for curing cells (Table VII; Barile, 1973a). Not only is it very difficult to eliminate mycoplasmas, but it is also very time-consuming. Prevention is much more effective than is elimination of contamination. However, if the cell culture is irreplaceable, we recommend the combined use of specific, high-titered neutralizing antisera and a pretested effective antibiotic.

TABLE VIII Recommended Procedure for Eliminating Mycoplasmas from Contaminated Cell Cultures

Antisera plus antibiotic
1. Antisera: a. Use specific, pretested, high-titered neutralizing sera. b. Add 5-10% sera to cell medium 2. Antibiotic a. Pretested & shown effective. b. Concentration predetermined to fall between effective dose for mycoplasma & toxic dose for cell culture. c. We have found 10 μg/ml tetracylcine & 100 μg/ml kanamycin, effective. 3. Procedure: a. Cells grown in medium containing 5-10% neutralizing antisera plus antibiotics at least 4 weeks with weekly subcultures & medium changes. b. Remove antisera and antibiotics, & test cells for mycoplasmas by direct culture and indirect staining procedures for 4 weeks. c. If contamination persist, retest antibiotic sensitivity & replace if necessary. d. Identify mycoplasma: 10% of contaminated cells have more than one contaminant. Replace antisera if necessary. Repeat procedure.

In brief, 5 to 10 percent of antisera and the pretested antibiotic are added to cell culture medium and the cells are grown for four weeks with weekly subcultures and medium changes. The specific antisera replace the serum component of the medium. After four weeks, the antibiotic and antisera are removed and the cells are examined for mycoplasmas for an additional four weeks. If contamination persists, the mycoplasma contaminant is retested for antibiotic sensitivity and for identity, because 10 percent of the cell cultures have mixed contamination (Table VIII, Barile, 1973a,b).

TABLE IX Procedures Used to Eliminate Mycoplasmas from Contaminated Virus Pools

1. Antibiotics:
   a. Must be pretested & shown to be effective.
   b. Use high concentrations: 1000-10,000 μg/ml or greater; incubate overnight at 2-8°C.
   c. Recommend: tetracyclines and/or kanamycin.
2. Filtration:
   a. Use 220 nM filter or smaller, when applicable.
   b. Filter twice.
   c. Avoid use of pressure which can force mycoplasmas through small porosity.
3. Combine antibiotic & filtration
   a. Extremely effective.
4. Ether or cloroform inactivation:
   a. Very effective, when applicable.

1. Eliminating Mycoplasma from Contaminated Virus Pools

In contrast, mycoplasmas can readily be eliminated from contaminated virus pools by the combined use of pretested effective antibiotics and gentle filtration through a 220 nM filter twice. In addition, because all mycoplasmas are very sensitive to ether and chloroform, these reagents are extremely effective when applicable (Table IX).

G. Preventing Mycoplasma Contamination

Contamination can be effectively prevented by using good, basic quality control procedures. Mycoplasma-free cell cultures of early passage should be stored frozen in a large number of vials. During study, the cells are monitored, and when found contaminated, discarded immediately. Preventive measures used successfully for reducing the risk of contamination are based on either eliminating the sources or on aborting the spread of contamination (Barile 1973a,b). For example, primary cell cultures are rarely contaminated and should be used whenever possible; avoid mouth-pipetting and use rigid sterile procedures. Bovine

TABLE X Preventive Measures for Reducing the Risk of Mycoplasma Contamination of Cell Cultures

1. Use primary cells when feasible.
2. Eliminate mouth pipetting.
3. Use rigid sterile procedures.
4. Treat commercial bovine sera, trypsin, etc.
5. Filter the air servicing work-room.
6. Use laminar flow hood for work-area.
7. Decontaminate work-area daily.
8. Monitor and treat working-reagents.
9. Discard contaminated cells immediately.
10. Store mycoplasma-free cells for future use.
11. Good luck!!

sera and other reagents are treated by gentle filtration twice through a 220 nanometer filter and/or heat inactivated at 56°C for 45 minutes (See G.1). Laminar-flow hoods are used and the working area and rooms are decontaminated routinely (Table X).

1. Pretesting and Pretreating Bovine Sera

Preventative measures have been used successfully to reduce the amount of mycoplasma contamination of cell cultures maintained in our laboratories. There has been no contamination with bovine mycoplasmas since we began pretesting and pretreating our supplies of commercial bovine sera as follows: several large lots of bovine sera are examined for mycoplasma contamination by the large volume broth culture procedure (Barile and Kern, 1971). If mycoplasmas do not grow, a large supply of the tested serum lot is purchased and stored frozen in 100 ml volumes. Cell cultures grown in medium with the serum lot are monitored weekly for mycoplasmas. If bovine mycoplasmas are isolated from these cell cultures, the lot of serum is replaced with another pretested mycoplasma-free lot. Every bottle of bovine serum is routinely pretreated by heat-inactivation at 56°C for 45 minutes and/or by filtration twice through a 220 nM filter. If heat-inactivation is unsuitable for growth of a particular cell culture, we recommend that the complete medium containing serum be filtered twice before use. Since high pressure filtration can force the plastic mycoplasma cells through the filter, high pressures should not be used. We have found that filtering complete medium containing 5 to 20% serum through a 220 nM filter can be accomplished without excess pressure, and that it can effectively remove contaminating mycoplasmas.

II. MYCOPLASMA-CELL-VIRUS INTERACTIONS

A. Mycoplasma-Cell Interactions

We are frequently asked: Can the investigator live with mycoplasma contamination, i.e., Does mycoplasma contamination affect the results of my study? The answer can be determined only experimentally by examining the effect of the particular mycoplasma contaminant on the test system under study, and because mycoplasmas are known to affect cell cultures in many ways (Barile, 1973a), the investigator is obliged to make the effort and to rule out a mycoplasma effect. Mycoplasmas can produce cytopathic effects, inhibit lymphocyte transformation, cause chromosomal aberrations, mimic viruses, and either decrease or increase virus yields. Some mycoplasmas, especially the pathogens, can produce hemolysis, adsorb to red blood cells, as well as many other types of cells and hamagglutinate (Barile, 1973a). Some can alter the antigenicity of the cell membrane resulting in the development of autoantibodies to lung and brain tissues, as well as cold agglutinins during infection.

B. Hemadsorption

Figure 2 illustrates hemadsorption of sheep red blood cells to colonies of *Mycoplasma pulmonis*. In addition, sheep or guinea pig red blood cells can also adsorb to mycoplasmas in infected or

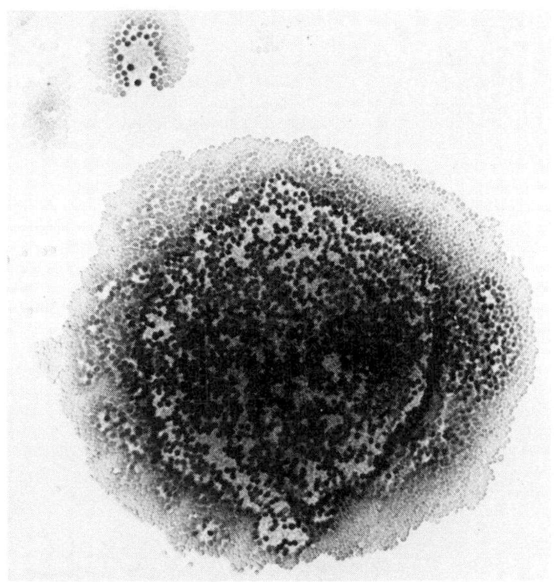

FIGURE 2 *Illustrates hemadsorption of sheep red blood cells onto a colony of M. pulmonis.*

contaminated cell cultures. The binding site is probably sialic acid because treating red blood cells with neuraminidase inhibits mycoplasma hemagglutination. Therefore, the investigator interested in hemagglutinating agents must be cognizant of mycoplasma contamination.

C. Mycoplasma Attachment to Cell Membranes

A number of mycoplasmas have been seen closely associated with or attached to the plasma membranes of animal cells using either histologic staining, immunofluorescence or electron microscopy (Barile, 1973a).

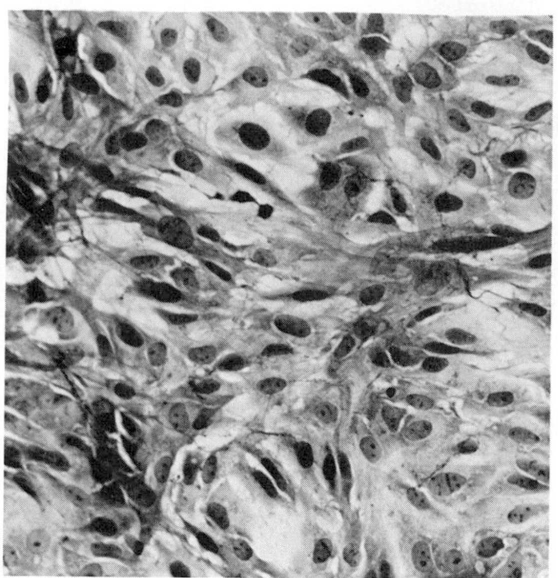

a

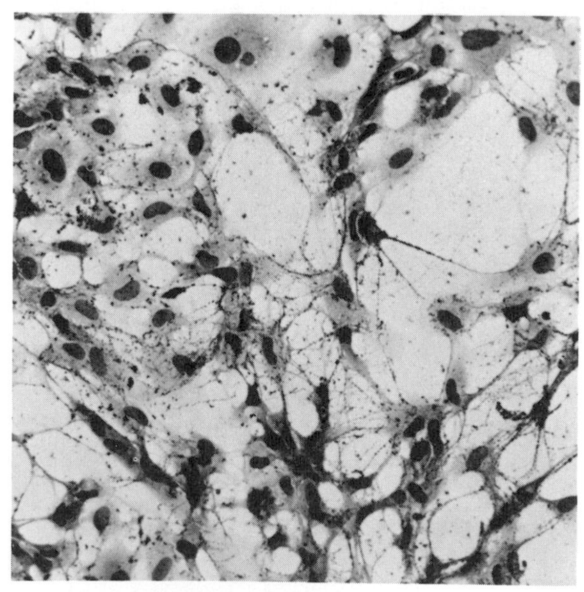

b

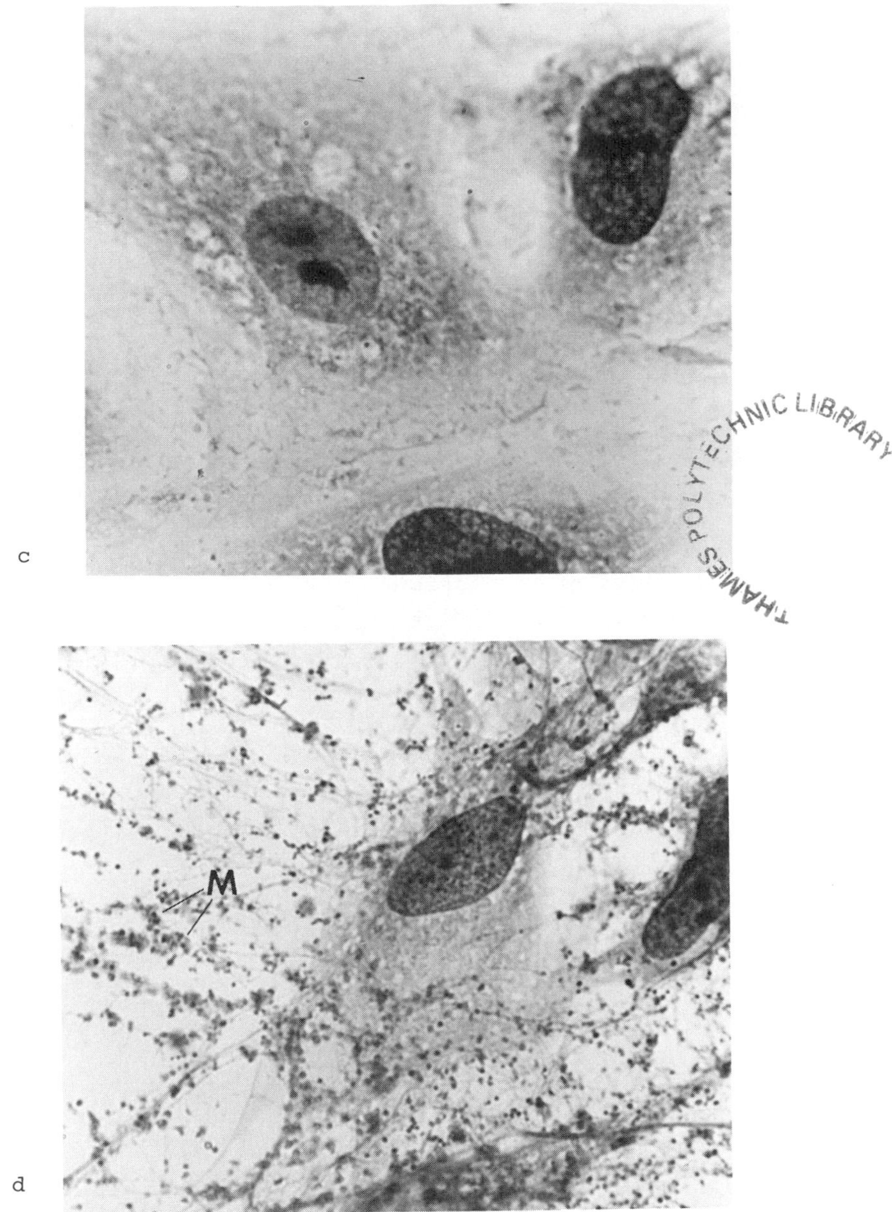

FIGURE 3 *Illustrates attachment of M. hyorhinis to a primary rabbit kidney cell culture at low (b) and high (d) magnifications stained by the Giesma procedure. Note multiple mycoplasmas (M) per infected cell (3d). Compare mycoplasma infected (b and d) to mycoplasma-free (a and c) cell cultures.*

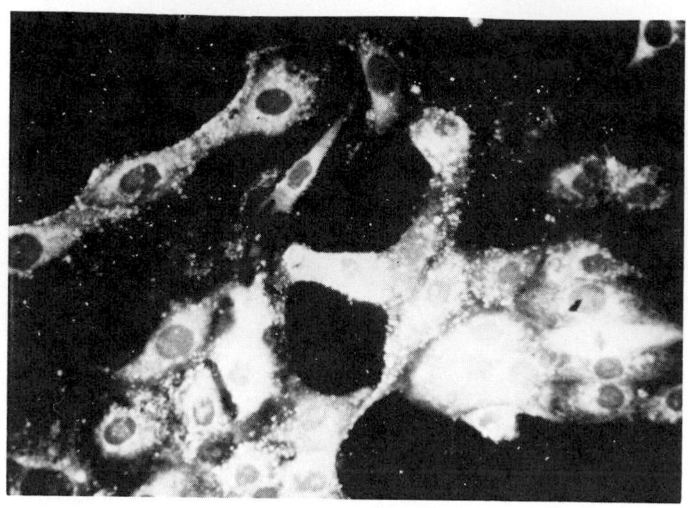

FIGURE 4 *Illustrates a positive homologous immunofluorescence reaction in a M. hominis contaminated HEp-2 cell culture.*

1. *Histologic Staining*

Figure 3 illustrates attachment of *Mycoplasma hyorhinis*, a common contaminant, to a HEp-2 cell culture. The mycoplasmas appear as pleomorphic bodies attached to the infected cell membrane. Note the number of mycoplasmas per infected cell and the distortion of the cell morphology. Since mycoplasmas attach to membranes, ordinary histologic staining procedures provide one useful, simple, indirect procedure for detecting mycoplasmas in contaminated cells (Barile, 1973a).

2. *Immunofluorescence*

Figure 4 illustrates a positive homologous immunofluorescent reaction with a mycoplasma contaminated HEp-2 cell culture. The fluorescent bodies appear very closely associated with the cell membrane. Rhodamine is used as a counterstain. We have shown that immunofluorescence is an extremely effective procedure for

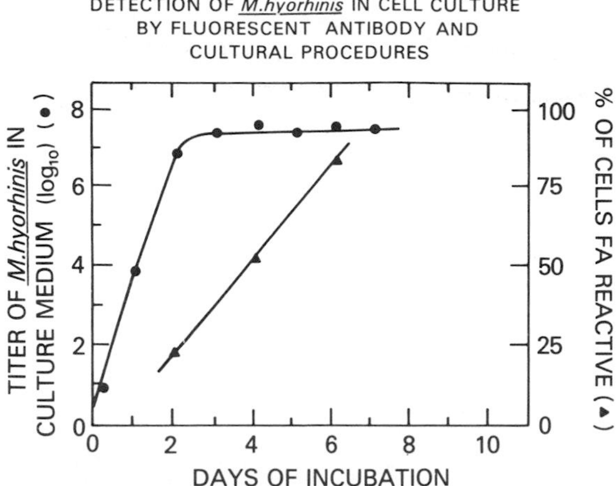

FIGURE 5 *Relative sensitivity for detecting mycoplasmas by comparing isolation and direct culture procedures with indirect fluorescent antibody procedures in M. hyorhinis contamination of cell cultures.*

detecting mycoplasma in chronically infected or contaminated cell cultures (Barile *et al.*, 1962; Barile and Del Giudice, 1972).

3. *Sensitivity of Immunofluorescence*

The relative sensitivity of the direct culture and the indirect fluorescent antibody procedures for detecting the presence of mycoplasmas in contaminated cell cultures is shown in Fig. 5. Cells were inoculated with small number of mycoplasmas and then tested daily by both procedures for 7 days. Whereas, very small numbers of mycoplasmas were isolated and detected by direct culture procedures, the fluorescent antibody reactions were positive only when the cells became *chronically* infected with large numbers, 7 logs/ml or greater, of mycoplasmas. In our studies, primary isolation and direct culture procedures were shown to be more sensitive than indirect procedures for detecting mycoplasma contamination. Because most contaminated cell cultures become chronically infected with mycoplasmas, immunofluorescence is a useful

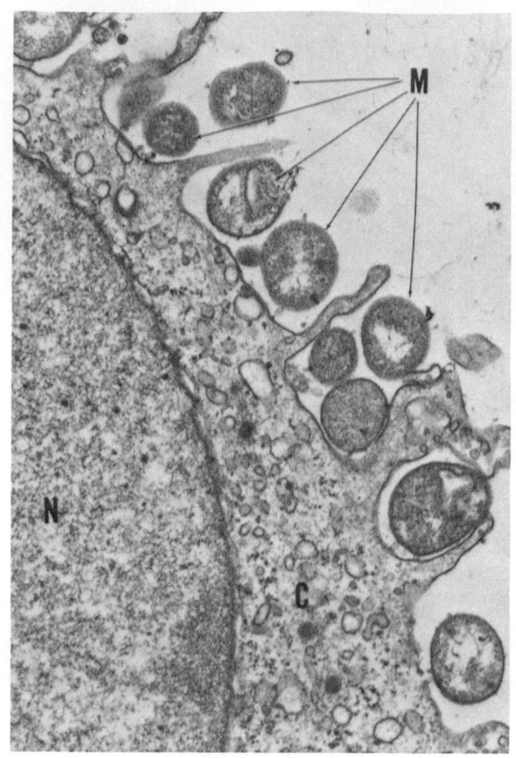

FIGURE 6 *Illustrates a thin-section preparation of a M. pulmonis infected HeLa cell culture. Note multiple mycoplasmas (M) attached to tissue cell membrane.*

indirect procedure and is effective for detecting contamination with the non-cultivable strains of *Mycoplasma hyorhinis*. These contaminants become cell-adapted, lose their ability to grow on artificial broth-agar medium and cannot be isolated using available culture procedures (Hopps *et al.*, 1973).

4. *Electron Microscopy*

Figure 6 illustrates a thin-section preparation of a mycoplasma contaminated HeLa cell culture. Occasionally, the electron microscopist is the first to detect mycoplasma contamination. Note the number of mycoplasmas per infected cell and their close

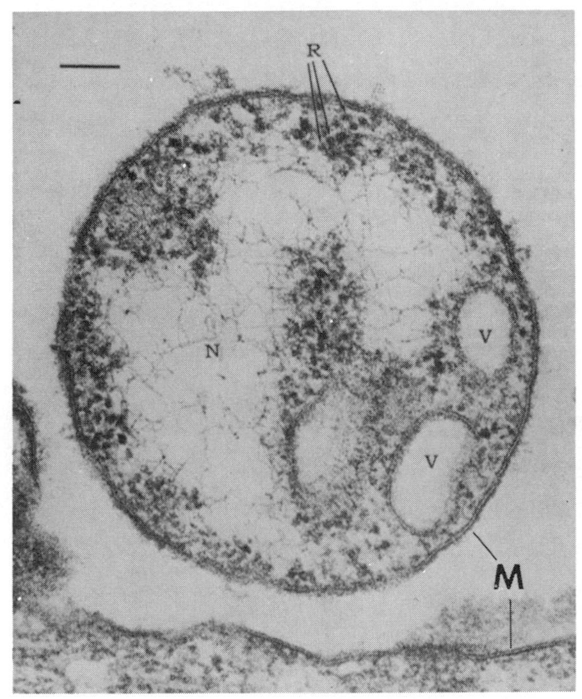

FIGURE 7 *Illustrates the ultrastructure of M. pulmonis in infected HeLa cell culture. Note and compare the three-layered unit membranes (M) of the mycoplasma and HeLa cell.*

association to the HeLa cell membrane. Some of the pathogenic species, such as *M. pneumoniae, M. gallisepticum,* and *M. pulmonis,* have well-organized terminal apparati used for attachment to tissue cell membranes. For example, *Mycoplasma pneumoniae* has a terminal spike (Collier, 1972), which permits the pathogen to anchor firmly to the infected tracheal cell membranes and to resist removal by the coughing and/or ciliary mechanisms. The ultrastructure of the mycoplasma cell is illustrated in Fig. 7. Unlike bacteria, mycoplasmas are devoid of cell wall and are enclosed by a typical three-layered unit membrane. Accordingly, mycoplasmas are susceptible to the same physical and chemical stresses which are injurious for the membranes of cell cultures (Barile, 1973) and thereby make elimination of mycoplasma contamination a challenging problem. Moreover, it is not surprising then, that mycoplasmas have now become very useful tools and a popular model for studying the chemical structure and function of membranes (Razin, 1975; Razin and Rottem, 1975).

D. Arginine Dehydrolase Activity

Another important property of mycoplasmas is their ability to break down arginine for energy. Schimke and Barile (1963) have shown that certain mycoplasmas catabolize arginine by a three-enzyme dehydrolase system with release of ATP. Since animal cells do not have this pathway, detection of arginine deiminase activity in cell cultures is a rapid chemical method for detecting mycoplasma contamination (Barile and Schimke, 1963).

The effect of arginine on the growth of the arginine-utilizing *Mycoplasma arginini* is shown in Fig. 8. Glucose and glutamine had no effect on growth, whereas the addition of arginine produced a marked stimulation of growth. Because mycoplasma contaminants require arginine for energy, they rapidly deplete arginine from the medium, depriving the cell culture of an essential amino acid, and producing profound effects on cell function and on virus propagation.

E. Covert Mycoplasma Contamination

Mycoplasma contamination may appear insidious and may go undetected because: (1) even high titers, 10^7 to 10^8 logs/ml of mycoplasmas do not produce the overt turbid growth commonly associated with bacterial and fungal contamination; (2) the initial cytopathic effects, and especially those produced by the arginine users, can be minimal and (3) mycoplasma contaminants compete for amino acids, sugars, pyrimidines and other nutrients causing nutritional and "toxic" cellular changes.

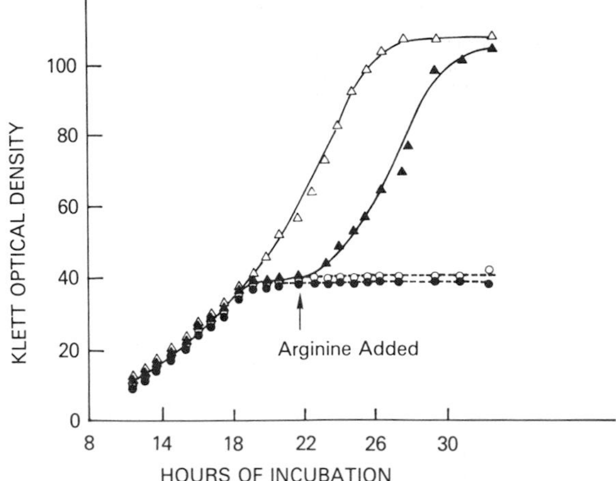

FIGURE 8 *Illustrates the effect of arginine on the growth of M. arginini, an arginine-utilizing contaminant. The addition of arginine to the original broth (△) or added during culture (▲) produced a marked stimulation of growth. The addition of glucose (●) and glutamine (o) to broth had no effect.*

These "toxic" changes can be reversed by changing the medium, providing the investigator with a false sense of security. To a large extent, the biochemical activity of the contaminant predetermines the cytopathic effect (CPE). The CPE depends on whether the mycoplasma utilizes arginine or glucose for energy. Some mycoplasmas deplete arginine and affect cell function, but may not produce a destructive CPE.

F. Cytopathic Effects

A number of mycoplasmas can produce very severe CPE, characterized initially by the development of microcolonies followed by microlesions and small foci of necrosis (Barile, 1973a). As a result, many mycoplasmas, such as *M. gallisepticum* (Fig. 9), produce plaque formation, a property commonly associated with viruses. As the chronic mycoplasma infection or contamination proceeds, the CPE becomes much more generalized, and large areas of

FIGURE 9 *Plaque formation in primary chick cell cultures produced by M. gallisepticum.*

the monolayer are destroyed. Because the mycoplasma can generate large amounts of acid, the cells detach from the glass, producing a macroscopic V-shaped CPE within one to two weeks (Fig. 10). Repeated media changes and stronger buffering systems can delay or reduce the CPE and contamination can go undetected. On the other hand, the macroscopic CPE can be exploited as a means of detecting contamination by mycoplasma fermenters (Barile, 1973a). In brief, tenfold dilutions of the suspect "pour-off" or spent tissue culture medium is inoculated into primary rabbit, monkey, or chick embryo cell cultures, incubated for 2 to 4 weeks without feeding and observed for macroscopic CPE (Fig. 10). Uninoculated cell cultures are maintained in the same manner and serve as negative controls.

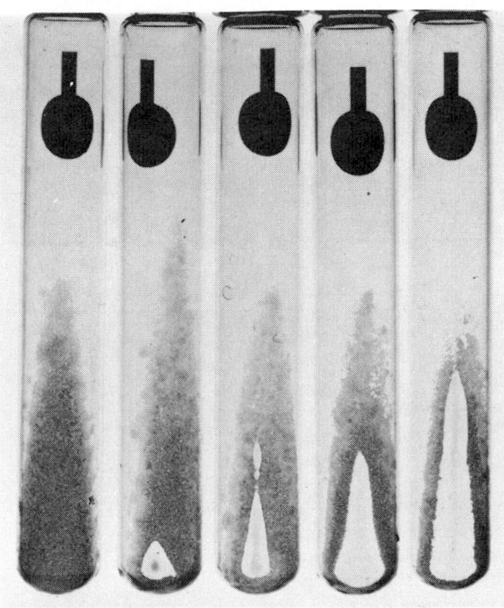

FIGURE 10 *Illustrates the typical inverted V-shaped, macroscopic cytopathic effect produced by mycoplasma fermenters. Note the progression of cell destruction in a M. hyorhinis-infected L-cell culture stained with one-tenth diluted Gentian Violet stain.*

G. Effects of Mycoplasma Contamination

However insidious mycoplasma contamination may appear to the unsuspecting investigator, mycoplasma can produce severe microscopic and macroscopic CPE, including plaque formation. They can also affect cells in many other ways (Barile, 1973a). Some mycoplasma inhibit lymphocyte blastformation and some cause chromosomal aberrations. Mycoplasma deplete nutrients from medium and can affect or influence protein and/or nucleic acid synthesis. Some mycoplasmas alter the antigenicity of cell membranes, causing the production of auto-antibodies, and therefore, mycoplasma have been incriminated as possible causative candidates for various auto-immune diseases (Barile, 1965). In addition, certain mycoplasmas can either decrease or increase virus yields.

TABLE XI The Effect of Mycoplasmas on Stimulation of Human Lymphocytes to Phytohemagglutinin*

Lymphocytes Inoculated with Mycoplasmas		Total ^{3}H-Thymidine Uptake cpm/culture (x 10^3)	
		Stimulant	
Species	Arginine Users	None	PHA
None		8.7	331.6
Broth medium		11.1	–
M.orale	+	–	1.5
M.arthritidis	+	–	2.1
M.hominis	+	0.55	18.7
M.pneumoniae	–	22.5	396.2
M.bovigenitalium	–	–	407.0
M.canis	–	–	498.3

M.hominis suppressed ^{3}H thymidine uptake.
M.hominis and other arginine users suppressed *PHA* stimulation. The non-arginine utilizing, fermenting mycoplasmas had no effect. on PHA stimulation.
* Barile & Leventhal (1968)

H. Effect on Lymphocyte Blastformation

Certain mycoplasmas can inhibit lymphocyte blastformation. In our studies, lymphocyte cultures were infected with either one of three fermenters or with one of three arginine users, including *Mycoplasma orale, M. arthritidis* and *M. hominis* (Table XI).

The data represent mean values from 10 patients and indicate that the arginine-using mycoplasmas markedly inhibited thymidine uptake of unstimulated cells and also inhibited the stimulatory effect of phytohemagglutinin (PHA). The mycoplasma fermenters, which do not utilize arginine for energy, had either no effect or a stimulatory effect. In subsequent studies, we examined the effect of arginine depletion on the inhibition of PHA by *M. hominis*. Whereas, *M. hominis* inhibited PHA stimulation (Table XII), feeding the cells with complete medium or supplementation with additional arginine reversed the *M. hominis* effect. On the other hand, arginine-free medium did not reverse the effect (Barile and Leventhal, 1968).

TABLE XII The Effect of Arginine-Utilizing Mycoplasmas on Stimulation of Human Lymphocytes to Phytohemaglutinin†

After 3 Days, Cells Were Treated as Follows:	Total Tritiated Thymidine Uptake (day 5) cpm/Culture (x10^3)		
	Mycoplasma-Free		*M. hominis** Infected & PHA
	Control	PHA	
None	15.2	446.0	8.0
Replace with complete Eagle's medium (MEM)	5.5	346.4	477.0
Replace with arginine-Free EMEM	3.6	145.5	4.6
Add 10mM arginine	9.4	128.4	237.5

**M. arginine, M. orale* & other arginine users showed similar effects. Mycoplasma fermenters (non-arginine users) had no effect.
† Barile & Levethal (1968)

I. Effect of Mycoplasmas on Virus Yields

We have shown that certain mycoplasmas can influence virus propagation. The effect produced is dependent on the particular mycoplasma, virus and cell cultures used. Whereas some mycoplasmas have no detectable effect, others were shown to either increase or decrease virus yields. Several mechanisms have been established. For example, the arginine-using mycoplasmas decrease yields of arginine requiring DNA viruses by depleting arginine from the medium. On the other hand, some mycoplasmas increase the yields of viruses by inhibiting interferon induction and/or activity (Barile, 1973a, Singer *et al.*, 1973).

1. Decreased Vaccinia Virus Yields by Arginine Depletion

Singer *et al.* (1970) have shown that arginine-utilizing mycoplasmas can decrease the titer of vaccinia virus (an arginine-requiring DNA virus) from 10^1 to 10^2 logs (Table XIII).

TABLE XIII Vaccinia Virus Yields from Mycoplasma-Free & Mycoplasma Infected Cells with and without Additional Arginine†

Mycoplasma infected	Arginine added	HA cells	Log_{10} differ from control	HEF cells	Log_{10} differ from control
Control (none)	No	5.5		5.5	
	Yes	5.8	+ 0.3	5.8	+ 0.3
M.arginini infected	No	3.3	– 2.2	4.1	– 1.4
	Yes	5.2	– 0.3	5.2	– 0.3
M.hyorhinis infected	No	5.5	± 0	5.5	± 0

HA = human amnion cells; HEF = human epithelial cells
Additional 0.4mM arginine v/v 0.1mM arginine
Data express mean values of several experiments
† Singer *et al* 1972

Addition of arginine reverses the effect. *Mycoplasma hyorhinis* (a fermenter which does not require arginine for energy) had no effect.

2. *Decreased Herpesvirus Yields by Arginine Depletion*

Manischewitz *et al.* (1975) showed that the arginine-utilizing mycoplasmas can also decrease titers of *Herpes simplex virus* (HSV), another arginine-requiring DNA virus (Fig. 11). In these studies, the titers of HSV grown in cells contaminated with *M. arginini* were 10^2 to 10^3 logs lower than titers obtained with mycoplasma-free cell cultures. *M. hyorhinis,* a fermenter, had no effect. Addition of arginine reversed the effect. We have suggested that the *M. arginini* effect may provide a simple procedure to establish the arginine requirement for other DNA viruses (Barile, 1973).

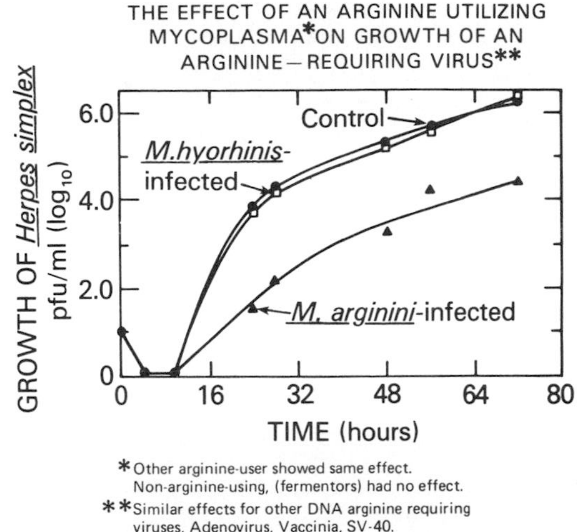

FIGURE 11 *Illustrates the growth of Herpes simplex virus in mycoplasma-free (●) and M. arginini (▲) and M. hyorhinis (□) infected Vero cell cultures. M. arginini, an arginine user, reduced HSV titers and the addition of arginine reversed the mycoplasma effect.*

3. *Reduction of Viral Plaques*

Singer *et al.* (1972) showed that mycoplasmas can also significantly reduce the morphology and number of Semliki Forest virus plaques. Manischewitz *et al.* (1975) showed that *M. arginini* can also reduce the number and size of HSV plaques. The reduction in number and size of HSV plaques was reversed by the addition of arginine. *A. laidlawii*, a non-arginine-utilizing mycoplasma, had no effect. To summarize, the arginine-utilizing mycoplasmas can reduce the number and plaque size of arginine-requiring DNA viruses, and additional arginine can reverse the effect.

4. *Increasing Virus Yields by Inhibiting Interferon Induction*

Singer *et al.* (1969a) have shown that mycoplasmas can also increase virus yields. These studies were initiated because two of the same cultures maintained in parallel gave different titers of

TABLE XIV Increased Semliki Forest Virus Yields from *M. Hyorhinis* Infected Hamster Cells‡

Cells	*Titer SFV log_{10}	Log_{10} difference from control	% control
Control	†(a) 6.4		
	(b) 6.4		
	(c) 6.4		
M. hyorhinis infected	(a) 7.3	(a) +0.9	(a) 790
	(b) 7.3	(b) +0.9	(b) 790
	(c) 7.1	(c) +0.7	(c) 500

*PFU/0.1 ml.

†a, b, and c represent three separate experiments on three lots of cells.

‡Singer *et al.*, 1969a

Semliki Forest virus (SFV). Subsequent studies showed that the cell culture producing the higher titers was contaminated with *Mycoplasma hyorhinis* (Table XIV). Increased SFV titers were seen only when low input was used (Table XV). With low SFV input, a small number of cells are infected, leaving most of the cultured cells available to accept the interferon generated, and to develop resistance to viral infection. Thus, even though mycoplasmas inhibit interferon induction, several cycles of virus multiplication are required to achieve amplification and to show a significant increase in virus yield. With high virus input, most or all of the cells are infected initially, and consequently the effect of mycoplasmas on interferon inhibition and on cell resistance cannot be demonstrated (Singer *et al.*, 1969b).

5. *Inhibiting Virus Induction of Interferon*

Although mycoplasmas do not directly induce interferon, they can inhibit interferon activity induced by either a virus, such as Vesicular Stomatitis Virus (VSV), or by synthetic RNA copolymers. In the VSV studies, mycoplasma-free and mycoplasma contaminated

TABLE XV VSV Replication in Control and *M. Arginini* Infected Hamster Embryo Fibroblasts†

Hours Post VSV Inoc.	Multiplicity of Infection 0.00003			Multiplicity of Infection 0.03		
	VSV TITER (Log_{10}/0.1 ml.)		Log_{10} Diff. from Control	VSV TITER (Log_{10}/0.1 ml.)		Log_{10} Diff. from Control
	Control Cultures	*M. Arginini* Inf. Cultures		Control Cultures	*M. Arginini* Inf. Cultures	
4	0	0	0	2.2	2.2	0
6	0.9	0.9	0	4.4	4.3	– .1
8	1.8	2.1	+ .3	5.2	5.4	+ .2
12	3.3	3.9	+ .6	6.4	6.5	+ .1
24	5.9	7.1	+ 1.2	N.D.*	N.D.	N.D.

* N.D. = not done

† Singer *et al* 1969b

cells were inoculated with high input of VSV and then assayed one and two days later for both virus and interferon titers. Whereas mycoplasma did not affect the virus titers, they did inhibit induction of interferon reducing titers from a high of 1:256 to less than 1:4 (Table XVI). Additional arginine had no effect (Singer *et al.*, 1969).

6. *Inhibiting Poly I·C Induction of Interferon*

Singer *et al.* (1969a) reported similar effects with the synthetic double-stranded Poly I·C, a potent inducer of interferon. The addition of Poly I·C stimulated interferon production and produced cell resistance with a significant decrease in virus yields in the mycoplasma-free cells. On the other hand, the mycoplasmas in contaminated cells inhibited interferon production and thereby nullified the Poly I·C effect, resulting in titers similar to the control cells without Poly I·C (Table XVII).

TABLE XVI Induction of Interferon by Semliki Forest Virus in Mycoplasma Free and Mycoplasma Infected Hamster Cells

Days Following Mycoplasma Infection	Cells‡	*Titer SFV (Log_{10}) After Interferon Induction		Titer of Interferon Induced	
		a†	b†	a	b
1	Control	7.6	7.3	1:128	1:64
	M. arginini Infected	7.4	7.3	1:8	1:8
	M. hyorhinis Infected	7.6	7.3	<1:4	<1:4
2	Control	7.3	7.5	1:256	1:64
	M. arginini Infected	7.5	7.7	1:8	1:4
	M. hyorhinis Infected	7.5	7.3	<1:4	<1:4

* PFU/0.1 ml. † a and b represent two separate experiments on two lots of cells.
‡ From: Singer *et. al*, (1969)

TABLE XVII Induction of Viral Resistance by Poly I·Poly C in Mycoplasma-Free and Mycoplasma Infected Hamster Cells Using Vesicular Stomatitis Virus

Cells Infected with Mycoplasmas	Poly I* Poly C Added	VSV Titers Log_{10}	Log_{10} differ from control
Control (mycoplasma-free)	No	6.8	
	Yes	5.2	– 1.6
M. arginini - infected	No	7.1	
	Yes	6.7	– 0.4
M. hyorhinis - infected	No	7.0	
	Yes	6.9	– 0.1

* 100 μ/ml Poly I· Poly C with 300μ/ ml neomycin

TABLE XVIII Sensitivity of Mycoplasma Infected Cells to Exogenously Applied Interferon†

Cells	Interferon titer		
	a*	b	c
Control	1:32	1:32	1:32
M. arginini infected	1:8	1:4	1:4
M. hyorhinis infected	1:4	1:4	1:4

*a, b, and c represent three different experiments on three separate cell lots.
†Singer *et al.*, 1969a

7. *Effect on Interferon Activity*

Singer *et al.* (1969a) also showed that mycoplasmas can affect interferon activity. In these studies, a standard preparation of hamster interferon was assayed for activity in mycoplasma-free and mycoplasma-contaminated hamster cells (Table XVIII). The findings show that mycoplasmas decreased interferon activity 4- to 8-fold. Therefore, mycoplasmas render cell cultures less sensitive to exogenously supplied interferon, and, consequently, contamination can affect the results of interferon assays.

Mycoplasma contamination could be used to advantage. Inhibition of interferon production and activity by mycoplasmas may be used to enhance virus yields and may play a useful role in the detection of latent viruses.

III. ISOLATION AND DETECTION OF MYCOPLASMAS

A. Detecting Cell Culture Contamination

We have used four basic approaches for primary isolation and detection of mycoplasmas (Barile, 1973a, 1974), including: (1) a standard culture procedure for routine specimens, cell cultures, biopsy tissues, throat cultures; (2) a semi-solid broth culture procedure for screening cell cultures for bacterial, fungal or mycoplasmal contamination; (3) the large specimen-broth culture procedure for isolation of mycoplasmas from contaminated commercial bovine sera; and (4) the use of primary and continuous cell cultures for the isolation and detection of fastidious, pathogenic mycoplasmas, and the cell-adapted strains of *M. hyorhinis*.

B. Direct Culture Procedures; Isolation of Mycoplasmas

Our standard medium is the Edward-Hayflick formula supplemented with arginine and dextrose as energy sources, phenol red as a dye indicator, thymic DNA and vitamins (Barile, 1974). Cell cultures are examined as follows: one ml or one-tenth ml of cell suspension is inoculated into broth or onto agar media, respectively. Agar plates are inoculated in duplicate: one is incubated aerobically, and the other in a 5 percent carbon dioxide-95 percent nitrogen atmosphere shown to be a superior environment for primary isolation of most mycoplasma contaminants (Barile *et al.*, 1962; Barile and Schimke, 1963). The cultures are incubated at 36°C for several weeks and examined periodically for growth of mycoplasmas. Detailed culture procedures are published elsewhere (Barile, 1973a, 1974; Barile *et al.*, 1973).

C. Efficacy of Culture Procedures

The pretesting of culture media and the standardization of culture procedures can influence successful isolation and growth of mycoplasmas (Table XIX). Because the quality of the medium

TABLE XIX Standardization of Culture Medium

I. Basal medium: pretest each batch
II. Horse serum: pretest for toxicity
  1. non-inactivate: *M. pneumoniae*
  2. inactivate (56°, 30') *M. arginini*
III. Fresh yeast extract
  1. 10% for *M. pneumoniae*
  2. 5% for *M. arginini*
IV. L-arginine-HCI
  1. >0.5% toxic for *M. hyosynoviae*
V. Thallium acetate (1:2000)
  1. toxic for Ureaplasmas ¢ some mycoplasmas
VI. Agar: pretest for toxicity
  1. use purified Ionagar #2 or Noble agar
  2. use optimal gel strength (too firm inhibits colony formation)
VII. pH optimal:
  1. Thermoplasmas, pH 2-3
  2. Ureaplasmas, pH 6.0
  3. Mycoplasmas pH 7.2-7.8
  4. *Achoeplasma laidlawii*, pH 8.5

Medium components: pretest each lot of each component for toxicity & growth promotion

Completed medium: pretest each batch for efficacy, using small numbers of fastidious mycoplasmas

components can vary, each lot of each component should be pretested for its toxicity and for its growth-promoting properties. For example, different lots of horse serum vary in their growth-promoting properties and in their toxicity. Thus, each lot must be pretested. In addition, whereas 10 percent fresh yeast extract stimulates growth of *Mycoplasma pneumoniae*, it may be toxic for certain strains of *M. arginini* which prefer 5% yeast extract.

Likewise, whereas arginine (0.2%) stimulates growth, too much arginine (>0.5%) can inhibit growth of certain mycoplasmas (Barile, 1974). The agar can also be toxic. Purified ionagar No. 2 (Oxoid) or Noble agar (Difco) are superior products and should be used. Too much agar can also inhibit colony formation and growth. In summary, there is no standard medium, but rather each medium must be standardized and each batch must be pretested for toxicity and for growth-promoting properties to assure successful isolation and propagation of mycoplasmas.

D. Optimal Culture Procedure

The selection of the medium and culture procedure is dependent on the particular need and the particular mycoplasma under study. For example, the ureaplasmas, spiroplasmas, and especially the thermoplasmas (which prefer to grow at 56°C and pH 2) have special requirements, media and growth conditions (Table XX). Another example of selecting the proper culture procedure is the use of the large specimen procedure for the isolation of mycoplasmas from contaminated bovine sera. This procedure is far superior to the standard agar culture procedure for this purpose. The best medium for primary isolation of the arginine-using mycoplasmas is semi-solid broth. However, the recommended medium for isolating mycoplasmas from contaminated cells is the standardized agar medium incubated in a 5% CO_2 and 95% nitrogen atmosphere.

1. *The Semi-Solid Broth Medium Procedure*

The semi-solid broth medium procedure is a simple broth medium for screening cells for microbial contamination (Barile, 1973a). In brief, a small amount of agar (0.05%) is added to our standard broth medium dispersed in screw-capped tubes. The agar-in-broth provides an oxygen gradient and permits mycoplasmas to produce turbid growth, microcolony formation, and metabolic changes in pH and color of the culture medium, facilitating the detection of growth by visual examination. The arginine-utilizing mycoplasmas produce an alkaline shift and the fermenters an acid shift in pH. This medium will also support the growth of many bacterial and fungal contaminants.

TABLE XX Preferred Medium for Growth of Mycoplasmas

1. Ureaplasmas: Hepes buffer, urea, acid pH (6.0)
2. Spiroplasmas: Bove's medium, 29-32°C
3. Thermoplasmas: acid pH (2.0), 56°C
4. Acholeplasmas: no serum, pH 8.5
5. Arginine users: semi-solid broth
6. Fermenters: agar medium, 5% CO_2 in N_2; broth for *M. hyopneumoniae*
7. "Non-cultivable" *M. hyorhinis* strains: cell cultures
8. Contaminated bovine serum: large specimen-broth culture procedure

2. *Summary*

In summary, (a) we have used culture procedures to isolate over 3000 mycoplasmas from contaminated cell cultures, commercial bovine sera, and biologic products, and also from various normal and infected tissues of man and animals; (b) the isolation and direct culture procedures are the most sensitive and the most effective methods available for detecting mycoplasmas, and are the recommended procedures for examining cell cultures for mycoplasma contamination; and (c) because some strains of *M. hyorhinis* are non-cultivable (Hopps *et al.*, 1973), an indirect procedure must be included in tests to detect mycoplasma contamination.

E. Principles of Indirect Procedures for Detecting Mycoplasma Contamination

The indirect procedures are reviewed elsewhere (Barile, 1973a). In brief, these procedures are based on the exploitation of several basic biological, biochemical, and antigenic properties of the mycoplasma contaminants: (a) because mycoplasmas attach to cell membranes and produce CPE, various staining procedures have been developed and used effectively to detect contamination, including acridine orange, orcein, histologic stains, such as hematoxylin and eosin or Giesma, fluorescent antibody and specific DNA binding stains, such as bisbenzimidazole; (b) some procedures are based on the presence of a particular enzymatic pathway in mycoplasmas, but not animal cells, such as arginine deiminase and uracil phosphoribosyl transferase activity; (c) other procedures are based on the presence of mycoplasma RNA or DNA in contaminated cell cultures.

F. Usefulness of Indirect Procedures

The usefulness of any particular procedure is dependent on at least three basic factors:

(a) the sensitivity of the test: What is the minimal number of mycoplasmas capable of detection? The minimal number or order of sensitivity in our experience is generally 7 logs/ml or greater of mycoplasmas;

(b) the specificity or validity of the basic idea: Do all mycoplasma contaminants possess the basic property under examination? For example, not all mycoplasma contaminants have arginine deiminase activity, and some do not have uracil phosphoribosyl transferase activity. Therefore, these indirect procedures and others have limitations which must be recognized. Moreover, some mycoplasmas attach poorly to cell membranes and are present mainly in the medium fluids. Consequently, these contaminants do not fix as efficiently as do the membrane-associated mycoplasmas in contaminated cell cultures and may miss detection;

(c) the presence of artifacts and other non-specific background activity: Some transformed cell cultures contain excessive amounts of nuclear materials which can affect interpretation of results using DNA stains and autoradiography.

Accordingly, we prefer to inoculate cell culture specimens into an indicator primary cell culture and to use uninoculated cell cultures as negative controls. In summary, it is not enough to have a good basic idea. The basic mycoplasma property must be capable of exploitation before an effective test can be developed. Nonetheless, indirect procedures have merit and should be used. Selection of the most suitable indirect procedure depends in part on the facilities and procedures available to the investigator; e.g., some laboratories routinely perform enzyme assays, electron microscopy, radioactive isotope studies or nucleic acid determinations, etc. In these cases, the investigator is well-advised to use procedures routinely available to him. However, it must be emphasized that ordinary histologic staining procedures are simple, inexpensive, sensitive, effective, and available to every investigator.

G. DNA Staining Procedure

One of the promising indirect procedures developed recently by T. R. Chen (Houston, Texas) is based on the use of a fluorescent bisbenzimidazole compound (Hoechst 33258) which specifically binds to DNA. A similar procedure was reported by Russell *et al.* (1975). In collaboration with Chen, we have examined cell cultures infected with 42 known distinct mycoplasmas species and the results indicate that the DNA-staining procedure provides an effective means for detecting mycoplasmas in infected cell cultures. The procedure is as follows: The cell culture specimen is inoculated into an indicator primary rabbit or monkey kidney cell culture and grown either on cover slips in Leighton tubes or in plastic tissue culture dishes (30 × 10 mm), Falcon, Oxnard, CA. Cell cultures grown in dishes are incubated at 36 ± 1°C in a 5% CO_2-in-air atmosphere until confluent growth is obtained; 2 to 7 days of incubation is generally satisfactory. An uninoculated cell culture serves as a negative control, and a *M. hyorhinis* contaminated cell culture serves as the positive control. The cell culture specimen is fixed at room temperature (RT°) by adding 2 to 3 ml of a 1:3 glacial acetic-acid-methanol fixative to the dish containing the cell culture and medium fluids. After 5 minutes, the fluids are removed and the fixation is repeated. The specimen is air-dried and stored at RT^{o} until stained. A stock concentration of the stain (bisbenzimidazole) is prepared in Hank's balanced salt solution to contain 50 μg/ml. The stain is stored at $4^{o}C$ in opaque bottles wrapped in aluminum foil because bisbenzimidazole is light-sensitive, and will deteriorate on storage. The test stain solution is pretested and used at a concentration of 0.05 to 1.0 μg/ml. For staining, 2 to 3 ml of freshly prepared staining solution is added to the fixed specimen and incubated at RT^{o} for 10 minutes. The stain is removed and the procedure is repeated. The specimen is rinsed 4 to 5 times with distilled water, air-dried, and examined by ultra-violet microscopy at 400 to 600 magnifications. Because the stain is more stable at an acid pH, the specimen is mounted in citric acid-disodium phosphate buffer, pH 5.5 (22.2 ml 0.1 M citric acid + 27.8 ml of 0.2 M Na_2 HPO_4 + 50 ml of glycerol) for examination. A cover slip is placed over the specimen and excess buffer is removed with blotting paper. On microscopic examination, the nuclear DNA of the cell culture appears as large (15-20 μ) spherical bodies and the mycoplasmal DNA appears as small (0.5 to 1 μ) plemorphic spherical, coccoidal or bacillary structures. Occasionally, large thin filamentous strands of mycoplasmal DNA are also present. Figure 12 illustrates a fluorescent reaction with the nucleus of a control mycoplasma-free culture of HEp-2 cells. Figure 13 illustrates a HEp-2 cell culture contaminated with small numbers of *M. cavie*. Note the many small plemorphic fluorescent bodies of mycoplasma DNA and compare them to the nucleus of the HEp-2 cell culture.

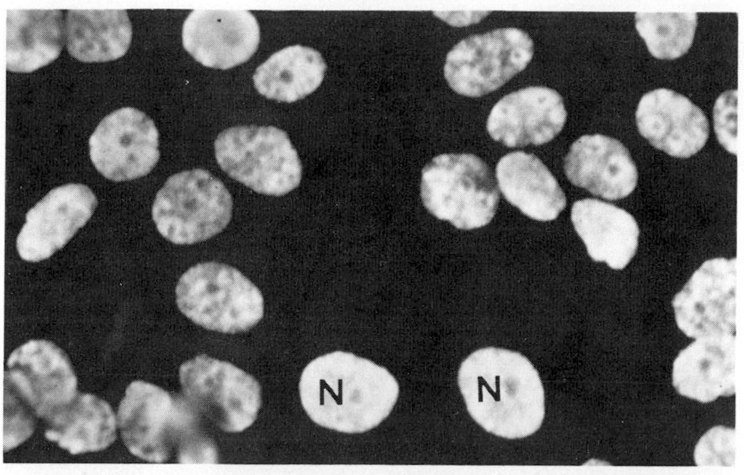

FIGURE 12

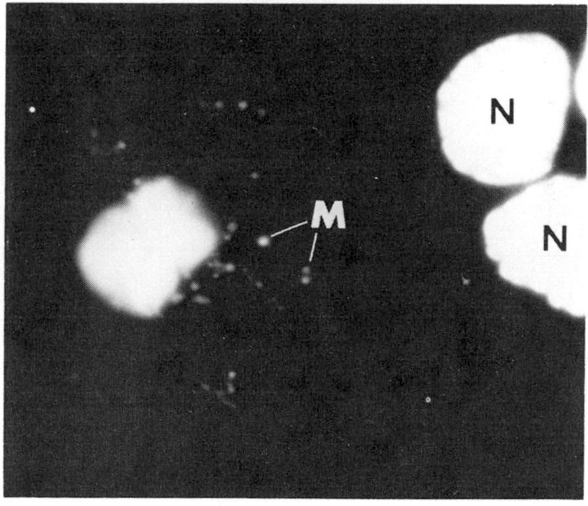

FIGURE 13

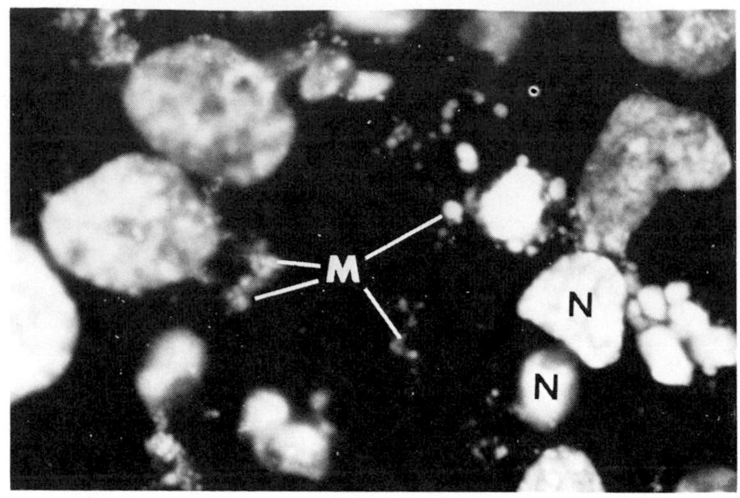

FIGURES 12-14 *Illustrates the use of a DNA-stain (bisbenzimidazole) as an indirect procedure for detecting mycoplasma contamination of cell cultures. Note the fluorescent reaction of the nucleus (N) of a mycoplasma-free, control HEp-2 cell culture (Fig. 12). Figure 13 illustrates a moderate and Fig. 14 a heavy M. caviae infection of an HEp-2 cell culture. The fluorescent mycoplasmal DNA (M) appear as small (0.5-1.0 μ), pleomorphic, spherical, coccoidal or bacillary structures. Note and compare the size of the fluorescent DNA of the HEp-2 (N) and mycoplasma cells (M).*

IV. CONCLUSION AND SUMMARY

A broad overview of our present knowledge of mycoplasma contamination of cell cultures is presented. Three general areas are discussed: (1) the epidemiology of contamination, i.e., incidence, prevalence, sources, prevention and elimination; (2) mycoplasma-cell-virus interactions with a discussion of some specific effects of mycoplasma contamination on cell function and virus propagation; and (3) some of the advantages as well as disadvantages of the direct culture and indirect procedures for primary isolation and detection of mycoplasma contamination. In conclusion, mycoplasmas have been shown to affect cell cultures in many ways. The investigator using cell cultures must be cognizant of mycoplasmas and their properties, must maintain a constant surveillance for mycoplasma contamination and must provide information in published reports on procedures used and data obtained on the presence or absence of mycoplasma contamination in order to properly interpret the results of study.

REFERENCES

Barile, M. F. (1965). *In* "Methodological Approaches to the Study of Leukemias" (V. Defendi, ed) pp. 171-181. Wistar Inst. Press, Philadelphia, PA.

Barile, M. F. (1973a). *In* "Contamination of Cell Cultures" (J. Fogh, ed) pp. 131-172, Academic Press, N.Y., N.Y.

Barile, M. F. (1973b). *In* "Methods and Applications of Tissue Cultures" (P. F. Kruse and M. K. Patterson, Jr. eds.) pp. 729-734, Academic Press, N.Y., N.Y.

Barile, M. F. (1974). *Inst. Nat. de la Sante Res. Med. 33*, 135-142.

Barile, M. F. and Schimke, R. T. (1963). *Proc. Soc. Exptl. Biol. and Med. 114*, 676-679.

Barile, M. F. and Leventhal, B. G. (1968). *Nature 219*, 751-752.

Barile, M. F. and Kern, J. (1971). *Proc. Soc. Exptl. Biol. and Med. 138*, 432-437.

Barile, M. F. and Del Giudice, R. A. (1972). *In* "Pathogenic Mycoplasmas," Ciba Foundation Symposium. pp. 165-188, Elsevier, Amsterdam.

Barile, M. F., Malizia, W. F. and Riggs, D. B. (1962). *J. Bacteriol. 84*, 130-136.

Barile, M. F., Del Giudice, R. A., Hopps, H. E., Grabowski. M. W. and Riggs. D. B. (1973). *N.Y. Acad. Sci. 225*, 251-264.

Collier, A. M. (1972). *In* "Pathogenic Mycoplasmas," Ciba Foundation Symposium. pp. 307-328, Elsevier, Amsterdam.

Hopps, H. E., Meyer, B. C., Barile, M. F. and Del Giudice, R. A. (1973). *N.Y. Acad. Sci. 225,* 265-276.

Manischewitz, J. E., Young, B. G. and Barile, M. F. (1975). *Proc. Soc. Exptl. Biol. and Med. 148,* 859-863.

Razin, S. (1975). *In* "Progress in Surface and Membrane Science" (Danielli, J. F., Rosenberg, M. D. and Cadenhead, D. A., eds.), Academic Press, N.Y., N.Y.

Razin, S. and Rottem, S. (1975). *In* "Biochemical Methods in Membrane Studies" (A. H. Maddy, ed.), Chapman and Hall Ltd., London.

Robinson, L. B., Wichelhausen, R. A. and Roizman, B. (1956). *Science 124,* 1147.

Russell, W. C., Newman, C. and Williamson, D. H. (1975). *Nature 253,* 461-462.

Stanbridge, E. (1971). *Bacteriol. Rev. 35,* 206.

Schimke, R. T. and Barile, M. F. (1963). *J. Bacteriol. 86,* 195-206.

Singer, S. H., Barile, M. F. and Kirschstein, R. L. (1969a). *Proc. Soc. Exptl. Biol. and Med. 131,* 1129-1134.

Singer, S. H. Kirschstein, R. L. and Barile, M. F. (1969b). *Nature 222,* 1087-1088.

Singer, S. H., Fitzgerald, E. A., Barile, M. F. and Kirschstein, R. L. (1970). *Proc. Soc. Exptl. Biol. and Med. 133,* 1439-1442.

Singer, S. H., Ford, M., Barile, M. F. and Kirschstein, R. L. (1972). *Proc. Soc. Exptl. Biol. and Med. 139,* 56-58.

Singer, S. H., Barile, M. F. and Kirschstein, R. L. (1973). *N.Y. Acad. Sci. 225,* 304-310.

RECENT STUDIES - RNA TUMOR VIRUSES

RAYMOND V. GILDEN

Viral Oncology Program
Frederick Cancer Research Center
Frederick, Maryland

The purpose of this presentation is to summarize information on interrelationships among mammalian RNA tumor viruses and to describe how this information has been used to evaluate recent putative human isolates. Immunochemical and molecular hybridization technology has provided definitive data for analysis. Some examples of a specific requirement for both techniques to achieve complete evaluation of certain viruses will be described. Recent developments in detection and distribution of "sarcoma specific" nucleic acid sequences will also be discussed briefly. Pertinent references can be found in reference (1) or as indicated.

A broad grouping can be made of mammalian type C viruses by analysis of the ∿30,000 dalton internal protein commonly referred to as the group specific (gs) antigen or p30 according to molecular weight. Based on both immunologic and NH_2 terminal sequence analyses, seven p30 groups have been recognized (Table I). The criteria for inclusion in the groupings are: (a) identity reactions in gel diffusion (Fig. 1, reference 1), (b) >95% relatedness based on quantitative complement fixation, and (c) minimal

TABLE I Grouping of Mammalian Type C Viruses[a] According to p30 Relationships

Species	Designation
Mouse (M. musculus)	MuLV
Rat	RaLV
Cat	FeLV
Syrian Hamster	HaLV
Cat-Baboon	RD_{114}, BaLV
Woolly Monkey, Gibbon Ape	WoLV, GaLV
M. caroli[b], Pig[b]	

[a]Viruses from cows and guinea pigs often referred to as type C viruses are clearly distinct from those listed in this table based on morphology.

[b]These viruses are easily distinguished from WoLV and GaLV based on molecular hybridization.

or no sequence differences (Fig. 6, reference 1). By these criteria, all mouse type C strains, whether ecotropic or xenotropic, share a common p30. In an analysis of six strains, only a single difference in the first 25-30 residues at the NH_2 terminal was found and peptide maps revealed only 3-4 maximal differences in the ∿40 tryptic peptides compared. In contrast either by immunoassays, by sequencing, or by peptide map analysis, all other type C virus isolates from other species are clearly distinct despite obvious homology, e.g., sharing of interspecies determinants (Fig. 2, reference 1). It is important to emphasize that diagnostic antisera are inherently group-specific or can be so rendered by absorption with heterologous p30's.

In our original studies, p30's of four species were found to be distinct. This led to the hypothesis that each species would possess an easily recognizable, distinctive type C virus and, explains the enthusiasm over detecting a unique p30 in a virus (RD_{114}) isolated in human tumor cells passaged through fetal cats. At present, we know that RD_{114} is representative of a second class of cat virus - in addition to FeLV - and is able to replicate only in heterologous cells. This important property has emphasized the need for appropriate substrates to isolate endogenous viruses, a point now well appreciated by investigators in the field.

Most surprising was the recent finding that the p30's of type C viruses isolated from baboons gave identity reactions with RD_{114} p30 in gel diffusion. NH_2 terminal sequence analysis showed no differences in the initial 14 residues. Matching these sequences to rat (RaLV), mouse (MuLV), hamster (HaLV), or cat (FeLV) p30's requires one gap event occurring presumably at position 5. The fact that this is a relatively rare event in protein evolution further emphasizes the close relationship among these viruses. Along with molecular hybridization data, this has suggested that an interspecies horizontal transmission event has occurred at some time in the remote past.

The sixth p30 grouping includes viruses isolated from a single woolly monkey, multiple gibbon apes, pigs, and the Asian mouse species, *Mus caroli*. If the results with the original primate isolates followed the pattern of genomic inheritance established for other type C viruses, they would lead inescapably to the conclusion that similar viruses are present in man. It is now clear that viruses of the woolly-gibbon (WoLV/GaLV) group are not endogenous in primates but are endogenous in pigs and *M. caroli* based on hybridization studies (2). This has given rise to speculation concerning horizontal transmission, again in the remote past, suggesting that the virus was transmitted horizontally from mouse ancestor to select primate species.

A seventh "group" is represented by the Chinese hamster type C virus. This possesses interspecies p30 determinants but does not react with any of the six group-specific antisera.

I. IMMUNOLOGIC RELATIONSHIPS - LOW MOLECULAR WEIGHT POLYPEPTIDES

It is now well established that the low molecular weight polypeptides (p12, p15) can be utilized to prepare immunoassays which permit type-specific differentiation among members of a p30 group. Thus, p12 assays easily differentiate among the two GaLV and one WoLV isolates (3). Such assays also subdivide MuLV's in precise agreement with molecular hybridization results. The p15 assays distinguish RD_{114} from baboon virus isolates and allow the type-specific differentiation of endogenous viruses from two species of baboon (Papio cynocephalus and P. Hamadryas) (4). These assays for the two groups of primate viruses were extremely useful in evaluating putative human virus isolates as described below.

II. NUCLEIC ACID SEQUENCE RELATIONSHIPS

In most instances viruses belonging to a p30 group will show extensive cross-hybridization. The most useful techniques at present are evaluation of extent of hybrid formation using stringent conditions and single-strand nucleases (e.g., S-1 nuclease from A. oryzae) and the degree of hybrid fidelity measured by thermal stability of heteroduplexes. It is important to recognize that reaction extent is a minimal estimate of relatedness and that thermal stability is a more accurate reflection of this parameter, especially if one compares sequences related on a evolutionary basis. Thus, 50% cross-hybridization with a $3^{o}C$ difference in Tm (midpoint of thermal dissociation curve) probably indicates $\sim$95% base sequence relatedness. It is obviously crucial in such comparisons to know virus history and to have supportive evidence for evolutionary divergence as opposed to the addition or deletion of new gene sequences. There seems little doubt that definitive studies will require genomic fragments (e.g., prepared by restriction enzymes) to evaluate all possibilities.

The high degree of type specificity achieved under stringent hybridization conditions has allowed the detection of contributions from different species to hybrid viruses produced *in vivo*. For examples the Kirsten (Ki) and Harvey (H) strains of "murine" sarcoma virus arose by passage of murine leukemia virus in rats. These viruses actually consist of MuLV-derived sequences and rat-derived sequences, the latter being smaller in subunit size (30S vs. 35S) and also detectable in normal rat cultured cells (5). Ki and HSV were thus laboratory creations. HSV also produces sarcomas in hamsters from which a virus (B-34), has been isolated with proteins of the hamster endogenous virus and nucleic acid sequences of hamster, rat, and mouse origin. The mammalian sarcoma viruses which are replication defective also contain a smaller RNA genome than the helper viruses which provide structural components. B-34, for example, has HaLV 35S RNA (normal helper subunit) and 30S RNA(s) with MuLV and rat-derived sequences; the latter two possibly both comprise a single subunit (6). Thus, the 35S subunit directs synthesis of structural components and the 30S RNA contains information for cell transformation. In the rat, this 30S RNA does not share sequences with 35S RNA of rat type C viruses, while in the mouse, the 30S RNA of Moloney sarcoma virus (no evidence of other species involvement in its origin) is related to the Moloney leukemia virus. The B-34 virus provides a dramatic case in point for the need for complete analysis by both hybridization and immunologic methods. The immunologist would call this a hamster virus, although preparation of cDNA, in general, gives mainly mouse-virus specificity (why this is so is unclear); however, analysis with cDNA's specific from mouse, hamster, and rat (30S) sequences shows all three to be present.

For our purposes it is important to note that hybridization methodology easily discriminates among the WoLV/GaLV group; at the same time it shows only intragroup cross-hybridization and, among the RD_{114}-baboon virus group, it is even capable of detecting slight differences between the two baboon species (7).

III. VIRUS-HOST RELATIONSHIP

Hybridization of viral nucleic acid sequence to host DNA has provided ample evidence for the concept that type C viruses are inherited in the host genome. The exceptions to this are the WoLV/GaLV viruses in primates. These viruses do hybridize with low fidelity to mouse DNA thus the hypothesis that transmission has occurred in the remote past. In similar fashion RD_{114} hybridizes only to the domestic cat and to several other cats from the Mediterranean basin. This is a highly significant distribution since unique sequence DNA's of all Felidae are highly related. It is also surprising to note that FeLV sequences are also distributed in this narrow fashion in the DNA of Felidae. This virus is transmitted horizontally in cats and, as a related virus, is also present in cat cell DNA. In terms of virus expression, all cat cells tested produce RD_{114} RNA but, thus far, FeLV RNA has not been found in uninfected cells. Baboon virus probes give reactions only with domestic or related cats in the same pattern as RD_{114} with other primate DNA's in a phylogenetic pattern (8). This has suggested the possibility of a baboon progenitor to cat ancestor transmission. We should note that the viral transcripts can be used to discriminate among closely related baboon species which is not possible with unique sequence baboon DNA. Some low-level positive results were obtained with human DNA and baboon viral cDNA; we emphasize that this is based at best on distantly related sequences.

One of the more important recent findings has been the isolation of a sarcoma virus-specific sequence from Rous sarcoma virus (RSV). In contrast to the mammalian virus, strains of RSV are competent for replication as well as for transformation. Mutants defective in transformation are readily isolated and these lack a portion of the RSV genome. By absorption procedures a cDNA specific for sarcoma viruses was prepared by Bishop and colleagues (9) which revealed homologous sequences in the DNA of all birds tested. In contrast, helper virus (non-sarcoma) sequences are found only in chickens and in a few closely related birds. Sarcoma sequences are found in all avian sarcoma viruses but in no leukemia viruses. This provides evidence for a highly conserved gene in Aves which is related to malignant transformation and not to viral structural genes. Thus, there seems to be a common thread between sarcoma-specific genes in mammalian and avian cells in terms of occurrence in normal cells although differences in expression may occur. These are obviously systems which will be exploited in the immediate future.

IV. PUTATIVE HUMAN VIRUSES

Gallagher and Gallo (10) described a virus isolate from human acute myelogenous leukemia (AML) cells which was subsequently grown in heterologous cells (11). In a collaborative study (7) it was shown that two viruses were present, one identical to woolly monkey virus and one identical to P. cynocephalus virus. Hybridization and immunoassay techniques gave completely concordant results. While the issue is not considered completely settled based on previous indication of WoLV/GaLV reverse transcriptase (12) and p30 (13) in leukemic cells, these data are consistent with contamination from laboratory viruses. By way of background we point out that, while positive reports of viral antigens in human tumors and normal tissue have appeared (14, 15), other laboratories (16, 17) have reported negative results using procedures of comparable sensitivity. Also cats, mice, and gibbons have antibodies (depending on history, strain, etc.) to protein of homologous viruses; no antibodies to such proteins in humans have been found (18). While testing will continue with well defined antigens and hybridization systems, evidence for a human type C virus, or expression of the same based on relationships to known viruses, is certainly not convincing to most. Even if the single AML case represents a true isolate, the overwhelming experience in terms of isolation is negative. For example, the same laboratory which claimed isolation from the AML case reports 200 negative attempts at isolation (R. C. Gallo, personal communication) and many other laboratories including our own, also report negative results.

Why then do some species yield type C virus easily (mice, cats, baboons) while others yield none at all? One possibility, based on long-term residence of viral genomes in host DNA, is that key genetic elements allowing complete synthesis of structural components are lost. The search may then be for the "oncogene" which seems a reality at least for birds. Similar studies with mammalian viruses may provide the desired probes (19).

REFERENCES

1. R. V. Gilden, *Advances in Cancer Research, 22:* 157-202, (1975).
2. M. M. Lieber, C. J. Sherr, G. J. Todaro, R. E. Benveniste, R. Callahan, H. G. Coon, *Proc. Natl. Acad. Sci, 72:* 2315-2319, (1975).
3. S. R. Tronick, J. R. Stephenson, S. A. Aaronson and T. G. Kawakami, *J. Virol. 15:* 115-120, (1975).
4. J. R. Stephenson, R. K. Reynolds, S. A. Aaronson, *J. Virol. 17:* 374-384, (1976).
5. N. Tsuchida, R. V. Gilden, M. Hatanaka, *Proc. Natl. Acad. Sci. 71:* 4503-4507, (1974).
6. N. Tsuchida, R. V. Gilden, M. Hatanaka, *J. Virol. 16:* 832-837, (1975).

7. H. Okabe, R. V. Gilden, M. Hatanaka, J. R. Stephenson, R. E. Gallagher, R. C. Gallo, S. R. Tronick, S. A. Aaronson, *Nature 260*: 264-266, (1976).

8. R. E. Benveniste, G. J. Todaro, *Proc. Nat. Acad. Sci., 71:* 4513-4518, (1974).

9. D. Stehlin, R. V. Guntaka, H. E. Varmus, J. M. Bishop, *J. Mol. Biol., 101*: 349-365, (1976).

10. R. E. Gallaher, R. C. Gallo, *Science, 187:* 350-353, (1975).

11. N. M. Teich, R. A. Weiss, S. Z. Zalahuddin, R. E. Gallagher, D. H. Gillespie, R. C. Gallo. *Nature, 256:* 551-555, (1975).

12. G. J. Todaro, R. C. Gallo, *Nature, 244:* 206-209, (1973).

13. C. J. Sherr, G. J. Todaro, *Science, 187:* 855-857, (1975).

14. M. Strand, J. T. August, *J. Virol. 14:* 1584-1596, (1974).

15. G. J. Sherr, G. J. Todaro, *Proc. Nat. Acad. Sci., 71:* 4703-4707, (1974).

16. H. P. Charman, M. H. White, R. Rahman, R. V. Gilden, *J. Virol., 17:* 51-59, (1976).

17. J. R. Stephenson, S. A. Aaronson, *Proc. Nat. Acad. Sci., 73*: 1725-1729, (1976).

18. H. P. Charman, N. Kim, M. White, H. Marquardt, R. V. Gilden, T. Kawakami, *J. Nat. Cancer Inst., 55:* 1419-1424, (1975).

19. E. M. Scolnick, R. S. Howk, A. Anisowicz, P. T. Peebles, C. D. Scher, W. P. Parks, *Proc. Nat. Acad. Sci., 72:* 4650-4654, (1975).

Workshop I

DIFFERENTIATION

CELL DIFFERENTIATION OF JUVENILE CHICK TESTIS

II. CONVERSION OF PROGESTERONE TO TESTOSTERONE IN VITRO

HAROLD H. LEE
*LORETTA LIANG-TANG**
*NORIMOTO KAZAMA***

Department of Biology
University of Toledo
Toledo, Ohio

ABSTRACT

Radioactive 17α-hydroxyprogesterone, androstenedione and testosterone were definitively identified by recrystallization to constant specific activity when monolayer cultures of testicular cells from juvenile cockerels were incubated with progesterone-^{3}H. These results suggested that the testicular cells retained their differentiated characteristics for as long as 11 days *in vitro* and that they possessed all the necessary enzymes for the conversion of progesterone to testosterone. Results from double-

*Present address: Department of Obstetrics and Gynecology, University of Rochester Medical School, Rochester, New York.
**Present address: Division of Reproductive Biology, C. S. Mott Center for Human Growth and Development, Wayne State University, Detroit, Michigan.

labelled experiments suggested that pregnenolone, dehydroepiandrosterone and andros-5-ene-3β, 17β-diol might be formed. However, recrystallization data did not support those of the chromatography. This study failed to demonstrate the reversibility of the 3β-hydroxysteroid dehydrogenase-isomerase reaction.

I. INTRODUCTION[1]

One of the advantages in using cell culture over in vivo approaches for endocrinological studies is the preclusion of endocrine contributions from other organs. However, one should be aware that cells in vitro may change, or dedifferentiate, although transformation from one differentiated cell type to another (Gurdon, 1974) is rare. Testis is not only an endocrine organ but also a developing system throughout most of the life span of the animal. The germinal population proliferate and differentiate continuously while the steroidogenic cells, Leydig and sertoli, produce androgens. These cellular characteristics are not shared by most organ systems whose proliferative abilities either cease or decrease to a level necessary for the maintenance of a stable population of differentiated cells.

Compared to investigations on the physiology of the testes in vivo, studies using cell cultures have been few. Cultures derived from mouse Leydig cell tumors failed to synthesize androgens (Shin, 1967). Monolayer cultures of immatured rat testes, previously primed in vivo with HCG, were unable to synthesize testost-

[1]The following trivial names and abbreviations are used: pregnenolone (Pnl), pregn-5-ene-3β-ol-20-one; progesterone (P), pregn-4-ene-3, 20 dione; 17α-hydroxyprogesterone (17αP), pregn-4-ene-17α-ol-3, 20 dione; androstenedione (Δ^4-dione), androst-4-ene-3, 17-dione; dehydroepiandrosterone (DHEA), androst-5-ene-3β-ol-17-one; testosterone (T), androst-4-ene-17β-ol-3-one; androstenediol (Δ^5-diol), androst-5-ene-3β, 17β-diol; TLC, thin-layer chromatogram; SA, specific activity; dpm, disintergrations per minute; cpm, counts per minute.

erone unless androstenedione, an immediate precursor of testosterone, was supplied as substrate (Steinberger *et al.*, 1970). Recently Steinberger and coworkers (1975) were able to isolate and maintain a viable, although not proliferative, population of sertoli cells. The steroidogenesis of these cultures have not been elucidated.

In view of the fact that testicular cultures would lend themselves in the investigation of not only steroidogenesis but also spermatogenesis, we established a primary monolayer culture system derived from juvenile cockerel testes. Previous publications showed that this cockerel testis culture had histoformative ability and were capable to synthesize androgens (Lee, 1971; Tang and Lee, 1973). The primary cultures reach confluency in about 10-12 days after seeding.

We now report that the primary cultures from cockerel testes are able to convert progesterone to testosterone via the Δ^4-pathway. Therefore, the differentiated characteristic of the steroidogenic cell population are retained without added hormonal stimulation.

II. MATERIALS AND METHODS

A. Cell Cultures

The method of cell culture has been reported previously in detail. Briefly, testes aseptically removed from 10-12-week-old fowl (White Leghorn variety) were decapsulated, minced and dissociated into single cell suspensions with 0.125% trypsin (Nutritional Biochemical Corp., grade 1:125) in saline. The cells, at an initial density of 2×10^6 per dish were cultured in 60-mm polystyrene plastic tissue culture dishes (Falcon Plastics) at 37^oC in an atmosphere of 5% CO_2 in air. Each dish contained 5 ml of modified Ham's F12 nutrient mixture supplemented with 1% horse serum, 10% tryptose phosphate broth (DIFCO Laboratories) and antibiotics. Monolayers with distinct reorganized patterns, i. e., 3 or 4 days after seeding, were used.

B. Incubation with Steroids

Radioactive steroids, 1, 2-^{3}H-progesterone (SA = 47.8 Ci/m mole or 8.3 Ci/m mole), progesterone-4-^{14}C (SA = 52.8 Ci/m mole), DHEA-7-^{3}H (SA = 25 Ci/m mole) were purchased from New England Nuclear Company. Their radiochemical purity was examined by thin layer chromatography (TLC) with a solvent system of benzene-heptane-ethyl acetate (5:2:3). For recrystallization studies, progesterone was purified prior to use by TLC with a solvent system of chloroform:acetone (92:8). Steroid substrates were added directly to the cultures in a volume of 10 μl containing various quantities of isotope as indicated in each experiment. The durations of incubations varied from 15 minutes to 24 hours depending on experiments.

C. Extraction of Steroids

At the end of the incubation, cells were scraped off the bottom of the culture dishes with a rubber policeman. The cells together with the culture medium were transferred to a 50 ml capped conical centrifuge tube. Steroids were extracted with either 9 volumes of chloroform-methanol (5:4) or 24 ml of methylene dichloride with or without carrier steroids. The organic phase was concentrated or dried over anhydrous sodium sulfate, both under N_2. Aliquots of these extracts were analyzed for recovery with a scintillation spectrometer. The recovery at this stage was 85-93% of the initially added radioactivity. Depending on subsequent steps, the above procedures varied slightly as indicated.

D. Fractionation by TLC

As a preliminary identification of the conversion products from progesterone, the concentrated samples were transferred to silica gel G TLC plates and subjected to ascending chromatography in benzene-heptane-ethyl acetate (5:2:3) according to Nugara and Edwards (1970). Fractions corresponding to authentic steroids (purchased from either Sigma Chemical or Steroloids, Inc.) were eluted with acetone and the radioactivities were determined.

E. Further Identification of Conversion Products by Acetylation

For the definitive identifications of products converted from progesterone, authentic steroid of P, 17αP, Δ^4-dione, Pnl, 17αPnl, DHEA or T at 40 μg each was added to the first extract. The mixture was then analyzed with TLC with a solvent of chloroform-acetone (92:8). TLC twice developed in the same direction effectively separated P and Δ^4-dione from each other and from other neutral steroids; other neutral steroids were also separated. The steroid spots, detected by a germicidal lamp and by Rhodamine 6G (0.015% in absolute ethanol), were aspirated onto a Pasteur pipette packed with glass wool. The steroids were then eluted with a mixture of chloroform-methanol (1:1).

The steroids were again dried and were acetylated by incubating with 200 μl of pyridine and 40 μl of acetic anhydride for 18 hours at room temperature (Dominquez *et al.*, 1963). The acetylated extracts were then chromatographed for further purification prior to recrystallization. The solvent systems were as follow:

I. Chloroform-acetone, 92:8
II. Heptane-benzene-ethyl acetate, 20:50:30
III. Hexane-ethyl acetate, 50:20
IV. Benzene-diethyl ether, 80:20
V. Heptane-benzene-ethyl acetate, 20:60:20
VI. Hexane-ethyl acetate, 50:50

The solvent systems used for individual acetylated derivative will be described in the Result section. The TLC plates used for this procedure were silica gel 60F-254 (Brinkman Instruments, Inc.).

F. Final Identification by Recrystallization

A final semi-purified fraction was mixed with 13-22 mg of authentic carrier compound. Repeated crystallizations were performed with appropriate solvent pairs (in Results) and the SA in both the crystals and mother liquor solids were determined. Crystals of approximately 600-700 μg were weighed on a Cahn electrobalance model G2. Radioactivities of each sample were again analyzed with a scintillation spectrometer. When the values of the SA of 3 successive recrystallizations were within ±5% of the average of the three values, the purity, i. e., the definitive identification, of an individual compound is established.

III. RESULTS

A. Time Course Studies

Progesterone-^{3}H (SA = 8.3 Ci/m mole) at 0.5 μCi/ml of medium was incubated with 11-day cultures for 7 intervals, 15 min., 30 min., 1, 2, 3, 6 and 24 hours. The first steroid extracts were concentrated and analyzed by TLC with a solvent system of benzene-heptane-ethyl acetate (5:2:3). The data (Fig. 1) was calculated as percent of total cpm. Two major metabolites appeared in the fractions corresponding to androstenedione and DHEA at the 15 minute incubate. With increase in time of incubation, there was a substantial increase of radioactivity in fractions corresponding to testosterone and an unidentified compound X. By 3 hours or longer, essentially all the added progesterone had been converted to testosterone, DHEA, androstenedione and compound X.

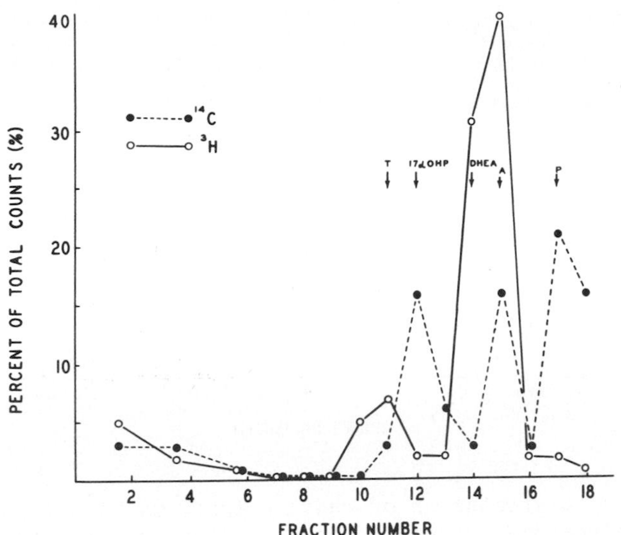

FIGURE 1 *Rate of metabolism of progesterone-1, 2-^{3}H by testicular cells in monolayer cultures. For details of incubation see text. Areas corresponding to authentic steroids on the TLC are indicated.*

B. Double-labelled with P-^{14}C and DHEA-^{3}H

Equimolar concentrations (2.5 µmole) of progesterone-^{14}C and DHEA-^{3}H both at 0.5 µCi/ml of culture medium were added to 5-day testicular cultures. Preliminary identifications of the metabolites were carried out with TLC by the solvent systems, benzene-heptane-ethyl acetate (5:2:3). The percents of radioactivities of either ^{14}C or ^{3}H present in fractions corresponding to authentic steroids were presented in Fig. 2. The 2 major metabolites labelled with ^{14}C were 17α-P and Δ^4-dione. The major metabolites with ^{3}H appeared to correspond to androstenedione and testosterone.

C. Identifications of Δ^5 Intermediates

Spots in solvent system II corresponding to Pnl was eluted, acetylated and chromatographed again in System III in the same

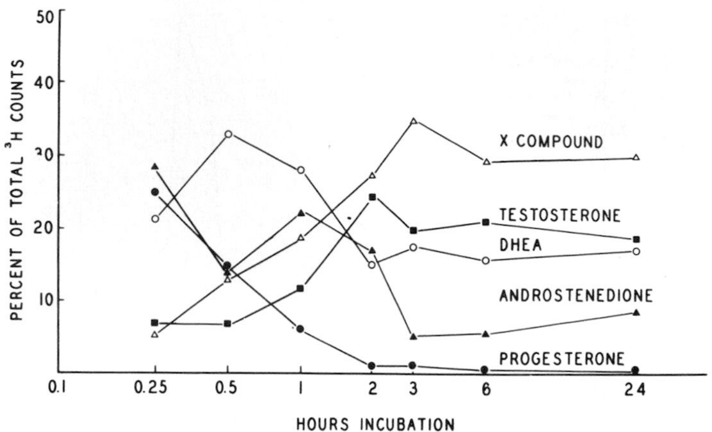

FIGURE 2 *Distribution of radioactivity on thin layer chromatograms of steroid extracts from monolayer cultures incubated with P-^{14}C and DHEA-^{3}H at 37°C for 3 hours. Solvent system in text. Areas corresponding to authentic steroids on the TLC are indicated by arrows.*

direction twice and then in System IV in the second dimension. The Pnl spot was then eluted and chromatographed with System V. However, recrystallization with authentic Pnl-acetate failed to achieve constant SA although the first crystal contained 2101 cpm/mg. The solvent pairs were acetone alone, acetone-heptane, chloroform-heptane, methanol-water and ethyl acetate-heptane, in that order.

For the identification of DHEA, the same solvent systems and procedures for TLC were used as for Pnl. The solvent pairs were acetone alone, acetone-70% aqueous methanol, chloroform-heptane, acetone-70% aqueous methanol, and methanol-50% aqueous methanol, in that order. Although the TLC purified fraction had SA = 1670 dpm/mg, the first crystal had a gross cpm less than twice the background.

To identify Δ^5-diol, the preliminary TLC spot correspond to authentic Δ^5-diol was eluted, acetylated and chromatographed in two dimension with solvent system VI. The area corresponding to authentic Δ^5-diol-diacetate was eluted and recrystallized five times with added authentic Δ^5-diol diacetate. The solvent pairs for recrystallization were acetone-70% aqueous methanol, methanol alone, methylene dichloride-heptane, chloroform-heptane, and acetone-50% aqueous methanol, in that order. Although the initial steroid mixture had SA = 24,696 dpm/mg, the first crystal had only a gross count less than twice the background.

Biochemical determinations to establish the identities of the possible Δ^5 intermediates, Pnl, DHEA and Δ^5-diol, leading to testosterone from progesterone have yielded negative results. The reverse reaction, i. e., P $\rightarrow$ Pnl, did not exist.

Attempts to identify the compound X in Fig. 1 were not successful. Radioactivity disappeared when a second TLC was carried out.

D. Identifications of Δ^4 Intermediates

After a 20 minute incubation with P-^{3}H, only 40% of the original radioactivity remained with the P as revealed by TLC with solvent system I. The areas corresponding to 17αP, Δ^4-dione, and testosterone were eluted for intensive purification as described below.

1. 17αP

Areas corresponding to authentic 17αP was eluted, acetylated and the acetylated compound was chromatographed again in solvent system VI two dimensionally. For recrystallization, authentic 17αP was added. Constant SA for both crystal and mother liquor were obtained for the last three solvent pairs (Table I).

2. Δ^4-dione

The area corresponding to authentic Δ^4-dione was eluted from the preliminary TLC and was acetylated. Although no acetylation occurred, a radioactive product corresponding to authentic Δ^4-

TABLE I Recrystallization of 17α-hydroxyprogesterone Produced from ^{3}H-progesterone by Juvenile Cockeral Testicular Cells

Recrystallization number	Solvent pairs	Crystals		Mother liquor solid	
		SA dpm/mg	%Deviation from mean	SA dpm/mg	%Deviation from mean
1	CH_2Cl_2/heptane	1074	+0.5	1330	
2	CH Cl_3/heptane	1066	-0.2	1145	+3.9
3	ethyl acetate/heptane	1071	+0.2	1088	-1.2
4	acetone/H_2O	1063	-0.5	1070	-2.8
Mean		1068		1101[a]	

[a]Mean for SA of mother liquor solid was derived from the last three specific activities.

TABLE II Recrystallization of Androstenedione Produced from ^3H-progesterone by Juvenile Cockeral Testicular Cells

Recrystallization number	Solvent pairs	Crystals		Mother liquor solid	
		SA dpm/mg	%Deviation from mean	SA dpm/mg	%Deviation from mean
1	acetone/heptane	10,817	+0.02	12,238	
2	CH_2Cl_2/heptane	10,795	-0.2	10,908	+0.7
3	ethyl acetate/heptane	10,867	+0.5	10,816	-0.1
4	acetone/isooctane	10,781	-0.3	10,761	-0.6
Mean		10,815		10,828[a]	

[a]Mean for SA of mother liquor solid was derived from the last three specific activities.

TABLE III Recrystallization of Testosterone (acetate)[a] Produced from ^{3}H-progesterone by Juvenile Cockeral Testicular Cells

Recrystallization number[b]	Crystals		Mother liquor solid	
	SA dpm/mg	%Deviation from mean	SA dpm/mg	%Deviation from mean
1	3564		7447	
2	3387	+2.6	4937	
3	3353	+1.5	4382	
4	3284	-0.6	3566	+1.9
5	3250	-1.6	3468	-0.9
6	3236	-2.0	3467	-0.9
Mean	$\overline{3302}$[c]		$\overline{3501}$[d]	

[a]The radioactive substance obtained from the area corresponding to authentic testosterone on the preliminary TLC was acetylated before crystallization.

[b]Crystallized from solvent pairs of acetone - 70% aqueous methanol.

[c]Mean for SA of crystals was derived from the last five specific activities.

[d]Mean for SA of mother liquor solid was derived from the last three specific activities.

dione was found. Recrystallization with authentic Δ^4-dione as carrier to constant SA was obtained for the last three solvent pairs (Table II).

3. Testosterone

The area corresponding to authentic testosterone on the preliminary TLC was eluted, acetylated and chromatographed in the same direction twice with system VI. Recrystallization with authentic T-acetate to constant SA with one solvent pair was obtained after six times (Table III).

Therefore, the 2 Δ^4 pathway intermediates and testosterone were definitively identified according to the accepted criteria (Axelrod *et al.*, 1965).

IV. DISCUSSION

Our present findings demonstrate that monolayer cultures of cockerel testes are capable of converting progesterone to testosterone. The double labelled experiment (Fig. 2) indicates the possibility that both Δ^4 and Δ^5 may exist in cockerel testis in vitro. Although multiple TLC analyses suggest that the Δ^5 pathway may be present with labelled progesterone as substrate, experiments on recrystallizations of several Δ^5 intermediates failed to yield constant SA. Therefore, unlike the rabbit testes and sheep adrenal (Rosner *et al.*, 1965; Ward and Engel, 1966) there is no reversibility of Δ^5-3β-hydroxysteroid dehydrogenase-isomerase in the cockerel testes *in vitro* converting P $\rightarrow$ Pnl. However, conversion of DHEA to Δ^4-dione is still feasible.

Conversion of P-^{3}H to testosterone via the Δ^4 pathway has been firmly established from the results of the present investigation. Authenticities of 17α-P, Δ^4-dione and T from H^3-P have been verified radiochemically according to the criteria of Axelrod (1965). Therefore, the 3 enzyme complexes, 17α-hydroxylase,

C_{17-20} lyase and 17βOHSD, for steroidogenesis are present in the primary monolayer cultures of cockerel testis. Unless the Δ^5 pathway exists in vivo, one can conclude thst the monolayers derived from cockerels retain the differentiated characteristics for as long as 11-days *in vitro*. The testes for the primary culture are at the stage prior to the onset of spermatogenesis when steroid synthesis of the testis is at the highest level. Prior to this stage, going back to embryonic gonadol development, the biochemical differentiation with respect to steroidogenesis might, however, differ. The difference might exist in the metabolic pathways whose influence on the gonadol development may be significant. The present culture system offers an opportunity to investigate this problem and that of the steroidogenic cell-germ cell interactions.

ACKNOWLEDGMENTS

This study was supported by a grant (HD 06725) from the National Institute of Child Health and Human Development, USPHS. One of us (Harold H. Lee) is recipient of a University of Toledo Trustee's Research Award (1974 and 1975).

REFERENCES

Axelrod, L. R., Matthijssen, C., Goldzieher, J. W. and Pulliam, J. E. (1965). *Acta Endocrinol* (Kbh) *Suppl*. 99, 3.

Dominquez, Q. V., Seely, J. R. and Gorski, J. (1963). *Anal. Chem. 35,* 1243.

Gurdon, J. B. (1974). "The Control of Gene Expression in Animal Development," Harvard University Press, Cambridge, Massachusetts.

Lee, H. H. (1971). *Develop. Biol. 24,* 322.

Nugara, D. and Edwards, H. M., Jr. (1970). *J. Nutr. 100,* 539.

Rosner, J. W., Hall, P. F. and Eik-Nes, K. B. (1965). *Steroids 5,* 199.

Shin, Seung-il. (1967). *Endocrin.* 81, 440.

Steinberger, A., Heindel, A. A., Lindsey, J. N., Elkington, J. S. H., Sanborn, B. M. and Steinberger, E. (1975). *Endo. Res. Comm. 3* (In Press).

Steinberger, E., Steinberger, A. and Ficher, M. (1970). *Recent Progr. Horm. Res. 26,* 547.

Tang, F. Y. and Lee, H. H. (1973). *Endocrin. 92,* 318.

Ward, M. G. and Engel, L. L. (1966). *J. Biol. Chem. 241,* 3147.

DIFFERENTIATION OF SENSORY CELLS IN CULTURES OF AGGREGATES OF DISSOCIATED EMBRYONIC CHICK OTOCYSTS

MARY FAITH ORR

Department of Anatomy
Northwestern University
Medical and Dental Schools
Chicago, Illinois

I. INTRODUCTION

Several studies have shown that the embryonic chick otocyst developed sensory epithelia of the inner ear in organ culture that were similar to the sensory epithelia that developed in vivo (Fell, 1928; Friedmann, 1956, 1965, 1968; Orr, 1968). Basilar papillae and cristae were frequently observed. These structures developed morphologically although there was no anatomical development in culture. The sensory epithelia were composed of fully differentiated sensory cells and supporting cells. In electron microscopic studies Friedmann (1959, 1968, 1969) and Friedmann and Bird (1961) demonstrated that sensory cells in the sensory epithelium of the basilar papillae have a surface structure (hair) composed of sterocilia; ocassionally a basal body was observed.

It was also shown by Friedmann that nerve fibers had formed synaptic contacts with the sensory cells. The sensory cells in the epithelium covering a ridge-shaped crista have a surface structure composed of a kinocilium and numbers of sterocilia.

A light microscopic study of organ cultures of aggregates of dissociated otocysts and associated tissues revealed that areas of reconstructed epithelium differentiated into sensory cells and supporting cells. The epithelium in these areas was innervated by nerve fibers. There were other areas of reconstructed epithelium in which a few poorly differentiated sensory cells were observed; no nerve fibers were demonstrated in these epithelia (Orr, 1968).

This is a report of the preliminary electron microscopic observations of the surface structures of sensory cells that have differentiated in reconstructed epithelium in organ cultures of aggregates of dissociated embryonic chick otocysts.

II. MATERIALS AND METHODS

A. Dissociation and Organ Culture Methods

1. Dissociation

Embryonic chick otocysts along with the surrounding mesenchyme and acoustic ganglion were dissected out of the embryo on the 4th day of incubation and dissociated according to the method of Moscona (1961). The otocysts were incubated for 15 minutes in Calcium- and Magnesium- free Hanks' balanced salt solution (Ca- and Mg- free HBSS) and then for 45 minutes in 1% trypsin (Difco 1:250) dissolved in Ca- and Mg- free HBSS. The tissue was washed three times in Ca- and Mg- free HBSS and then placed in one ml of nutrient medium or regular HBSS. The otocysts were dissociated by flushing through a fine pipette and the cellular suspension was allowed to aggregate for one hour in a test tube that was turned by hand at frequent intervals or by rotation in a roller drum (revolving 10 times per hour).

2. Organ Culture

Aggregates were placed in organ culture chambers according to the method previously described (Orr, 1968, 1975b). Control otocysts were cultivated in the same manner. Two nutrients were used with equal results: (1) The supernatant from clotted chicken plasma and embryo extract (equal parts), or (2) equal parts fetal calf serum and embryo extract. The pH was adjusted to 7.4-7.6. The medium was changed at 2 or 3 day intervals for 12 days.

B. Preparation of Histologic and Electron Microscopic Specimens

1. Histologic Preparations

Otocysts and cultures of otocysts and aggregates were fixed, processed, embedded in paraffin and serially sectioned according to the methods described in previous reports (Orr, 1968, 1975b). Some of the sections were stained with periodic acid-Schiff technique (PAS) followed by hematoxylin, while others were impregnated with silver according to Bodian's protargol method followed by the periodic acid-Schiff technique. Details of these staining methods were described by Orr (1975b).

2. Preparation of Electron Microscopic Specimens

Cultures of aggregates and intact otocysts were fixed for electron microscopy in 1% osmium tetroxide in Veronal acetate buffer (Palade, 1952) adjusted to pH 7.2-7.4. Specimens were also fixed in 1% glutaraldehyde in HBSS followed by post-fixation in 1% osmium in HBSS. Tissues were processed in graded ethanols, cleared in propylene oxide, infiltrated in equal parts propylene oxide and Araldite 502, and embedded in Araldite 502 (Orr, 1975 a and b). Light microscopic sections (0.5 to 1.0 micron in thickness) were stained with a mixture (equal volumes) of 1% methylene blue in 1% borax solution and 1% azure II (Richardson *et al.*, 1960). Ultrathin sections were cut on a Porter-Blum MT I ultramicrotome, stained with saturated uranyl acetate followed by lead citrate (Reynolds, 1963) and examined with a Hitachi 11 F electron microscope.

III. RESULTS

As noted in the introduction certain observations have been reported that were made on serial sections of histologic preparations of organ cultures of aggregates and intact otocysts (Orr, 1968). In order to make a clear presentation of the results of this study it was considered necessary to illustrate similar observations in this report. The photomicrographs shown here have not appeared in any other publications.

A. Embryonic Chick Otocyst

Figure 1 is a photomicrograph of an embryonic chick otocyst on the 4th day of incubation. The pseudostratified epithelium in the ventral portion of this sac-like structure merged with a cuboidal epithelium on the dorsal surface. The epithelium contained cells undergoing cell division and cells with a cilium. Underlying the epithelium was a thin basement membrane and the acoustic ganglion was adjacent to the pseudostratified epithelium (Orr, 1968, 1975a).

B. Sensory Epithelia in Twelve-day-old Cultures

A basilar papilla that developed and differentiated in a twelve-day-old culture is shown in figure 2. The basilar papillae were composed of rows of sensory cells; each sensory cell was surrounded by supporting cells. The presence of hair structures on the surface of sensory cells was considered one characteristic

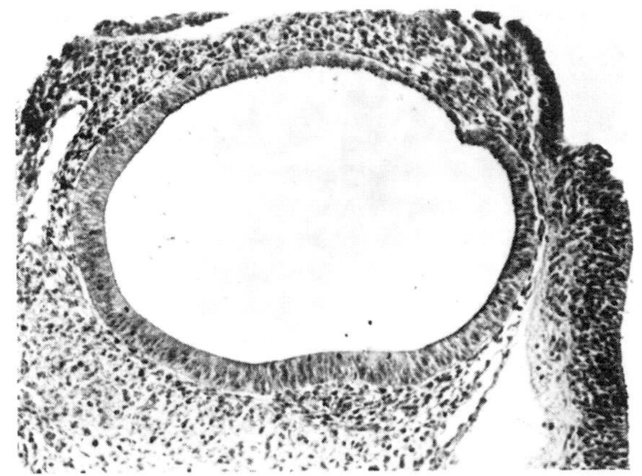

FIGURE 1 *This is a photomicrograph of a histologic section of an embryonic chick otocyst on the 4th day of incubation. The acoustic ganglion is not shown in this section. PAS and hematoxylin. X75.*

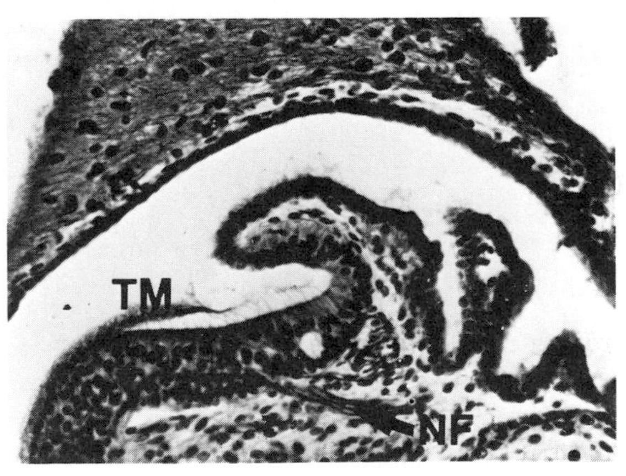

FIGURE 2 *This is a section of a basilar papilla in a 12-day-old culture. Note the hairs extending from the surface of the hair cells towards the tectorial membrane (TM). Nerve fibers (NF, arrow) can be observed innervating the sensory epithelium. Bodian and PAS. X650.*

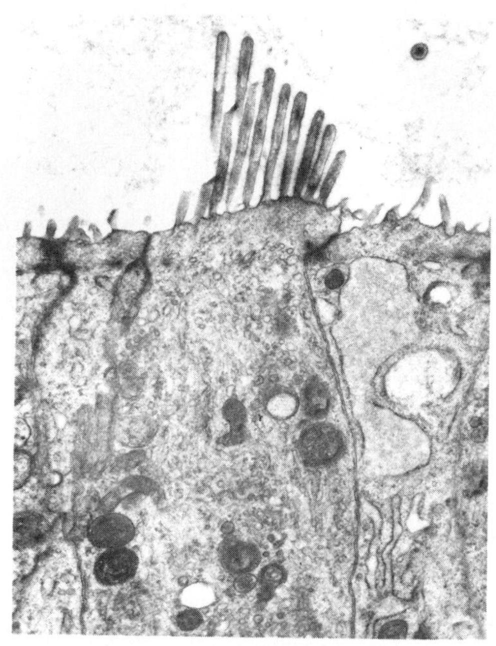

FIGURE 3 *A sensory cell in a basilar papilla that developed in a 12-day-old culture is shown in this electron micrograph. Note the sterocilia on the surface; the gradation from the tallest to the shortest sterocilia can be seen in this section. X11,750.*

that indicated complete cytodifferentiation of the sensory cells. Hairs extended from the surface of sensory cells towards the tectorial membrane. Nerve fibers impregnated with silver were observed innervating the sensory epithelium. An electron microscopic examination of a basilar papilla revealed that there were numbers of sterocilia on the surface of the sensory cells (Figure 3). Basal bodies of kinocilia were also observed in some cells.

The sensory cells in the sensory epithelium that covered a crista were 'tear-drop' shaped; two sensory cells are shown in Figure 4. An electron microscopic examination of a cross section of the surface structure (hair) of a similar sensory cell demonstrated that the sterocilia were arranged in orderly rows with a kinocilium located in a constant position (Figure 5).

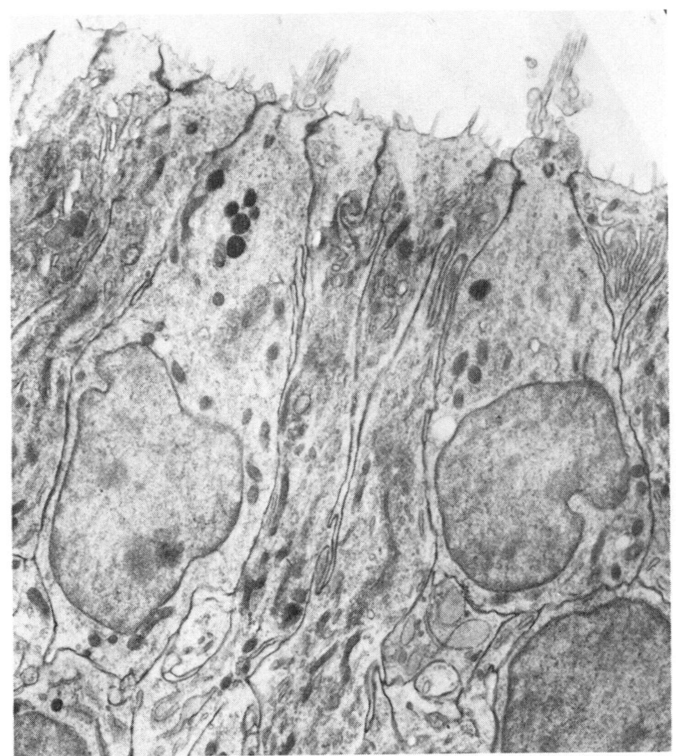

FIGURE 4 *Two sensory cells in the sensory epithelium of a crista can be seen in this electron micrograph. X11,750.*

C. Sensory Epithelia in Twelve-day-old Cultures of Aggregates

Cellular suspensions (that contained undissociated groups of cells) of embryonic otocysts reaggregated into clumps of cells within an hour. After three days in culture reconstructed epithelium had formed in organ cultures of aggregated clumps of cells. Although morphologic development did not occur, areas of the reconstructed epithelium differentiated into sensory cells and supporting cells (Figure 6). Figure 7 is an example of a sensory epithelium in which sensory cells with demonstrable hairs are shown in a histologic section. An examination of histologic preparations impregnated with silver revealed that nerve fibers were present in these sensory areas of epithelium (Orr, 1968). Furthermore, with electron microscopy it was shown that the

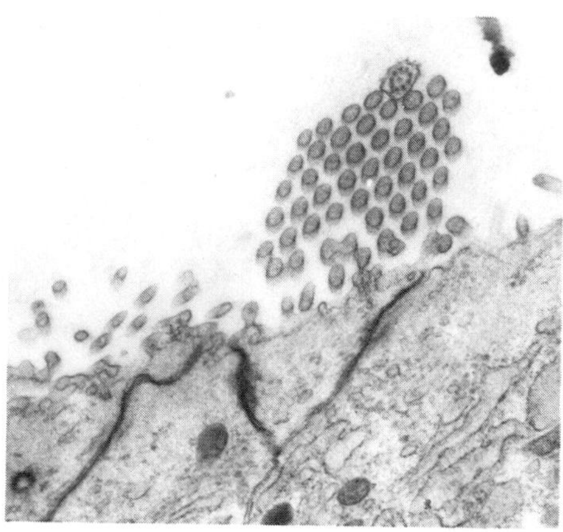

FIGURE 5 *An electron micrograph of a cross section of the surface structure of a sensory cell similar to those depicted in Figure 4. The orderly arrangement of the sterocilia and the location of the kinocilium are demonstrated in this micrograph. X7,050.*

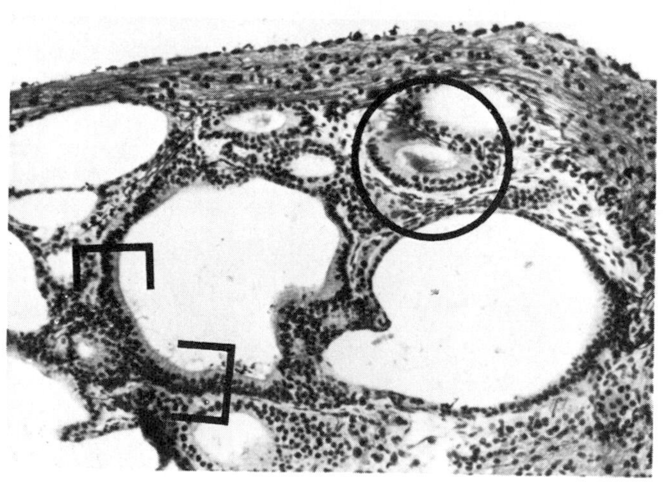

FIGURE 6 *This is a photomicrograph of a histologic section of a 12-day-old culture of aggregates of dissociated otocysts. One area of sensory epithelium is within brackets. Another sensory epithelium is circled; PAS positive material is present in the lumen of this structure. Bodian and PAS. X250.*

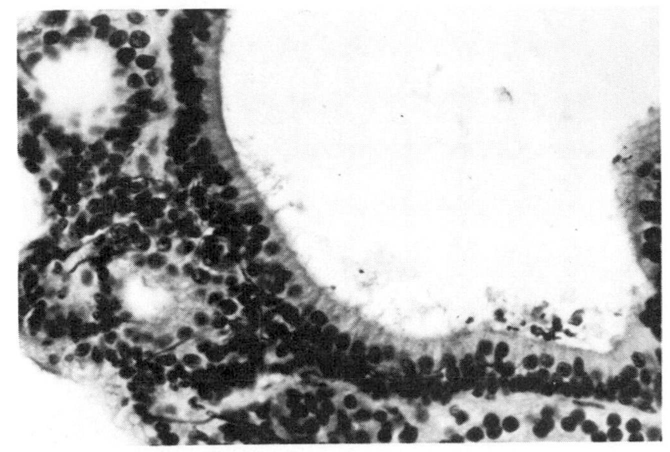

FIGURE 7 *The sensory epithelium within brackets in Figure 6 is shown here at higher magnification. Note the well-aligned row of sensory cells with hair structures extending into the lumen. The hairs have been distorted by the preparation procedure. Bodian and PAS. X400.*

sterocilia of the sensory cells had developed and were comparable with the sterocilia of sensory cells that had developed in cultures of intact otocysts (Figure 8).

There were other areas of reconstructed epithelium that contained a few sensory cells that were poorly differentiated (Figure 9). Nerve fibers were not observed innervating these areas of epithelium. The electron microscopic examination showed that there were several abnormal configurations of the sterocilia on the surface of the sensory cells. In some electron micrographs of longitudinally cut sensory cells the sterocilia were disorganized and some appeared enlarged (Figure 10). Sterocilia that were abnormally bent were observed in other sections (Figure 11). In Figure 12 there is one sensory cell cut in longitudinal section with a few sterocilia that appeared normal. A cross section of a surface structure in this same electron micrograph revealed a kinocilium and numbers of sterocilia, however, the arrangement

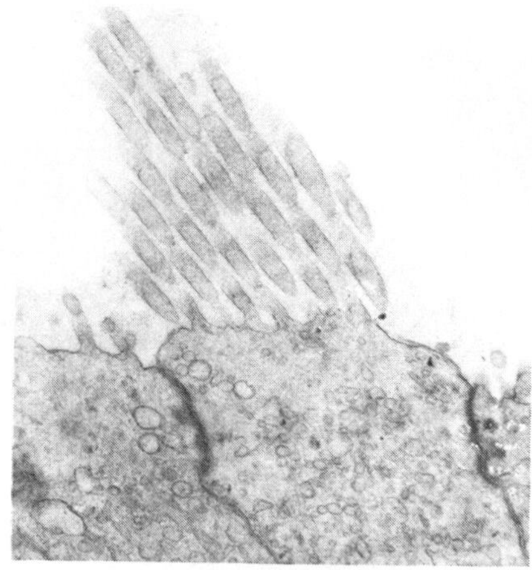

FIGURE 8 *An electron micrograph of the sterocilia of a sensory cell that differentiated in a 12-day-old culture of aggregates. These sterocilia are comparable with those observed on the surface of sensory cells that differentiated in cultures of intact otocysts. X7,050.*

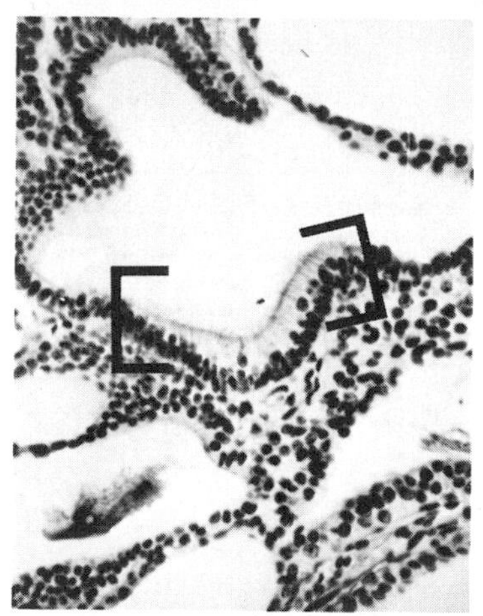

FIGURE 9 *This photomicrograph of a histologic section of a 12-day-old culture of aggregates reveals a sensory epithelium (within brackets) in which only two poorly differentiated sensory hair cells can be observed. Bodian and PAS. X250.*

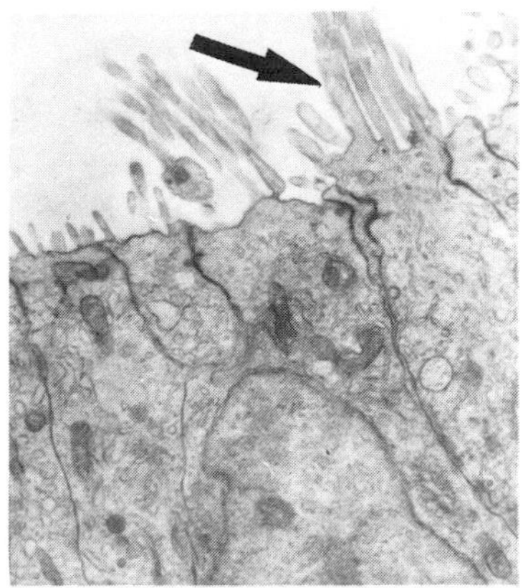

FIGURE 10 *The sterocilia shown in this electron micrograph are abnormal in development and distribution. Note that one sterocilium in this electron micrograph is enlarged (arrow). X7,000.*

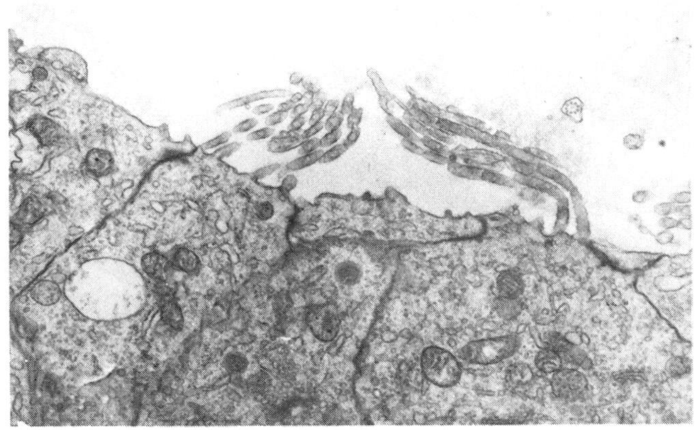

FIGURE 11 *In this electron micrograph the sterocilia of the sensory cell on the right are abnormally bent. X7,000.*

of the sterocilia was disorganized. Compare this cross section with the cross section in Figure 5.

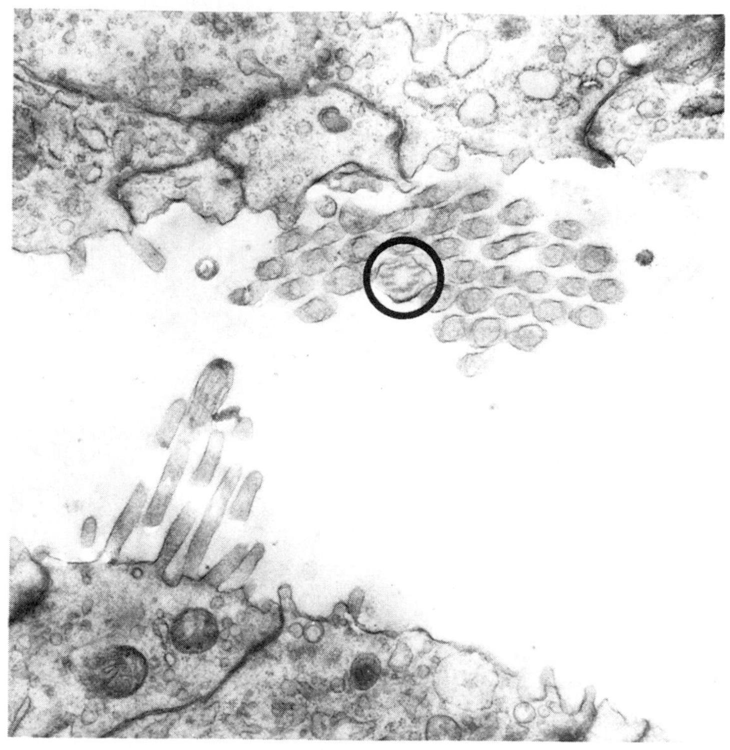

FIGURE 12 *In this electron micrograph there is one cell cut in longitudinal section with a few sterocilia that appear to be normal; the tip of a kinocilium can be seen. A kinocilium (circled) and numbers of disorganized sterocilia can be observed in this tangential section across a hair structure in the same micrograph. X7,000.*

IV. DISCUSSION

This electron microscopic study has shown that there were well-developed sterocilia on the surface of sensory cells that had developed in organ cultures of aggregates of dissociated embryonic chick otocysts. It has been shown in a study of histologic preparations of cultures of aggregates that nerve fibers innervated areas of epithelium that had differentiated into sensory hair cells and supporting cells (Orr, 1968). Synaptic contacts between nerve fibers and sensory cells in reconstructed epithelia have not been explored with the electron microscope.

The findings in this study have also confirmed that the cytodifferentiation of the sensory cells in some reconstructed epithelia was affected by the lack of some factor. The sterocilia of some of the sensory cells were revealed to be abnormal in distribution, development and physical structure; and, there was a definite indication that the presence of a kinocilium did not prevent disorganization of the sterocilia. Based on light microscopic studies of histologic preparations it was suggested that complete cytodifferentiation of sensory cells was dependent on innervation of epithelium (Orr, 1968). Friedmann (1968) had also noted that "the inclusion of certain neurogenic elements plays an essential role in the differentiation of the sensory areas" of the inner ear of the embryonic chick in culture. Sher (1971) reported that innervation of the sensory areas of the inner ear of the mouse preceded differentiation of the epithelium.

The most likely explanation for the poor differentiation of the surface structures of the sensory cells was the lack of innervation of the sensory epithelium. However, it should be noted that there may be other factors involved in some instances. Some of the undissociated groups of cells in the cellular suspension were probably responsible for some of the sensory epithelium in the reconstructed epithelium in cultures of aggregates. It has been shown that there were groups of cells in the embryonic chick otocyst that have cohesive apical junctions that were not disrupted by trypsin although the desmosomes were affected (Orr, 1975a). The poor differentiation of some of the sensory cells may be due to incompletely reformed intercellular connections.

This study was restricted to the formation of sterocilia on the surface of the sensory hair cells since this was considered one characteristic of cytodifferentiation that would be most revealing. The influence of the innervating nerve fibers has been discussed. However, it should be noted that a previous study also indicated that histodifferentiation of the epithelium into sensory epithelia required the proximity of mesenchymal and epithelial tissue (Orr, 1968). This requirement would be particularly important in the early stages of localization and development of the sensory epithelia.

V. CONCLUSION

This study with electron microscopy has shown that the sterocilia were well-formed on the surface of sensory cells in some of the reconstructed epithelium in cultures of aggregates of dissociated embryonic chick otocysts. The sterocilia of other sensory cells were abnormal in distribution, development and physical structure. The most probable cause was considered to be the lack of innervation of these epithelia.

ACKNOWLEDGMENTS

The author wishes to thank Mrs. Ethel Golliday for technical assistance and Ms. Sue Decker for assisting in preparing the prints. This investigation was supported by a United States Public Health Service Grant NS 08569.

REFERENCES

Fell, H. B. (1928). *Arch. Exptl. Zellforsch. 7*, 69-81.

Friedmann, I. (1956). *Ann. Otol. Rhinol. Laryngol. 65*, 98-109.

Friedmann, I. (1959). *J. Biophys. Biochem. Cytol. 5*, 263-268.

Friedmann, I. (1965). *In* "Cells and Tissues in Culture" (E. N. Willmer, ed.), Vol. II, pp. 521-548. Academic Press, New York.

Friedmann, I. (1968). *J. Laryng. and Otol. 82*, 185-201.

Friedmann, I. (1969). *Acta Oto-laryngologica 67*, 224-238.

Friedmann, I., and Bird, E. S. (1961). *J. Ultrstruct. Res. 20,* 356-365.

Moscona, A. A. (1961). *Exptl. Cell Res. 22,* 455-475.

Orr, M. F. (1968). *Devel. Biol.* 17, 39-54.

Orr, M. F. (1975a). *Devel. Biol. 47,* 325-340.

Orr, M. F. (1975b). *In* "Handbook of Auditory and Vestibular Research Methods" (C. A. Smith and J. A. Vernon, eds.), Chapt. 5, pp. 127-174. C. C. Thomas, Springfield, Illinois.

Palade, G. E. (1952). *J. Exptl. Med. 95,* 285-298.

Reynolds, E. S. (1963). *J. Cell Biol. 17,* 208-212.

Richardson, K. C., Jarell, L., and Fincke, E. H. (1960). *Stain Technology 35,* 313-323.

A MODEL FOR THE "PERMISSIVE" EFFECT OF GLUCOCORTICOIDS ON THE GLUCAGON INDUCTION OF AMINO ACID TRANSPORT IN CULTURED HEPATOCYTES

MICHAEL W. PARIZA[1]
ROLF F. KLETZIEN[2]
VAN R. POTTER

McArdle Laboratory for Cancer Research
Medical School
University of Wisconsin
Madison, Wisconsin

I. INTRODUCTION

The ability of glucocorticoids to potentiate cyclic AMP (cyclic adenosine 3', 5'-monophosphate) mediated processes has been termed "permissive" (Ingle, 1952; Thompson and Lippman, 1974; Kletzien *et al.*, 1975). However, the biochemical phenomena involved in glucocorticoid "permissive" effects are poorly under-

[1]Present address: Department of Food Microbiology and Toxicology, Food Research Institute, University of Wisconsin, Madison, Wisconsin 53706
[2]Present address: Sidney Farber Cancer Center, 35 Binney Street, Boston, Massachusetts 02115

stood. Schaeffer *et al.* (1969) reported that one site of glucocorticoid "permissive" action on glycogenolysis was to increase the pool of inactive liver glycogen phosphorylase, and Butcher *et al.* (1971) found that long-term exposure of H-35 hepatoma cells to hydrocortisone increased the basal level of tyrosine aminotransferase (TAT) activity as well as TAT induction by dibutyryl cyclic AMP. Zahlten (1974) has commented that glucocorticoids could induce many enzymes involved in cyclic AMP-mediated processes.

We recently reported (Kletzien *et al.*, 1975) the identification of an amino acid transport system as a site of glucocorticoid "permissive" action in cultured hepatocytes at the membrane level. This finding may be of fundamental importance in the process of gluconeogenesis from amino acids. During periods of prolonged fasting, gluconeogenesis by liver and kidney cells from amino acids (especially alanine) mobilized from muscle protein (Exton, 1972, but see also Odessey *et al.*, 1974) is the principal mechanism whereby normal blood glucose levels are maintained. Additionally, Park (1974) has commented that *in vivo* the concentration of amino acids is rate-limiting for their conversion to glucose. Thus, our finding that glucocorticoids interact with glucagon in the regulation of the transport of substrates for gluconeogenesis has important implications.

II. MATERIALS AND METHODS

Figure 1 summarizes the methods and procedures which we use to prepare *non-proliferating* monolayer cultures of adult rat liver parenchymal cells. Methods for determining amino acid transport in cultured hepatocytes using the non-metabolizable alanine analog, α-aminoisobutyric acid (AIB), have also been previously reported (Kletzien *et al.*, 1976a).

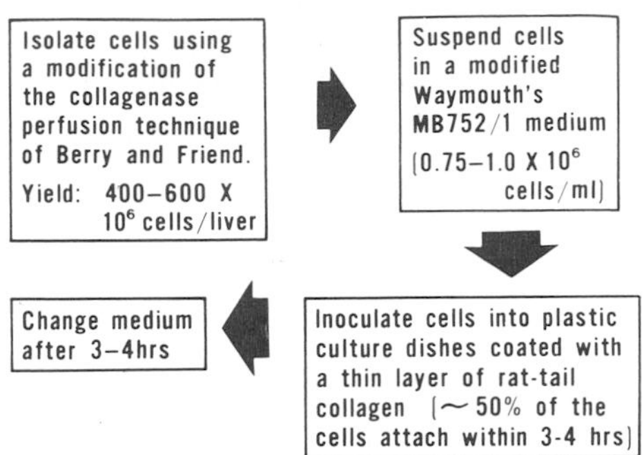

FIGURE 1 *Procedure for isolating and culturing liver parenchymal cells from adult rats (Berry and Friend, 1969; Bonney et al., 1973; Pariza et al., 1974, 1975, 1976a; Kletzien et al., 1976a).*

III. RESULTS AND DISCUSSION

A. Characteristics of Hepatocytes in Primary Monolayer Culture

Previously we have shown that isolated rat liver parenchymal cells placed in primary monolayer culture do not degenerate but rather improve biochemically during the first 48 hours in culture (Pariza *et al.*, 1974, 1975; Kletzien *et al.*, 1976a. Additionally, we have emphasized the desirability of following more than one criterion for viability, and in this regard autoradiography with ^{3}H-leucine has proven particularly useful in estimating the fraction of viable cells present in monolayer cultures (Pariza *et al.*, 1975). Autoradiograms prepared with labeled precursors of RNA (such as ^{3}H-cytidine) would also be highly relevant. However, cultured hepatocytes *do not proliferate* in serum-containing media currently in use (Bohney *et al.*, 1974; Pariza *et al.*, 1974, 1975) and for this reason autoradiography with labeled precursors of DNA is not available as a marker for viability of cultured hepatocytes. This particular property is considered to be significant, because liver parenchymal cells do not divide in the adult rat under normal circumstances. Thus, retention of this property by liver parenchymal cells in culture is to be expected if the adult differentiated state is to be maintained (Bissell *et al.*, 1973; Bonney *et al.*, 1974; Pariza *et al.*, 1975).

There are certain undeniable difficulties which accompany any attempt to study "normal" cells in culture. This is particularly true when the cells under study are from an adult tissue and are therefore highly differentiated. We have discussed many of these problems in previous reports as well as our attempts to determine through biochemical analysis exactly how "adult" non-proliferating cultured hepatocytes really are (Pariza *et al.*, 1974, 1975). While one cannot expect these cells to perform every liver cell function at once (Pariza *et al.*, 1974), it is nonetheless true that the primary culture system offers many advantages not available with other systems for studying liver. One of these advantages is the ability to study the effect of hormones added singly, simultaneously or sequentially using a large number of cultures simultaneously prepared from the liver of one rat (Pariza *et al.*, 1976b). Another advantage is the simplicity with which membrane transport systems can be studied (Kletzien *et al.*, 1976a,b).

B. Hormonal Induction of Amino Acid Transport

Table I shows the effect of glucagon, dibutyryl cyclic AMP, dexamethasone, and insulin, alone and in combination, on the transport of AIB in adult rat liver parenchymal cells in primary culture. Glucagon, insulin and dibutyryl cyclic AMP stimulated the transport of AIB whereas dexamethasone (a synthetic glucocorticoid) alone or in combination with insulin has no affect on AIB transport. However, treatment with dexamethasone in combination with glucagon or dibutyryl cyclic AMP resulted in a 2-fold stimulation of AIB transport over that observed with either glucagon or dibutyryl cyclic AMP alone (Table I). This is to our knowledge the first published demonstration of a glucocorticoid

TABLE I The Effect of Glucagon, Dibutyryl Cyclic AMP, Dexamethasone, and Insulin Alone and in Combination on α-aminoisobutyric Acid (AIB) Transport in Cultured Hepatocytes (Kletzien *et al.*, 1975; Kletzien *et al.*, 1976a)

Hormone	AIB transport (n moles/4 minutes/mg protein)
None	0.71 ± 0.02
Dexamethasone	0.70 ± 0.03
Glucagon	1.33 ± 0.06
Dexamethasone and glucagon	2.77 ± 0.11
Dibutyryl cyclic AMP	1.66 ± 0.06
Dexamethasone and dibutyryl cyclic AMP	3.43 ± 0.10
Insulin	1.54 ± 0.05
Dexamethasone and insulin	1.55 ± 0.05

"permissive" effect on an amino acid transport system.

Extensive characterization of the "permissive" action of glucocorticoids on the glucagon induction of amino acid transport in cultured hepatocytes is in progress and will be published separately (Pariza *et al.*, 1976a; Kletzien *et al.*, 1976b). However, we would like to present in this communication a summary of relevant findings to date.

1. The glucocorticoids dexamethasone, hydrocortisone, and corticosterone all produce a "permissive" effect on the glucagon induction of AIB transport.

2. Glucocorticoids increase the V_{max} without affecting the K_m of the glucagon-induced transport system.

3. A "permissive" effect is observed when glucocorticoids are given simultaneously with glucagon or before glucagon addition (glucocorticoid added, removed, and then glucagon added). However, in the latter case the "permissive" response appears af-

fected by the rate at which the glucocorticoid is metabolized as dexamethasone gives a more prolonged effect than does hydrocortisone.

4. Cyclic AMP levels increase by more than 50-fold within 10-12 minutes after glucagon addition, and this peak value is followed by a slow decline in cyclic AMP levels. The cyclic AMP response to glucagon appears unaffected by prior exposure to dexamethasone, thus supporting the data of Table I that the glucocorticoid "permissive" effect is beyond cyclic AMP.

5. There is a "lag period" of 1-2 hrs following glucagon addition before the initial increase in AIB transport is detected. Moreover, exposure to glucagon for only 30 minutes results in a marked induction in AIB transport when assayed at 120, 240 or 360 minutes after glucagon removal. Additionally, a "permissive" effect on transport is seen in cells pretreated with glucocorticoid and then handled as above.

6. In contrast to the results presented above, a "permissive" transport effect fully induced by treatment with dexamethasone and glucagon together for 12 hrs decays with a half-life of about 60 minutes when glucagon is removed, whether or not dexamethasone is removed simultaneously. Moreover, the half-life of a fully induced "permissive" transport system in glucagon alone (dexamethasone removed) is several hours.

7. Cycloheximide or puromycin block decay of a fully induced transport system when dexamethasone and glucagon are removed.

A model consistent with our findings is shown in Fig. 2. The basic tenet of the model is the existence of a transport system in an *inactive* form which can be converted to an *active* form by a process initiated by glucagon. While there is circumstantial evidence that cyclic AMP may be involved in this process since

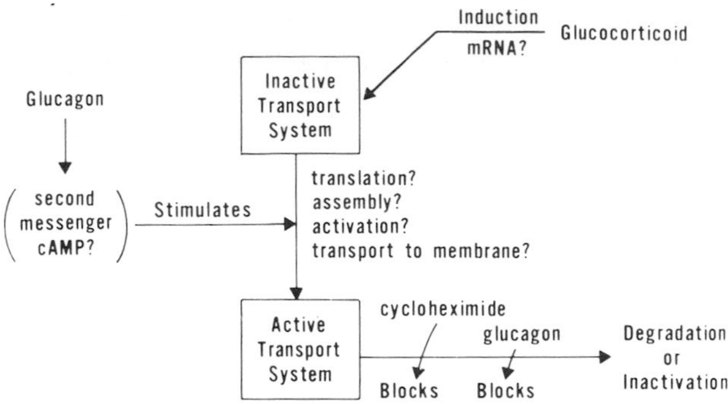

FIGURE 2

DBcAMP acts in the absence of glucagon (Table I), we do not yet have unequivocal data on this point. Tolbert *et al.* (1973) reported a lack of correlation between cyclic AMP accumulation and gluconeogenesis in suspensions of isolated liver cells treated with catecholamines and Michell (1975) has proposed a role for phosphatidylinositol as a "second messenger" independent of cyclic AMP. It is clear that the mechanism whereby the inactive transport system is converted to an active transport system (Fig. 2) is dependent upon the nature of the glucagon-indicated process, and therefore we can not yet stipulate how this conversion occurs. However, some possibilities which could be involved are listed in Fig. 2: translation of mRNA, assembly of inactive sub-units, activation of a critical sub-unit, or transport to the "proper spot" in the membrane.

The model (Fig. 2) further proposes that glucocorticoids do not directly interact with the glucagon-initiated process, but rather increase the pool of inactive transport system, possibly through induction of mRNA synthesis. This would explain why glucocorticoids alone do not increase AIB transport, and why they cannot alone support a fully-induced transport system.

Our model (Fig. 2) also includes a mechanism for degradation or inactivation of the active transport system. There is strong evidence that cycloheximide or puromycin inhibits or retards this mechanism (Pariza *et al.*, 1976a). Moreover, the data conform to the proposition that the mechanism for inactivation or degradation must itself be induced: short exposure to glucagon (30 minutes) permits the appearance of an active transport system at later time points in cells pre-incubated in dexamethasone or no hormones, whereas removal of glucagon from cells pre-treated 12 hours with glucagon plus hydrocortisone results in a rapid decay of the active transport system.

The model shown in Fig. 2 is at present a working hypothesis. We propose it to convey our current thinking concerning the presently available data (Potter, 1964). Elsewhere (Pariza *et al.*, 1976b) we have discussed how the "permissive" effect of glucocorticoids on the glucagon induction of amino acid transport in liver *in vivo* may depend upon a sequence of hormones modulated independently by external physiological stimuli. While it is very difficult to investigate complex hormonal interactions in the whole animal, rat liver parenchymal cells in primary monolayer culture are well-suited for such studies.

ACKNOWLEDGMENTS

The authors are grateful to Miss Joyce Becker and staff for cell culture media and facilities, and to Jon Shaw and Mark Kleinschmidt for technical assistance.

Financial support was provided in part by Training Grant T01-CA-5002 and Grants CA-07175 and R01-CA-17334 from the National Cancer Institute.

REFERENCES

Berry, M. N. and Friend, D. S. (1969). *J. Cell Biol. 43,* 506-520.

Bissell, M. D., Hammaker, L. E. and Meyer, U. A. (1973). *J. Cell Biol. 59,* 722-734.

Bonney, R. J., Walker, P. R. and Potter, V. R. (1973). *Biochem. J. 136,* 947-954.

Bonney, R. J., Becker, J. E., Walker, P. R. and Potter, V. R. (1974). *In Vitro 9,* 399-413.

Butcher, F. R., Becker, J. E. and Potter, V. R. (1971). *Exptl. Cell Res. 66,* 321-328.

Exton, J. H. (1972). *Metabolism 21,* 945-990.

Ingle, D. J. (1952). *J. Endocrinology.* xxxiii - xxxvii.

Kletzien, R. K., Pariza, M. W., Becker, J. E. and Potter, V. R. (1975). *Nature 256,* 46-47.

Kletzien, R. K., Pariza, M. W., Becker, J. E. and Potter, V. R. (1976a) *J. Biol. Chem., 251,* 3014-3020.

Kletzien, R. F., Pariza, M. W., Becker, J. E., and Potter, V. R. (1976b). *J. Cell. Physiol., in press.*

Michell, R. H. (1975). *Biochim. Biophys. Acta 415,* 81-147.

Odessey, R., Khairallah, E. A. and Goldberg, A. L. (1974). *J. Biol. Chem. 249,* 7623-7629.

Pariza, M. W., Becker, J. E., Yager, Jr., J. D., Bonney, R. J. and Potter, V. R. (1974). *In* "Differentiation and Control of Malignancy of Tumor Cells" (W. Nakahara, Ono, T., Sugimura, T. and Sugano, H., eds.), pp. 267-284. University Park Press, Baltimore.

Pariza, M. W., Yager, Jr., J. D., Goldfarb, S., Gurr, J. A., Yanagi, S., Grossman, S. H., Becker, J. E., Barber, T. A. and Potter, V. R. (1975). *In* "Gene Expression and Carcinogenesis in Cultured Liver" (L. E. Gerschenson and E. B. Thompson, eds.), pp. 137-167. Academic Press, New York.

Pariza, M. W., Butcher, F. R., Kletzien, R. F., Becker, J. E., and Potter, V. R. (1976a) *Proc. Nat. Acad. Sci. U.S.A. 73,* 4511-4515.

Pariza, M. W., Kletzien, R. F., Butcher, F. R. and Potter, V. R. (1976b). *Adv. Enzyme Regulation, 14,* 103-115.

Park, C. R. (1974). *In* "Regulation of Hepatic Metabolism" (F. Lundquist and N. Tygstrup, eds.), p. 112. Academic Press, New York.

Potter, V. R. (1964). *Nat. Cancer Inst. Monograph 13,* 111-116.

Schaeffer, L. D., Chenoweth, M. and Dunn, A. (1969). *Biochim. Biophys. Acta 192,* 292-303.

Thompson, E. B. and Lippman, M. E. (1974). *Metabolism 23,* 159-202.

Tolbert, M. E. M., Butcher, F. R. and Fain, J. N. (1973). *J. Biol. Chem. 248,* 5686-5692.

Zahlten, R. N. (1974). *New England J. Medicine 290,* 743-744.

CYTOPLASMIC MESSENGER RIBONUCLEOPROTEINS AND TRANSLATIONAL CONTROL DURING GROWTH AND DIFFERENTIATION OF EMBRYONIC MUSCLE CELLS

SATYAPRIYA SARKAR
JNANANKUR BAG

Department of Muscle Research
Boston Biomedical Research Institute
Department of Neurology
Harvard Medical School
Boston, Massachusetts

I. INTRODUCTION

Although a large amount of information about the physicochemical properties of the myofibrillar proteins and their role in the contraction-relaxation cycle of muscle tissue is currently available (for a review see Taylor, 1972), very little is known about myofibrillogenesis and its regulation during the growth and differentiation of muscle cells. Recent studies from a number of laboratories have suggested that translational control may exert a subtle regulation of protein synthesis during myogenesis of the skeletal and cardiac muscle fibers. In cultures of skeletal muscle cells, which are used by many investigators as a model system for studying various aspects of terminal differentiation, the mononucleated rapidly proliferating muscle cells undergo several cycles of cell division, then withdraw from the mitotic cycle and fuse to form a multinucleated embryonic myotube (for a review see Holtzer *et al.*, 1972). It is at this stage, namely the time of cell fusion and the formation of the myotube, that the intensive synthesis of myosin and a number of muscle-specific enzymes (reviewed by Yaffe and Dym, 1972) is first observed. In cultures of rat myoblast cells treatment with actinomycin D at, or just prior to, fusion does not block the immediate appearance of myosin and other muscle-specific proteins (Yaffe and Dym, 1972). Buckingham *et al.* (1974) have recently reported that during the transition from non-differentiated myoblasts to the differentiated myotube stage, rapidly labeled poly (A)-containing RNAs of different sizes are stabilized with a significant increase in their half-lives. They have shown that a 26 S RNA species, considered as the putative mRNA coding for the 200,000 dalton large subunit of myosin (also referred to as the myosin heavy chain), accumulated as a free messenger ribonucleoprotein (mRNP) particle sedimenting as 110-120 S in sucrose gradients of cell lysates. After cell fusion this RNA species was located by pulse-chase experiments in a group of very large polysomes (40-60 ribosomes) which are known to synthesize myosin heavy chain in embryonic muscles (Sarkar and Cooke, 1970; Heywood and Rich, 1968). Chacko and Xavier (1974) have shown that when precardiac mesodermal cells, which are the progenitors of cardiac tissue, were treated with 5-bromodeoxyuridine (BrdU), a thymidine analog or actinomycin D at a non-differentiated stage of growth in culture, they were able to continue growth and express their differentiated phenotype. These studies suggest that critical regulatory controls of gene expression may operate at a post-transcriptional level during the terminal differentiation of skeletal and cardiac muscle cells. However, the precise biochemical nature of these regulatory mechanisms remains to be understood.

Actin and myosin or myosin-like proteins are also present in a wide variety of non-muscle eukaryotic cells (for a review see Pollard and Weihing, 1974). It has been suggested that non-muscle (also referred to as cytoplasmic) actin and myosin may play important roles in processes such as cellular motility, adhesion, contraction and cytokinesis in eukaryotic cells. Rubinstein *et al.* (1974) have recently suggested that the cytoplasmic contractile system found in almost all eukaryotic cells, including myoblasts and BrdU-suppressed myogenic cells (Holtzer *et al.*, 1972), may be controlled by a genetic program which is insensitive to BrdU. In contrast, the myofibrillar contractile system, which is specific for differentiated muscle cells, may arise from the readout of a separate genetic program whose expression is sensitive to BrdU (Rubinstein *et al.*, 1974). If indeed this is the case, the mechanisms by which the expression of these different genetic programs is regulated in a myogenic cell remain to be understood.

The polymorphism of various muscle proteins adds another complexity in the problem of differentiation of muscle tissue. Recent evidence suggests that at least in the case of myosin, a major heteropolymeric myofibrillar protein, different sets of structural genes coding for myosin subunits (also referred to as the light and heavy chains) are expressed in various types of differentiated muscle fibers (e.g. slow, fast, cardiac, smooth) resulting in different isozymic forms of myosin characteristic of each muscle type (Sarkar *et al.*, 1971; Sarkar, 1972; Lowey and Risby, 1971; Weeds and Frank, 1972; for a review see Taylor, 1972). The expression of such a "family of genes" coding for the polymorphic forms of muscle proteins must be precisely regulated during the life cycle of the muscle cell.

In order to elucidate the regulatory mechanisms involved in myogenesis we have selected the myosin heavy chain and actin as two suitable muscle-specific markers and probed whether the biosynthesis of these proteins is regulated at a cytoplasmic level. We report here the isolation and characterization of cytoplasmic nonpolysomal mRNP particles containing actin and myosin heavy chain mRNAs from embryonic muscle cells. The significance of these results in relation to some of the cellular processes in muscle cells, outlined above, is also discussed.

II. RESULTS

A. Subribosomal mRNP Particles Containing Actin mRNA

Since a large fraction of cellular mRNAs coding for 15,000-35,000 dalton polypeptides sediment in sucrose gradients between 8-20 S (for a review see Brawerman, 1974), the expected size of the cytoplasmic mRNP particles, which may contain such mRNAs, should be within the range 20-40 S (Spirin, 1972). A crude subribosomal pellet of embryonic muscle cells was, therefore, used to search for nonpolysomal mRNP particles. The subribosomal pellet was obtained by centrifugation of the postribosomal supernatant of homogenates of 14-day old chick embryonic leg and breast muscles at 255,000 g for 4 hr and details of the preparative procedures are described in our published reports (Bag and Sarkar, 1975; Bag *et al.*, 1975). When subribosomal particles were centrifuged through linear 5-20% sucrose gradients, about 60% of the uv-absorbing material sedimented as a broad peak of about 8-10 S (Fig. 1). The remaining material showed a heterogeneous distribution sedimenting between 16-40 S (tubes, 9-25, Fig. 1). Based on two considerations, size and heterogeneity of the sedimentation profile, this 16-40 S fraction of the gradient runs was pooled to look for cytoplasmic mRNP particles.

When the RNA isolated from the 16-40 S particles was analyzed by centrifugation through 5-20% linear sucrose gradients, the profile shown in Fig. 2 was obtained. The major part of the RNA sedimented as a broad peak of about 18-20 S. However, a significant amount of the uv-absorbing material showed a very heterogeneous profile with S values equal to and larger than that of 28

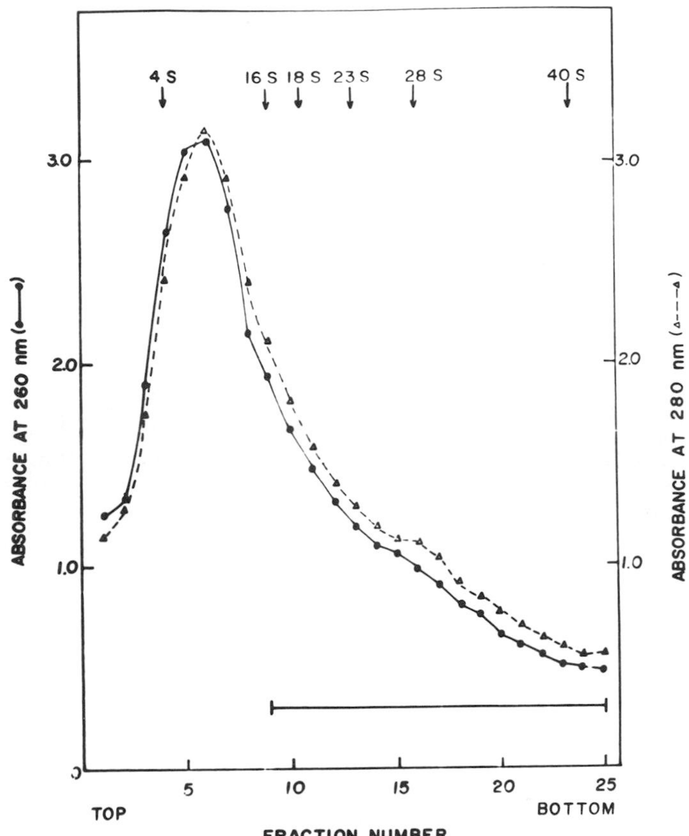

FIGURE 1 *Sedimentation profile of subribosomal particles in 5-20% linear sucrose gradients in 0.01 M Tris-HCl, pH 7.6 containing 0.5 M KCl, 0.002 M EDTA and 500 μg/ml of heparin. Centrifugation was done for 17 hr at 24,000 rpm at 2° in a Spinco SW 25.1 rotor. For details see Bag and Sarkar (1975). Fractions indicated by the bar (16-40 S) were pooled for the isolation of RNA fractions with messenger activity. Reproduced from Bag and Sarkar (1975).*

S rRNA run as a marker. Morris *et al.* (1972) have previously reported that a 10-17 S RNA fraction isolated from muscle polysomes can direct the synthesis of actin in a heterologous cell-free system. We, therefore, pooled an 8-20 S fraction (indicated by the bar in Fig. 2) from the gradient runs in order to test whether

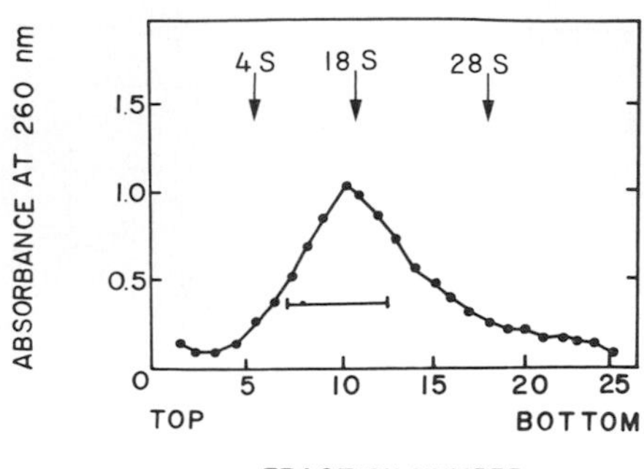

FIGURE 2 *Sedimentation profile of RNA fractions isolated from the pooled 16-40 S subribosomal particles (Fig. 1). Centrifugation was done at 2°; for 17 hr at 24,000 rpm in a Beckman SW 25.1 rotor. For details see Bag and Sarkar (1975). The bar indicates the pooled fractions (8-20 S), used for in vitro translation. Reproduced from Bag and Sarkar (1975).*

this nonpolysomal RNA fraction indeed programs the synthesis of actin or any other muscle protein of similar (40,000 - 50,000 dalton) size.

The preincubated S-30 system prepared from wheat germ embryos according to the method of Roberts and Paterson (1973) was used to test the mRNA activity of the RNA fractions isolated from the 16-40 S particles. Both the heterogeneous RNA isolated from these particles (tubes 9-25 in Fig. 1) and the 8-20 S RNA fraction (Fig. 2) strongly stimulated amino acid incorporation, about 12-15 fold, in this system (Table I). This stimulatory activity was comparable to that obtained with globin mRNA isolated and partially purified from rabbit reticulocyte polysomes. These results indicate that the RNA fraction associated with the 16-40 S subribosomal particles exhibits a strong messenger activity in an *in vitro* cell-free system.

TABLE I Stimulation of Amino Acid Incorporation by RNA Fractions Isolated from 16-40 S Subribosomal Particles in Wheat Germ Embryo Cell-Free[a] System

Source of RNA	μg of RNA used	$[^{35}S]$ met incorporated (cpm/assay)	Degree of stimulation
Minus RNA		2,400	Control
Rabbit globin mRNA	2.5	30,000	12.5-fold
Chicken 4S RNA	80	2,600	
Chicken 28S rRNA	10	2,300	
Total RNA from pooled tubes 9-25 (Fig. 1)	12	29,000	12-fold
8-20S RNA fraction (Fig. 2)	8	36,000	15-fold

[a]Assays were performed as described by Bag and Sarkar (1975).

B. In Vitro Translation of actin mRNA isolated from subribosomal Particles

Incubation mixtures using wheat germ embryo S-30 system programmed with the 8-20 S RNA fraction, as described above, were used for the identification of actin as one of the possible products of cell-free translation. After addition of highly purified chicken actin as carrier and dialysis to remove excess radioactive amino acids, the *in vitro* products were copurified by three cycles of reversible salt-dependent transformation of globular (G) to fibrillar (F) actin. Polymerization of the products to F-actin was done by dialyzing the sample against 5 m*M* N-2-hydroxyethylpiperazine-N'-2-ethane sulfonic acid (Hepes), pH 7.0, 0.1 *M* KCl and 0.003 *M* $MgCl_2$. The F-actin and ribosomes were pelleted by centrifugation at 160,000 g for 2 hr. The sample was then dialyzed against a depolymerizing medium, 5 m*M* Hepes, pH 7.0, 0.2 m*M* $CaCl_2$ and 0.2 m*M* ATP and the ribosomes were pelleted at 160,000 g for 2 hr leaving the soluble G-actin in the supernatant. Details of these steps are described elsewhere (Bag and Sarkar, 1975; Bag

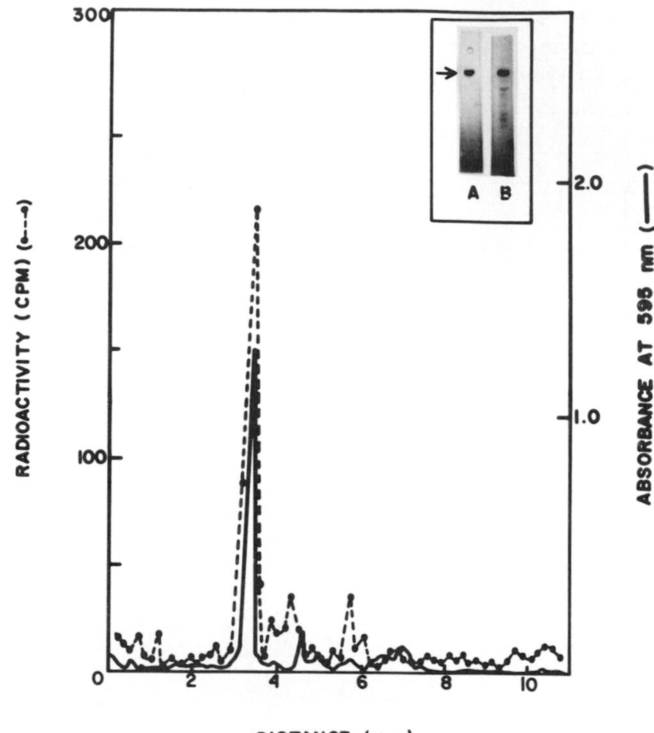

FIGURE 3 *Sodium dodecyl sulfate polyacrylamide gel electrophoresis of labeled products of in vitro translation of 8-20 S RNA copurified with chicken actin. F-actin samples containing 1400 cpm (specific activity 25,000-30,000 cpm/mg) were analyzed using 10.5 cm of 10% gels at 4 mA/gel for 4 hr. The gels were stained with Coomassie brilliant blue, scanned, sliced and processed for radioactivity (Bag and Sarkar, 1975). Insert: Electrophoretograms of gel runs of purified actin (gel A) and the purified labeled product (gel B). Reproduced from Bag and Sarkar (1975).*

et al., 1975). About 60% of the actin, initially added as carrier, was recovered in the final pellet of F-actin after three cycles of copurification. The final pellet contained about 10% of the initial nondialyzable radioactivity incorporated in the cell-free system.

When portions of the F-actin pellet, purified as described above, were analyzed by acrylamide gel electrophoresis in the

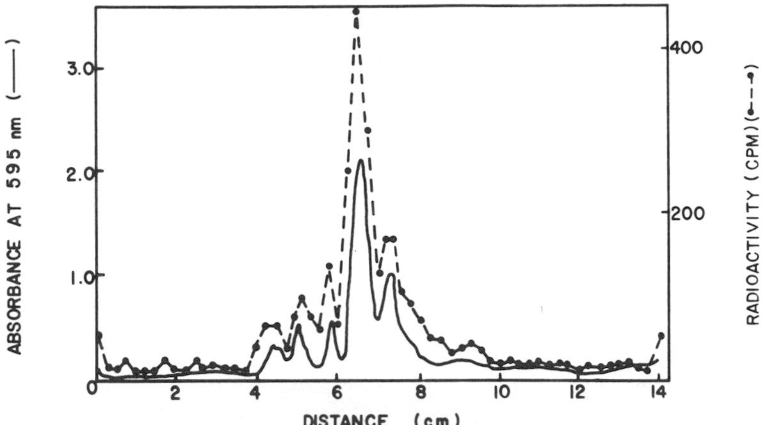

FIGURE 4 *Sodium dodecyl sulfate polyacrylamide gel electrophoresis of CNBr peptides of purified products of in vitro translation of 8-20 S RNA. For details see Bag and Sarkar (1975). Samples containing 3000-3500 cpm were analyzed using 14 cm 12.5% polyacrylamide high cross-linked gels (bisacrylamide ratio of 1:10) in the presence of 8 M urea and 1% dodecyl sulfate. Gels were run at 2 mA/gel for 20 hr, and then stained, scanned at 595 nm, sliced and counted. Reproduced from Bag and Sarkar (1975).*

presence of sodium dodecyl sulfate, about 70% of the radioactivity applied to the gel was recovered in a densitometric peak of the actin band (Fig. 3). A number of additional minor bands, also observed in the radioactivity profile, may be due to other polypeptides synthesized in the *in vitro* system, such as tropomyosin and troponin components of the thin filaments. These proteins are copurified with F-actin due to strong mutual interactions and migrate in the same region of the gel where the minor radioactive peaks (Fig. 3) are observed (Potter and Gergely, 1974).

When portions of the *in vitro* products, copurified with F-actin as described above, were electrophoresced in the presence of 8 *M* urea and absence of sodium dodecyl sulfate, the only major radioactive peak obtained in the gel run comigrated with the actin band (data not presented here; Bag and Sarkar, 1975). About 80% of the radioactivity applied to the gel was recovered in the peak corresponding to actin band. These results indicate that as judged by mobilities in two kinds of polyacrylamide electrophoretic systems, one based on charge and the other based on size, the *in vitro* products showed properties similar to that of muscle actin.

In order to prove that the 8-20 S RNA fraction indeed programs the *de novo* synthesis of the 42,000 dalton polypeptide chain of actin, the *in vitro* products were labeled with a mixture of ^{14}C-labeled amino acids and then copurified with chicken actin, as described above. The sample was aminoethylated and cleaved with CNBr and the resulting peptides were analyzed by sodium dodecyl sulfate polyacrylamide gel electrophoresis using high cross-linked gels, as described by Swank and Munkres (1971). The results are shown in Fig. 4. The densitometric scan shows that one major and five minor peaks were resolved in the gel run. Some of these peaks presumably represent mixtures of peptides which are not completely resolved (Elzinga, 1970). The correspondence of radioactive peaks from the *in vitro* translated products with the densitometric peaks and the absence of any significant amount of radioactivity in other regions of gel indicate that total synthesis of the polypeptide chain was achieved. Control experiments in which the 8-20 S RNA was omitted did not give any detectable amount of radioactivity above the background level in the actin band. These results indicate that the 8-20 S RNA fraction indeed directs the *de novo* synthesis of muscle actin in a heterologous translation system.

C. Myosin Heavy Chain mRNP in Embryonic Muscle Cells

The mRNA coding for the 200,000 dalton chick embryonic myosin heavy chain has been recently purified and characterized in our laboratory. The purified mRNA has a molecular weight of about 2.23×10^6 consisting of about 6400 nucleotides with a poly(A) segment of about 170 adenylic acid residues at the 3' end (Sarkar *et al.*, 1973; Mondal *et al.*, 1974; Sarkar, 1974, 1975). The myosin heavy chain mRNA sediments in sucrose gradients as 25-27 S

(Heywood and Nwagwu,1969) but migrates in polyacrylamide gels as 30-32 S (Sarkar *et al.*, 1973; Mondal *et al.*, 1974). Due to the very large size of myosin heavy chain mRNA, the expected size of an mRNP particle containing this mRNA should be much larger than the 40 S ribosomal subunit (Spirin, 1972). Therefore, the post-polysomal supernatant of embryonic muscle cells containing 80 S monosomes, 60 S and 40 S ribosomal subunits was selected in order to look for myosin heavy chain mRNP. This choice was justified by our initial experiments which showed that an RNA fraction, 24-32 S, isolated from the post-polysomal supernatant of embryonic muscle cells, showed a strong mRNA activity in the wheat germ embryo S-30 system.

The procedure used for the isolation of myosin heavy chain mRNP includes the following steps: centrifugation of the post-polysomal supernatant (Bag and Sarkar, 1975) at 255,000 g for 2 hr to obtain a pellet of oligosomes, monosomes, ribosomal subunits and myosin heavy chain mRNP; treatment of the post-polysomal particles with EDTA and separation of the ribosomal subunits from myosin heavy chain mRNP by centrifugation in sucrose gradients containing EDTA. Since ribosomal subunits are unfolded in the presence of EDTA whereas mRNP particles are resistant to EDTA treatment, the myosin heavy chain mRNP sedimented in the bottom one-fifth of the sucrose gradients where particles larger than 80 S were located. The myosin heavy chain mRNP was detected by its characteristic A_{260}/A_{280} ratio of about 1.2-1.3, which is lower than the corresponding value (1.6-1.8) of ribosomal subunits. Details of these preparative methods are described elsewhere (Bag *et al.*, 1975; Bag and Sarkar, 1976). When samples of myosin heavy chain mRNP, pooled from the gradient fractions, were recentrifuged through a second sucrose gradient in the presence of 0.5 *M* KCl, the mRNP was further purified, free from the contaminating ribosomal subunits. The results are shown in Fig. 5. About 60-65% of the uv-absorbing material applied to the sucrose gradient was recovered in a distinct peak which

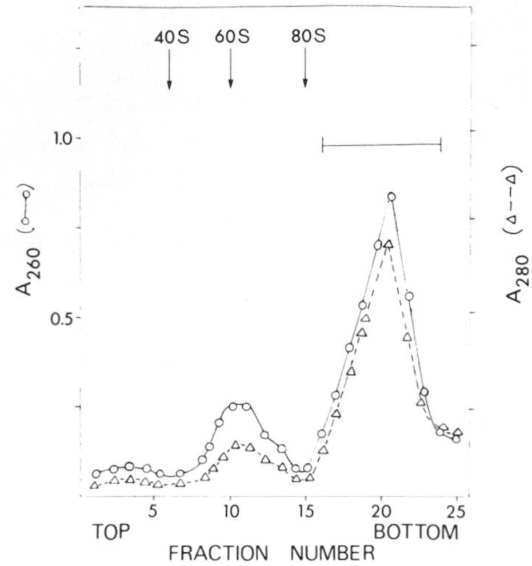

FIGURE 5 *Purification of myosin heavy chain mRNP by sucrose gradient centrifugation. Crude samples of myosin heavy chain mRNP obtained by sucrose gradient centrifugation of EDTA-treated postpolysomal particles were pooled from gradient fractions and centrifuged through a 15-40% linear sucrose gradient in 20 mM Tris-HCl, pH 7.6; 0.5 M KCl; 0.005 M magnesium acetate at 47,000 rpm for 90 min in a Spinco SW 50.1 rotor. For details see Bag et al. (1975). Fractions (16-25), indicated by the bar, were pooled and processed further for the isolation of myosin heavy chain mRNA.*

showed an A_{260}/A_{280} ratio of about 1.2. The approximate S value of this peak calculated by the method of Martin and Ames (1961) was about 120. When the RNA isolated from the 120 S particles was assayed for mRNA activity in the wheat germ embryo S-30 system, a strong stimulatory activity, about 10-12 fold, was observed (Bag *et al.*, 1975).

The RNA isolated from the 120 S particles (Fig. 5) was analyzed by agarose-polyacrylamide gel electrophoresis using chick embryonic muscle ribosomal RNAs (28 S and 18 S) and myosin heavy chain mRNA purified from the myosin-synthesizing polysomes (Sarkar *et al.*, 1973; Mondal *et al.*, 1974; Sarkar, 1974, 1975) as markers. The results are shown in Fig. 6. Densitometric scans

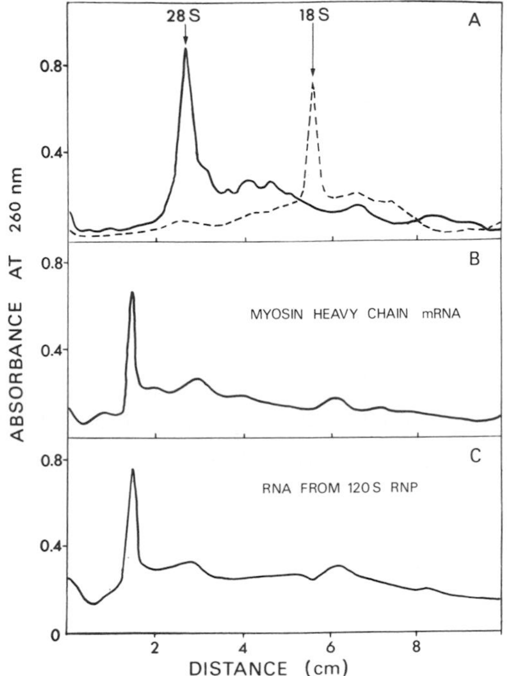

FIGURE 6 *Densitometric scans of polyacrylamide gel runs of RNA isolated from 120 S mRNP particles. For details see Bag et al. (1975). Panel A:* —— *30 μg of 28 S chick embryonic muscle rRNA;* ---- *25 μg of 18 S rRNA. Panel B:* —— *28 μg of myosin heavy chain mRNA purified from myosin synthesizing polysomes (Sarkar et al., 1973; Mondal et al., 1974). Panel C:* —— *32 μg of RNA isolated from 120 S mRNP particles (Fig. 5). Electrophoresis was carried out using 2.5% gels containing 0.5% agarose cross-linked with 0.175% bisacrylamide. For details see Sarkar et al. (1973). The gels were scanned at 260 nm in a Gilford recording spectrophotometer fitted with a linear transport device for scanning gels.*

of the gel runs indicate that purified myosin heavy chain mRNA isolated from muscle polysomes migrated slower than the 28 S rRNA (panels A and B). This is in agreement with the published reports on the electrophoretic mobility and size of myosin heavy chain mRNA (Mondal *et al.*, 1974; Sarkar, 1974, 1975; Morris *et al.*, 1972). The RNA isolated from the 120 S particles gave a densitometric peak which corresponded to that of purified myosin heavy chain mRNA (panels B and C). These results indicate that a unique species of RNA, which is larger than 28 S ribosomal RNA and is identical to myosin heavy chain mRNA in electrophoretic mobility, is present in the 120 S particles.

D. In Vitro Translation of mRNA Isolated from 120 S Myosin Heavy Chain mRNP Particle

Based on the results presented in the preceding section, it was of interest to test whether the RNA isolated from the 120 S particles can indeed direct the synthesis of myosin heavy chain in a heterologous cell-free system. The rabbit reticulocyte lysate system, previously used by us for *in vitro* translation of highly purified myosin heavy chain mRNA (Mondal *et al.*, 1974; Sarkar, 1974, 1975) was used for these studies. The products, labeled with a mixture of [^{14}C] amino acids, were subjected to the following purification steps: coprecipitation at low ionic strength with highly purified chick embryonic myosin added as carrier after removal of excess radioactive amino acids by dialysis; chromatography of the myosin on a column of DEAE-cellulose; and polyacrylamide gel electrophoresis in the presence of sodium dodecyl sulfate of the column-purified myosin. Details of these methods have been previously reported by us (Sarkar and Cooke, 1970; Sarkar *et al.*, 1973; Mondal *et al.*, 1974).

The radioactivity profile and the densitometric scan of the sodium dodecyl sulfate-gel runs of the *in vitro* products, purified by DEAE-cellulose chromatography, are shown in Fig. 7. About 70% of the counts applied to the gel was recovered in a peak which corresponded to the densitometric peak of the 200,000 dalton myosin heavy chain. No detectable amount of radioactivity above the background level was found in the myosin light chain regions of the gels (indicated as LC_1 and LC_2 in Fig. 7). This indicates that the mRNAs coding for the light chains (average molecular weight of 20,000) was absent in the 120 S particles.

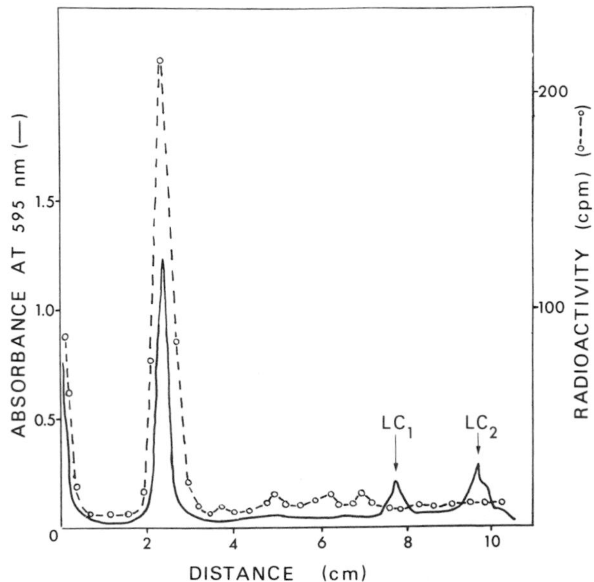

FIGURE 7 *Sodium dodecyl sulfate polyacrylamide gel electrophoresis of labeled products of in vitro translation of RNA isolated from 120 S mRNP particles. The products were copurified with chick embryonic leg muscle myosin added as a carrier by low ionic strength precipitation and DEAE-cellulose chromatography. For details see Bag et al. (1975). The purified products containing 1200-1400 cpm (specific activity 25,000 cpm/mg) were analyzed using 10.5 cm of 5% gels at 4 mA/gel for 2.5 hr. After staining with Coomassie brilliant blue the gels were scanned at 595 nm, sliced and counted as described by Bag and Sarkar (1975). The migration of the light chains present in the carrier myosin are indicated with arrows and marked as* LC_1 *and* LC_2.

Control experiments using the reticulocyte lysate alone, incubated in the absence of exogenously added mRNA, indicated that globin was predominantly synthesized. As judged by the absence of any detectable level of counts above the background in the myosin heavy band (data not shown), it is concluded that the reticulocyte lysate alone does not synthesize any myosin heavy chain. These results indicate that the translated products of the RNA from 120 S particles behaved like chick embryonic myosin heavy chain as judged by a number of properties.

E. Proteins Associated with Actin and Myosin Heavy Chain mRNP Particles

In order to prove that the cytoplasmic mRNP particles described in the previous section are true cellular entities, the nature of the protein moieties of these particles was investigated using two approaches. The relative protein and RNA contents were estimated by determining the buoyant densities after fixing the mRNP particles with formaldehyde (Spirin, 1972). The protein components of these particles were also analyzed by sodium dodecyl sulfate polyacrylamide gel electrophoresis.

When samples of formaldehyde-fixed 120 S particles and 16-40 S subribosomal particles were centrifuged to equilibrium in preformed CsCl gradients, the absorbance profile gave a single peak at a density of 1.40 g/cm^3 (Fig. 8). The protein content of these particles, calculated from this value by the method of Spirin (1972), amounted to 75%. In contrast, the 80S monosomes prepared from embryonic chick muscles gave a peak at a density of 1.57 g/cm^3 (Fig. 8). The buoyant densities of embryonic muscle ribosomal subunits usually ranged between 1.58-1.585 for 60 S and 1.52-1.53 g/cm^3 for 40 S subunits, respectively (data not shown here; Bag and Sarkar, 1976). These values are in good agreement with those reported in the literature for eukaryotic ribosomes and ribosomal subunits (for a review see Spirin, 1972). As judged by the absorbance profile and the absence of any detectable peak in the density range of 1.5-1.6 g/cm^3, the mRNP particles were free of contaminating ribosomal subunits. Furthermore, the characteristic buoyant density of 1.4 g/cm^3 appears to be a common property of muscle mRNP particles irrespective of their sizes, namely 120 S versus 16-40 S and the nature of the mRNA present, e.g. actin mRNA versus myosin heavy chain mRNA. The buoyant density of the cytoplasmic muscle mRNP particles reported here is in good agreement with the published values in the literature which were obtained mainly by pulse-labeling of RNA in order to detect free mRNPs in a variety of eukaryotic systems

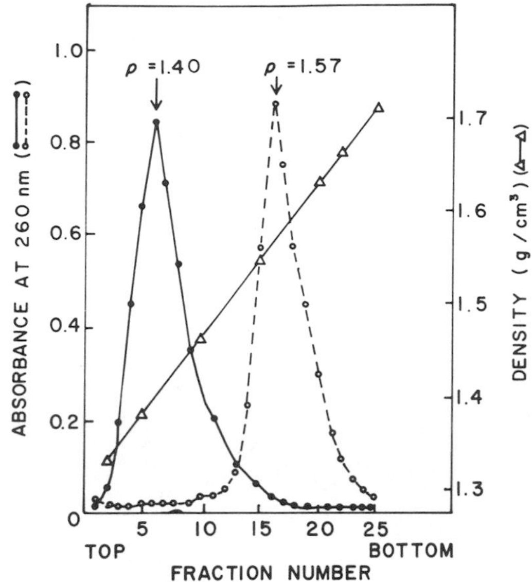

FIGURE 8 *Buoyant densities of cytoplasmic mRNP particles and 80 S chick embryonic muscle monosomes. For details see Bag and Sarkar (1975). The pooled subribosomal fractions (16-40 S) as indicated in Fig. 1 and the 120 S particles (Fig. 5) were used as mRNP particles. Centrifugation was done at 35,000 rpm for 40 hr in a Beckman SW 50.1 rotor at 20°* *using preformed CsCl gradients.* ●, *absorbance of mRNP particles. The same profile was obtained with both 16-40 S and 120 S particles.* o, *absorbance of monosomes.* Δ, *density of gradient fractions.*

(for a review see Spirin, 1972). Since RNA is pelleted in CsCl gradients, the absence of any detectable pellet in the CsCl gradient runs of mRNP particles makes it highly unlikely that the mRNA activity reported in the previous sections is due to any protein-free mRNA species present in the preparations.

We have also analyzed the carboxymethylated protein samples of ribosomal subunits and mRNP particles by sodium dodecyl sulfate polyacrylamide gel electrophoresis. The electrophoretograms are shown in Fig. 9. The majority of typical ribosomal proteins in the molecular weight range of 15,000-30,000, which migrated in the middle region of the gel runs of 60 S and 40 S subunits (gels A and B), could not be detected in the mRNP particles (gel C). A

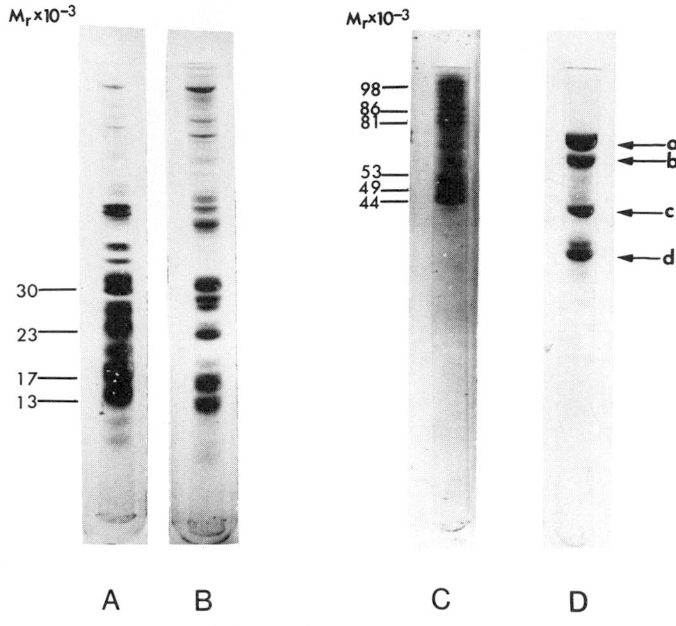

FIGURE 9 *Sodium dodecyl sulfate polyacrylamide gel electrophoresis of mRNP particles and chick embryonic muscle ribosomal subunits. For details see Bag and Sarkar (1975) and Bag et al. (1975). Electrophoresis of carboxymethylated samples of mRNP particles and ribosomal subunits was carried out using 10 cm long 10% gels at 4 mA/gel for 4 hr. The gels were stained with Coomassie brilliant blue. The amounts of samples applied to the gels are: (A) 0.3 A_{280} unit of 60 S ribosomal subunit; (B) 0.4 A_{280} unit of 40 S ribosomal subunit; (C) 0.2 A_{280} unit of 120 S mRNP (Fig. 5); (D) a mixed sample of the following markers: 12 μg of conalbumin (molecular weight 78,000); 8 μg of bovine serum albumin (molecular weight 68,000); 6 μg of chicken actin (molecular weight 42,000) and 8 μg of chicken tropomyosin (molecular weight 34,000). The bands corresponding to these 4 marker proteins are indicated with arrows (a-d) in decreasing order of molecular weight.*

number of distinct polypeptides, seven to eight, ranging in molecular weight from 44,000-100,000, which do not appear to represent the typical ribosomal proteins, were present in the upper one-third segment of the electrophoretograms of the mRNP particles (gel C). The major band of these proteins is a 44,000 component which was also absent in the ribosomal subunits. As judged by these characteristic gel patterns and the buoyant densities, it is concluded that the mRNP particles are cytoplasmic entities unrelated to and distinct from monoribosome, ribosomal subunits and mRNA-containing "initiation complexes" of embryonic muscle cells.

F. Are the Protein Moieties Identical in All mRNP Particles

In order to test whether cytoplasmic mRNP particles containing mRNAs coding for different myofibrillar proteins contain the same or different polypeptide components, the proteins of two mRNP particles of unequal size, one containing myosin heavy chain mRNA and the other containing actin mRNA, were compared by dodecyl sulfate-polyacrylamide gel electrophoresis. The results are shown in Fig. 10. The electrophoretograms of actin mRNP (gel 2) and a mixed sample of actin and myosin heavy chain mRNPs (gel 3) show that the three faster components in the 44,000-53,000 dalton range (see also Fig. 9 for assignment of molecular weights) are present in both mRNP particles. Three distinct and a number of minor protein bands in the molecular weight range of 44,000-53,000 (see also Fig. 9) were consistently found in gel runs of myosin heavy chain mRNP. Although protein bands of similar mobilities were also found in electrophoretograms of actin mRNP particles, the relative amounts of these high molecular weight components were always lower than those present in the myosin heavy chain mRNP (panel A, gels 1 and 2). When mixed samples of the two mRNPs were electrophoresed, the three faster components co-migrated together (panel A, gel 3). This pattern did not change even when electrophoresis was run for a longer time (panel B), or a combination of decreased sample load and increased time of electrophoresis was used to facilitate resolution of the bands (panel C). The results indicate that the protein components of different mRNP particles are very similar. However, the subtle variation in the intensity and number of protein bands in the 81,000-86,000 dalton range of actin and myosin heavy chain mRNP

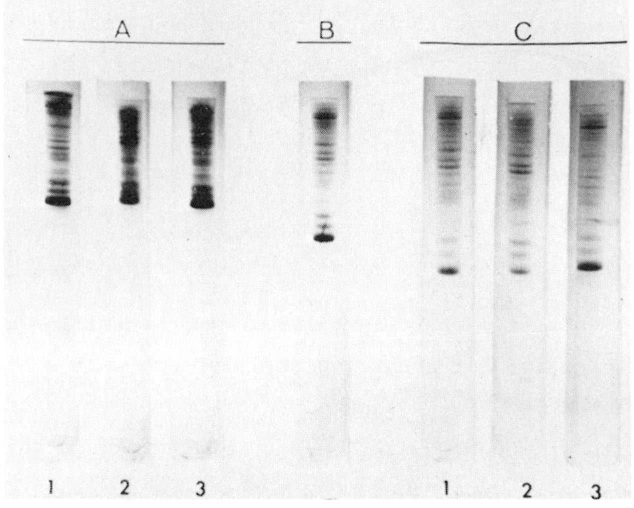

FIGURE 10 *Electrophoresis of carboxymethylated proteins of myosin heavy chain mRNP and actin mRNP particles. For details see Bag and Sarkar (1975) and legends to Fig. 9. The following amounts of samples were used: Panel A: 0.08 A_{280} unit of actin mRNP (gel 1); 0.1 A_{280} unit of myosin heavy chain mRNP (gel 2); and a mixture of 0.04 A_{280} unit of actin mRNP and 0.05 A_{280} unit of myosin heavy chain mRNP, Panel B: a mixture of 0.04 A_{280} unit of actin mRNP and 0.05 A_{280} unit of myosin heavy chain mRNP. Panel C: 0.03 A_{280} unit of actin mRNP (gel 1); 0.04 A_{280} unit of myosin heavy chain mRNP (gel 2); and a mixture of 0.03 A_{280} unit of actin mRNP and 0.04 A_{280} unit of myosin heavy chain mRNP (gel 3). Conditions for electrophoresis was increased to 5 hr for the mixed sample of mRNPs shown in panel B and 6.5 hr for the gels shown in panel C.*

particles suggests that there may be subpopulations of mRNP particles containing additional or different proteins. In order to understand the molecular basis of this phenomenon we are continuing these studies using mRNP particles prepared and purified by different methods.

G. In Vitro Translation of mRNP Particles

Since mRNP particles have been implicated in the process of stabilization of mRNAs during the transition of myoblasts to myotube stages (Buckingham *et al.*, 1974), it is of interest to study whether or not the mRNP particles represent "non-translatable"

TABLE II Translation of Myosin Heavy Chain mRNP Particle and Protein-free mRNA in Rabbit Reticulocyte Lysate System[a]

mRNA added μg	mRNP added (equivalent μg RNA)	% of total cpm incorporated in myosin heavy chain band[b]
-	-	Not detectable
-	6	2.65
-	10	5.10
-	15	5.70
-	20	5.84
-	30	5.71
5	-	2.95
10	-	5.84
15	-	7.43
20	-	8.31
30	-	8.37

[a]Translation of 120 S myosin heavy chain mRNP (Fig. 5) and protein-free mRNA isolated from the mRNP were done as described by Bag *et al.* (1975).

[b]The *in vitro* products copurified with carrier myosin (for details see Mondal *et al.*, 1974; Bag *et al.*, 1975) were electrophoresed and the radioactivities in the myosin heavy chain bands were determined.

forms of mRNAs. If the mRNP particles are found to be inactive in an *in vitro* translation system, this would suggest that they may regulate the biosynthesis of myofibrillar proteins through a negative control, i.e., the mRNP particles in the myoblast stage are blocked in translation. Removal of this negative control, e.g., in the myotube stage, would lead to an accelerated rate of cell-specific protein synthesis. In order to test this possibility we have compared the relative efficiency of *in vitro* translation of myosin heavy chain mRNP with that of protein-free myosin heavy chain mRNA using the rabbit reticulocyte lysate system.

The products from the *in vitro* incubations were copurified with carrier myosin and analyzed by sodium dodecyl sulfate gel electrophoresis as described in the previous section. The radioactivity recovered in the myosin heavy chain band is calculated as per cent of the total radioactivity incorporated in the cell-free system. The results presented in Table II indicate that the myosin heavy chain mRNP particle was able to program the *de novo* synthesis of the myosin heavy chain in a heterologous system. Further, at low concentrations of mRNA or equivalent amount of mRNP the relative efficiencies of translation did not show any significant difference. However, at saturating concentration about 30% higher incorporation was achieved with protein-free mRNA. These results clearly indicate that the mRNP particle, at least in the rabbit reticulocyte lysate system, does not exhibit any stringent negative control in translation. Since it is quite likely that the *in vivo* regulation of translation of mRNAs may be different from the *in vitro* conditions, other possible interpretations regarding the role of mRNP particles in myogenesis during the terminal differentiation of muscle cells should also be carefully considered. Further discussion along these lines is presented in the next section.

III. DISCUSSION

The studies presented here show that mRNAs coding for two important myofibrillar proteins, actin and myosin heavy chain, are present as nonpolysomal cytoplasmic mRNP particles in embryonic muscle cells. These particles are true cytoplasmic entities unrelated to ribosomes, as shown by the absence of typical ribosomal proteins, by their characteristic buoyant density and the presence of a unique class of nonribosomal protein moieties firmly bound to them. The mRNAs isolated from these particles direct the synthesis of polypeptides in cell-free systems which are indistinguishable from authentic skeletal muscle actin and myosin heavy chain. These results, considered together with the fact that the myosin heavy chain mRNP itself can be translated in heterologous system, indicate that the mRNAs are present in these particles in a form which can be translated without further processing.

It is currently believed that mRNAs exist as ribonucleoprotein particles in the nucleus and cytoplasm of a large number of eukaryotic cells (for a review see Spirin, 1972). However, the nature and possible role of these proteins bound to these particles is not thoroughly understood. Cytoplasmic polyribosomal mRNPs isolated by dissociation of polysomes from a number of species and tissues contain, as judged by sodium dodecyl sulfate gel electrophoresis, two major protein bands of approximate molecular weights 52,000 and 78,000 and a large number of minor bands ranging from 6-13 (for a review see Blobel, 1973). The 78,000 dalton component, which is present in both nuclear and polysomal mRNPs, is specifically bound to the poly (A) segment at the 3'-end of eukaryotic mRNAs (Blobel, 1973; Quinlan *et al.*, 1974; Firtel and Pederson, 1975). The results presented here indicate that the cytoplasmic muscle mRNP particles contain seven to eight distinct polypeptides including a prominent 44,000 dalton component. Among these major protein components are bands corresponding to the 52,000 and 78,000 dalton components previously reported in many eukaryotic mRNPs. The electrophoretic pattern of the cytoplasmic mRNA-associated proteins described in these studies supports the recent view that proteins associated with eukaryotic mRNPs are much more complex with respect to size and number than previously thought (for a review see Barrieux *et al.*, 1975).

Previous reports have shown that specific mRNP particles, viz. globin α-chain mRNP, globin α and β chain mRNPs and histone mRNP, are present in rabbit reticulocytes (Jacobs-Lorena and Baglioni, 1972), chick erythrocytes (Spohr *et al.*, 1972) and unfertilized sea urchin eggs (Gross *et al.*, 1973) respectively. While the sea urchin embryo during the early stages of development may be considered as an undifferentiated cell, the reticulocyte and erythrocyte represent a specific type of cell engaged primarily in the synthesis of only one protein, viz. globin, which accounts for about 80% of the cellular proteins. Embryonic muscle cells, on the other hand, are a terminally differentiated eukaryotic system which exhibit many unique characteristics. These include a post-mitotic and non-replicating chromosome, a highly ordered contractile apparatus formed by the post-translational assembly of a large number of myofibrillar proteins and translational control of gene expression during differentiation. The work presented here describes the first successful isolation and partial characterization of actin and myosin heavy chain mRNP particles. Assuming that highly purified myosin heavy chain mRNP contains 1 mole of mRNA per mole, we have calculated the molecular weight of the mRNP from the known molecular weight of the purified mRNA and the RNA/protein ratio of the mRNP. This estimate gives a value of 8.11×10^6 which is in the expected range for an mRNP of this size (Spirin, 1972). Previous reports in the literature on myosin heavy chain mRNP were based on detection using either pulsed RNA (Buckingham *et al.*, 1974; Buckingham and Gross, 1975) or nonpolysomal sedimentation of the mRNA in gradient runs (Heywood *et al.*, 1975).

The ability of the myosin heavy chain mRNP to program the *in vitro* synthesis of myosin heavy chain, as reported here, indicates that the binding of specific proteins to the mRNA *per se* does not exert a stringent negative control on the translation process. In view of these results the stabilization of muscle-specific mRNAs as cytoplasmic mRNPs during the myoblast to myotube transition of cultured muscle cells (Buckingham *et al.*, 1974) may be interpreted as a mechanism by which the concentration of "translatable" mRNAs of the cell is increased. According to this view the mRNAs are stored as biologically active rather than inactive species in a cytoplasmic pool. The entry of the mRNAs from this compartment to the polysomes may be a key regulatory event in translation during the terminal differentiation of muscle cells. This model of translational control involving mRNPs also explains the accelerated rate of cell-specific protein synthesis observed in the post-fusion stages of cultured muscle cells. The details of the transit of mRNAs from mRNP particles to polysomes is now being investigated by us. The 44,000 dalton protein, which we have observed in all classes of mRNPs and which seem to be absent in polysomal mRNPs derived from embryonic muscle cells (manuscript in preparation) may be involved in some aspect of this translational control.

Bester *et al.* (1975) have recently reported the isolation of a dialyzable, low molecular weight RNA species from the postpolysomal supernatant of embryonic muscles. They have postulated that this RNA species exerts a negative control of translation on cytoplasmic mRNP particles. In view of the results presented here, the regulation of translation by biologically active cytoplasmic mRNP particles during the growth and differentiation of muscle cells through a positive control, as discussed above, is favored by us.

ACKNOWLEDGMENTS

The authors thank Drs. John Gergely, Henry Paulus and Helga Boedtker for many helpful discussions; Dr. Paul C. Leavis for his advice on the isolation and characterization of actin; and Dr. Raman K. Roy for his help in the preparation of globin mRNA and wheat germ embryo S-30 cell-free system. It is a pleasure to acknowledge the skillful technical assistance of the Misses Ann Sutton and Ven-jim Chen. This work was supported by grants from the National Institutes of Health (AM 13238), the American Heart Association (No. 71-915), and the Muscular Dystrophy Associations of America, Inc. This work was initiated during the tenure of a research fellowship to J. Bag from the American Heart Association, Massachusetts Affiliate, Inc.; more recent work has been carried out during the tenure of a fellowship of the Medical Foundation, Inc., Boston, Massachusetts.

REFERENCES

Bag, J. and Sarkar, S. (1975). *Biochemistry 14*, 3800-3807.

Bag, J. and Sarkar, S. (1976). *J. Biol. Chem. 251*, 7600-7609

Bag, J., Roy, R. K., Sutton, A. and Sarkar, S. (1975). *In* "Regulation of Growth and Differentiated Function in Eukaryote Cells" (G. P. Talwar, ed.), pp. 111-127. Raven Press, New York.

Barrieux, A., Ingraham, H. A., David, D. N. and Rosenfeld, M. G. (1975). *Proc. Nat. Acad. Sci. U.S.A. 72*, 1523-1527.

Blobel, G. (1973). *Proc. Nat. Acad. Sci. U.S.A. 70*, 924-928.

Brawerman, G. (1974). *Ann. Rev. Biochem. 43*, 621-642.

Buckingham, M. E., Caput, D., Cohen, A., Whalen, R. G. and Gross, F. (1974). *Proc. Nat. Acad. Sci. U.S.A. 71*, 1466-1470.

Buckingham, M. E. and Gross, F. (1975). *FEBS Lett. 53*, 355-359.

Chacko, S. and Xavier, J. (1974). *Dev. Biol. 40*, 340-354.

Elzinga, M. (1970). *Biochemistry 9*, 1365-1374.

Firtel, R. A., and Pederson, T. (1975). *Proc. Nat. Acad. Sci. U.S.A. 72*, 301-305.

Gross, K. W., Jacobs-Lorena, M., Baglioni, C. and Gross, P. R. (1973). *Proc. Nat. Acad. Sci. U.S.A. 70*, 2614-2618.

Heywood, S. M. and Rich, A. (1968). *Proc. Nat. Acad. Sci. U.S.A. 59*, 590.

Heywood, S. M. and Nwagwu, M. (1969). *Biochemistry 8*, 3839-3845.

Heywood, S. M., Kennedy, D. S. and Bester, A. J. (1975). *FEBS Lett. 53*, 69-72.

Holtzer, H., Sanger, J. W., Ishikawa, H. and Strahs, K. (1972). *Cold Spring Harbor Symp. Quant. Biol. 37*, 549-566.

Jacobs-Lorena, M. and Baglioni, C. (1972). *Proc. Nat. Acad. Sci. U.S.A. 69*, 1425-1428.

Lowey, S. and Risby, D. (1971). *Nature 234*, 81-85.

Martin, R. G. and Ames, B. N. (1961). *J. Biol. Chem. 236*, 1372-

1379.

Mondal, H., Sutton, A., Chen, V. and Sarkar, S. (1974). *Biochem. Biophys. Res. Commun. 56,* 988-995.

Morris, G. E., Buzash, E. A., Rourke, A. W., Tepperman, K., Thompson, W. C. and Heywood, S. M. (1972). *Cold Spring Harbor Symp. Quant. Biol. 37,* 535-541.

Pollard, T. D. and Weihing, R. R. (1974). *Crit. Rev. Biochem. 2,* 1-65.

Potter, J. D. and Gergely, J. (1974). *Biochemistry 13,* 2697-2703.

Quinlan, T. J., Billings, P. B. and Martin, T. E. (1974). *Proc. Nat. Acad. Sci. U.S.A. 71,* 2632-2636.

Roberts, B. E. and Paterson, B. M. (1973). *Proc. Nat. Acad. Sci. U.S.A. 70,* 2330-2334.

Rubinstein, N. A., Chi, J. C. H. and Holtzer, H. (1974). *Biochem. Biophys. Res. Commun. 57,* 438-445.

Sarkar, S. and Cooke, P. H. (1970). *Biochem. Biophys. Res. Commun. 41,* 918-925.

Sarkar, S., Sreter, F. A. and Gergely, J. (1971). *Proc. Nat. Acad. Sci. U.S.A. 68,* 946-950.

Sarkar, S. (1972). *Cold Spring Harbor Symp. Quant. Biol. 37,* 14-17.

Sarkar, S., Mukherjee, S. P., Sutton, A., Mondal, H. and Chen, V. (1973). *Prep. Biochem. 3,* 583-604.

Sarkar, S. (1974). *In* "Exploratory Concepts in Muscular Dystrophy II" (A. T. Milhorat, ed.), pp. 172-184. Excerpta Medica, Amsterdam.

Sarkar, S. (1975). *In* "Eukaryotes at the Subcellular Level: Development and Differentiation" (J. Last, ed.), pp. 315-368. Marcel Dekker, New York.

Spirin, A. S. (1972). *In* "The Mechanism of Protein Synthesis and Its Regulation" (L. Bosch, ed.), pp. 515-537. North-Holland Publishing, Amsterdam.

Spohr, G., Kayibanda, B. and Scherrer, K. (1972). *Eur. J. Biochem. 31,* 194-208.

Swank, R. T. and Munkres, K. D. (1971). *Anal. Biochem. 39*, 462-477.

Taylor, E. W. (1972). *Ann. Rev. Biochem. 41*, 577-616.

Weeds, A. G. and Frank, G. (1972). *Cold Spring Harbor Symp. Quant. Biol. 37*, 9-14.

Yaffe, D. and Dym, H. (1972). *Cold Spring Harbor Symp. Quant. Biol. 37*, 543-547.

EXPLANTS OF NEWBORN RAT BRAIN CORTEX SHOW MUTUAL INFLUENCE OF FATTY ACID PATTERNS AFTER COMBINATION AT DIFFERENT AGES IN ORGANOTYPIC CULTURES

M. GIESING
F. ZILLIKEN

Institute for Physiological Chemistry
University of Bonn
Bonn, West Germany

SUMMARY

Explants of neonatal rat brain cortex were allowed to form fibrous connections with neocortex cultures after cultivation for 8 or 14 days. The determination of fatty acids revealed specific changes of the age dependent pattern. Palmitic and palmitoleic acid decreased in neonatal cultures after combination. Polyenoic fatty acids were reduced after 4 days. Explants which had been maintained in vitro prior to combination, showed a marked increase of polyunsaturated fatty acids. Several mechanisms involved in the mutual influence are discussed.

I. INTRODUCTION

Organotypic cultures of central and peripheral nervous tissues have been increasingly employed during the last two decades as model systems to study favorably cell differentiation and maturation. Tissue matures in vitro following dissection from the animal at a prenatal stage, i.e. dorsal root ganglia or shortly after birth, i.e. cortex cerebri and cerebellum. The process of maturation is paralleled by the onset of functional junctions between different cell types, synaptogenesis, myelin formation, spontaneous bioelectric activity and adequate responses on stimulatory and inhibitory manipulation as reviewed by Bornstein, Model (1972); Crain (1952); Crain, Bornstein (1972); Model *et al.* (1971). Cultures of rat neocortex form mature synapses within two weeks in vitro. The genesis of synapses is concomitant with the demonstration of propagated neuronal impulses and synaptic transmission, the presence of complex organotypic discharge patterns and pharmacological sensitivites (Crain, Bornstein 1972). Turnover studies of fatty acid metabolism revealed a threephasic maturation period of rat neocortex as we recently reported (Giesing, Zilliken 1975, a). The age dependency of lipids and fatty acids in developing cortex cerebri cultures is correlated with a posttraumatic repair phase, a period of synapse maturation and a postsynaptogenic stage. The availability of an analytical system to evaluate the maturation of nervous tissues implied a model system of in vitro recombination of different parts of the central nervous system, i.e. thalamus - hypothalamus, neocortex-cerebellum, neocortex - neocortex cultures. This paper deals with some aspects of combined neocortex cultures. Cultures which were maintained in vitro for different periods of time, were allowed to form connections with immature cortex explants. Fatty acids were assayed in order to differentiate inductive changes and mixtures of cells and cell processes of different age which have migrated from the explants.

II. MATERIALS AND METHODS

A. Tissue Culture

The Maximow-double-coverslip assembly was used to maintain cortex cerebri in organotypic formation. The tissue was dissected from rats on the second day post partum. Two explants grew on each collagen coated coverslip at 34.5°C in a properly sealed tissue chamber (Bornstein 1958). Each culture unit was fed with 60 µl of a nutrient solution composed of 32 Vol % Eagle's minimum essential medium, 26 Vol % fetal bovine serum, 34 Vol % Simm's balanced salt solution diluted by 8 Vol % of bidistilled water free of ammonia [final concentrations: 2.3×10^{-6} *M* D-Glucose; 1.2×10^{-3} *M* Insulin without zinc (a gift from Schering AG, Berlin, GFR) and 9.0×10^{-7} *M* L-Glutamine). No antibiotics were used. The volume of the feeding solution allowed the cultures to flatten out within 5 - 6 days. The medium was changed twice a week. After 8 and 14 days in vitro (DIV) respectively one freshly dissected explant was taken from animals two days post partum and added on the same coverslip. The distance between the cultures did not exceed 2 mm. The volume of the medium was not changed. The development of the cultures was examined by light microscopic observation.

B. Fatty Acid Analysis

Total lipids were extracted according to Folch *et al.* (1957) and quantitated gravimetrically after Egge *et al.* (1970). Total phospholipids were assayed by the determination of phosphorus after a method devised by Fiske, Subbarow (1935). Fatty acids were converted to methyl esters by transesterification in methanol containing 5% KOH after incubation for 16 h at 0 - 4°C. Free fatty acids were derivatized by diazomethane. Separation of individual fatty acids was accomplished on a S.C.O.T. column coated with DEGS (15 m) following isolation of the methyl ester moiety on silica gel columns (60 HR nach Stahl; Fa. Merck, Darmstadt, GFR). Nitrogen was used as carrier in a Perkin Elmer Gashchromatograph 900. The identification of individual components was carried out by combined gas liquid chromatography - mass spectrometry (LKB 9 000) with or without derivatization.

III. RESULTS

A. Morphological Development of Cortex Cultures after Combination

The cortex cultures we used were composed of two explants which had been adapted to the tissues' environment for a cultivation period of 8 or 14 days (A cultures). A cultures formed extensive fibrous connections within this time. At these two points neonatal explants were positioned on the same coverslip (N cultures). Interactions (I) between both cultures became visible by a continuous outgrowth of fibre bundles. Figures 1-7 show combined cortex cultures with an age difference of 14 DIV. Figures 1-4 show the succesive establishment of connections between NI and AI cultures by a serial observation every 24 hours. Connections were promoted by cellular outgrowth from the NI culture. The origin of junctions was in close proximity to the AI culture as marked by arrows in Fig. 1. Three days after combination (Fig. 3) first fibrous contacts were to be observed. Figures 5 and 6 show the contact area after 4 DIV. The pictures were taken from the same cultures which are to be seen in Figs. 1-4 on the same day as in Fig. 4. NI fibres insert at numerous points of the outgrowth area of the AI cultures. Figure 6 shows AI neurons contacting a NI fibre. The bundles connecting AI and NI cultures remained during continued cultivation. Figure 7 shows combined cortex cultures after 8 DIV. Fibre bridges are to be seen. The outgrowth pattern of N and NI cultures was dependent on the tissues' environment. As it is to be seen in Fig. 4 radial outgrowth from a NI culture followed the formation of contacting elements

1

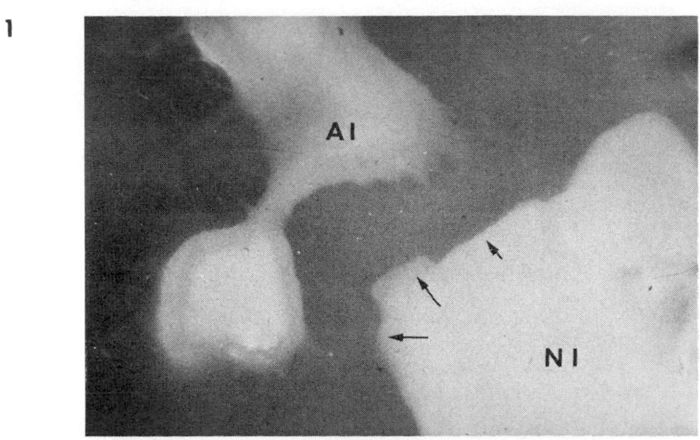

2

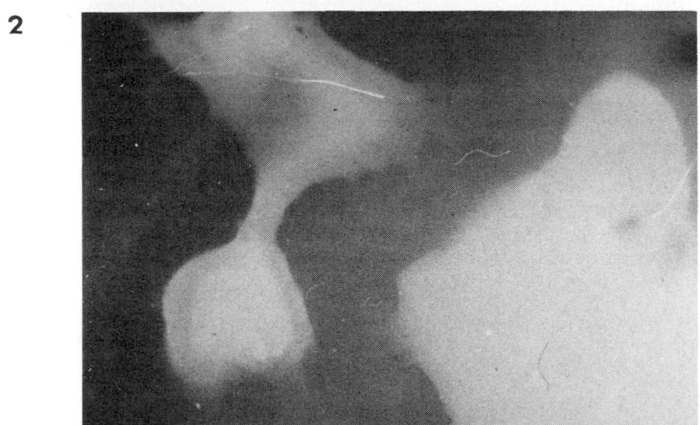

3

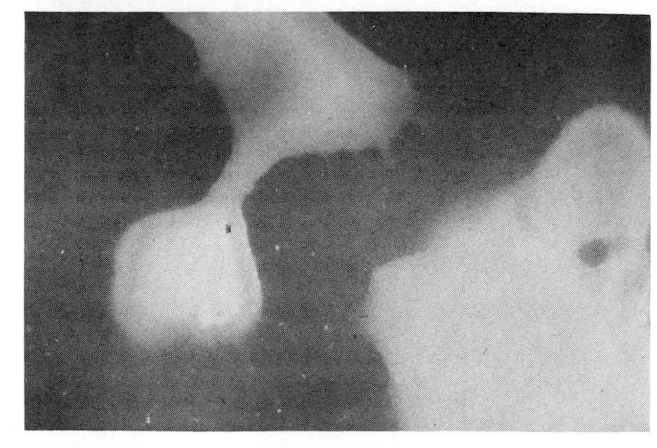

4

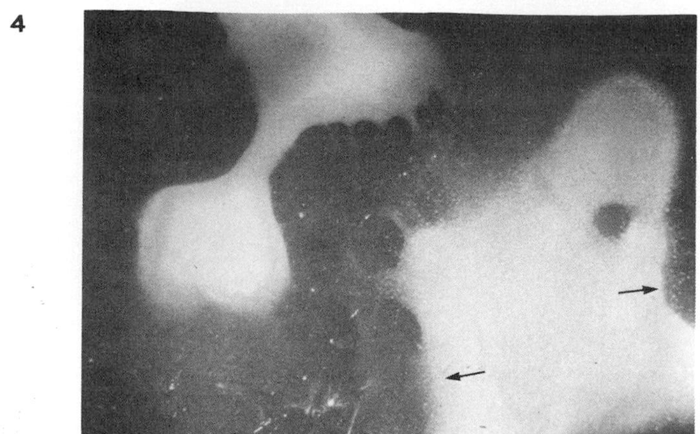

5

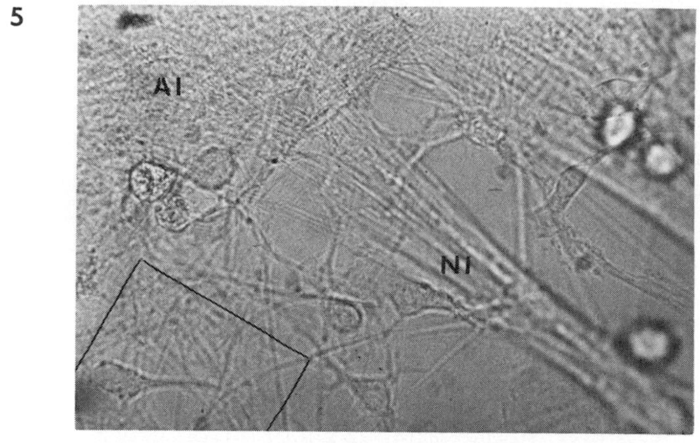

6

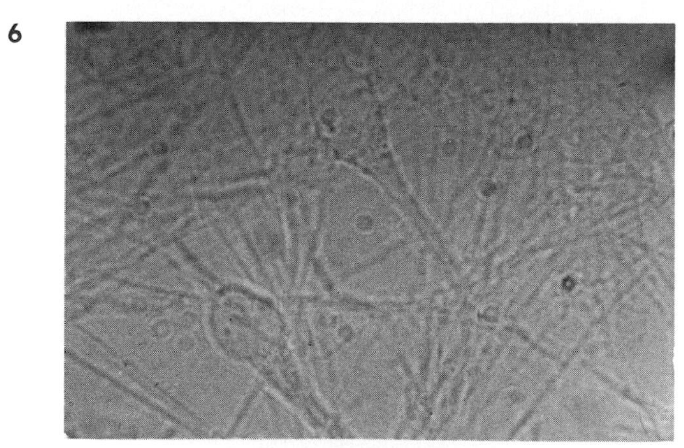

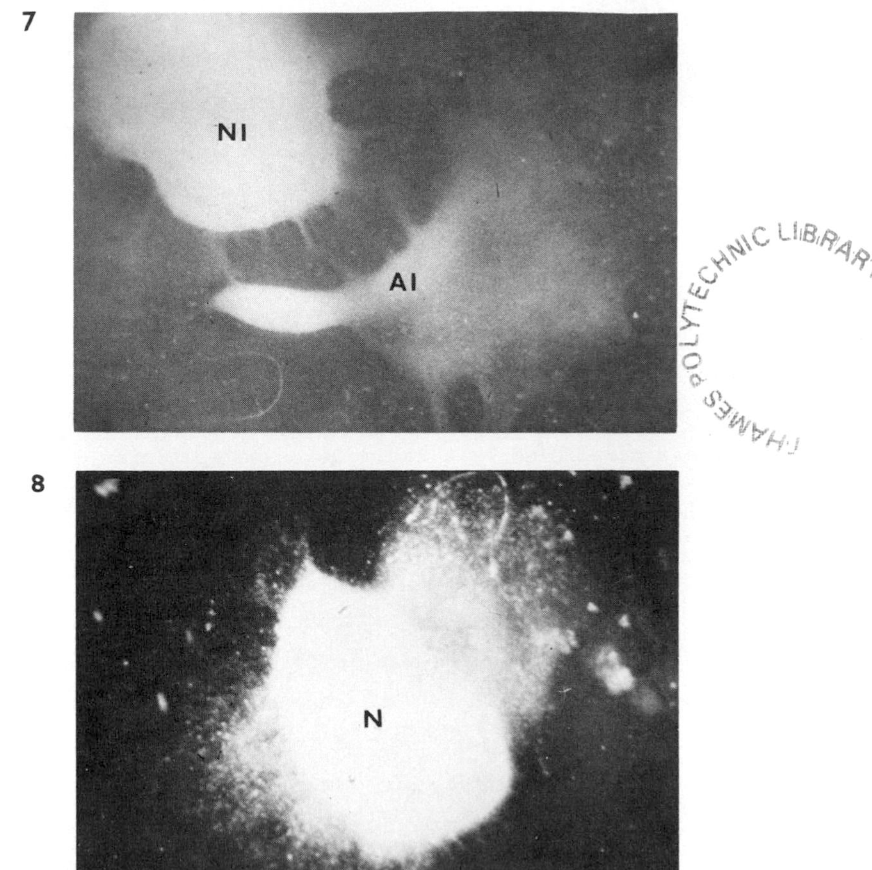

FIGURES 1-8 *Morphological development of combined rat neocortex cultures. AI = neocortex explants which have been adapted to the tissue's environment for 14 days in vitro (DIV) prior to the addition of a NI explant. NI = neonatal explant added to an AI culture. Figures 1-4 show the formation of morphological connections between AI and NI cultures by a serial observation every 24 h starting from the 1st day after combination; contact is established within 4 DIV (Fig. 4) and remains stable for at least 8 DIV (Fig. 7). Locations of outgrowth are marked by arrows in Fig. 1. Radial outgrowth from NI cultures follows the formation of connections (arrows in Fig. 4). Radial outgrowth is to be seen in N cultures (neonatal cultures without AI cultures) after 6 DIV (Fig. 8). The cell morphology of connections between AI and NI is shown in Figs. 5 and 6. Bundles of fibres grow out from the NI explant of Fig. 4 inserting the area of migrated cells of the corresponding AI culture. Figure 7 shows a section of*

Fig. 6. An AI neuronal cell process contacts a NI fibre. All pictures have been taken from living tissue. Magnification: × 30 (× 800 in Fig. 5; × 1260 in Fig. 6).

(arrows). If neonatal cultures are grown in the absence of A cultures the radial outgrowth is clearly visible after 6 DIV (Fig. 8).

B. Fatty Acids in Neocortex Cultures after Combination

Total lipid concentration in developing neocortex in vitro varied in relation to the cultivation time. The first four days in vitro were paralleled by a rapid decrease of lipids due to numerous dying cells which were traumatized by the dissection procedure or could not be sufficiently nourished. The minimum weight was observed after 12 DIV followed by an increase reaching a maximum after 3 weeks. If neonatal cultures were added after 8 or 14 DIV the initial decrease was less rapid and followed by a marked increase (Fig. 9 a, b). The succeeding raise of detected total lipid in NI cultures during the synaptogenic stage (Δ DIV 8) reached the level of dissection after 8 DIV.

Total lipids in the corresponding AI cultures decreased. AI lipids in combined neocortex cultures during the postsynaptogenic period were markedly reduced within 4 DIV. However light microscopic observation of AI cultures did not show a number of dead cells which are necessary to explain the reduce of lipids. Phospholipids remained constant in all combination experiments.

Individual fatty acids in combined neocortex cultures (Δ DIV 14) showed a complex pattern. As shown in Fig. 10 a loss of palmitic acid in NI cultures was observed. The loss of other fatty acids differed gradually. Palmitoleic acid showed a moderate decrease compared with C 16 ÷ 0 (Fig. 11). NI stearate and oleate were preferentially abated during the first 4 DIV followed by an increase (Fig. 12, 13). NI linoleate (Fig. 14), arachidonate

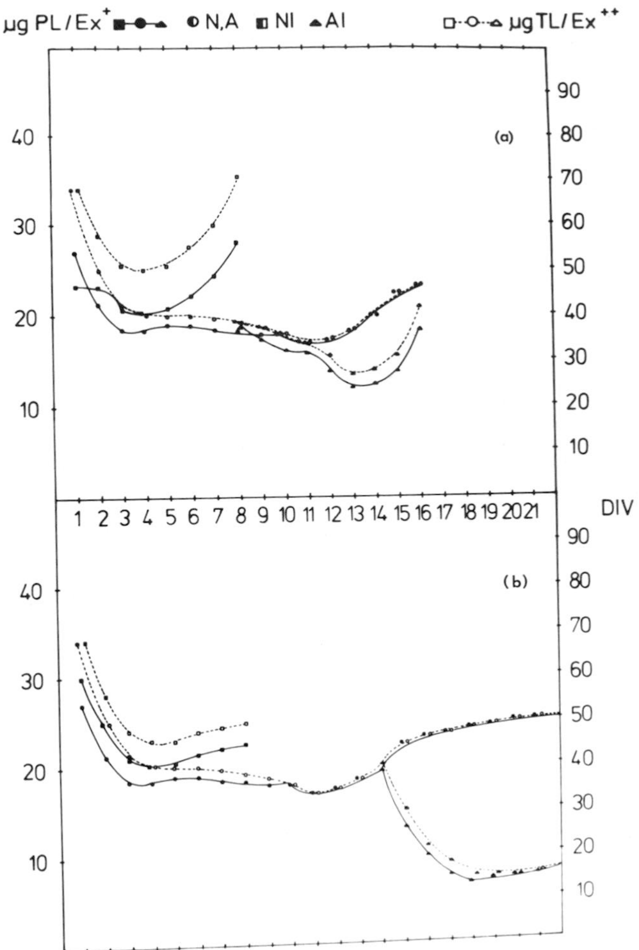

FIGURE 9 a, b *Lipid and phospholipid development in N, A, NI and AI cultures.* [+]*µg PL/Ex = µg total phospholipids per explant.* [++]*µg TL/Ex = µg total lipid per explant. (a) Combination at stage Δ DIV 8. (b) Combination at stage Δ DIV 14. For further comments see text! Each point is the average of triplets with at least 6 cultures.*

(Fig. 15) and docosahexaenoate (Fig. 16) seemed to be reduced rather from DIV 4 to 8.

AI fatty acids differed markedly from the NI patterns. Palmitic and palmitoleic acid were not influenced. Stearate and

10

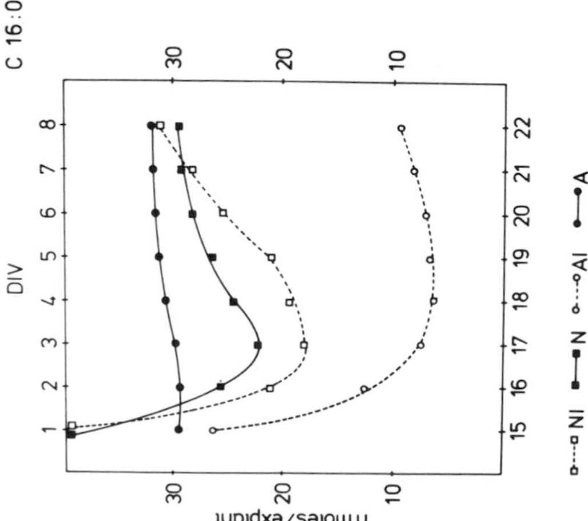

C 16:0
DIV
1 2 3 4 5 6 7 8
n moles/explant
30
20
10
15 16 17 18 19 20 21 22
NI N AI A

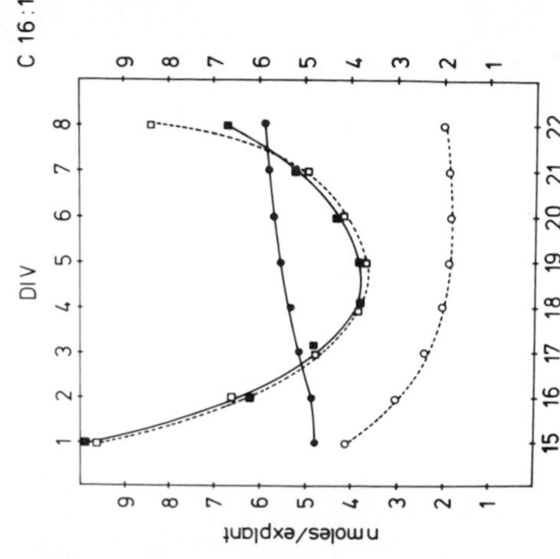

C 16:1
DIV
1 2 3 4 5 6 7 8
n moles/explant
9 8 7 6 5 4 3 2 1
15 16 17 18 19 20 21 22

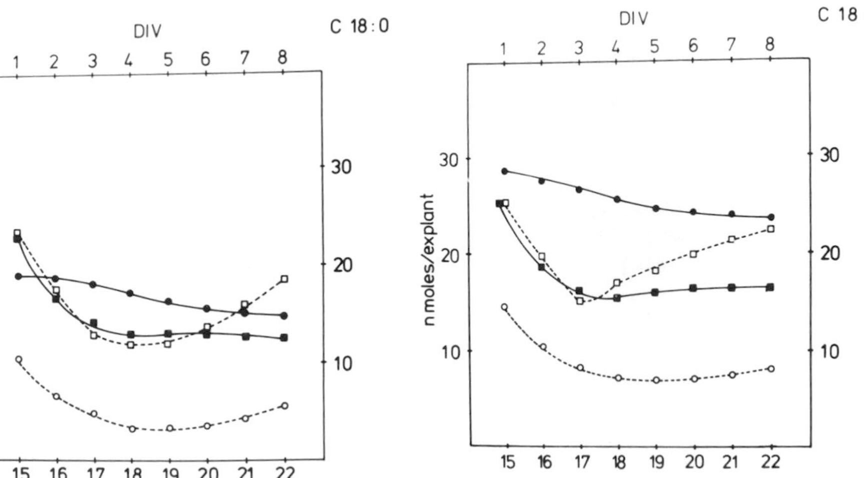

FIGURES 10-13 *Saturated and monoenoic fatty acids in combined neocortex cultures. NI cultures show a decrease of palmitic and palmitoleic acid (Figs. 10, 11). NI stearate and oleate are enhanced from the 4th day on after combination (Fig. 12, 13). AI saturated and monoenoic fatty acids are not influenced except that stearate and oleate are slightly increased from DIV 18 to 22. Each point is the average of triplets with at least 10 cultures per assay.*

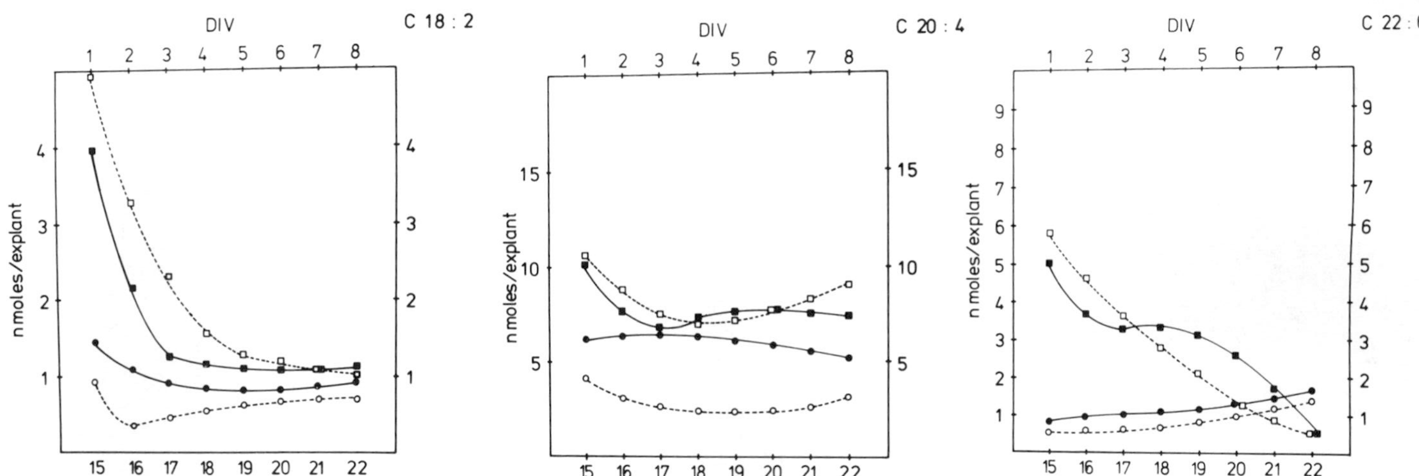

FIGURES 14-16 *Polyunsaturated fatty acids in cortex cultures after combination. AI linoleate and docosahexaenoate are rapidly increased (Figs. 14, 16). AI arachidonate is increased from DIV 20 to 22 at a low rate. NI linoleate and docosahexaenoate are lowered from DIV 4 to 8. Total number of cultures (same in Figs. 10-13) : 510.*

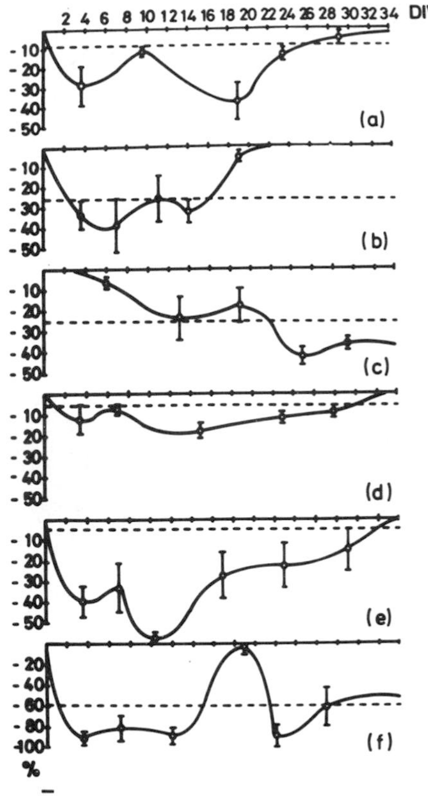

FIGURE 17 *Depletion of medium fatty acids in N and A cultures. (a) Palmitate; (b) palmitoleate; (c) stearate; (d) oleate; (e) linoleate; (f) arachidonate. The depletion lines were plotted on on the basis of the highest value of each individual fatty acid which was detected in incubated medium at 34.5°C after 4 DIV in the presence of cells. This value marks the absolute zero consumption lying on the abszissa (solid line). The concentration of individual fatty acids in freshly prepared medium is represented by the dotted line. Any depletion beyond this line corresponds to absolute or net consumption. The zone between the lines is related with relative depletion which might result from uptake, release or relative distribution changes. Fetal bovine serum as lipid source of the medium was replaced by equal parts of human cord serum and an aqueous extract of 9 days old chicken in ovo to a final medium concentration of 46 Vol %. Assays were performed as triplets of at least 120 cultures varying ± 2 DIV. Values are means ± S.E.*

oleate showed a slight increase during the last 4 days of the observation time. The concentration of fatty acids with 18 carbon atoms in A cultures was continuously diminished during the first weeks in vitro. After combination C 18 fatty acids were raised in the range : C 18 : 2 > C 18: 1 > C 18:0. AI docosahexaenoate was even more enhanced as compared with A cultures.

In order to evaluate a mediation of the mutual influence in combined neocortex cultures provided by a differential release of individual components into the medium, analysis of medium fatty acids of N and A cultures after incubation in the presence of the tissue was carried out. Results are set out in Fig. 17. All major medium fatty acids in N cultures (DIV 1 - 8) were found to be depleted. Only stearate might have been segregated by the tissue. A cultures (DIV 15 - 22) might have released palmitoleic, stearic and docosahexaenoic acid since the concentration in incubated medium was higher than in the freshly prepared nutrient solution.

IV. DISCUSSION

Rat brain neocortex cultures have been combined with neonatal cortex cerebri after a cultivation period of 8 and 14 days respectively. Massive connections were made by outgrowing fibre bundles within 3 to 4 days. Lipid analysis in NI cultures revealed a stabilizing effect since the decrease of total lipids per explant was less rapid. Total lipids in AI cultures were reduced. NI cultures showed a differential reduce of some relatively short chained fatty acids, namely C 16 : 0 and C 16 : 1 during the cultivation period of 8 days whereas polyunsaturated fatty acids decreased rather after 4 DIV. AI cultures showed a raise of di- and polyenoic fatty acids. The effect of the mutual influence in combined neocortex cultures cannot be explained by a release of components into the medium in A and N cultures. Combination of both cultures is essential. However the results presented do not prove the necessity of morphological connections. The selectivity of differences in cultures after combination rules out a possible synchronisation of the age dependent fatty acid pattern on account of a dilution effect provided by contaminating cells which have migrated from the explant. Differential changes of fatty acids in brain cortex cultures following combination with freshly dissected explants seem to be rather induced by activated or inhibited chain elongating and desaturating enzyme systems and/ or altered metabolic rates than by a mututal transport of components via the aqueous phase of the medium or along the connecting fibre bridges. Nevertheless the impuls which induces metabolic changes in AI and NI cultures, ought to be mediated and consequently effectuated on one of the two ways.

Cantrill, Carey (1975) reported on the differential timing of enzymes involved in brain fatty acid metabolism. Palmitic acid is preferably synthesized by de novo synthesis after Sun, Horrocks (1971). The decrease of palmitic acid in NI cultures might be due to a reduce of de novo synthesis or inactivation of the microsomal palmitoyl-CoA synthetase (Cantrill, Carey, 1975). NI palmitate might be preferably elongated to stearate from DIV 4 to 8 after combination and consequently desaturated. The changes of fatty acid pattern in AI cultures were correlated with a raise of polyenoic fatty acids. Saturated fatty acids were less influenced. The increase of linoleate and docosahexaenoate might be provided either by enhanced uptake from the medium or by activation of linolenate desaturase systems (Giesing- Zilliken, 1975, b).

Further experiments are in progress to study the way of mutual influence in combined brain cortex cultures. The culture model presented allows promising studies on the exchange of different species of effectuators between different parts of the brain. It might be usefull as well to investigate metabolic events in relation to stimulatory or inhibitory manipulations of spontaneous bioelectric activity.

ACKNOWLEDGMENTS

We are grateful to Mrs. Ursula Gerken for taking care of the cultures and to Mrs. Marlene Kühne for her assistance in the analytical work. The program has been supported in part by the Deutsche Forschungsgemeinschaft.

REFERENCES

Bornstein, M. B. (1958). *Lab. Invest. 7,* 134-137.

Bornstein, M. B. and Model, P. G. (1972). *Brain Res. 37,* 287-293.

Cantrill, R. C. and Carey, E. M. (1975). *Biochim. Biophys. Acta. 380,* 165-175.

Crain, St. M. (1952). *Proc. Soc. exp. Biol. (N.Y.). 81,* 49-51.

Crain, St. M. and Bornstein, M. B. (1972). *Science. 176,* 182-184.

Egge, H., Murawski, U., Müller, J. and Zilliken, F. (1970). *Z. Klin. Chem. u. Klin. Biochem. 8,* 488-491.

Fiske, C. H. and Subbarow, Y. (1935). *J. Biol. Chem. 66,* 375-400.

Folch, J., Lees, M. and Sloane Stanley, G. H. (1957). *J. Biol. Chem. 226,* 497-509.

Giesing, M. and Zilliken, F. (1975, a). *Nutr. Metabol. 19,* 251-262.

Giesing, M. and Zilliken, F. (1975, b). *Hoppe Seylers' Z. Physiol. Chem. 356,* 234.

Model, P. G., Bornstein, M. B., Crain, St. M. and Pappas, G. D. (1971). *J. Cell Biol. 49,* 362-371.

Sun, G. Y. and Horrocks, L. A. (1971), *J. Neurochem. 18,* 1963-1969.

USE OF *IN VITRO* CELL COLONY ASSAYS FOR MEASURING THE CYTOTOXICITY OF CHEMOTHERAPEUTIC AGENTS TO HEMATOPOIETIC PROGENITOR CELLS COMMITTED TO MYELOID, ERYTHROID AND MEGAKARYOCYTOID DIFFERENTIATION

ALEXANDER NAKEFF
FREDERICK A. VALERIOTE

Section of Cancer Biology
Mallinckrodt Institute of Radiology
Washington University School of Medicine
St. Louis, Missouri

I. INTRODUCTION

A major consideration in optimizing the use of various chemotherapeutic agents is to minimize their effects on the hematopoietic system since the efficacy of the agents is often limited by their cytotoxicity to normal blood cell progenitors.

Investigators have employed morphological as well as functional criteria from simple changes in cellularity to extent of destruction of the pluripotential hematopoietic stem cells (Bruce and Valeriote, 1968).

With regard to experimental techniques for assaying those hematopoietic cell progenitors already committed to differentiation, sophisticated approaches have been attempted including the use of ^{59}Fe incorporation, both following a standard dose of erythropoietin in hypertransfused recipients to measure erythropoietin-responsive cells (ERC) (Gurney *et al.*, 1962) and following marrow transplantion into supralethally-irradiated recipients to measure erythroid repopulating ability (ERA) (Blackett *et al.*, 1964; Hellman *et al.*, 1969), and the measurement of granulocyte repopulating ability (GRA) based on an endotoxin cell-mobilization technique for assessing granulocyte production (Hellman *et al.*, 1969). While Hellman *et al.* (1969) initially noted little difference in the effect of radiation on these functional assays, the response to chemotherapeutic agents revealed differences between the different cell populations (Lamerton and Blackett, 1974). Later studies by Hellman and Grate (1971) have demonstrated a consistant difference between these two populations with the erythrocytic series having been found to be more sensitive than the granulocytic series not only to cyclophosphamide and nitrogen mustard, but also to radiation. This difference is not confined only to the inherent difference between two distinct cell populations, but has been extended to individual compartments within the erythrocytic series which have been assessed by different functional assays and shown to respond differently to various drugs (Millar and Blackett, 1974). Presumably, this reflects mostly the different proliferative states at the time of drug exposure for that portion of the erythroid progenitor population being assayed.

The usefulness of these functional assay systems is limited by several factors including the fact that they are indirect assays which are interpreted as reflecting events occurring in specific progenitor compartments and, perhaps most importantly, are assays which cannot be applied to studies of human bone marrow. The *in vitro* growth of hematopoietic cell colonies with expression of differentiation as well as proliferation characteristics provides a more direct approach to the assay of hematopoietic progenitor cells. Using a methyl cellulose culture technique for the progenitor to the myeloid line, the colony-forming unit, culture (CFU-C), Brown and Carbone (1971) have carried out the most comprehensive study of the effect of anti-cancer agents on human marrow CFU-C and while the agents act in agreement with the classification of Bruce and Valeriote (1968), the absolute sensitivities do differ. Studies comparing the drug response of CFU-S and CFU-C in agar culture of mouse marrow have also been carried out (Chen and Schooley, 1970).

Techniques have been developed more recently for culturing colonies of erythroid and megakaryocytoid cells. The first successful growth of colonies of erythroid cells was reported by Stephenson *et al.* (1971) using mouse marrow in plasma cultures, followed by reports of Gregory *et al.* (1973) then Iscove *et al.* (1974) and Cooper *et al.* (1974) using both mouse and human marrow in methyl cellulose cultures. McLeod *et al.* (1975) have most recently developed an *in vitro* erythropoietic cell colony assay in plasma cultures of mouse marrow for a class of erythropoietin-dependent cell progenitor committed to erythropoiesis (CFU-E). The growth of megakaryocyte colonies has been reported in agar cultures of mouse marrow (Nakeff *et al.*, 1975; Metcalf *et al.*, 1975) and a plasma culture system has been described for a class of progenitor in mouse marrow committed to the megakaryocytic cell line (CFU-M) (Nakeff and Daniels-McQueen, 1976). These assays for CFU-M, CFU-E, CFU-C and CFU-S make it possible to study the effect of chemotherapeutic agents on both pluripotent and committed progenitor cell populations. However, since both culture technique and culture conditions have been shown to affect the growth and differentiation of cell colonies from hematopoietic tissue (*In Vitro* Culture of Hemopoietic Cells, 1971), it was imperative to standardize these factors for the three committed cell progenitors. In our experience, the plasma culture technique seemed most ideal since it was the only technique that permitted not only the growth of colonies from each progenitor but also, more importantly, allowed for their identification *in situ* using specific cyto- and histochemical properties.

This report will describe the microtiter plasma culture system as modified from that described by McLeod *et al.* (1975) which we have used successfully to assay for colony-forming units specific for erythropoiesis (CFU-E), granulopoiesis (CFU-C) and megakaryocytopoiesis (CFU-M) under almost identical culture conditions from the same pool of bone marrow. In addition, some preliminary data will be presented on the time-response of these committed cell progenitors, as well as the CFU-S, to cyclophosphamide.

II. MATERIALS AND METHODS

A. Mice

Twenty week-old $B6D_2F_1$ (C57Bl/6 × DBA/2) female mice from Jackson Laboratories were used throughout this study.

B. Collection and Preparation of Bone Marrow

A mono-dispersed cell suspension of bone marrow was obtained from mice by flushing each femur with 1 ml of cold modified Eagle's medium (HMEM) (McLeod *et al.*, 1975) supplemented with 2% heat-inactivated fetal calf serum (FCS) (Gibco) and passing the resulting suspension once through a 25G needle.

After determining the total nucleated cell count of the pooled marrow sample by electronic particle counting (Celloscope, Particle Data) in cetrimide, cells were concentrated by centrifugation (350 × g for 10 min at 4°C) then resuspended in an appropriate volume of HMEM for addition to the cultures.

C. Plasma Culture of CFU-E, C and M

The basic microtiter plasma culture technique was used as described by McLeod *et al.* (1975) for assay of CFU-E and CFU-C

TABLE I Composition of Cultures for CFU-E, C and M

Constituents	CFU-E ml	CFU-C ml	CFU-M ml
FCS[a]	0.3, HI[b]	0.2	0.2
Cond. medium	0.1, Ep[c]	0.1, L-CM[d]	0.1, PWM-CM[e]
Other	0.1, L-asparagine[f]	-	-
Culture medium	0.2, NCTC-109[g]	0.4, α-MEM[h]	0.4, L-15[i]
BEE[j]	0.1	0.1	0.1
BM cells	0.1 (3×10^6/ml)	0.1 (3×10^6/ml)	0.1 (1.5×10^7/ml)
BCP[k]	0.1	0.1	0.1
	1.0 ml	1.0 ml	1.0 ml

[a]Fetal calf serum (Gibco).
[b]Heat-inactivated (56°C for 45 min).
[c]Erythropoietin Step III from Connaught Laboratories, Toronto (3004-6) at 2.5 U/ml.
[d]Conditioned medium (CM) from mouse L-cells.
[e]CM from mouse spleen cells cultured at 1×10^6 cells/ml in α-MEM with 10% FCS for 4 days in the presence of a 1:320 final dilution of pokeweed mitogen (PWM) (Gibco).
[f]Schwarz-Mann at 0.2 mg/ml.
[g]Microbiological Assoc., Bethesda
[h]Flow lab., Bethesda
[i]Gibco, Buffalo.
[j]Bovine embryo extract (1:4 in medium) (Gibco).
[k]Bovine citrated plasma (Gibco)

and Nakeff and Daniels-McQueen (1976) for CFU-M. The particular culture conditions for CFU-E, C and M are presented in Table I. In brief, 0.1 ml of bone marrow (BM) cells at the appropriate cell concentration was added to 0.8 ml of the culture constituents (Table I) excluding the bovine citrated plasma (BCP). Finally, 0.1 ml of BCP was added, the culture mixed thoroughly to insure a mono-dispersed cell suspension and aliquots of 0.1 ml placed in wells of sterile, polyvinyl microtiter plates (Cooke Engineering, Alexandria, Va.) and permitted to clot at room temperature (10 min). Plates were then placed inside a 100 mm culture dish (Falcon) with a water reservoir to maintain humidity and incubated at 37°C at 100% humidity. Cultures for CFU-E were incubated at 5% CO_2 in air which was found to be optimal for their growth whereas cultures for CFU-C and CFU-M were incubated in 7.5% CO_2 in air.

D. Identification of CFU-E, C and M

After various times in culture, clots were removed after rimming and groups of 3 transferred to a 25 mm × 75 mm microscope slide. The clots were covered with filter paper soaked in 0.1 *M* sodium phosphate, pressed tightly to the slide and fixed with 5% glutaraldehyde (Electron Microscopy Sciences, Fort Washington, Pennsylvania) for 10 min.

1. CFU-E

Plasma cultures were stained for hemoglobin as described by McLeod *et al.* (1975). Briefly, after 2 min incubation in 1% benzidine, slides were placed in 2.5% hydrogen peroxide (in 70% ethanol) for 1 min, rinsed in distilled water for 1 min then counterstained in Harris' hematoxylin (Fisher) for 1 to 2 min. After "blueing" in running tap water for 2 to 3 min, slides were air-dried and mounted in Eukitt for microscopic examination at 400X. Colonies of more than eight benzidine-positive cells were scored as erythroid (CFU-E) and the total number counted in each culture.

2. CFU-C

Cultures were stained for 3 min in Harris' hematoxylin, "blued" for 2 to 3 min in running tap water, dried and mounted with Eukitt for microscopic examination at 200X. Colonies of greater than 50 granulocytic cells were scored as CFU-C. Occasional macrophage colonies of greater than 50 cells were also counted.

3. *CFU-M*

The "direct-coloring" thiocholine method of Karnovsky and Roots (1964) was used to stain for acetylcholinesterase activity following a 3 hour incubation at room temperature in a solution consisting of

10 mg acetylthiocholine iodide (Sigma, St. Louis)
15 ml 0.1 *M* sodium phosphate
2 ml 30 m*M* copper sulfate
1 ml 0.1 *M* sodium citrate
2 ml 5 m*M* potassium ferricyanide

Following a 1 min rinse in 0.1 *M* sodium phosphate and postfixation in absolute methanol for 10 min and 50% methanol for 30 sec, slides were counterstained in Harris' hematoxylin for 3 min, "blued" for 2 to 3 min in running tap water then dried and mounted in Eukitt. CFU-M were scored as colonies of more than 4 acetylcholinesterase-positive cells at a magnification of 400X.

E. CFU-S Assay

The spleen colony assay of Till and McCulloch (1961) was used to measure pluripotent stem cells. An appropriate fraction of the pooled marrow cell suspension was injected via the tail vein into groups of 10 syngeneic recipients receiving a total dose of 1000 rads of ^{137}cesium gamma rays. The dose was delivered in two 500 rad exposures administered three hours apart as described by Hellman and Grate (1971). Nine days later, the mice were killed, their spleens harvested, placed in Bouin's solution and the number of macroscopic spleen colonies counted. The number of CFU-S determined in this manner was expressed per femur without the consideration of an "f" factor.

III. RESULTS

A. Appearance of CFU-E, C and M as a Function of Time in Culture

In order to determine the optimal time of assay for CFU-E, C and M, plasma cultures of normal marrow were cultured at cell concentrations of 3×10^4 per 0.1 ml culture for CFU-E and C and 1.5×10^5 per 0.1 ml for CFU-M. After various periods of incubation, cultures were harvested and stained as described above and the number of CFU-E, C and M determined in each culture.

As shown in Fig. 1, the maximum number of CFU-E was obtained after 2 days in culture with most of the colonies having disappeared by the fifth day. With regard to the CFU-M, no colonies were observed over the first two days in culture. Colonies were observed, however, by day 3 with a peak incidence on day 4 and an almost complete disappearance by day 7. CFU-C appeared in culture with a peak incidence on day 4 and, as with the CFU-M, almost completely disappeared by day 7. On the basis of this data, CFU-E were routinely assayed after 2 days in culture with CFU-C and CFU-M being assayed after 4 days in culture.

B. Effect of Cyclophosphamide on CFU-S, E, C and M

Mice were injected intraperitoneally with a single dose of 2 mg of cyclophosphamide and at various times thereafter groups of four mice were killed and their marrow pooled and assayed for CFU-S, E, M and C.

As can be seen in Fig. 2, the number of CFU-S per femur decreased by about 30% by 2 hours after drug injection. There was a further decrease of about 30% over the next 22 hours which we attribute mostly to the loss of the initial surviving fraction of CFU-S through increased differentiation. Cellular recovery after this point was rapid and essentially complete by 4 days.

With regard to the number of CFU-E, there was a rapid and substantial loss of CFU-E by 2 hours at which point we were unable to

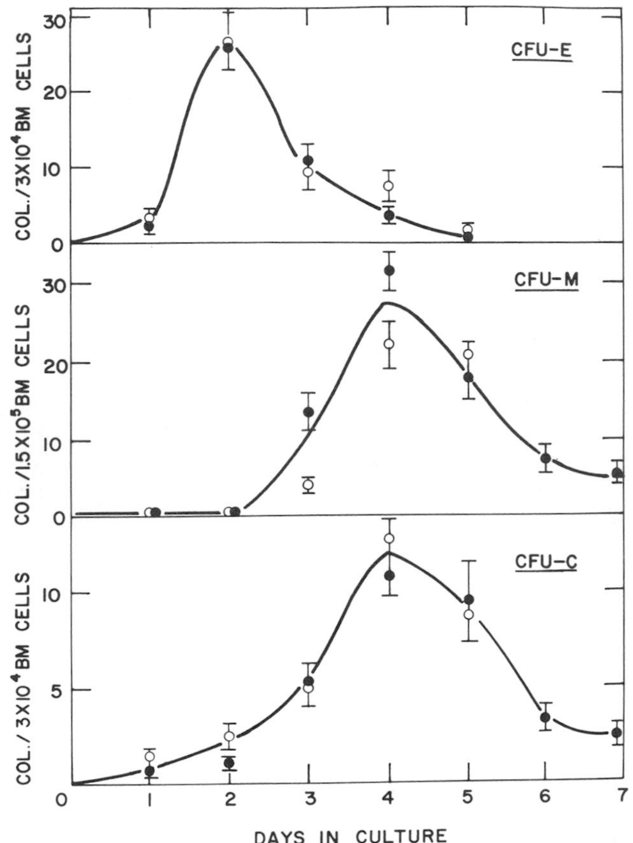

FIGURE 1 *The number of CFU-E, C and M per 0.1 ml plasma culture of normal mouse bone marrow as a function of the number of days in culture. Each set of symbols represents a separate experiment and each point is the average of nine cultures. Errors shown represent ± 1 SE.*

culture any erythroid colonies at the cell concentration plated. This was true also at 24 hours. By 2 days, however, recovery was observed with a rapid return to pretreatment levels by day 4.

The initial decrease in CFU-C was also rapid, reaching levels of about 5% of control by 2 hours. Cell recovery commenced thereafter so that by day 4 CFU-C colonies were confluent with the number of CFU-C probably being above the initial starting

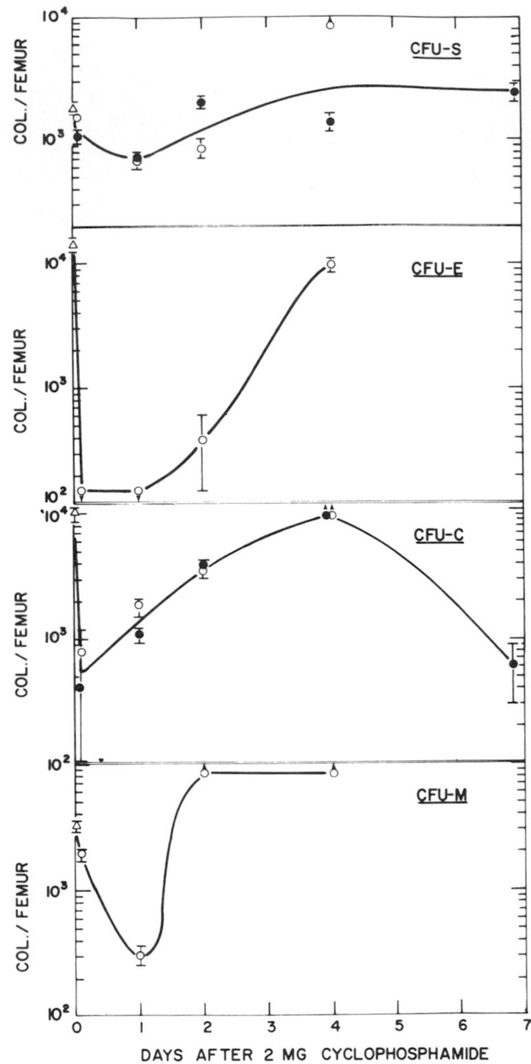

FIGURE 2 *The number of CFU-S, E, C and M per femur in mice as a function of time following a single intraperitoneal injection of 2 mg of cyclophosphamide. Each set of symbols represents a separate experiment and each point for CFU-E, C and M is the average of six cultures. Points with arrows down denote zero colony formation and those with arrows up denote confluent colonies. Errors shown represent ± 1 SE.*

values. There was a subsequent decrease in CFU-C per femur observed on day 7.

The decrease in the number of CFU-M per femur was more gradual with the largest decrease to about 10% of the pretreatment value being observed at 24 hours. Recovery after this time was extremely rapid with confluent colonies on day 2 being indicative of values above pretreatment.

IV. DISCUSSION

In vitro cell clonal assays have been described using the microtiter plasma culture technique for measuring hematopoietic progenitor populations committed to erythrocytic, granulocytic and megakaryocytic differentiation. In addition, preliminary data is presented of early changes induced in these populations, as well as in CFU-S, as a function of time following a single dose of cyclophosphamide.

In our experience, the microtiter plasma culture system offers several advantages for studying clonal ancestors of differentiated hematopoietic cells. This one system provides a standardized method of culture that permits not only the full expression of the potentials for proliferation and differentiation of the clonal progenitors of the three major blood cell lines but also the use of specific cyto- and histochemical assays with which to identify and thus quantitate *in situ* the total number of colonies derived from each differentiated progenitor. In addition, the relatively small size of the cultures permits the economic use of ingredients such as conditioned media which, in some instances, may be difficult (or expensive) to obtain and store under conditions ideal for maintaining their potency. Most importantly, perhaps, is the potential application of this culture system to the use and study of human marrow which has been demonstrated for human CFU-E by Prchal *et al.* (1974).

The time of appearance of CFU-E, C and M and their plating efficiences at optimal culture times correspond with those presented by McLeod *et al.* (1975) and Nakeff and Daniels-McQueen (1976) for both CFU-E and C and CFU-M, respectively.

Although the data on the effects of cyclophosphamide on CFU-S, E, C and M is preliminary, the procedures applied have demonstrated the utility of using the described cell clonal assays for experimentally determining hematopoietic cytotoxicities to both pluripotent stem cells and committed clonal cell progenitors.

The initial decrease of 30% in the number of CFU-S per femur 2 hours after the injection of 2 mg of cyclophosphamide is in agreement with previous data (Bruce and Valeriote, 1968) and reflects mostly the direct cell killing by the drug. The subsequent decrease observed over the next 22 hours we attribute mostly to the loss of surviving CFU-S from the pluripotent stem cell compartment through increased differentiation in order to repopulate more mature cell compartments (e.g. CFU-E and C) which are depleted more completely as a result of their greater sensitivity to cyclophosphamide. The rapid recovery in CFU-S observed after this time probably reflects the relatively large surviving fraction of CFU-S present two hours after the administration of cyclophosphamide and the increased production of new CFU-S following the subsequent recruitment of the surviving fraction of CFU-S into cell cycle.

Cyclophosphamide showed a differential cell killing of CFU-E, C and M. CFU-E were the most sensitive showing the largest decrease at two hours and a somewhat delayed recovery, although once recovery was underway by day 2, the return to pretreatment values was rapid. This decrease was even more profound considering that the marrow cellularity was relatively unchanged by two hours after cyclophosphamide with the largest decrease to about 25% of normal observed on day 1 with a return to normal cellularity by day 2. In subsequent studies, we were still unable to detect CFU-E at two hours after 2 mg of cyclophosphamide even when more than three times as many cells were plated. The differential effects we observed for CFU-E and C are supported by the studies of Hellman and Grate (1971) and Lamerton and Blackett (1974) who have shown using functional assays that the process of erythropoiesis is generally more sensitive than granulopoiesis to a number of chemotherapeutic agents, including cyclophosphamide. This difference may result from a biochemically-inherent greater sensitivity of CFU-E to cyclophosphamide or a difference in proliferative rate which has been shown to modify cyclophosphamide sensitivity (van Putten and Lelieveld, 1970). Both Axelrad *et al.* (1973) and Iscove and Sieber (1975) have described clonal assays for an erythropoietic progenitor that is more primitive than the CFU-E and referred to as the burst-forming unit (BFU-E). Although the lack of linearity between the number of BFU-E formed and the number of cells plated in the plasma culture system (Axelrad *et al.*, 1973) may prevent its use at this time as a quantitative assay for "erythroid-committed" progenitor cells, it should be noted that the CFU-E which we assay here are not the most primitive of the committed progenitors in the erythropoietic line.

The position occupied by the CFU-M in the megakaryocytic pathway of differentiation and its precise relationship to the CFU-S is being elucidated at this time (Nakeff, in press). It is known that megakaryocytes in agar colonies of mouse marrow are polyploid (Metcalf *et al.*, 1975) so that colonies must arise by a complex series of both endomitoses to increase ploidy and mitoses to increase the number of daughter megakaryocytes from a single progenitor cell. The extent to which the interplay of these processes reflects the observed response of the CFU-M to cyclophosphamide is certainly not clear from the data presented; however, it seems that this population of progenitors is not as sensitive to killing by cyclophosphamide as either CFU-E or CFU-C and that cellular recovery, once commenced, is rapid.

We have shown that the response of CFU-S, E, C, and M to a single dose of 2 mg of cyclophosphamide is unique to each cell line. The degree of cell loss and the subsequent pattern of recovery within each class of progenitor probably reflects a complex interrelationship among several factors including, the proliferative state of the progenitor cells, the demands for differentiation placed upon them and their inherent cell sensitivity to cyclophosphamide, the size of the progenitor cell pools and their reserve capacity, and, ultimately, the relationship of each of the "committed" progenitors to the pluripotential stem cell. Although the data presented is insufficient to answer these questions adequately at this time, we are accumulating more data on dose- and time-response relationships for each cell system to a variety of agents in order to better classify anticancer agents in terms of their relative toxicities to the various progenitors as well as to use these same agents to better understand the complex interrelationships among these four important hematopoietic cell progenitors.

ACKNOWLEDGMENTS

We wish to thank Ms. Marianne Schmidt for her excellent technical assistance. This investigation was supported by grant number 5P01CA13053-05, and awarded by the National Cancer Institute, DHEW.

REFERENCES

Axelrad, A. A., McLeod, D. L., Shreeve, M. M. and Heath, D. S. (1974). *In* "Hemopoiesis in Culture" (W. A. Robinson, ed.) pp. 226-237. U.S. Government Printing Office, Washington.

Blackett, N. M., Roylance, P. J. and Adams, K. (1964). *Brit. J. Haemat. 10*, 453-467.

Brown, C. H. III and Carbone, P. P. (1971). *Cancer Res. 31*, 185-190.

Bruce, W. R. and Valeriote, F. A. (1968). *In* "The Proliferation and Spread of Neoplastic Cells" pp. 409-422. Williams and Wilkins, Philadelphia.

Chen, M. G. and Schooley, J. C. (1970). *J. cell. Physiol. 75,* 89-96.

Cooper, C. C., Levy, J., Cantor, L. N., Marks, P. A. and Rifkind, R. A. (1974). *Proc. Nat. Acad. Sci. 71,* 1677-1680.

Gregory, C. J., McCulloch, E. A. and Till, J. E. (1973). *J. cell. Physiol. 81,* 411-420.

Gurney, C. W., Lajtha, L. G. and Oliver, R. (1962). *Brit. J. Haemat. 8,* 461-466.

Hellman, S., Grate, H. E. and Chaffey, J. T. (1969). *Blood 34,* 141-156.

Hellman, S. and Grate, H. E. (1971). *Blood 38,* 174-183.

In Vitro Culture of Hemopoietic Cells (1971). (D. W. van Bekkum and K. A. Dicke, eds), Radiobiological Institute TNO, The Netherlands.

Iscove, N. N., Sieber, F. and Winterhalter, K. H. (1974). *J. cell. Physiol. 83,* 309-320.

Iscove, N. N. and Sieber, F. (1975). *Exp. Hemat. 3,* 32-43.

Karnovsky, M. J. and Roots, L. (1964). *J. Histochem. Cytochem. 12,* 219-221.

Lamerton, L. F. and Blackett, N. M. (1974). *In* "Control of Proliferation in Animals Cells" (B. Clarkson and R. Baserga, eds), pp. 973-984. Cold Spring Harbour Symp.

McLeod, D. L., Shreeve, M. M. and Axelrad, A. A. (1974). *Blood 44,* 517-534.

Metcalf, D., MacDonald, H. R., Odartchenko, N. and Sordat, B. (1975). *Proc. Nat. Acad. Sci. 72,* 1744-1748.

Millar, J. L. and Blackett, N. M. (1974). *Brit. J. Haemat. 26,* 535-541.

Nakeff, A. Dicke, K. A. and Noord, M. J. van (1975). *Ser. Haemat. 8,* 4-21.

Nakeff, A. and Daniels-McQueen, S., (1976). *Proc. Soc. Exp.*

Biol. Med. *151*, 587-590.

Nakeff, A. (In press). *In* "Experimental Hematology Today" (S. J. Baum, ed), Springer-Verlag, New York.

Prchal, F. J., Axelrad, A. A. and Crookston, J. H. (1974). *Blood* *44*, 912.

Stephenson, J. R., Axelrad, A. A., McLeod, D. L. and Shreeve, M. M. (1971). *Proc. Nat. Acad. Sci.* *68*, 1542-1546.

Till, J. E. and McCulloch, E. A. (1961). *Radiat. Res.* *14*, 213-222.

Van Putten, L. M. and Lelieveld, P. (1971). *Europ. J. Cancer* *7*, 11-16.

MELANOGENESIS IN SOMATIC CELL HYBRIDS

FUNAN HU

Department of Cutaneous Biology

LINDA M. PASZTOR

Department of Pathology

Oregon Regional Primate Research Center
Beaverton, Oregon

I. INTRODUCTION

Genetic control of a differentiated function such as melanogenesis has been extensively studied in the phenotype of hybrids which were produced by the fusion of pigmented cells and nonpigmented fibroblasts derived from longterm cell lines (Davidson *et al.*, 1966; Silagi, 1967). Like other "luxury" functions, melanogenesis was suppressed in cell hybrids which contained one genetic complement but was expressed in hybrids which had two genomes from the pigmented cell parent (Fougere *et al.*, 1972; Davidson, 1974). We contend that some cells that have never functioned melanogenically and have been cultured in vitro for long periods of time may be regulated by negative control mechanisms (Davidson *et al.*, 1968), but that such factors as the ploidy and origin of the so-called undifferentiated cell parent are extremely critical in determining whether cell hybrids can express the differentiated function.

In earlier studies one of the parental cells in differentiated x nondifferentiated hybrids was a mouse fibroblast line. Unlike melanoma cells which were differentiated, these cells were undifferentiated in terms of melanogenesis. Strictly speaking, however, the differentiated function of the fibroblast is collagen, not melanin, synthesis. Should fibroblasts that are capable of collagen synthesis be designated undifferentiated?

Melanoma cells, on the other hand, form melanin, but enzymatically (Lerner and Fitzpatrick, 1953; Fitzpatrick and Kukita, 1956; Chang *et al.*, 1959) as well as ultrastructurally (Mishima, 1965; Cesarini, 1971) do not behave like normal melanocytes in all respects. Conceivably, then, the control of differentiation in normal and neoplastic cells may differ somewhat.

In this report, we describe our observations of several cell hybrids in vitro derived from (1) diploid melanocytes x pigmented and nonpigmented cells from long-term cell cultures and (2) melanoma x nonpigmented cells from long-term cell cultures as well as in vivo hybrids produced by the fusion of melanoma cells from C57BL/6 mice with diploid cells from the host (Hu and Pasztor, 1975).

II. METHODS AND MATERIALS

A. Cells

Two types of melanogenic cells were used: (a) primary cultures of terminally differentiated, diploid melanocytes prepared from adult rhesus monkey choroid (Hu *et al.*, in preparation) and (2) a neoplastic melanocyte cell line (PAZG) (Pasztor *et al.*,

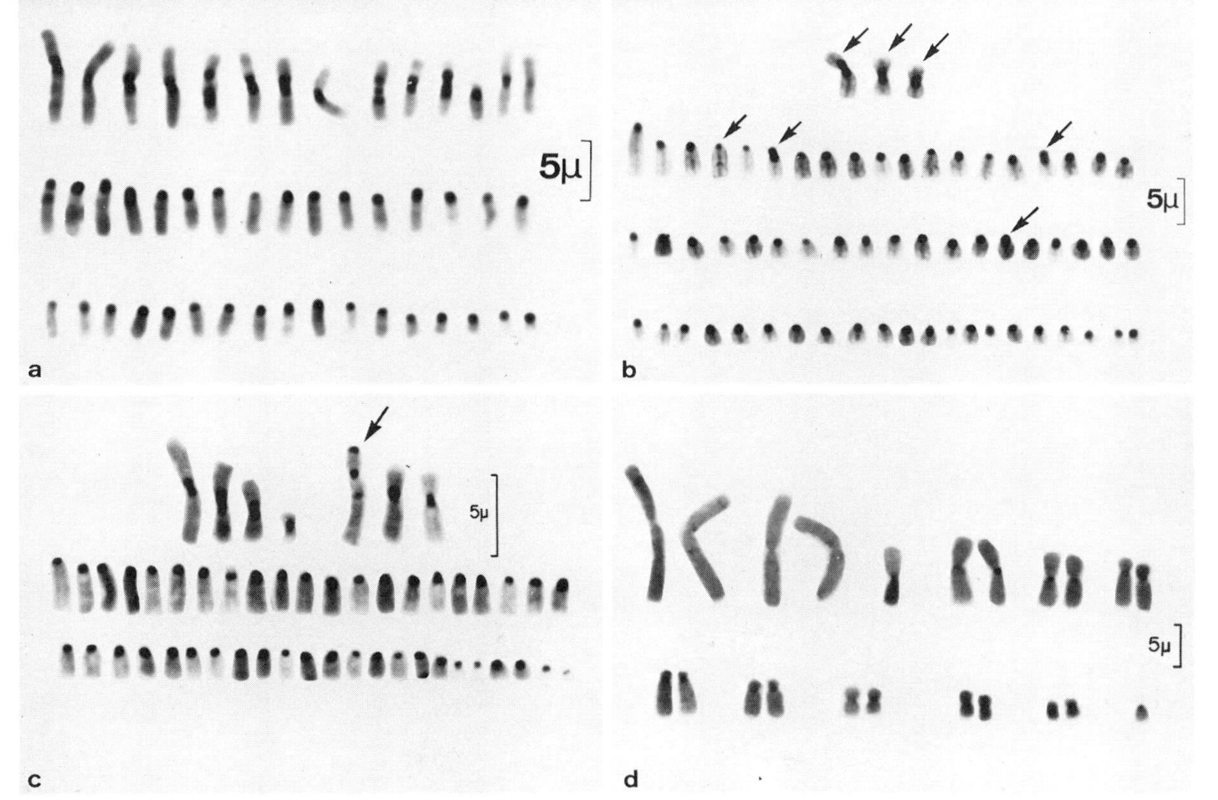

FIGURE 1 *Chromosomes stained to show C-banding. Bar = 5μ. (a) Karyotype of LM(TK$^-$) cells. (b) Karyotype of RAG cells. Arrows indicate the marker chromosomes. (c) Karyotype of PAZG cells. Arrow indicates the marker chromosome. (d) Karyotype of CH cells. (Reproduced with permission of S. Karger, A. G., Basel, from Proceedings of the IX International Pigment Cell Conference, 1977, Vernon Riley, editor).*

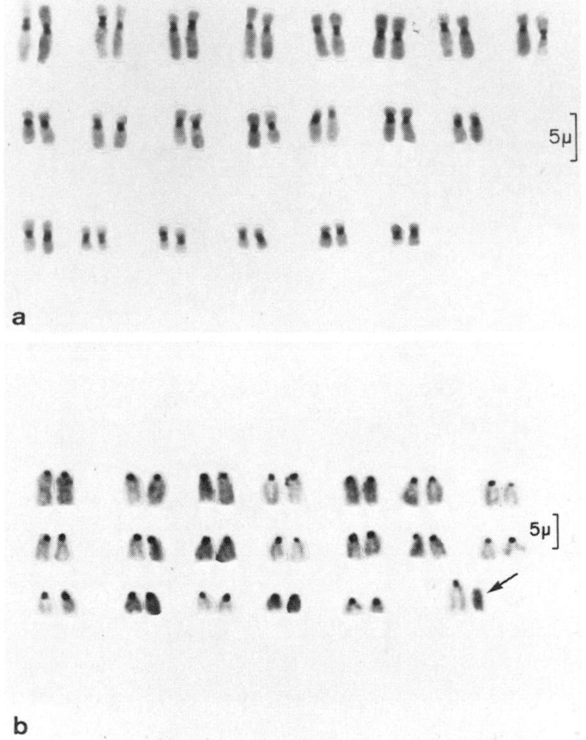

FIGURE 2 *Chromosomes stained to show C-banding. Bar = 5μ. (a) Karyotype of diploid rhesus cells. (b) Karyotype of male diploid C57BL/6 mouse cells. Arrow indicates the Y chromosome.*

1974). The nonmelanogenic parental cells were murine renal adenocarcinoma cells (RAG) (Klebe *et al.*, 1970), mouse fibroblasts [LM (TK^-)] (Kit *et al.*, 1963), Chinese hamster peritoneal cells (CH) (line B14 FAF 28-G3, obtained from American Type Culture Collection, CCL 14.1), and diploid cells from a male C57BL/6 mouse, most probably fibroblasts (Hu and Pasztor, , 1975). (Fig. 1a-d, Fig. 2a-b).

B. Hybridizations

All cells were grown in Eagle's minimal essential medium (MEM) supplemented with 10% fetal calf serum. The medium for the rhesus choroidal melanocytes also contained 75 μg/ml cytosine arabinoside, which killed any cell contaminants such as proliferating fibroblasts or endothelial cells and ensured a pure melanocyte population.

The PAZG cells were grown in MEM containing 200 μg/ml 8-azaguanine (AZG); the RAG, in 3 μg/ml AZG; the CH, in 50 μg/ml 5-iododeoxyuridine; and the LM(TK$^-$) in 30μg/ml 5-bromodeoxyuridine. All fusions in vitro were done either in suspension or monolayer cultures with β-propiolactone-inactivated Sendai virus.

A tumor produced when 10^6 PAZG cells were injected into a normal C57BL/6 male mouse was excised, trypsinized, and grown in HATG medium to suppress the proliferation of the PAZG cells and to select for cell hybrids produced by the fusion of the host with the PAZG cells. Cell hybrids received the genetic information for HGPRT which was lacking in PAZG cells but was required for survival in HATG medium.

III. RESULTS

Rhesus choroid (RC) x long-term cell line hybrids contained a reduced number of rhesus chromosomes (2n = 42). Most hybrid metaphases contained from 2 to 16 rhesus elements and two genomes contributed by the other parent. RC x LM(TK$^-$), F96 (Fig. 3a) and RC x RAG, F100 (Fig. 3b) hybrids were nonpigmented. Similarly, PAZG x CH, F57 (Fig. 4a) and PAZG x LM(TK$^-$), F34 (Fig. 4b) hybrids were nonpigmented and contained a reduced number of PAZG chromosomes and of total chromosomes respectively (Hu *et al.*, in press).

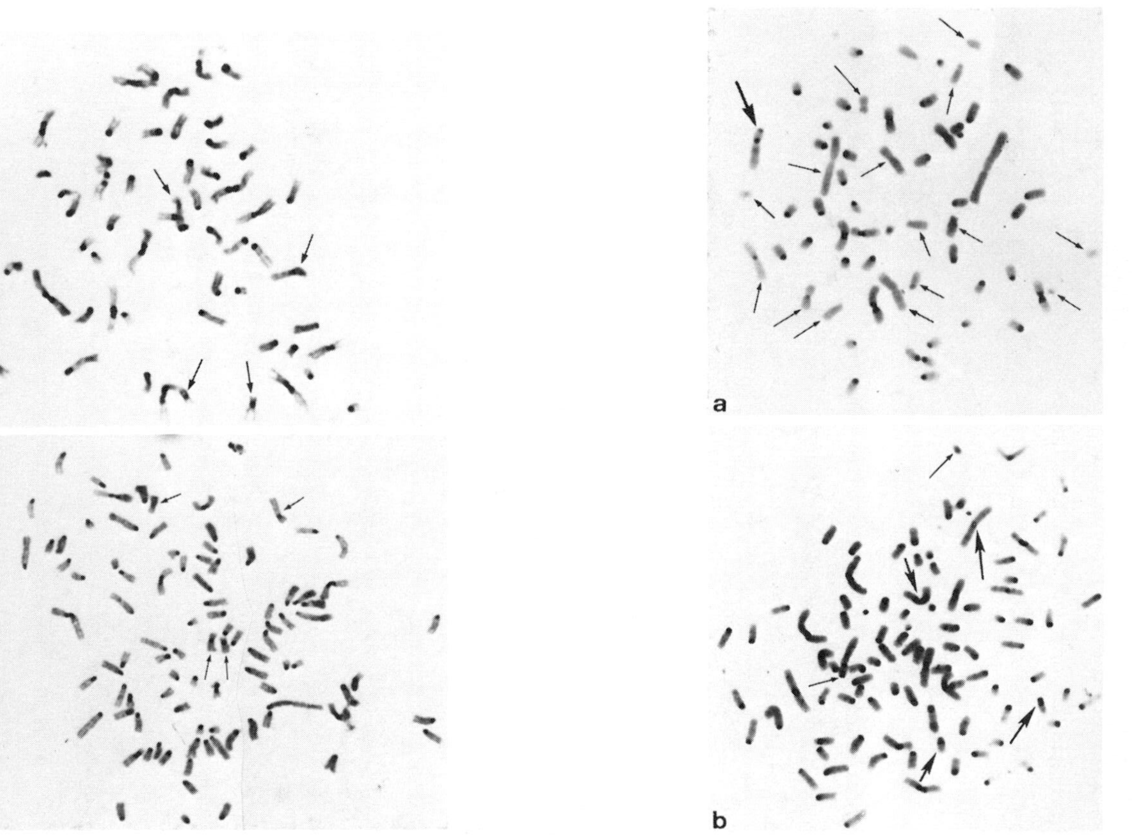

FIGURE 3 *Metaphase spreads, C-bands. Arrows indicate rhesus chromosomes. (a) RC x LM (TK^-) (F 96). (b) RC x RAG (F 100).*

FIGURE 4 *Metaphase spreads, C-bands. (a) PAZG x CH (F 57). Thin arrows indicate CH chromosomes; thick arrow, the PAZG marker. (b) PAZG x LM(TK^-) (F 34). Thin arrows indicate PAZG markers; thick*

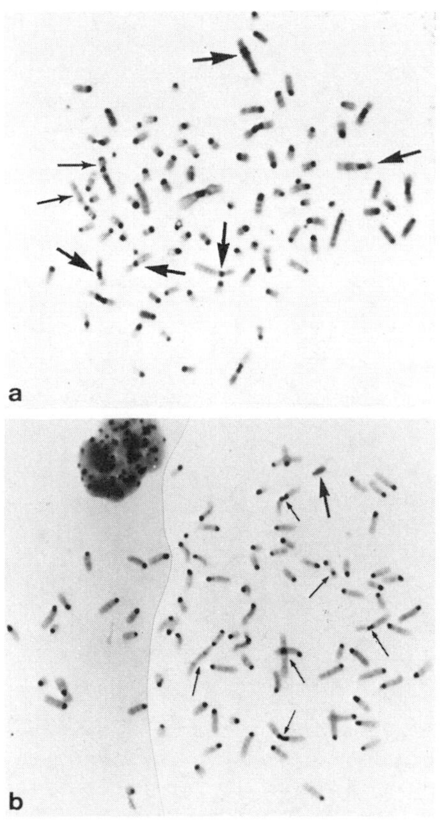

FIGURE 5 *Metaphase spreads. C-bands. (a) PAZG x RC (F 85). Thin arrows indicate PAZG markers; thick arrows, rhesus chromosomes. (b) PAZG x C57BL/6 diploid cells (MP), in vivo hybrids. Thin arrows indicate PAZG markers; thick arrow, the Y chromosome.*

RC x PAZG, F85 (Fig. 5a) and PAZG x C57BL/6 host cell, MP hybrids (Fig. 5b) were pigmented. Indeed, these two hybrid lines were more pigmented than the parental PAZG cells (Fig. 6a-d).

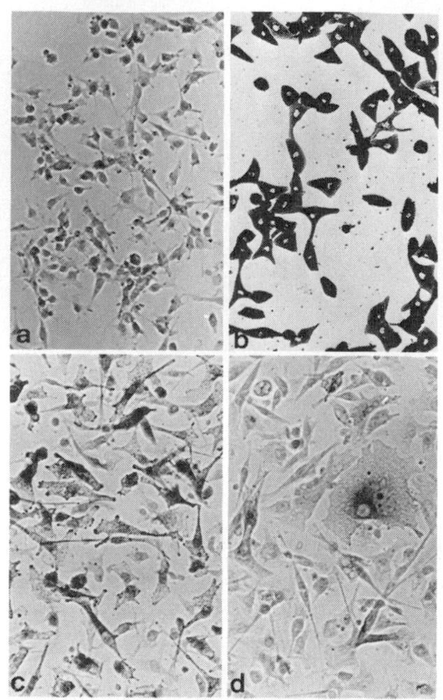

FIGURE 6 *Monolayer cell cultures. (a) Mouse melanoma cells (PAZG), DOPA reaction. Note only a few scattered cells darkened by the dopa reagent. Mag. 256 X. (b) Rhesus monkey choroidal melanocytes (RC). Living culture, unstained. All cells filled with dark granules. Mag. 160 X. (c) PAZG x diploid host cells (MP). DOPA reaction. Note larger and more dopa-reactive cells than (a). Mag. 256 X. (d) PAZG x RC (F 85). DOPA reaction. Note one large epithelial-like cell which contains dark dopa-reactive granules and several smaller dopa-reactive cells. Mag 256 X.*

IV. DISCUSSION

Our experiments were designed to answer two questions: (1) Are hybrids derived from the fusion of two melanogenic parental cells (e. g., F85) different from hybrids derived from the fusion of one melanogenic with one nonmelanogenic cell? (2) Does the control of differentaition differ in neoplastic and normal melanocytes?

Our results indicate that when both parental cells were melanogenic, the hybrid was melanogenic and that when only one parental cell was melanogenic, the hybrids were, with one exception, nonmelanogenic. Each of the nonmelanogenic hybrids was characterized by chromosome loss, however, and melanogenesis was not re-expressed in them. These results were the same whether the melanogenic parent was normal or neoplastic. The exceptional hybrid, MP, was initially characterized by a chromosome number which was consistent with its derivation from the fusion of one PAZG cell with one diploid host cell without chromosomal loss (Hu and Pasztor, 1975).

According to Harris (1972) and Wiener *et al.* (1971, 1974), malignancy develops when cells lose the gene(s) which control normal cell growth and can be corrected by the fusion of malignant with normal cells. Undoubtedly, the expression of melanogenesis depends upon whether the genetic information for this function is present. Davidson (1974) discussed a gene dosage effect in melanoma hybrids and suggested that a 2S genetic complement from the pigmented parental cell can overcome the negative control contributed by the nonpigmented parental cell. Since melanogenesis was expressed in our MP hybrid cells (an intraspecific fusion of melanoma and host cells from the same mouse strain), this requirement was unnecessary.

We agree with Davidson's observation (1974) that to use different species as parental cells complicates the interpretation of the results since some alterations in gene expression in the hybrids could be due to the combination of genomes of different species, not to the mechanisms of gene regulation. Silagi's hybrids (1967) were intraspecific (C57BL/6 melanoma x C3H cells from a long-term cell line) but unlike our MP hybrids, the undifferentiated parental cell was not diploid and not derived from the same mouse strain.

What control mechanism accounts for the enhancement of melanogenesis in MP hybrids? If we could be absolutely sure that no diploid pigment cell was involved in this fusion, we could postulate that once the amelanotic cell is removed from its natural environment, it can be derepressed or activated to synthesize its own melanin. This would be similar to the induction of human albumin in mouse hepatoma x human leukocyte hybrids (Darlington, 1974). This hypothesis is supported by Brumbaugh's (this Congress) finding of melanin formation in heterokaryons derived from amelanotic melanocytes x erythrocyte nuclei from the pink-eyed mutant and wild-type fowl respectively.

ACKNOWLEDGMENTS

We are indebted to Dr. H. G. Coon of the National Cancer Institute, who generously supplied the β-propiolactone inactivated Sendai virus used in our fusion experiments; to Coral Jean Cotterell, Cathy Taylor and Dinah Teramura for expert technical assistance; and to Margaret Barss for careful editing of the manuscript.

Publication No. 816 of the Oregon Regional Primate Research Center supported in part by Grants CA 08499 of the National Cancer Institute, RR 00163 and RR 05694 of the Animal Resources Branch, Division of Research Resources, AM 08445 of the National Institute of Arthritis and Metabolic Diseases, National Institutes of Health; and by funds from the Cammack Trust, Portland, Oregon.

REFERENCES

Cesarini, J. P. (1971). *Eur. J. Clin. Biol. Res. 16,* 316-322.

Chang, J. P., Speece, A. J. and Russell, W. O. (1959). *In* "Pigment Cell Biology" (M. Gordon, ed.), pp. 359-370. Academic Press, New York.

Darlington, G. (1974). *In* "Somatic Cell Hybridization" (R. L.

Davidson and F. de la Cruz, eds.), pp. 159-162. Raven Press, New York.

Davidson, R. L. (1974). *In* "Somatic Cell Hybridization" (R. L. Davidson and F. de la Cruz, eds.), pp. 131-146. Raven Press, New York.

Davidson, R. L., Ephrussi, B., and Yamamoto, K. (1966). *Proc. Natl. Acad. Sci. 56,* 1437-1440.

Davidson, R. L., Ephrussi, B., and Yamamoto, K. (1968). *J. Cell Physiol. 72,* 115-128.

Fitzpatrick, T. B. and Kukita, A. (1956). *J. Invest. Dermatol. 26,* 173-183.

Fougere, C., Ruiz, F., and Ephrussi, B. (1972). *Proc. Natl. Acad. Sci. USA. 69,* 330-334.

Harris, H. (1972). *J. Natl. Cancer Inst. 48,* 851-864.

Hu, F. and Pasztor, L. M. (1975). *Differentiation 4,* 93-97.

Hu, F., Pasztor, L. M., White, R., and Wilson, B. J. *In* "Proc. IX International Pigment Cell Conference." S. Karger, Basel. (In press).

Kit, S., Dubbs, D. R., Piekarski, L. J., and Hsu, T. C. (1963). *Exp. Cell Res. 31,* 297-312.

Klebe, R. J., Chen, T., and Ruddle, F. (1970). *J. Cell Biol. 45,* 74-82.

Lerner, A. B. and Fitzpatrick, T. B. (1953). *In* "Pigment Cell Growth" (M. Gordon, ed.), pp. 319-333. Academic Press, New York.

Mishima, Y. (1965). *Arch. Dermatol. 91,* 519-557.

Pasztor, L. M., Hu, F., Stankova, L., and Bigley, R. (1974). *J. Natl. Cancer Inst. 52,* 1143-1150.

Silagi, S. (1967). *Cancer Res. 27,* 1953-1960.

Wiener, F., Klein, G., and Harris, H. (1971). *J. Cell Sci. 8,* 681-692.

Wiener, F., Klein, G., and Harris, H. (1974). *J. Cell Sci. 15,* 177-183.

MORPHOLOGICAL TRANSFORMING FACTOR: INFLUENCE ON HISTOTYPIC ORGANIZATION OF EPITHELIAL BRAIN CELLS

RAMON LIM
DAVID E. TURRIFF
SHUANG S. TROY
*KATSUSUKE MITSUNOBU**

Departments of Surgery (Neurosurgery) and Biochemistry
and the Brain Research Institute
University of Chicago
Chicago, Illinois

I. INTRODUCTION

Tissue culture is the only artificial system where cells in the living state can be brought under direct observation for a relatively long period of time. Since brain cells during maturation undergo striking changes in morphology and histotypic pattern, they serve as a useful model for the study of morphogenesis at the cellular level and its relationship with chemical differentiation. In this article we summarize our experience with fetal brain cells in a monolayer culture as well as the effects of a tissue factor on the cells.

*Present address: Department of Neuropsychiatry, University of Okayama, Okayama City, Japan.

II. BASIC OBSERVATIONS

17-day Sprague-Dawley rat fetuses are used. The cerebrums and cerebellums are dissected, pooled and dissociated with trypsin in Tyrode solution free of calcium and magnesium. After seeding in a tissue culture flask, the cells form numerous small aggregates out of which two types of cells emerge: the flat epithelial cells which migrate rapidly onto the flask surface forming a thin layer of cell carpet; the bipolar cells (presumably neuroblasts) which migrate slowly and grow only on top of the flat cells (Fig. 1a). Before completely migrating out of the aggregates, cable-like interconnections between groups of neuroblasts are common. Upon subculture and by changing the medium within 24 hours, the neuroblasts are differentially eliminated leaving a culture of homogeneous epithelial cells (Fig. 1b). In the experiments reported in this article we deal exclusively with these epithelial brain cells. If permitted to reaggregate, the epithelial cells eventually mature into glial cells. However, if left in a monolayer, they will remain flat for an indefinite time.

In January, 1972, this laboratory (Lim *et al.*, 1972) made the observation that if an extract from adult rat brain is added to a culture of epithelial brain cells, dramatic morphological changes take place. The changes involve (1) the rounding up of individual cells and (2) the outshoot of numerous cell processes which are branched and interconnected between cells (Fig. 1c). Their appearance is characteristic of mature astrocytes. The morphological changes reach a peak at about 20 hours, after which further process outgrowth occurs if the medium containing the brain extract is renewed every other day (Fig. 1d). We routinely carry out the experiment on plastic culture flasks although the cells behave similarly on glass surface. The phenomenon is operationally designated as morphological transformation and the factor as the Morphological Transforming Factor. Care should be taken in

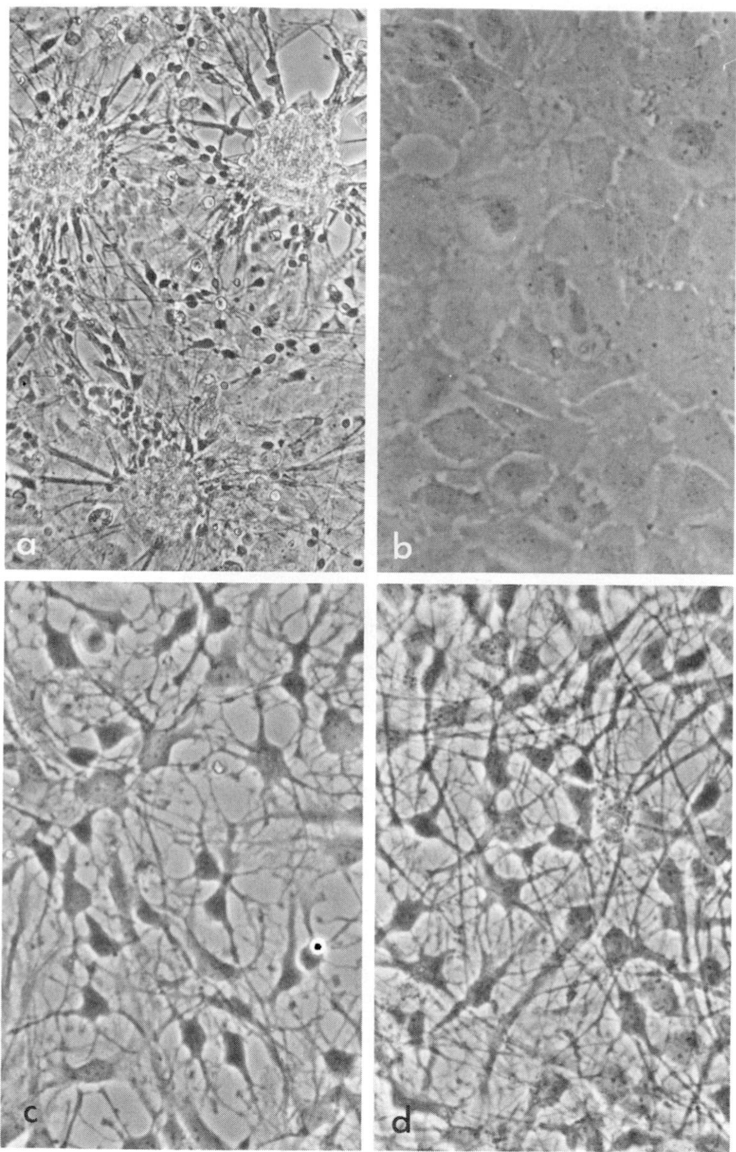

FIGURE 1 *Live phase-contrast photographs demonstrating morphological transformation in fetal brain cells. (a) Primary culture showing a mixture of epithelial cells and neuroblast-like cells (X 150). (b) Secondary culture showing the epithelial cells before exposure to brain extract (X 300). (c) Secondary culture showing the epithelial cells after exposure to brain extract for*

not confusing the term "transformation" in this article with the other connotation of the word as in "malignant transformation."

The brain extract is prepared by dialyzing the 100,000 × g supernatant of a brain homogenate. In order to confine ourselves to the macromolecules the extracts are always dialyzed before subjecting them to any study reported here, although the undialyzed extract produces essentially the same morphological effect.

III. NATURE OF THE EPITHELIAL CELLS

A. Are the Epithelial Cells Indeed Brain Cells?

The epithelial cells used for our studies make up about 50% of the total cell population dissociated from the brain tissue, the other 50% being neuroblasts. Since in a 17-day rat embryo most of the neuroblasts have already appeared whereas glial cells normally come into existence two weeks postnatally, it is reasonable to assume that a great proportion of these ill-defined cells are glial precursors. Such an assumption has previously been made by other investigators (Shein, 1965; Varon and Raiborn, 1969). We estimate that contamination by fibroblasts does not exceed 5%, based on the abundance of blood vessels embedded in the embryonic brain parenchyma; these blood vessels are the only source of fibroblasts since we meticulously peel off the meninges and the superficial vascular linings of the brain before dissociation. This estimate has been corroborated by our inability to obtain the same epithelial cells from other organs of the same embryos, even from those that are rich in fibroblasts such as heart and skeletal muscle. Furthermore, collagen fibers are never observed in the brain cell cultures with electron microscopy. That the

20 hours (X 300). (d) Secondary culture showing the epithelial cells after exposure to brain extract for a week, the medium containing the extract being changed every other day (X 300).

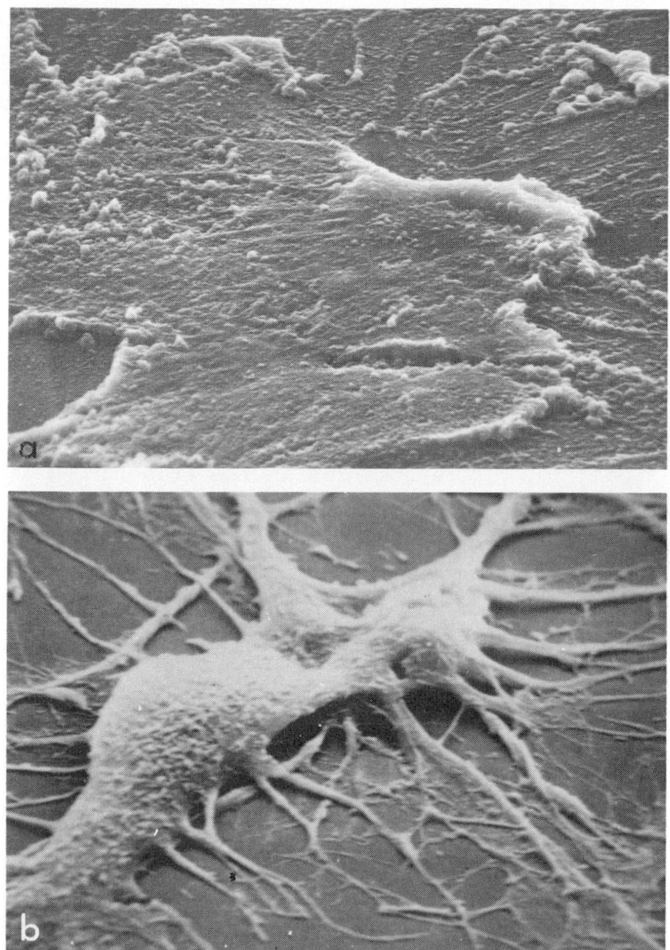

FIGURE 2 *Scanning electron microscopy of brain epithelial cells (a) before and (b) after morphological transformation by brain extract (X 2500).*

epithelial brain cells are indeed developing cells is supported by our unsuccessful attempts to obtain them from the adult brain.

B. Fine Structure of the Cells

With the scanning electron microscope (Fig. 2) a confluent monolayer of epithelial brain cells appears as a monotonous continuous sheet. Except for some thickened areas which probably represent cell borders, the cells are devoid of recognizable surface landmarks. A great contrast is observed with cells exposed to brain extract. These cells possess a body of definite volume, from which numerous processes and ramifications originate.

Transmission electron microscopy (Figs. 3 and 4) reveals differences in intracellular ultrastructure before and after transformation. Sheath microfilaments[1] (fibrillar structures that are 50 Å in diameter and arranged in bundles) are abundant in the flat cells but disappear following transformation by brain extract. Tonofilaments[1] (100 Å in diameter) are found in both types of cells but appear to be more numerous in the transformed cells. Those seen in the flat cells are usually randomly coiled without a definite orientation whereas those in the transformed cells are arranged in parallel arrays. One distinctive feature of the transformed cells is that their processes are packed with tonofilaments running along the axis. Microtubules (200 Å in diameter) are a minor component in both cells, but the difference in arrangement and orientation before and after transformation is similar to the tonofilaments. The flat and the transformed cells also differ in cell junctions. The desmosome-like junctions (complete with tonofilaments converging on both sides) characteristic of the flat cells are replaced by junctions similar to *zonula adhaerentes* described by Farquhar and Palade (1963) (see also Peters *et al.*, 1970).

C. Sequential Events During Morphological Transformation

We have studied the morphological changes with cinematography. Upon exposure of the epithelial cells to brain extract, a latent period covering the first 8 hours is observed, within which the only noticeable change is the clearer separation of the cell bor-

[1]The classification of the intracellular fibrillar structures in this article follows that of Wessells (1973).

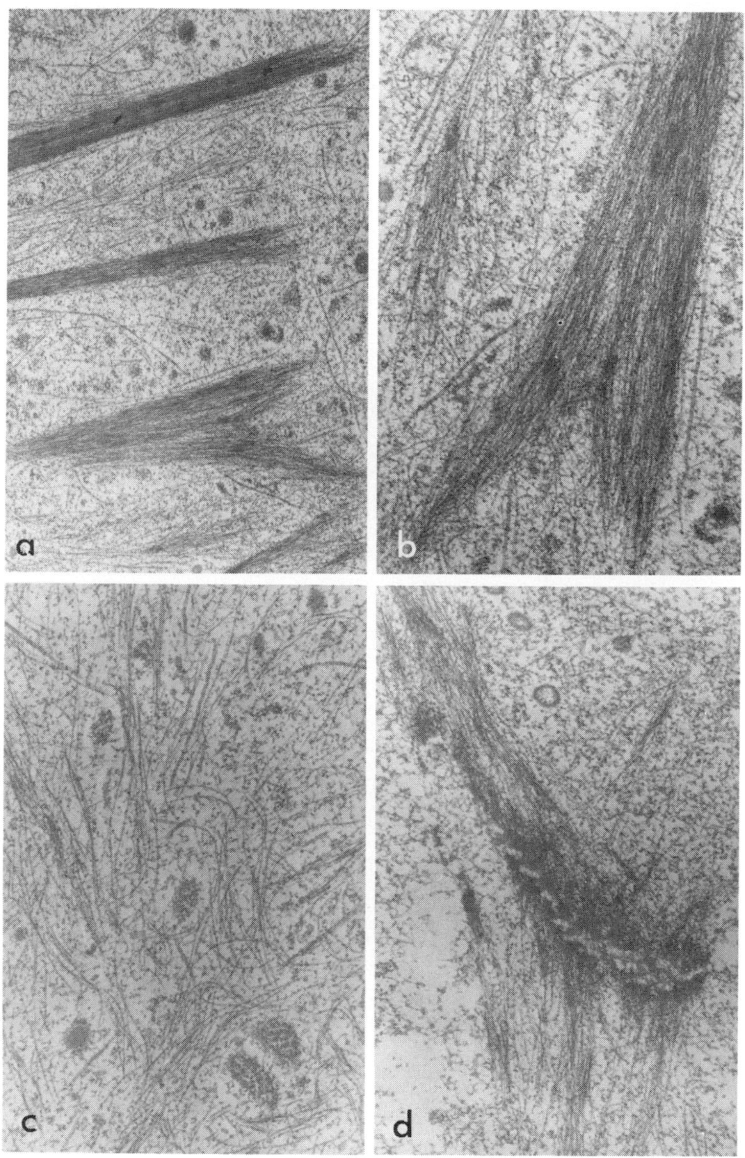

FIGURE 3 *Transmission electron microscopy of brain epithelial cells before transformation, showing: (a) sheath microfilaments (X 12,500); (b) detail of a bundle of sheath microfilaments (X 25,000); (c) tonofilaments with some microtubules (X 25,000); (d) a desmosome-like junction (X 25,000). All stained with phosphotungstic acid.*

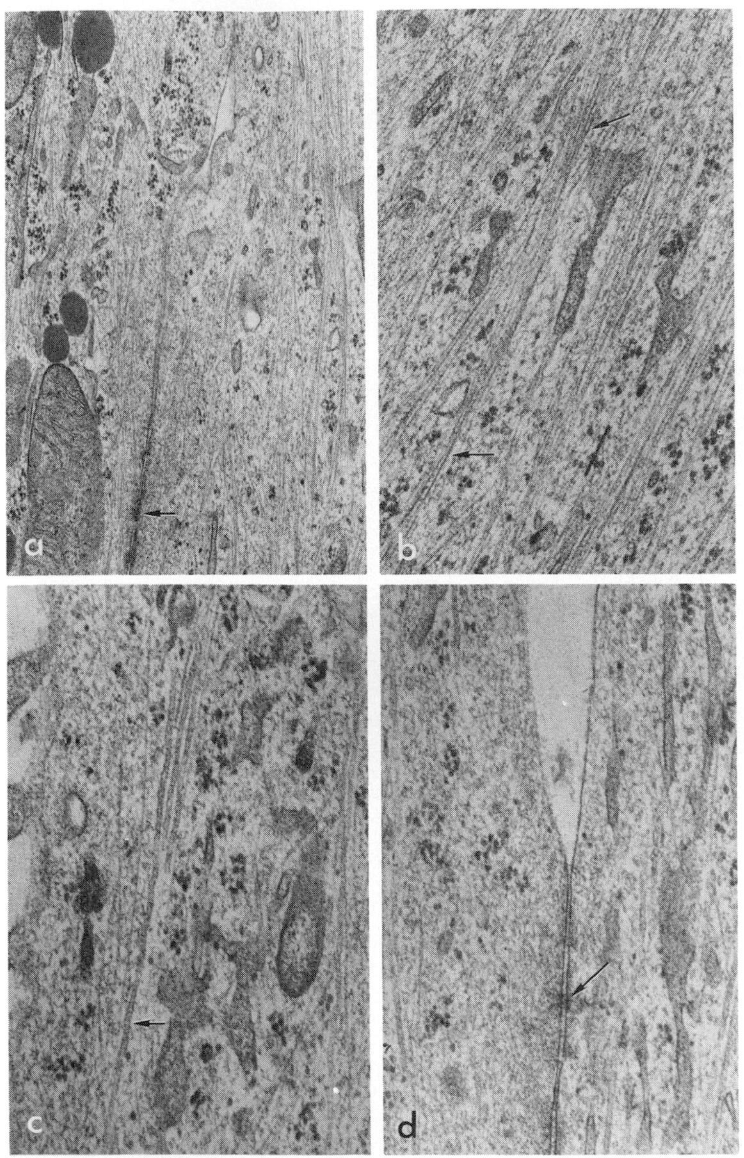

FIGURE 4 *Transmission electron microscopy of brain epithelial cells after transformation, showing: (a) numerous tonofilaments in two apposing cell processes; arrow indicates a junction complex between the two processes (X 25,000); (b) detail of tonofilaments (arrows) (X 50,000); (c) mixture of microtubules (arrow) and tonofilaments (X 50,000); (d) detail of a junctional complex (arrow)*

ders. Between the 12th and the 20th hours, dramatic changes occur. Many cells suddenly undergo mitotic activity after which the cells maintain a contracted appearance and extrude processes. All the cells become very active, constantly extending and retracting their processes. Since the cells are interconnected, they appear to be engaged in a tug-of-war, as reported earlier by Lumsden and Pomerat (1951) to be characteristic of glia. On the other hand, actual locomotion by the cells from one place to another is minimal. The general histotypic pattern is that of an interconnected cell net showing coordinated movement, as if being pulled from one end of the net (this dynamic relationship is common to glial cell cultures, as opposed to fibroblasts which exhibit no visible intercellular coordination). Three days after exposure to the extract, movement and mitosis gradually subside while the processes continue to extend. At the end of a week one sees a crisscrossing network with little movement or cell division.

In a parallel study using isotope incorporation (Table I), we found an increase in the rate of DNA synthesis in cultures exposed to brain extract over the control cells at the 12th hour and the 24th hour points. At the latter time point RNA and protein synthesis also increase. The series of events reflects cell division. A dramatic rise in intracellular cyclic AMP level occurs 4 days after stimulation by the brain extract. The time course of cyclic AMP changes, when taken together with the cinematographic observations on the cells, is consistent with the known inhibitory effects of the agent on cell division (Abell and Monahan, 1973) and motility (Johnson *et al.*, 1972).

Another interesting observation that distinguishes the flat and the transformed cells lies in the ability of some of the latter cells (about 20% of the total population) to concentrate

(X 50,000). All stained with osmium tetroxide, uranyl acetate and lead citrate.

TABLE I Chemical Differences[a] between Experimental and Control Cells

Time[b]	DNA synthesis[c]	RNA synthesis[c]	Protein synthesis[c]	Cyclic AMP
½ day	+63%	-19%	-15%	-
1 day	+110%	+53%	+58%	-18%
4 days	-43%	-60%	-54%	+270%
7 days	-50%	-75%	-43%	+370%

[a]The results are expressed as percentage increment or decrement in the experimental cells (exposed to brain extract) with respect to the control cells (not exposed to brain extract).

[b]Days after the initial exposure to brain extract.

[c]One-hour pulse labeling with a precursor.

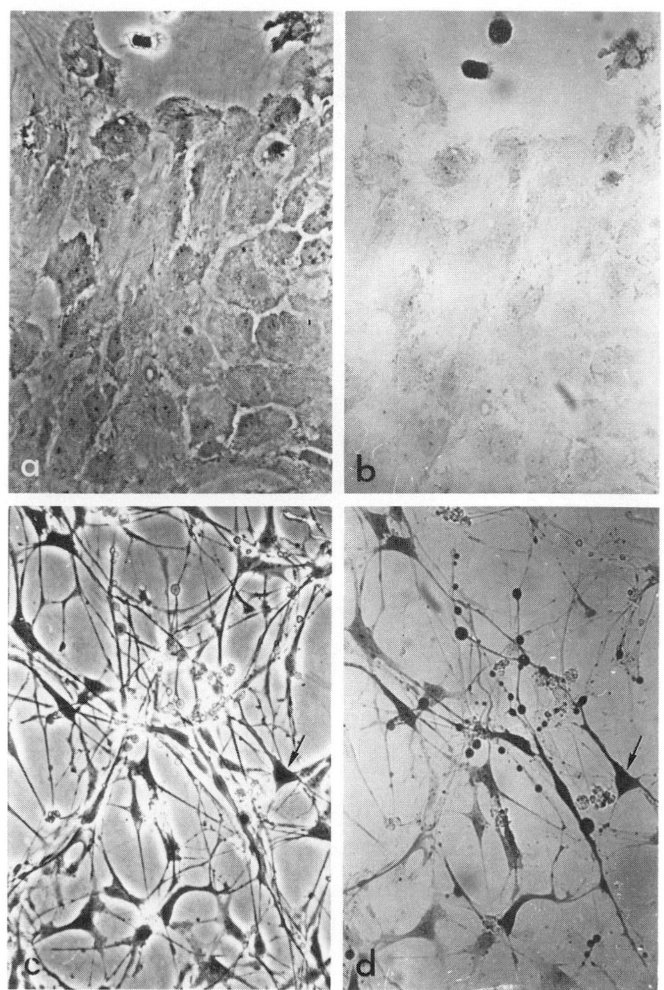

FIGURE 5 *Difference in methylene blue uptake by brain epithelial cells before and after transformation by brain extract. The dye was added to the media of live cultures to make a final dilution of 1/50,000 (w/v). (a) and (b): Phase-contrast and light microscopy, respectively, of the flat cells 4 hours after addition of the dye. (c) and (d): Phase-contrast and light microscopy, respectively, of the transformed cells 1 hour after addition of the dye; arrow indicates one of the cells with "silhouette" cell bodies and beaded cell processes as a result of excessive dye uptake. (X 200).*

methylene blue from the medium when used as a supravital stain (Fig. 5). Since Costero and Pomerat (1951) indicated that the uptake of methylene blue is more of a neuronal property than glial, the significance of this observation is yet to be clarified.

D. Dependency of Morphological Transformation on Cell Passage

The transformability of the epithelial brain cells decreases as the number of cell passage increases. On the 5th passage the flat cells are no longer responsive to stimulation by the brain extract. It is not known whether this decreasing transformability is due to the selection in favor of the non-responsive cells or is a result of progressive de-differentiation of all the epithelial cells.

E. Reversibility of Transformation and Temporal Summation of Transforming Effect

Withdrawal of brain extract from the medium leads to the eventual reversion of the transformed cells to the flat cells. On the other hand, renewing the medium containing the brain extract every other day not only sustains but also augments the outgrowing of cell processes. A temporal summation of transforming effect is particularly evident when the cells happen to respond poorly after the first exposure to the brain extract. The need of the continuous presence of the factor indicates a sustaining rather than inducing role with regard to differentiation.

F. Response to Dibutyryl Cyclic AMP and β-Adrenergic Agents

The epithelial cells can be transformed by 1 mM dibutyryl cyclic AMP. However, unlike the brain extract effect which shows a long latency, the effect of dibutyryl cyclic AMP is rapid (starting within 5 minutes of exposure and reaching a height at about 4 hours) and rather transient. Furthermore, whereas with brain extract cellular retraction and process outgrowth are equally prominent, with dibutyryl cyclic AMP cellular retraction is the main feature. Cyclic AMP at 1 mM has no effect.

1-Isoproterenol (0.01 mM) transforms the cells in a manner and time course similar to dibutyryl cyclic AMP. The effect is blocked by the β-blocker dichloroisoproterenol (0.1 mM) but not by the α-blocker phentolamine (0.1 mM). 1-Norepinephrine at 0.1 mM shows only a weak activity compared to isoproterenol. Thus the cells appear to have β-adrenergic receptors which utilize cyclic AMP as the mediator, a view consistent with earlier reports that β-adrenergic agents produce a transient surge in cyclic AMP level in astrocytoma cultures (Clark and Perkins, 1971; Gilman and Nirenberg, 1971), brain slices (Rall and Gilman, 1970; Schultz and Daly, 1973a and 1973b), and dissociated brain cells (Gilman and Schrier, 1972).

G. Response to Variations in Serum Concentration

Complete absence of serum may lead to a transient retraction of some of the epithelial cells in a manner similar to the effect of dibutyryl cyclic AMP. At serum concentrations of 2% or higher, this spontaneous change is not seen.

The influence of serum on the transforming effect of brain extract is complex. In the presence of 10% serum or higher, the transforming effect of brain extract normally observable at the 20th hour is suppressed or reduced. However, if the same culture is observed over a period of 4 or more days, with renewal of the medium containing brain extract every other day, the response eventually shows up to the same degree in high serum as in low serum. This is true even with serum concentrations as high as 17%.

H. Agents Blocking Morphological Transformation

The following agents block the transforming effect of brain extract: cycloheximide (0.1 μg/ml); colchicine (0.1 μg/ml); vinblastine (0.5 μg/ml); cytochalasin B (5 μg/ml).

IV. NATURE OF THE TRANSFORMING FACTOR

A. Some Initial Doubts

The question arose as to whether the observed morphological transformation is caused by some chemical message contained in the brain extract or simply by changes in physical conditions introduced by the extract, such as pH of the medium and the surface texture of the flask. The pH effect is ruled out by our ability to demonstrate the same transformation by brain extract under various pH conditions (Lim and Mitsunobu, 1975). The texture effect is ruled out by the observation that growing the epithelial brain cells in flasks pretreated (one-day preincubation) with brain extract does not result in morphological transformation (unless fresh brain extract is concomitantly present) and that transformed cells, if transferred to a new flask not currently or previously exposed to brain extract, retain the transformed morphology, thus indicating that the direct contact between the cells and the extract, and not between the flask surface and the extract, is essential for the observed morphological changes. Furthermore, other pieces of evidence against the flask surface effect are available: (1) growing the epithelial brain cells in plasma clot or collagen-coated surface, or the direct addition of gelatin and fibrinogen to the culture, does not result in process outgrowth; (2) the brain is one of the organs with the lowest collagen content yet its transforming effect is the highest among organs assayed; (3) extracts prepared from rat brains whose plasma fibrinogen has been eliminated by saline perfusion do not show a lower transforming activity compared with extracts from unperfused brains; (4) no fibrous deposit has ever been detected by scanning electron microscopy and no collagen fiber has ever been observed with transmission electron microscopy on cultures exposed to brain extract; (5) preincubation of the brain extract with collagenase or direct addition of the enzyme into the culture does not block morphological transformation.

Can the transforming effect be attributed to some special nutritional factor contained in brain extract? Since the small molecules have been routinely dialyzed out, this question directs itself to the macromolecules as possible nutrients. We have demonstrated that if the factor is a protein, it is the native molecule and not the amino acid constituents that are responsible for the activity. To prove this we first boiled the dialyzed brain extract for 10 minutes to coagulate the proteins (the mixture was now devoid of activity). With the coagulum in place, pronase (2 mg/ml) was introduced and the mixture incubated for 4.5 hours at 37°C until the coagulum cleared up. The solution was again boiled for 10 minutes to coagulate the pronase which was subsequently eliminated by centrifugation. The resulting protein digest was completely inactive for transformation.

The protein digest experiment described above also suggests that the putative transforming factor cannot be a heat-stable small molecule (such as cyclic AMP or prostaglandin) that might have adhered to the proteins, since the pronase digestion would have released them and they would have remained active in the supernatant fluid. The presence of contaminating cyclic AMP that might have exerted the transforming effect was also ruled out by the inability of cyclic AMP phosphodiesterase to suppress the transformation by brain extract when either pre-incubated with the extract or directly added to the culture.

B. What the Factor Is Not

Numerous commercially available proteins (Lim and Mitsunobu, 1974) were assayed; none transformed the cells. Other synthetic and natural products free of transforming activity (when tested at 1.2 mg/ml medium) include: polyvinylpyrrolidone, Ficoll, polyethylene glycol, hyaluronic acid, chondroitin sulfate, heparan sulfate, yeast RNA, and poly-L-aspartic acid. The following factors, known to affect growth and/or differentiation of some other cells, show no morphological effect on the epithelial brain cells: Nerve Growth Factor, phytohemagglutinin, concanavalin A, wheat germ agglutinin and pokeweed mitogen. Of these, Nerve Growth Factor was tested on three separate occasions: once with a sample purchased from Wellcome Research Laboratories, once with a sample obtained in our laboratory directly from the submaxillary glands of an adult male mouse, once with a gift supplied by Drs. Ing-Ming Jeng and Ralph A. Bradshaw of Washington University. All gave negative results when tested at a concentration optimum for neurite outgrowth in dorsal root ganglia. Furthermore, a partially purified Morphological Transforming Factor (see below) sent from our laboratory to Drs. Jeng and Bradshaw for radioimmunoassay and bioassay for Nerve Growth Factor was negative by either criterion. Cyclic AMP at 1 mM and prostaglandin E_1 at concentrations ranging from 0.01 μg/ml to 10 μg/ml are without effect on our cells (Lim *et al.*, 1973).

C. Biological Distribution of the Transforming Factor

Extracts from rat and pig brains are equally active on fetal rat brain cells. Other rat and pig organs, such as liver, kidney, heart and muscle, also contain significant amounts of the activity. 17-day rat embryos (brain or whole body extract) contain only 1/5 of the activity found in adult brain. The factor is also low in brain tumors such as rat astrocytoma and mouse neuroblastoma, where less than 1/10 of the activity in adult brain can be detected. Human red blood cells contain only 1/25 of the rat brain activity whereas the activity in microorganisms such as E. coli, Bacillus subtilis and yeast is practically nil (less than 1/200 of adult brain). All the comparisons above are made on the basis of extractable protein. The following body fluids, when used at concentrations up to 15% (v/v) in the medium, do not possess any transforming activity: fetal calf serum, bovine serum, horse serum, rat serum, chicken serum and human cerebrospinal fluid. Thus, the general distribution indicates that the factor is limited to solid organs of normal adult animals. Conditioned medium from the epithelial brain cells, even after a 5-fold concentration, does not show any transforming activity.

D. Chemical Properties of the Transforming Factor

The factor has an apparent molecular weight of 350,000 (Fig. 6). It is resistant to DNAase and RNAase. Incubation with trypsin does not destroy nor reduce the activity. However, the factor is completely inactivated by pronase (0.1 mg enzyme/ml) after an overnight incubation under aseptic conditions, indicating that it is a protein. There is no indication as to whether it is a pure or a conjugated protein. Periodate oxidation does not affect the transforming activity, suggesting that any carbohydrate moiety, if present, probably does not play an essential role in its function. The factor is resistant to reducing agents and alkylating agents. It has an isoelectric point below neutrality. Other protein properties include its denaturation by the extremes of pH, by 8 M urea, and by prolonged heating. We have achieved a 400-fold partial purification (Lim and Mitsunobu, 1975) of the factor by ethanol treatment, trypsinization and column chromatography with Sephadex G-200 and Sepharose 4B. The purest sample is active at 3 μg protein/ml medium, equivalent to a concentration of 1×10^{-8} M.

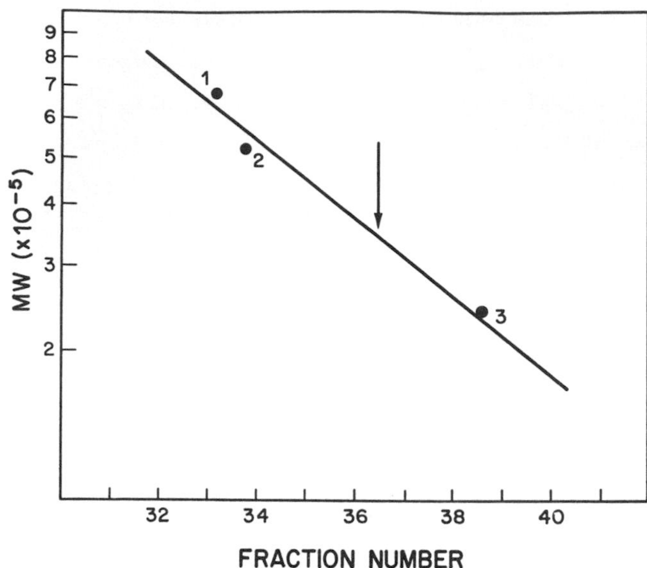

FIGURE 6 *Molecular weight determination for the Morphological Transforming Factor on a Sepharose 4B column. 1 = thyroglobulin; 2 = β-galactosidase; 3 = catalase. Arrow indicates the peak of the transforming activity.*

V. SUMMARY AND CONCLUSIONS

(1) Epithelial brain cells from rat fetuses can be stimulated by a tissue factor to grow out processes after a latency of about 8 to 12 hours.

(2) Most of the cells are glial cells.

(3) The effect of the tissue factor is not due to alteration of the physical conditions of the culture.

(4) The factor is a high molecular weight protein which is active when used in minute amounts.

(5) The cells stimulated by the factor show an increase in DNA, RNA, and protein synthesis at the early time point and an increase in cyclic AMP content at the late time point.

(6) The reversibility of transformation and the presence of the factor in adult rather than embryonic tissue suggest that the

factor plays a maintenance role in cellular differentiation and morphogenesis.

(7) The factor is present in the brain and several other organs of the mature animal.

ACKNOWLEDGMENTS

Dr. Mitsunobu was, and Dr. Troy is Research Associate in Dr. Lim's laboratory. Dr. Turriff is a Public Health Service Postdoctoral Fellow (NS-05017). This work was supported by U.S. Public Health Service grants NS-09228, CA-14599 and NS-07376.

REFERENCES

Abell, C. W. and Monahan, T. M. (1973). *J. Cell Biol. 59,* 549-558.

Clark, R. B. and Perkins, J. P. (1971). *Proc. Nat. Acad. Sci. USA. 68,* 2757-2760.

Costero, I. and Pomerat, C. M. (1951). *Am. J. Anatomy. 89,* 405-439.

Farquhar, M. G. and Palade, G. E. (1963). *J. Cell Biol. 17,* 375-412.

Gilman, A. G. and Nirenberg, M. W. (1971). *Proc. Nat. Acad. Sci. USA. 68,* 2165-2168.

Gilman, A. G. and Schrier, B. K. (1972). *Molecular Pharma. 8,* 410-416.

Johnson, G. S., Morgan, W. D. and Pastan, I. (1972). *Nature. 235,* 54-56.

Lim, R., Li, W. K. P. and Mitsunobu, K. (1972). *Abstracts of 2nd Annual Meeting, Society for Neuroscience, Houston, Texas.* p. 181.

Lim, R., Mitsunobu, K. and Li, W. K. P. (1973). *Exp. Cell Res., 79,* 243-246.

Lim, R. and Mitsunobu, K. (1974). *Science. 185,* 63-66.

Lim, R. and Mitsunobu, K. (1975). *Biochim. Biophys. Acta, 400,* 200-207.

Lumsden, C. E. and Pomerat, C. M. (1951). *Exp. Cell Res. 2,* 103-114.

Peters, A., Palay, S. L. and Webster, H. de F. (1970). "The Fine Structure of the Nervous System" Harper and Row, Publishers, New York.

Rall, T. W. and Gilman, A. G. (1970). *Neurosciences Res. Prog. Bull. 8,* 239-242.

Schultz, J. and Daly, J. W. (1973a). *J. Neurochem. 21,* 1319-1326.

Schultz, J. and Daly, J. W. (1973b). *J. Biol. Chem. 248,* 860-866.

Shein, H. M. (1965). *Exp. Cell Res. 40,* 554-569.

Varon, S. and Raiborn, C. W. Jr. (1969). *Brain Res. 12,* 180-199.

Wessels, N. K. (1973). *Neurosciences Res. Prog. Bull. 11,* 24-27.

ALTERATIONS IN THE PHENOTYPIC EXPRESSION OF CULTURED ARTICULAR CHONDROCYTES BY A PITUITARY GROWTH FACTOR (CGF)*

CHARLES J. MALEMUD
DAVID P. NORBY

Department of Pathology
Health Sciences Center
State University of New York at Stony Brook
Stony Brook, New York

I. INTRODUCTION

Recent advances have implicated serum (Todaro *et al.*, 1967; Holley and Kiernan, 1968, 1971, 1974; Leffert, 1974; Fryklund *et al.*, 1974; Turkington and Majumder, 1973) and organ-derived (Armelin, 1973; Gospodarowicz, 1974; Jones and Gospodarowicz, 1974; Lim and Mitsunobu, 1974; Revoltella *et al.*, 1974; Cohen and Carpenter, 1975) factors as playing major roles in the control of proliferation and phenotypic expression of cultured cells. Such factors have been studied in attempts to find serum fractions which can be used as substitutes for whole serum (Pierson and Temin, 1972; Dulak and Temin, 1973; Holley and Kiernan, 1974), to study the mechanisms by which these factors interact with the

*Abbreviations used: CGF, Chondrocyte Growth Factor; MSA, Multiplication - Stimulating Activity; EGF, Epidermal Growth Factor; FGF, Fibroblast Growth Factor; NIH-TSH, Thyrotropin; MEM, Minimal Essential Medium; DMEM, Dulbecco's Modified Eagle's Medium; BSA, Bovine serum albumin; DG-2-deoxy-D-glucose; 2-AIB, 2-amino*iso*butyric acid; Td-thymidine.

cell (Revoltella *et al.*, 1974; Rudland *et al.*, 1974a) and most recently to investigate bovine pituitary and brain molecules which profoundly augment the growth of cells not necessarily derived from these organs (Gospodarowicz, 1974, 1975). One of these pituitary factors has been designated chondrocyte growth factor (CGF) because of its relative selectivity for promoting the growth of cultured articular chondrocytes (Corvol *et al.*, 1972). CGF has not been purified and is recognized as a contaminant of several pituitary glycoprotein hormones.

II. METHODS AND MATERIALS

The majority of experiments performed with CGF have utilized NIH-TSH, lots B5, B6, and B7 (2.47-3.53 U/mg protein) and its effectiveness in promoting growth has been tested primarily on rabbit articular chondrocytes grown under monolayer conditions as previously described (Sokoloff *et al.*, 1970; Green, 1971; Corvol *et al.*, 1972). The concentration of TSH in these cultures has generally been 64-70 μg/ml. Recently we have tested various precursor fractions of TSH for CGF activity. These fractions designated LER-1853-B2 (a crude pituitary glycoprotein fraction) and LER-1874 (a fraction derived from CG-50 chromatography of LER-1853-B2) generously supplied by Dr. L. E. Reichert, are quite effective in promoting the growth of lapine chondrocytes; the growth response towards 1853-B2 being in the same range as NIH-TSH (e.g. 200-250% above control values) while that of LER-1874 was slightly higher (333%).

III. RESULTS AND DISCUSSION

The effects of CGF as NIH-TSH on cultured chondrocytes have been summarized in Table I. The role of CGF as a growth-promoting factor was first demonstrated by an increase in total DNA

TABLE I Alterations in the Chondrogenic Phenotype of CGF-treated Chondrocytes in Monolayer Culture

Cell property	Control	CGF-treated (64-70 μg/ml)	Reference
Growth response (Δ%)	-	218	Corvol *et al.*
^{3}H-Td incorporation-48 h. (cpm/μg DNA)	300	679	Malemud & Sokoloff
Generation time (hrs)	14-16	10	Malemud & Sokoloff
G_1 period (hrs)	6.7	0.1	Malemud & Sokoloff
Protein content (μg/μg DNA)	25-30	10-15	Malemud & Sokoloff
RNA content (μg/μg DNA)	4.7-5.2	3.7-4.1	Malemud & Sokoloff
$^{35}SO_4$ incorporation (dpm × 10^4/μg DNA)	1.67 ± 0.11	0.53 ± 0.02	Corvol *et al.*
Per cent of $^{35}SO_4$ incorporated into chondroitin sulfate	68	64	Malemud & Sokoloff
^{14}C-glucosamine incorporation (cpm/μg DNA)	657	230	Malemud & Sokoloff
Hyaluronate synthesis (nmoles/μg DNA)	0.60	0.57	Malemud & Sokoloff
2-deoxy-glucose transport-vmax (pmol/μg DNA/min)	76	20	Malemud & Sokoloff (to be published)
2-aminoisobutyric acid transport vmax (pmol/μg DNA/min)	41	8	Malemud & Sokoloff (to be published)

content/culture flask of CGF-treated cells. CGF augmented DNA synthesis which was maximal at 48 hours after subculture of primary cultures and declined thereafter. CGF also shortened the generation time from 16 to 10 hours. Three of the cell cycle phases, S, G_2, and M were identical with respect to time in control and CGF-treated cultures. The G_1 phase of CGF-treated chondrocytes, however, was almost totally absent, thus accounting for the shortened generation time.

In addition to promoting the growth of chondrocytes, CGF had a negative effect on RNA, protein and sulfated glycosaminoglycan synthesis. Thus, the protein, RNA content and net synthesis of sulfated glycosaminoglycans was reduced in CGF-treated chondrocytes. However, the profile of the glycosaminoglycan species common to rabbit articular cartilage and cultured chondrocytes (Srivastava *et al.*, 1974) was not altered. The per cent of $^{35}SO_4$ incorporated into chondroitin 4 and 6 sulfates or doubly-sulfated chondroitin was similar in control and CGF-treated chondrocytes (Table I). Although glucosamine incorporation was similarly reduced, synthesis of hyaluronic acid was unaffected by CGF (Malemud and Sokoloff, 1974). Thus, the reduction in glycosaminoglycan synthesis was specific for the sulfated types.

The effect of CGF on sulfated glycosaminoglycan synthesis as well as on proliferation was reversible. When chondrocytes were initially grown in CGF-containing medium and CGF was then withdrawn and replaced with control medium, increments in growth and reduction in $^{35}SO_4$ incorporation were lessened when CGF was withdrawn at earlier intervals in the culture period.

Collagen synthesis was also reduced by CGF. Chondrocytes were cultured in the presence of LER-1874 (70 μg/ml) for 5 days. With the last medium change, the cells were labeled with ^3H-glycine (0.5 μCi/ml) in MEM with Earle's Salts supplemented with sodium ascorbate (100 μg/ml) and 10% fetal bovine serum for 20 hours. The medium was dialyzed against 0.2 M NaCl for 24 hours and the retentate counted for radioactivity. The cell pellet was

TABLE II Effect of CGF (as LER-1874) on ^{3}H-glycine Incorporation by Articular Chondrocytes in Monolayer Culture

		^{3}H-glycine incorporation dpm × 10^3/μg DNA[a]		
		Cellular		Medium
Group	DNA (μg/flask)	TCA-insol.	TCA-sol.	non-dialyzable
Control	24.2	20.7 ± 1.3	5.48 ± 0.20	3.26 ± 0.16
LER-1874 (70 μg/ml)	104.8	2.98 ± 0.12	1.15 ± 0.06	0.64 ± 0.03

[a]Mean ± s.e. (6 determinations).

precipitated with 6% TCA and the soluble and insoluble fractions counted for radioactivity. The results of this experiment are shown in Table II. They indicate that the total ^{3}H-glycine incorporation into the medium and cell pellet compartments were reduced in the presence of LER-1874. These results reflected the overall reduction in total protein synthesis in the presence of CGF, but further experiments (see below) indicated that the proportion of ^{3}H-glycine incorporated into collagen as distinct from other cellular protein was also quantitatively reduced.

An analysis of the ratio of α1 to α2 collagen chains was carried out primarily on the cell pellet fraction, since 80-85% of the ^{3}H-glycine incorporated into collagen in 20 hours appears in this fraction (Norby *et al.*, 1977). After treatment of this fraction according to methods to be described elsewhere (Norby *et al.*, 1977), collagen was separated from proteoglycan by DEAE-cellulose chromatography (Miller, 1971). The chain composition of the collagen was then analyzed by CM-cellulose chromatography (Miller, 1971). Absorbance was monitored at 280 nm and 6.5 ml fractions collected. Alternate fractions were counted for radioactivity. Radioactivity appearing in fractions corresponding to carrier collagen peaks were pooled for each peak to give α1/α2 ratios.

The total ^{3}H-glycine incorporated into the cell pellet fraction of collagen was reduced in CGF-treated cultures (Control 15.3×10^{4} dpm/mg DNA; LER-1874, 0.35×10^{4} dpm/mg DNA). The $\alpha 1/\alpha 2$ ratio of control cultures was 5.37, suggesting a mixture of Type I and Type II collagens (Norby *et al.*, 1977), while that of CGF-treated cultures was 1.4. The greatly reduced ^{3}H-glycine incorporation into the collagen of CGF-treated cultures made it difficult to conclude whether LER-1874 induced an alteration in the $\alpha 1/\alpha 2$ ratio. Unlike the control cultures, more ^{3}H-glycine incorporation into collagen was found in the medium than in the cell pellet of the CGF-treated cells. When the medium of LER-1874 cultures was analyzed by a modification of the method described above, the $\alpha 1/\alpha 2$ ratio was 2.96, suggesting the presence of a mixture of Type I and Type II collagens.

The above studies have suggested that in addition to affecting chondrocyte proliferation in a positive manner, CGF suppressed the phenotypic expression of protein, glycosaminoglycan and collagen synthesis. These results are quite unlike that reported for other growth-promoting agents such as MSA (Pierson and Temin, 1972; Dulak and Temin, 1973), EGF (Cohen and Carpenter, 1975; Hollenberg and Cuatrecasas, 1975) and FGF (Gospodarowicz, 1974, 1975). These latter growth factors stimulate not only proliferation, but also most synthetic cell processes. Thus, the likelihood existed that CGF exerted its pleiotypic effect (Hersko *et al.*, 1971; Rudland *et al.*, 1974b) by limiting the transport of precursors required for incorporation into synthetic processes. Conversely, augmentation of growth might have resulted from increased transport of nutrients which by an unknown mechanism were shunted into the pathways directed towards growth.

Table I shows that the transport of 2-DG (a glucose analogue) and 2-AIB (an amino acid analogue) was reduced in the presence of

CGF (as NIH-TSH). These measurements were made 66 hours after introducing CGF to the cells. An antecedent effect of CGF appeared to be a requirement for reduced 2-DG transport, inasmuch as treatment of control cultures with CGF only at the time of measuring uptake failed to have any effect on transport of 2-DG. Thus, the effect of CGF on reducing 2-DG and AIB transport may not be directly involved in altering membrane transport, but may instead be involved in producing changes in the cell membrane which reduce the capacity of the CGF-treated cell to transport these small molecules. It should also be noted in this regard that CGF does not support the growth of chondrocytes in the absence of serum. When cells were grown in BSA (3 mg/ml) or BSA + CGF (20 μg/ml), growth either did not occur or was not sustained. (Figure 1). This suggested the possibility that CGF activated a serum component in order to augment growth.

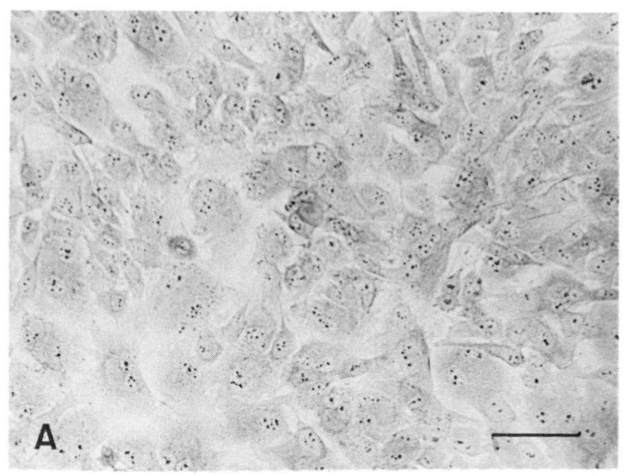

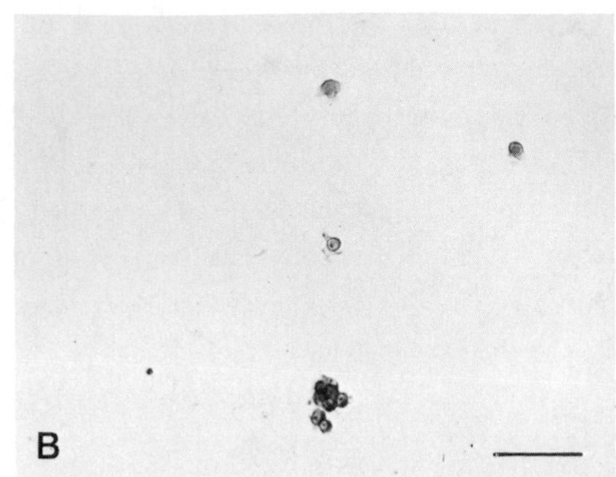

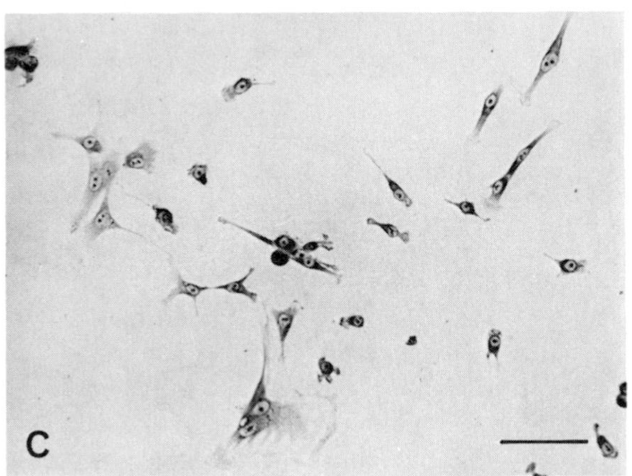

FIGURE 1 *Failure of cultured articular chondrocytes to proliferate in the absence of fetal bovine serum. 2A. Normal control chondrocytes grown for 4 days in DMEM + 10% FBS. 2B. Chondrocytes grown in the absence of fetal bovine serum, with 3 mg/ml BSA as protein source. Cells have attached, with greatly reduced plating efficiency, and have not spread out. 2C. Chondrocytes grown in BSA, 3 mg/ml + CGF (20 μg/ml). Some cells have attached*

ACKNOWLEDGMENT

We wish to thank the Hormone Distribution Officer of The National Institute of Arthritis, Metabolism and Digestive Diseases for supplying NIH-TSH and Dr. Leo E. Reichert, Emory University School of Medicine for the LER-1853-B2 and 1874 fractions. Supported by a grant from The National Institute of Arthritis, Metabolism and Digestive Diseases (AM 17258-01).

REFERENCES

Armelin, H. (1973). *Proc. Natl. Acad. Sci. 70*: 2702-2706.

Cohen, S. and Carpenter, G. (1975). *Proc. Natl. Acad. Sci. 72*: 1317-1321.

Corvol, M-T., Malemud, C. J. and Sokoloff, L. (1972). *Endocrinology 90*: 262-271.

Dulak, N. C. and Temin, H. M. (1973). *J. Cell. Physiol. 81*: 153-160.

Fryklund, L., Uthne, K. and Sievertsson, H. (1974). *Biochem. and Biophys. Res. Comm. 61*: 950-956.

Gospodarowicz, D. (1974). *Nature 249*: 123-127.

Gospodarowicz, D. (1975). *J. Biol. Chem. 250*: 2515-2520.

Green, W. T., Jr. (1971). *Clin. Orthop. 75*: 248-260.

Hershko, A., Mamont, P., Shields, P. and Tomkins, G. (1971). *Nature* (New Biol.) *232*: 206-211.

Hollenberg, M. D. and Cuatrecasas, P. (1975). *Fed. Proc. 34*: 1556-1563.

Holley, R. and Kiernan, J. (1968). *Proc. Natl. Acad. Sci. 60*: 300-304.

Holley, R. and Kiernan, J. (1971). *In* "Ciba Foundation Symposium on Growth Control in Cell Cultures" (G. E. Wolstenholme and J.

and spread out, but growth is not sustained. Giemsa (Bar equals 0.1 mm).

Knight, eds.), pp. 3-10. Churchill Livingston, London.

Holley, R. and Kiernan, J. (1974). *Proc. Natl. Acad. Sci. 71*: 2908-2911.

Jones, K. and Gospodarowicz, D. (1974). *Proc. Natl. Acad. Sci. 71*: 3372-3376.

Leffert, H. L. (1974). *J. Cell. Biol. 62*: 767-779.

Lim, R. and Mitsunobu, K. (1974). *Science 185*: 63-66.

Malemud, C. J. and Sokoloff, L. (1974). *J. Cell. Physiol. 84*: 171-180.

Malemud, C. J. and Sokoloff, L. (to be published).

Miller, E. J. (1971). *Biochemistry 10*: 1652-1659.

Norby, D. P., Malemud, C. J. and Sokoloff, L. (1977). *Arthritis Rheum. 20*: 709-716.

Pierson, R. W., Jr. and Temin, H. M. (1972). *J. Cell. Physiol. 79*: 319-330.

Revoltella, R., Bertolini, L., Pediconi, M. and Vigneti, E. (1974) *J. Exp. Med. 140*: 437-451.

Rudland, P. S., Gospodarowicz, D. and Seifert, W. (1974a) *Nature 250*: 741-774.

Rudland, P. S., Seifert, W. and Gospodarowicz, D. (1974b) *Proc. Natl. Acad. Sci. 71*: 2600-2604.

Sokoloff, L., Malemud, C. J. and Green, W. T., Jr. (1970). *Arthritis Rheum. 13*: 118-124.

Srivastava, V. M. L., Malemud, C. J. and Sokoloff, L. (1974). *Conn. Tiss. Res. 2*: 127-136.

Todaro, G., Matsuya, Y., Bloom, S., Robbins, A. and Green, A. (1967). *In* "Growth Regulating Substances for Animal Cells in Culture" (V. Defendi and M. Stoker, eds.) pp. 87-98. Wistar Institute Press, Philadelphia, Pa.

Turkington, R. W. and Majumder, G. C. (1973). IRCS (International Research Communication System) March, Pg. 8.

RESTORATION OF PIGMENT SYNTHESIS IN MUTANT MELANOCYTES AFTER FUSION WITH CHICK EMBRYO ERYTHROCYTES

D. G. SCHALL
J. A. BRUMBAUGH

School of Life Sciences
University of Nebraska
Lincoln, Nebraska

I. INTRODUCTION

When chick erythrocyte nuclei are fused with various established mammalian cell lines, they often resume DNA and RNA synthesis (Harris, 1967; Johnson and Harris, 1969). When a reactivating erythrocyte nucleus develops a nucleolus and synthesizes RNA (Sidebottom and Harris, 1969), it begins the synthesis of chick specific proteins (Harris *et al.*, 1966; Cook, 1970). When the host nucleus of a heterokaryon enters the cell cycle, the chick nucleus may also begin DNA synthesis, but out of phase with the host nucleus. This asynchrony of DNA synthesis causes "premature chromosome condensation" or "chromosome pulverization" in the chick nucleus (Schwartz *et al.*, 1971; Johnson *et al.*, 1970; Kato and Sandberg, 1968). It is possible for these chromatin fragments to become incorporated into the nucleus of the host cell (Schwartz *et al.*, 1971) but remain karyotypically undetectable (Boyd and Harris, 1973). Cook (1970), Schwartz *et al.* (1971) and Klinger and Shin (1974) showed that mouse A9 cells defective for inosinic acid pyrophosphorylase (IMP: pyrophosphate phosphoribosyl transferase, E. C. 2.4.2.8), can produce clones of IMP-producing cells after fusion with chick embryo erythrocyte nuclei. Chick specific IMP was produced by these stable, reproducing, postfusion cells.

In this study chick embryo neural crest melanocytes, defective in pigment production, were fused with inactive erythrocyte nuclei from normally pigmented embryos. Colonies of pigment producing cells were recovered indicating restoration of normal pigment synthesis.

II. MATERIALS AND METHODS

A. Melanocyte Cultures from Neural Crest

The posterior two-thirds of the somites of 72 hour chick embryos were dissected, trypsinized and plated out in 60 mm culture dishes at 2.0-3.0 $\times$ 10^5 cells per dish. They were grown in medium F-12 (GIBCO) with 1% bovine serum albumin, 2% fetal calf serum, and antibiotics at 38.5^{o}C in a 5% CO_2 incubator. On the third and fourth culture days, small colonies of melanocytes became evident which proliferated so that on days 5 and 6 melanocytes accounted for approximately 70-90% of the cell population. In this study wild type (normal; $+^{Pk}/+^{Pk}$) melanocytes produced dark pigment granules as early as days 4 and 5 while pinkeye (*pk/pk*) melanocytes never produced overt pigment. Pinkeye melanocytes both *in vivo* and *in vitro* fail to deposit melanin upon the forming pigment granule (premelanosome) matrices (Brumbaugh, 1968; Brumbaugh *et al.*, 1973; Brumbaugh and Lee, 1975). Some cultures were labeled with ^{3}H-thymidine (1.0 μ Ci/ml) during the 24 hours prior to harvesting for fusion. Fertile eggs for the cultures were from genetic stocks maintained at the School of Life Sciences, University of Nebraska-Lincoln.

B. Erythrocytes

Erythrocytes were collected from 9-10 day chick embryos of the wild type (pigmented; $+^{Pk}/+^{Pk}$) genotype by cutting the allantoic blood vessels and subsequently collecting the allantoic fluid (Bolund *et al.*, 1969). The cells were centrifuged and washed twice in Hank's solution without glucose and counted. This preparation contains essentially only erythrocytes since granulocytes do not normally enter the circulation until the 14th day of incubation. Thrombocytes are rarely encountered (Lucas and Jamroz, 1961).

C. Cell Fusion

Five or 6-day melanocyte cultures were trypsinized, counted, and pelleted into a centrifuge tube. They were then mixed thoroughly with 0.5 ml of the erythrocyte preparation so that the ratio of erythrocytes to melanocytes was 10:1. U. V. inactivated Sendai virus preparation (0.5 ml; Connaught Laboratories, Willowdale, Ontario) was then added so that a final concentration of 12-16,000 HAU's per ml was achieved. A typical experiment involved $1.5\text{-}2.0 \times 10^6$ melanocytes and $1.5\text{-}2.0 \times 10^7$ erythrocytes. The remainder of the fusion procedure was a modification of Harris *et al.* (1966). After fusion the cells were replated in standard medium on the basis of 1.0×10^5 melanocytes per dish.

D. Microscopy

At 24 and 48 hours post-fusion sample dishes were fixed and stained with Giemsa for light microscopic determination of fusion rate and plating efficiency.

Living, post-fusion cultures were periodically screened for heterokaryons and/or pigment production with the light microscope. When desired cells or colonies were located they were fixed *in situ* for electron microscopy and embedded. Ultrathin sections made parallel to the plane of each culture dish were stained with uranyl acetate and lead citrate, then viewed and photographed using a Philips 201C electron microscope.

Electron microscopic autoradiography was performed according to the method of Brumbaugh and Lee (1975) using Ilford L-4 emulsion and an exposure time of 12 days.

III. RESULTS

Twenty-four hours after fusion, Giemsa stained cultures showed that approximately 13.0% of the melanocytes contained one or more erythrocyte nuclei. Erythrocyte nuclei were readily discerned because of their small size and condensed chromatin. Figure 1 shows a heterokaryon between a wild type pigmented melanocyte containing a few pigment granules and an erythrocyte nucleus (Fig. 1, arrow). Figure 2 shows a similar heterokaryon but the melanocyte is of the amelanotic pinkeye genotype, hence no pigment granules. Further verification of erythrocyte fusion is shown in the high resolution autoradiograph (Fig. 3) prepared 24 hours after fusion. The larger, less condensed, labeled melanocyte nucleus is clearly distinguished from the smaller, condensed, unlabeled erythrocyte nucleus (*, Fig. 3). The erythrocyte nucleus is completely surrounded by melanocyte cytoplasm and is clearly an integral part of the cell.

Melanocyte homokaryons were also classified in the Giemsa stained preparation at a frequency of 32.5%. Not all of these are true homokaryons since suspected multinucleated cells frequently contained separating cell membranes when observed with the electron microscope.

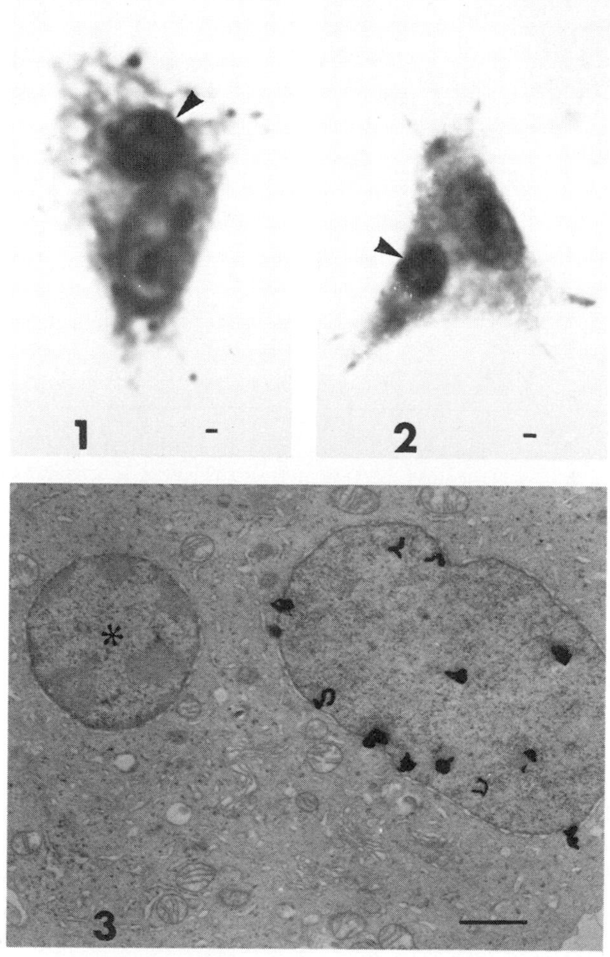

FIGURE 1 *Heterokaryon of a wild type ($+^{Pk}/+^{Pk}$) chick melanocyte and a chick embryo erythrocyte nucleus (arrow) 24 hr post fusion. A few pigment granules are present in the cytoplasm. 2225X. Scale equals 1 micron.*

FIGURE 2 *Heterokaryon of a pinkeye (pk/pk) chick melanocyte and a chick embryo erythrocyte nucleus (arrow) 48 hr post fusion. No pigment granules are present as is characteristic of the pinkeye genotype. 2500X. Scale equals 1 micron.*

FIGURE 3 *High resolution autoradiograph of a ^{3}H-thymidine labeled melanocyte and a chick embryo erythrocyte nucleus (*) 24 hr post fusion. Silver grains are evident over the melanocyte nucleus but not the erythrocyte nucleus. 14,000X. Scale equals 1 micron.*

TABLE I Summary of Restoration Experiments

Treatment and genotype	Melanocyte phenotype: Premelanosomes	Melanocyte phenotype: Melanin Deposition
Erythrocyte donor - wild type; $+^{Pk}/+^{Pk}$	Yes	Yes
Melanocyte host - pinkeye; *pk/pk*	Yes	No
Post-fusion restored cells	Yes	Yes

Two fusion experiments using a total of 3.4 × 10^6 melanocytes of the amelanotic pinkeye (*pk/pk*) genotype and 3.4 × 10^7 erythrocytes from the pigmented, wild type ($+^{Pk}/+^{Pk}$) genotype were conducted. Three pigment producing colonies, one in the first experiment and two in the second, were detected after visual screening. Each colony occurred in a separate dish. One colony was detected on the fourth post-fusion day and was followed for 32 additional days during which time it divided several times. All of the daughter cells appeared to be pigment producers. The second and third colonies were detected on days 10 and 24 after fusion and were subsequently fixed for electron microscopy. Table I summarizes the genetic design of the experiments and the results.

Figure 4 is a portion of the cytoplasm from a typical, synthetically active, pinkeye melanocyte. Even though the cell has been subjected to the cytochemical test for dopa oxidase (notice the darkened Golgi system), the forming pigment granules (premelanosomes) show no sign of melanin deposition (see arrow, Fig. 4). This cell is to be contrasted with the 24 day, post-fusion cell in Fig. 5, which contains several frank pigment granules in various stages of melanin deposition.

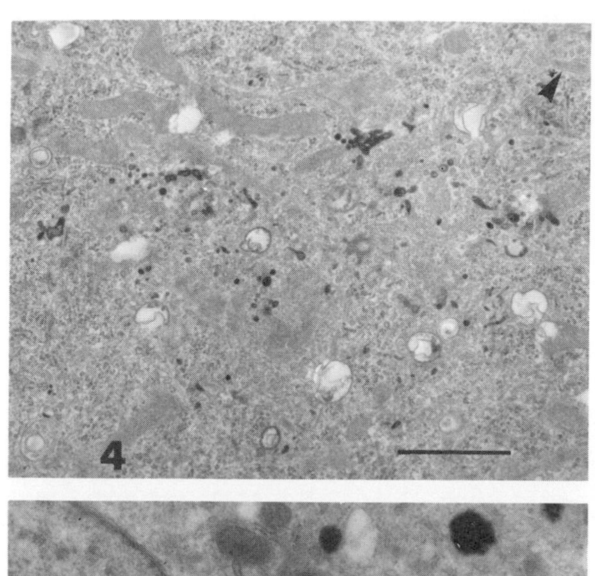

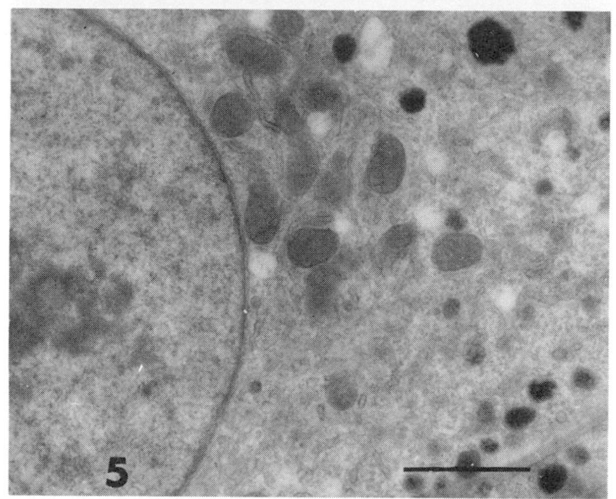

FIGURE 4 *Cytoplasm of a pinkeye (pk/pk) melanocyte in culture. The Golgi system has been cytochemically stained. The arrow locates a solitary unpigmented premelanosome. 21,000X. Scale equals 1 micron.*

FIGURE 5 *A melanogenically active cell from a colony of such cells present in a pinkeye (pk/pk) culture 24 days after fusion with wild type erythrocyte nuclei. Note the several electron dense pigment granules. 22,750X. Scale equals 1 micron.*

The 24 hour post-fusion plating efficiency, as determined by counting attached cells in Giemsa stained preparations, was approximately 20%. This means that only 20% of the 3.4×10^6

melanocytes fused (6.8×10^5) were recovered. Thus approximately one pigment colony was produced per 2.3×10^5 recovered melanocytes.

IV. DISCUSSION

There are several ways of explaining the restoration of pigment synthesis in the pinkeye melanocytes. Since only mononucleated dividing pigment cells were recovered, it is not possible to suggest that a reactivated erythrocyte nucleus in a persisting heterokaryon produced the factor necessary for pigment synthesis. The rate of restoration (1 in 2.3×10^5) seems too high to be due to back mutation. This is corroborated by the fact that pinkeye adult feathers do not show wild type flecks or "ticks". Integration of the erythrocyte genome and the melanocyte genome to produce a tetraploid cell is a plausible explanation since karyotypes of the pigmented clones were not determined. Such integrations have not been reported in chick erythrocyte-mammalian cell fusions, however (Boyd and Harris, 1973; Schwartz, *et al.*, 1971; Klinger and Shin, 1974.

The differentiating melanocytes used in these experiments were mitotically very active, dividing both before and after fusion. This lends credence to the hypothesis that "premature chromosome condensation" or "chromosome pulverization" (Johnson, *et al.*, 1970) occurs in the erythrocyte nucleus of the heterokaryons as they attempt to divide in synchrony with the melanocyte nucleus. An erythrocyte-derived chromosome fragment, bearing the wild type allele of the pinkeye lesion could explain the restoration of pigment production if it became incorporated into the melanocyte nucleus. A similar hypothesis is used to explain the production of chick IMP in IMP deficient mouse A9 cells after fusion with chick erythrocytes (Schwartz, *et al.*, 1971; Boyd and Harris, 1973). Like Boyd and Harris (1973) and Schwartz (1971) we also obtained a phenotypically stable, dividing population of restored cells.

Several investigators (Cook, 1970; Schwartz, *et al.*, 1971; Boyd and Harris, 1973; Klinger and Shin, 1974) have shown that it is possible to correct genetically defective mammalian cells by fusion with chick erythrocytes. The correcting molecules were usually chick specific. All of these systems depended on some type of cell selection and usually involved ubiquitous enzymes necessary for cell survival and/or growth. In this study restoration required the synthesis of a specific product of differentiation-melanin. Thus a pigment gene normally inactive in the erythrocyte is apparently activated after exposure to the nuclear or cytoplasmic environment of the melanocyte.

V. SUMMARY

Five to 6-day cultures of chick embryo melanocytes of the amelanotic pinkeye genotype were fused with 9-10 day chick embryo erythrocytes from pigment-producing, wild type embryos. Three colonies of pigment producing cells were produced in two different experiments at a rate of one per 2.3×10^5 melanocytes recovered after fusion. These results strongly suggest that the wild type pigment gene from the erythrocyte nucleus is activated after exposure to the nuclear and/or cytoplasmic environment of the melanocyte.

REFERENCES

Bolund, L., Ringertz, N. R., Harris, H., (1969), *J. Cell Sci. 4,* 71-87.

Boyd, Y. L., Harris, H., (1973), *J. Cell Sci. 13,* 841-861.

Brumbaugh, J., (1968), *Devep. Biol. 18,* 375-390.

Brumbaugh, J., Lee, L., (1975), *Genetics 44,* 333-347.

Brumbaugh, J., Bowers, R., Lee, K., (1973), *Yale Journal of Biology and Medicine 46,* 523-534.

Cook, P. R., (1970), *J. Cell Sci.* 7, 1-3.

Harris, H., (1967), *J. Cell Sci.* 2, 23-32.

Harris, H., Watkins, J. F., Ford, C. E., Schoefl, G. I., (1966), *J. Cell Sci. 4,* 499-525.

Johnson, R. T., Harris, H., (1969), *J. Cell Sci. 5,* 625-643.

Johnson, R. T., Rao, P. N., Hughes, H. D., (1970), *J. Cell. Physiol. 76,* 151-158.

Kato, H., Sandberg, A. A., (1968), *J. Nat. Cancer Inst. 41,* 1117-1123.

Klinger, H. P., Shin, S., (1974), *Proc. Nat. Acad. Sci. U.S.A. 71,* 1398-1402.

Lucas, A. M., Jamroz, C., (1961), "Atlas of Avian Hematology," Agriculture Monograph 25, United States Department of Agriculture.

Schwartz, A. G., Cook, P. R., Harris, H., (1971), *Nature New Biology 230,* 5-8.

Sidebottom, E., Harris, H., (1969), *J. Cell Sci. 5,* 351-364.

ACKNOWLEDGMENTS

This investigation was supported by NIH research grant GM18969 from the National Institute of General Medical Sciences.

Note added in proof: Leung, W. C., Chen, T. R., Dubbs, D. R., Kit, S., (1975), *Exp. Cell Res. 95,* 320-326, have restored TK^- mouse cells by fusing them with chick erythrocytes. The functioning thymidine kinase (TK) was electrophoretically of the chick type and was accompanied by chick microchromosomes.

Workshop II

NUTRITIONAL REQUIREMENTS FOR CELL GROWTH

THE SERINE AND GLYCINE REQUIREMENTS OF CULTURED HUMAN LYMPHOCYTE LINES

J. R. BIRCH
D. W. HOPKINS

Searle Research Laboratories
High Wycombe, England

I. INTRODUCTION

Although human lymphocyte cell lines are now widely used, frequently in large quantities, surprisingly little is known about their precise qualitative and quantitative nutritional requirements. As part of a project aimed at defining the nutrient requirements of lymphocyte lines, we have studied the amino acid requirements of a range of lines from normal individuals. In particular we have looked at the requirement for amino acids which are generally considered to be non-essential for cultured cells.

The culture medium most commonly used for lymphocyte culture is R.P.M.I. 1640. In addition to the 13 amino acids usually considered essential for growth (Eagle, 1959), this medium contains 7 'non-essential' amino acids as well as the tripeptide glutathione (Table I).

TABLE I Medium RPMI 1640

'Essential' amino acids	'Non-essential' amino acids
L-Arginine	L-Asparagine
L-Cystine	L-Aspartic acid
L-Glutamine	L-Glutamic acid
L-Histidine	(Glutathione, reduced)
L-Isoleucine	Glycine
L-Leucine	L-Hydroxyproline
L-Lysine	L-Proline
L-Methionine	L-Serine
L-Phenylalanine	
L-Threonine	
L-Tryptophan	
L-Tyrosine	
L-Valine	

II. MATERIALS AND METHODS

A. Culture Methods

R.P.M.I. 1640 (pH 7.4) supplemented with 10% v/v dialysed foetal calf serum was used for these experiments. The medium contained the following modified phosphate concentrations - Na_2HPO_4-$2H_2O$, 500 mg/litre and $NaH_2PO_4.2H_2O$ 440 mg/litre. Trace metals were also added (3.6 μM $FeSO_4.7H_2O$, 2.0 μM $CuSO_4.5H_2O$ and 3.5 μM $ZnSO_4.7H_2O$). Cells were grown as 10 ml static cultures in 8 oz prescription bottles. Agitated cultures were grown in 10 ml medium in 50 ml Erlenmeyer flask shaken at 200 r.p.m.

B. Cell Types

The lymphocyte lines used and their characteristics are summarised in Table II.

TABLE II Human Lymphocyte Cell Lines

Cell designation	Donor	Chromosome analysis	Sex
MICH	Spleen cells from a patient with idiopathic thrombocytopaenia	Subtetraploid	F
BEC 11	Lymphoid cells from tonsils of a patient with tonsilitis	Diploid	F
BRI 7	Peripheral lymphocytes from a normal patient	Diploid	M
BRI 8	Peripheral lymphocytes from a normal patient	Diploid	M
RPMI 1788	Peripheral lymphocytes from a normal patient	Diploid	M

C. Cell Counting

Growth was measured as viable cell count on a haemocytometer slide, using nigrosin (50 mg/100 ml phosphate buffered saline) to stain non-viable cells.

III. RESULTS

A. Omission of Individual 'Non-Essential' Amino Acids

The effect of omitting individual supposedly non-essential amino acids on the growth of MICH cells is shown in Table III. Only serine had a significant effect on growth. In its absence the duration of the lag phase was increased, the population doubling time was increased and the maximum population density was decreased. Omission of glutamic acid (not shown in Table III) and

TABLE III Effect of 'Non-Essential' Amino Acids on Growth of Cultured Human Lymphocytes (MICH)

Amino acid omitted	Duration of lag phase (h)	Minimum population doubling time (h)	Maximum cell population density × 10^5/ml (inoculum deducted)
Control	10	23	8.2
Asparagine	14	21	9.3
Aspartic acid	16	21	8.5
Glutathione	12	23	8.4
Glycine	10	23	7.4
Hydroxyproline	11	22	8.2
Proline	12	22	8.6
Serine	28	42	3.1

the whole group of 'non-essential' amino acids other than serine had no effect on growth of MICH.

B. Effect of Glycine on Serine Requirement

If glycine was omitted from the culture medium the requirement of MICH for serine was absolute. This is shown in Fig. 1. In the presence of glycine alone growth rates were reduced. Glycine was slightly stimulatory even in the presence of serine. Interestingly, in agitated culture as opposed to static culture, there was no growth with glycine alone. This may indicate that the cells are leaky to glycine as suggested by Lockhart and Eagle (1959). Cell lines BRI 8 and BEC 11 gave essentially the same results as MICH.

C. Results With Other Cell Lines

Other cell lines differed in their response to glycine and serine. Cell line R.P.M.I. 1788 did not grow in the absence of serine even when glycine was present (Fig. 2). As with MICH

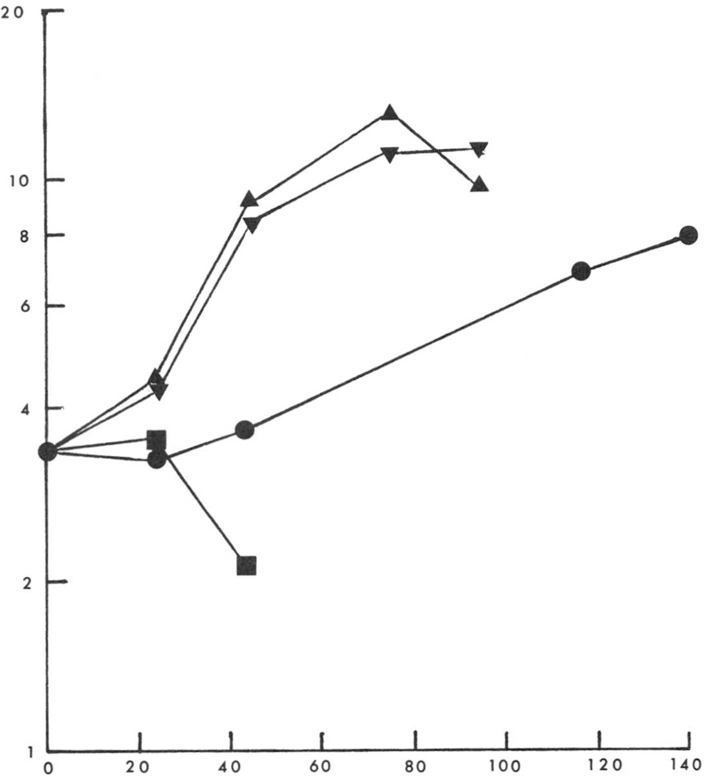

FIGURE 1 *Effect of glycine and serine on growth of human lymphocytes (MICH). Abscissa, incubation period (h). Ordinate, viable cells/ml × 10^5. , 1640 without glycine and serine; ●, 0.13 mM glycine; ▼, 0.29 mM serine; ▲, 0.13 mM glycine + 0.29 mM serine.*

growth rate and maximum population density were slightly increased when glycine was added to medium already containing serine.

A third type of response is shown in Fig. 3 for cell line BRI 7. This cell line grew slowly even in the complete absence of serine. Glycine increased growth rate and maximum population density.

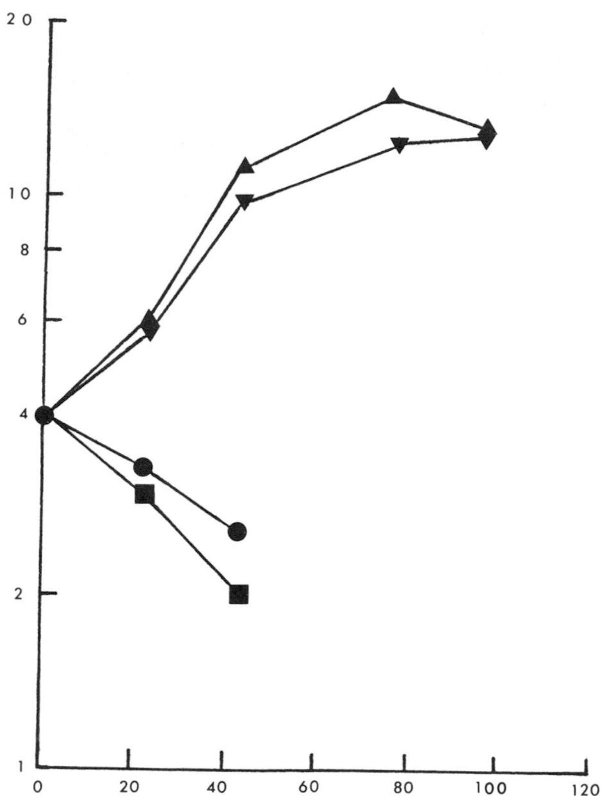

FIGURE 2 *Effect of glycine and serine on growth of human lymphocytes (RPMI 1788). Abscissa, incubation period (h). Ordinate, viable cells/ml × 10^5.* ■, *1640 without glycine and serine;* ●, *0.13 mM glycine;* ▼, *0.29 mM serine;* ▲, *0.13 mM glycine + 0.29 mM serine.*

D. Quantitative Serine Requirement

The quantitative requirement of MICH for serine in the absence of glycine was determined. Figure 4 shows the influence of serine concentration on maximum population density. From the slope of this graph the growth yield coefficient (amount of cell produced per unit weight of nutrient) is 0.35 × 10^5 cells/μg serine. Growth rates as opposed to maximum population densities were not influenced by serine concentrations in the range 10-

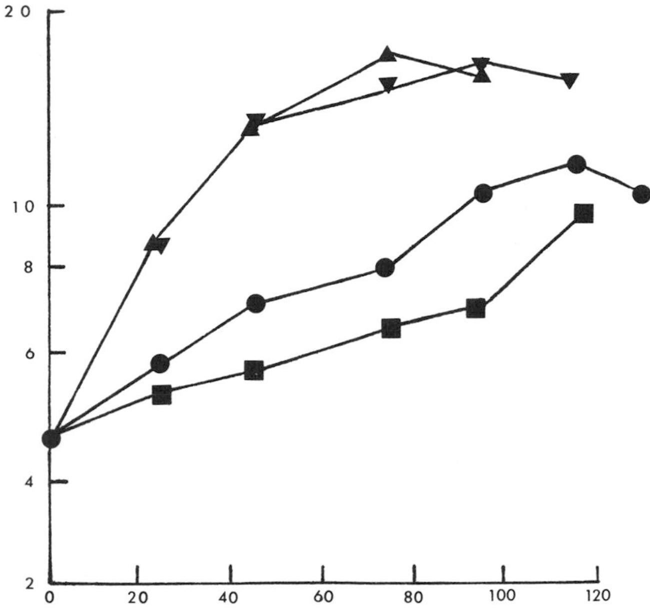

FIGURE 3 *Effect of glycine and serine on growth of human lymphocytes (BR17). Abscissa, incubation period (h). Ordinate, viable cells/ml × 10^5. ■, 1640 without glycine and serine; ●, 1.3 mM glycine; ▼, 0.29 mM serine; ▲, 0.13 mM glycine + 0.29 mM serine.*

100 µg/ml but were reduced at concentrations below 10 µg/ml. The normal serine concentration of 30 µg/ml would be close to a growth limiting level in the absence of glycine. However, glycine had a marked sparing effect on the quantitative requirement and in the presence of glycine 2.5 µg serine/ml was sufficient to restore growth rates and maximum population densities to the normal levels.

IV. DISCUSSION

With certain exceptions, serine has been considered non-essential except at low cell concentrations (Eagle, 1959). However,

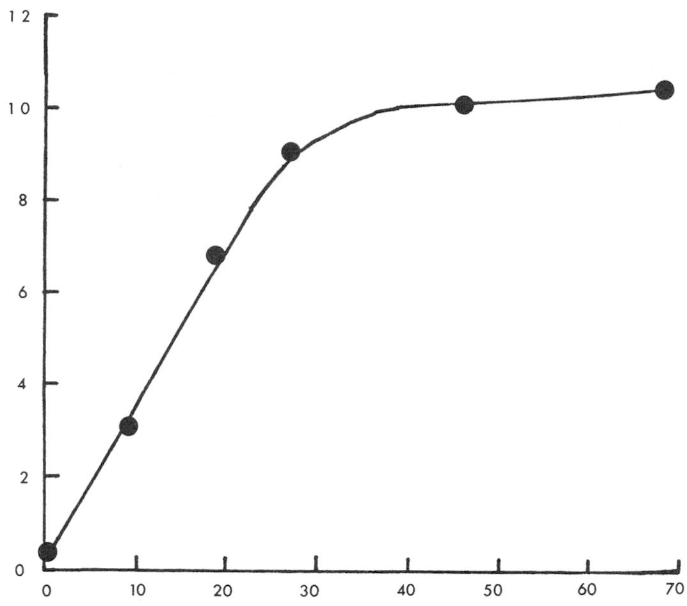

FIGURE 4 *Effect of serine concentration on growth of human lymphocytes (MICH). Abscissa, serine concentration in culture medium (μg/ml). Ordinate, maximum cell population/ml* × 10^5.

a requirement has been shown for short-term cultures of normal and leukaemic blood cells and phytohaemagglutinin stimulated normal peripheral blood lymphocytes (Regan *et al.*, 1973). These requirements were shown to be due to reduced enzyme levels in the 'phosphorylated pathway' for serine biosynthesis (Pizer and Regan 1972). We assume that a similar enzyme deficiency may account for the serine requirement of our permanent cell lines. The sparing action of glycine is not surprising since the amino acids are interconvertible through the action of serine transhydroxymethylase (L-Serine: tetrahydrofolate 5, 10-hydroxymethyltransferase E.C.2. 1.2.1.). The varying responses of different cell lines may reflect variations in the level of this enzyme and of the enzymes of serine biosynthesis. This may prove to be of value in the biochemical 'finger-printing' of cell lines. Ultimately studies of the nutritional requirements of cells isolated from various sources may be useful in designing chemotherapeutic agents aimed at inhibiting specific cell populations in vivo (Dubrow *et al.*, 1973).

REFERENCES

Dubrow, R., Pizer, L. I. and Brody, J. I. (1973). *J. Nat. Cancer Inst. 51*, 307-311.

Eagle, H. (1959). *Science 130*, 432-437.

Lockhart, R. Z. Jr. and Eagle (1959). *Science 129*, 252.

Pizer, L. I. and Regan, J. D. (1972). *J. Nat. Cancer Inst. 48*, 1897-1900.

Regan, J. D., Vodopick, H., Takeda, S., Lee, W. H. and Faukon, F. M. (1969). *Science 163*, 1452-1453.

AUTOCLAVABLE LOW COST CELL CULTURE MEDIA

LEONARD KEAY

Department of Microbiology and Immunology
Division of Biology and Medical Sciences
Washington University School of Medicine
St. Louis, Missouri

I. INTRODUCTION

For the last twenty-five years animal cells have been grown in chemically-defined media supplemented with some type of serum. These chemically-defined media consist of inorganic salts, vitamins, amino acids and sometimes other organic compounds (Morgan *et al.*, 1950; Eagle 1955a; Healy *et al.*, 1954; Waymouth 1955; Ham 1963). Even in these media the term chemically-defined (that is knowing all of the components present and their precise amounts) may be questioned since the purity of many components is generally not completely established.

The supplementation of media with serum creates several problems (Fedoroff *et al.*, 1972). The most obvious is economic. The supplementation of medium with 10% serum accounts for 75-90% of the material costs (based on the cost of dry powder media).

However, more recently, serum has been implicated as a source of mycoplasmal and viral contamination of cell cultures (Molander *et al.*, 1972; Hopps, 1974). The final reason is the variation between various samples of sera (Boone *et al.*, 1972) and the desirability of a totally chemically-defined medium.

The precise role(s) of serum is unknown. It may be a source of hormones, macromolecules required for protecting cell surfaces, free fatty acids, sterols, trace minerals or vitamins, and attempts to replace serum usually involve substitutions for these possible serum components (Higuchi, 1973; Taylor, 1974).

The problem of mycoplasma and virus contamination is the reason for seeking an autoclavable medium. Serum or other native protein cannot be autoclaved without coagulation and filtration will not remove mycoplasma or viruses. Therefore any completely autoclavable medium will have to also be serum-free medium.

Other desirable characteristics of any serum-free autoclavable medium are that it be totally chemically-defined, that it support the growth of a wide range of cells, that cells grow in it without an extensive period of adaptation, that high cell yields be obtained without repeated refeeding, and that the cells remain unchanged in karyotype, cell surface antigens, virus susceptibility, etc., when grown in such a medium over an extended period of time.

Since the pioneer work of Eagle on the amino acid and vitamin requirements of HeLa and L-cells (Eagle 1955b and c) leading to the formulations of BME and MEM, few detailed nutritional studies have been carried out, although many formulations have been published.

Serum-free formulations ML 192/2 and MB 752/1 in which bovine serum albumin and bactopeptone were used as supplements to a salts/amino acids/vitamins medium were developed by Waymouth (1956, 1959). The more complex chemically-defined medium NCTC109 supported the growth of adapted L-cells (NCTC 2071) in stationary systems but not in agitated systems. Cell disintegration rapidly occurred unless methylcellulose was added (Bryant *et al.*, 1961). Macromolecules such as methylcellulose, polyvinylpyrrolidone, protamine sulfate and serum albumin have been added to various serum-free media, although it has been shown that exhaustively dialyzed serum does not support cell growth, i.e. it is more than just "macromolecular."

The nutritional requirements for cell growth have been found to be dependent on the density of the cell inoculum and even in the case of the complex medium NCTC109, cell growth does not occur in the absence of serum if the inoculum is too small (Bryant *et al.*, 1961; Fioramonte *et al.*, 1958; Eagle and Piez, 1962). A complex chemically defined medium supporting the growth of L-cells was also developed by Healy *et al.* (1955). At the other extreme, attempts were made to cultivate HeLa and L-cells in systems containing bacteriological media (Mayyasi and Schuurmans, 1956; Ginsberg *et al.*, 1955), but the addition of serum was necessary and the growth and quality of the cells rather poor.

Other additives formulated into serum-free systems include unsaturated fatty acids, putrescine, insulin, thyroxine, pyruvate, ethanol, gluconolactone, and nucleotide precursors. Most serum-free media contain high levels of amino acids (both essential and nonessential), usually 3-10 fold higher than in MEM. Also, the vitamin levels are higher especially inositol and choline and other vitamins such as vitamin B_{12}, biotin and lipoic acid often included (Ham, 1965; Nagle *et al.*, 1963; Higuchi, 1963; Tribble and Higuchi, 1963; Blaker *et al.*, 1961; Lieberman and Ove, 1959; Higuchi and Robinson, 1973; Lasfargues *et al.*, 1973; Morrison and Jenkin, 1972; Birch and Pirt, 1970; Takaoka and Katsuka, 1971; Holmes and Wolfe, 1961).

The role of insulin is uncertain, but it has been shown (Temin, 1967) that the growth of chick embryo fibroblasts is stimulated by insulin in serum-free systems and low molecular weight fractions have been isolated from serum which have this insulin-like activity but are not suppressed by antiinsulin sera (Pierson and Temin, 1972). Insulin has been shown to markedly stimulate the growth of HeLa cells in serum-free systems, as have vitamin B_{12} and biotin (Higuchi and Robinson, 1973; Blaker *et al.*, 1971).

Bactopeptone has been used as part of several serum-free media (Waymouth, 1956, 1959), either alone or in conjunction with other additives such as lactalbumin hydrolysate, polyvinylpyrrolidone, and yeast extract (Lasfargues *et al.*, 1973). Bactopeptone possesses the advantages that it is both low cost and autoclavable, but it is not chemically-defined. In fact, probably less is known about the composition of bactopeptone than is known about serum. Recently the growth promoting activity of bactopeptone has been shown to be in the low molecular weight dialysate fraction (Taylor *et al.*, 1972b) and some peptide fractions shown to promote cell growth (Taylor *et al.*, 1974; Hsueh and Moskowitz, 1973a, 1973b) although no full characterization of the components have been made. Other peptones including soybean derivatives have been shown to support cell growth (Taylor *et al.*, 1974; Hsueh and Moskowitz, 1973a, 1973b) although no full characterization of the components have been made. Other peptones including soybean derivatives have been shown to support cell growth (Taylor *et al.*, 1972a; Healy and MacMorine, 1972).

Autoclavable media have been less extensively studied than serum-free media. Nagle developed a heat stable medium for the suspension culture of L-cells, HeLa and cat kidney cells (Nagle, 1968). The medium contained high levels of amino acids and after autoclaving, the medium was completed by the addition of sodium bicarbonate, methyl cellulose and glutamine, each autoclaved separately (the glutamine autoclaved as a dry powder). This medium was subsequently modified by increasing the level of choline 50-fold (Nagle, 1969), and then by elimination of the glutamine by replacement with a series of amino acids (Nagle and Brown, 1971).

Yamane *et al.*, 1968 found that autoclaved MEM would support the growth of baby hamster lung and L-cells provided that the autoclaving was carried out at pH 4.0-4.5 (maintained by a succinate buffer) and that glutamine and serum were added after autoclaving (Yamane *et al.*, 1968).

Pumper *et al.* (1965) found that autoclaved bactopeptone or peptone dialysate when combined with autoclaved medium 199 or MEM (without glutamine) supported a 7 fold increase in rabbit heart myocardial cells in 7 days.

It is interesting to note that only a few of these serum-free or autoclavable media are commercially available. A formulation of Waymouths 752/1, bovine serum albumin and linoleic acid is available in liquid form ($HW10_5$) and autoclavable BME and MEM are available in dry powder form.

The approach taken in this laboratory is that the most useful medium would be based on a simple medium commercially available in dry powder form which with a minimum of additions or manipulations would be autoclavable and support the growth of many commonly used cell types. The starting point was a simple medium (MEM) plus a supplement (bactopeptone), the latter being chosen because of low cost and autoclavability.

Initial studies were on the growth of commonly used cell lines in this serum-free system, followed by an investigation of the autoclavability of those serum-free systems shown to support cell growth when sterilized by filtration.

Finally some studies will be carried out on the problems of scale up of these systems for cell and virus production.

II. MATERIALS AND METHODS

Dry powdered media and serum were obtained from Microbiological Associates, Bethesda, Md., and KC Biologicals, Lenexa, Kansas. Bactopeptone was obtained from Difco Laboratories. Inorganic salts (ACS Reagent grade) were obtained from Fisher Scientific Co., St. Louis, Mo. Darvan #2, a polyalkylbenzene sulfonate detergent, was obtained from R. T. Vanderbilt Co., New York, N. Y., and other additives and organic compounds were obtained from Sigma Chemical Co., St. Louis, Mo. Unless otherwise stated, all solutions were sterilized by membrane filtration.

Growth studies in monolayer or stationary suspension cultures were generally carried out in 75 cm^2 plastic flasks containing 25 ml medium and incubated at 37° in a 5% CO_2/95% air atmosphere. Agitated suspension cultures were carried out at 37° in glass spinners with teflon coated magnetic stirring bars. Cell counts were determined microscopically with a hemocytometer (after trypsinization in the case of monolayers) and cell viability determined by the trypan blue dye exclusion method. Media and serum substitutes were autoclaved at 10 psi for 10 minutes. Bactopeptone was prepared as a 20% w/v solution and generally 5% v/v added to the medium to give a final concentration of 1% w/v. SSK7 (Lasfargues, 1973) contained bactopeptone 5%, PVP-360 3.5%, lactalbumin hydrolysate 2.5%, yeastolate 0.5%, glucose 0.5%. Insulin was generally prepared as a 1 mg/ml solution in phosphate-buffered saline.

III. RESULTS AND DISCUSSION

A. The Growth of Established Cell Lines in Serum Free Media

Initial experiments were carried out with L-cells (L-929 and a derivative L-60TM which grows more readily in suspension) and then extended to other established cell lines.

The results obtained with L-cells and BHK have been published (Keay, 1975) and will therefore only be summarized here.

A number of peptones and protein hydrolysates were examined for their ability to support the growth of L-cells when they were added to supplement the simple MEM medium. The results (see Table I) show that a number of peptones support the growth of L-929 cells, whereas the more complete hydrolysates do not support cell growth. In fact the addition of hydrolysates reduces the growth

TABLE I Growth of L-929 Cells on Peptones and Protein Hydrolysates

Additive (0.5% w/v)	Cell counts (10^6 cells)[a]
Bactopeptone (Difco)	36.3
Prime meat peptone (Sheffield)	37.5
Proteose peptone (Difco)	17.2
Bacto-tryptone (Difco)	38.2
Bacto-tryptose (Difco)	33.5
Soy peptone (Sheffield)	23.3
Edamin - S (Sheffield)	2.6
NZ-amine AS (Sheffield)	8.1
NZ-amine AW (Sheffield)	7.2
Trypticase Soy Broth (BBL)	3.2
Acidase #2 (BBL)	4.0

[a]Initial cell count was 1.15×10^6.

stimulating effect of the peptones. The addition of 0.5% lactalbumin hydrolysate reduces the growth on bactopeptone by more than 50%. It was observed that as cell growth continued more of the cells detached and grew in suspension until the whole system consisted of bunches of floating cells. Although this may be a disadvantage if a monolayer system is necessary, it is an advantage if a suspension culture is desired. It was found that the addition of insulin or oleate had no effect on the growth of L-cells in MEM supplemented with 0.5% bactopeptone.

Using 0.5% bactopeptone as a supplement to Joklik-modified MEM, excellent growth was obtained L-60TM cells in spinner culture.

The L-cells grew at once in this simple medium without any period of adaptation, and have been cultured in this serum-free system for several months at a time.

When attempts were made to culture BHK cells using bactopeptone as a serum substitute two facts were quickly noted; the cells did not grow well on MEM or Joklik-modified MEM, but grew

TABLE II The Growth of BSC-1 Cells in Serum-free Systems

Medium	10^6 cells/75 cm^2 flask	
	Expt A[a]	Expt B[b]
MEM, 10% FCS	-	19.0
MEM, 1%BP	-	18.0
MEM, 1%BP, insulin[c]	-	12.2
MEM-α 10%FCS	27.0	-
MEM-α insulin[c]	4.2	-
MEM-α 1%BP	24.0	23.0
MEM-α 1%BP, insulin[c]	19.0	17.5

[a]In experiment A, monolayers were refed after 5 days, counted at 8 days.
[b]In experiment B, monolayers were refed after 3 and 6 days, counted at 9 days.
[c]Insulin concentration 0.23 I.U./ml.

better on richer media such as MEM-alpha (without the ribosides and deoxyribosides, Stammers *et al.*, 1971), F-12, or RPMI 1640; and the cells grew as rather nodular clumps some of which detached from the surface and grew in suspension. The addition of a poly-alkylbenzene sulfonate detergent Darvan #2 (300 μg/ml) was sufficient to convert the system to a completely suspension culture even in stationary systems, although clumping still occured especially in the MEM-alpha and RPMI 1640 systems. Monolayer systems of BHK cells could be produced but were difficult to maintain because of subsequent cell detachment. It was found that growth of BHK cells on F-12 supplemented with 0.5% bactopeptone was unaffected by the addition of insulin.

The methods and media developed with the L and BHK cells were used in studies on the growth of five other cell lines (CHO, 3T3, BSC-1, HeLa and KB) (Keay, 1976).

In the case of BSC-1 cells (an established African Green Monkey Kidney Line) it was found that the cells would grow with bac-

TABLE III The Growth of CHO Cells in Serum-free Media[a]

	Cells/ml			
	MEM, BP	MEM, BP insulin[b]	MEM-α, BP	MEM-α, BP, insulin[b]
Day 4	230,000	275,000	145,000	200,000
Day 5	185,000	365,000	180,000	340,000
Day 6	165,000	850,000	135,000	675,000
Day 7	120,000	930,000	100,000	810,000

[a]The initial cell density was 50,000 cells/ml.
[b]The insulin concentration was 0.11 I.U./ml.

topeptone (1%) as a serum substitute (see Table II). The cells remained an anchorage dependent line with very few detached cells observed. The confluent monolayers appeared to be healthier when MEM-alpha was used than when MEM was used, although the cell numbers are not too different. At confluency the monolayers developed the whorls observed in confluent monlayers grown in MEM + 10% FCS. The addition of insulin appears to be unnecessary for cell growth. It was found that BSC-1 cells grown to confluency with bactopeptone as a serum substitute could be infected with SV40 and that the yield of virus produced was about the same as in systems containing serum.

Chinese Hamster Ovary cells (CHO) were found to grow with bactopeptone (1%) as a serum substitute in MEM or MEM-α provided that insulin was added (see Table III). The optimal level of insulin has not been determined, but little difference was observed between 0.11 and 0.56 I.U./ml. It was observed that whereas with MEM-α + 10% FCS the CHO cells grew as a monolayer until close to confluency, when bactopeptone was used as a serum substitute the cells grew in stationary suspension in grape-like clusters. It was observed that the cell viability dropped rapidly as the cells

TABLE IV The Growth of Balb/c 3T3 Cells in Serum-free Systems

Medium	Day 5	Cells/ml[c] Day 7[b]	Day 12
MEM, BP	55,000	85,000	100,000
MEM, BP, Insulin[a]	130,000	275,000	360,000
MEM-α, BP	45,000	100,000	100,000
MEM-α, BP, Insulin[a]	100,000	235,000	190,000

[a]Insulin concentration was 0.11 I.U./ml.
[b]Cells centrifuged down and refed with fresh medium.
[c]The initial cell density was 25,000 cells/ml.

multiplied to about 10^6 cells/ml and immediate subculturing by centrifugation and resuspension in fresh medium was necessary. The CHO cells were cultured in this manner in the system MEM-α + BP + insulin for 3 months after which they were recultured in MEM-α + 10% FCS whereupon they immediately resumed the monolayer growth and previous morphology.

Balb/c 3T3 cells are a mouse line which has only previously grown as a contact-inhibited monolayer. However, when the 3T3 cells were subcultured into MEM-alpha supplemented with 1% bactopeptone and insulin it was observed that although most of the cells attached initially, they subsequently dissociated and grew in stationary suspension in grape-like clusters. The clumps grew quite large and were difficult to break up for cell counting which with the initial attachment made quantitative growth data less easy to obtain. This tendency to attach initially was only observed when new plastic flasks were used but not when used flasks were reused. Table IV shows the effect of insulin on the growth of 3T3 cells in the serum-free systems. Little difference was observed between growth in MEM and MEM-α. 3T3 cells were grown in a spinner bottle in MEM-α, bactopeptone and insulin, polyvinylpyr-

TABLE V The Effect of Medium on the Growth of HeLa Cells in the Presence of Bactopeptone[a] and Insulin[b]

		Cells/ml	
	Day 0	Day 2	Day 4
MEM	200,000	400,000	820,000
1st	200,000	405,000	930,000
Expt. MEM-alpha	200,000	700,000	1,100,000
	200,000	500,000	1,150,000

	Day 0	Day 3	Day 4	Day 5	Day 6
MEM	100,000	310,000	390,000	600,000	660,000
2nd	100,000	360,000	460,000	600,000	730,000
Expt. MEM-alpha	100,000	480,000	880,000	1,060,000	1,120,000
	100,000	510,000	750,000	1,010,000	900,000

[a]The bactopeptone supplement was 5% of a 20% solution.
[b]The insulin level was 0.06 I.U./ml.

rolidone being added as a possible macromolecular protective agent. In a 100 ml spinner, the cell density rose from 1×10^5 to 4.8×10^5 in 3 days. However, sometimes the cell density plateaued at 3×10^5 cells/ml and the system does not appear to have any practical value.

HeLa cells have been grown in several serum-free systems and it was found that they grew readily in the system MEM-alpha plus bactopeptone (1%) provided that insulin is added. The cells grew as a stationary suspension. The results shown in Table V show that growth in MEM-α was considerably better than in MEM. This is probably because the MEM-α contains the vitamin B_{12} and biotin which have been shown to be necessary for the growth of HeLa cells. There was little cell growth unless insulin was added, but as little as 0.011 I.U./ml insulin had a pronounced effect on cell growth.

TABLE VI The Growth of KB Cells in Serum-free Systems[a]

Medium	Day 4	Day 7	Day 10
MEMα, BP	80,000	50,000	-
MEMα, BP, 0.01 I.U./ ml insulin	140,000	440,000	746,000
MEMα, BP, 0.06 I.U./ ml insulin	150,000	370,000	755,000
MEMα, BP, 0.23 I.U./ ml insulin	110,000	300,000	620,000
MEM, BP, 0.23 I.U./ ml insulin	150,000	270,000	200,000
MEM, BP, 0.23 I.U./ ml insulin, B_{12}, biotin	190,000	360,000	592,000

[a]Initial cell density was 70,000 cells/ml. Bactopeptone concentration was 1% w/v.

HeLa cells grew well in agitated suspension in the system MEM-α, bactopeptone, insulin and polyvinylpyrrolidone the density rising from 2×10^5 to 1.4×10^6 in 5 days.

Results similar to those obtained with HeLa cells were obtained with KB cells. Insulin was required for cell growth, and apparently due to its vitamin B_{12} and Biotin content, MEM-α was much more effective than MEM when bactopeptone was used as a serum substitute (see Table VI). The cells grew as a stationary suspension in large grape-like clusters.

HeLa and KB cells were grown in spinner cultures in MEM-α, bactopeptone, PVP and insulin and used to prepare adenovirus. The yield and quality of the virus produced was the same as that produced in Joklik-MEM supplemented with horse serum.

B. Growth of Primary and Normal Cells in Serum-Free Media

Much of the work on cell nutritional requirements has been criticized because it has all been carried out with HeLa or L-cells (Ham, 1974). Several of the serum-free media developed support the growth of a wide range of established cell lines, but not primary or normal cells (Higuchi and Robinson, 1973; Lasfargues *et al.*, 1973).

Recently some progress has been made on extending our work to primary and secondary chick embryo fibroblasts. Chick embryo primary cells or secondary cells (grown in MEM + 3% FCS) (Keay and Schlesinger, 1974) were centrifuged and washed with PBS to remove residual fetal calf serum and trypsin, and resuspended in PBS for dilution into selected media. The results (Table VII) show that the complex serum substitute SSK7 (3% v/v) (Lasfargues, 1973) or the simpler bactopeptone, lactalbumin hydrolysate, polyvinylpyrrolidone combination (3% v/v) (Keay, 1975) promote cell growth as well as does fetal calf serum provided that insulin is also added. Insulin alone does not promote cell growth in this simple MEM system.

No attempts have yet been made to continuously culture the chick embryo cells in this serum-free system or to extend the method to other diploid or normal cells.

C. Autoclavable Low Cost Cell Culture Media

As described above, it is clear that many established cell lines will grow continuously on simple commercially available media supplemented with an autoclaved bactopeptone solution (or a more complex combination), insulin having to be added in some cases.

The question is whether an autoclavable medium can be developed from this system. Previously Yamane (1968) has grown cells on MEM autoclaved at low pHs provided glutamine is added after autoclaving. The autoclaving of glutamine and insulin represent

TABLE VII The Growth of Chick Embryo Fibroblasts in Serum-free Media[a]

	Expt A	Expt B	Expt C	Expt D	Expt E
Type of cell	Primary	Primary	Primary	Secondary	Secondary
10^6 cells inoculum	10.0	12.0	11.6	2.0	2.0 M
No. of days incubation	5	5	6	2	5
MEM	0.8	3.6	1.0	-	0.8
MEM, insulin	2.4	-	2.9	-	3.1
MEM, FCS	17.0	8.4	6.1	4.9	8.0
MEM, BP	4.6	4.9	-	-	-
MEM, BP, insulin	9.2	15.4	0.7	2.4	-
MEM, (BP-, LH-, PVP)	-	-	-	-	3.0
MEM, (BP-, LH-, PVP) insulin	-	-	5.4	5.3	4.2
MEM, SSK7	6.6	6.0	3.0	-	2.3
MEM, SSK7, insulin	14.4	7.2	6.0	4.9	4.7

[a]The insulin concentration was 0.46 I.U./ml.

specific problems. It was originally claimed that glutamine was a requirement for the groth of L cells and HeLa cells, but subsequently several glutamine-free media have been devised. Of course one problem is that when undefined additives such as serum or peptones are used, it is not clear whether glutamine or a glutaminine-substitute is being added. It is interesting to note that the glutamine-free media contain pyruvate without which cell growth is reduced.

Initial experiments were carried out with L-60TM cells in stationary suspension in plastic flasks. The cells had been growing in MEM supplemented with autoclaved bactopeptone solution (final concentration of bactopeptone 1%). The medium used was AutoPow^R MEM (Flow Laboratories) and autoclaving was carried out at 10 psi for 10 min with and without prior addition of solid bactopeptone and with and without prior neutralization with sodium bicarbonate. Glutamine was added to all media after autoclaving.

The results (see Table VIII) show clearly that Autopow^R MEM and bactopeptone do not support the growth of L-60TM cells as well when autoclaved together as when autoclaved separately and then combined. Further experiments were carried out with regular MEM (i.e., with glutamine and without succinate buffer). After neutralization with sodium bicarbonate, the medium was autoclaved at 10 psi for 10 min, cooled, the insoluble material solubilized by acidification with HCl, followed by immediate reneutralization with NaOH solution. Bactopeptone (20% w/v, autoclaved separately) was added to give a final concentration of 1% w/v, and glutamine (292 mg/l) was also added to compensate for any loss of glutamine during autoclaving.

L-60TM cells have been grown in stationary suspension in this system for several months by subculturing several times weekly by centrifugation. Preliminary experiments show that BHK cells will grow in autoclaved F-12 medium supplemented with bactopeptone and Darvan and this is being investigated further.

TABLE VIII The Growth of L-60TM Cells in Autoclaved Media

Medium	pH at autoclaving	Additives after autoclaving	Cell counts, 10^5 cells/ml[a]			
			Day 2	Day 5	Day 8	Day 10
Autoclaved (MEM, BP)	4.0	gln	2.4	4.5	4.0	1.2
Autoclaved (MEM, BP)	7.0	gln	3.0	4.2	2.7	2.0
Autoclaved MEM	4.0	BP, gln	2.8	9.0	9.8	3.1
Autoclaved MEM	7.0	BP, gln	2.5	8.8	8.0	2.6
Unautoclaved MEM, BP, gln (control)			2.5	8.7	8.2	2.7

[a]Initial cell density was 1.5×10^5 cells/ml.
[b]On Day 8, all were centrifuged and subcultured to 2.0×10^5 cells/ml.
[c]Medium was AutoPow[R] MEM (Flow Laboratories).

How does the above system compare with the ideal medium? It meets essentially all of the requirements except one, namely that it is not chemically defined. The problem of chemical definition will probably be solved when the precise composition and roles of both serum and bactopeptone can be determined and both replaced by their purified growth promoting components.

It has not yet been established how long various types of cells will continue to proliferate in these autoclaved media or what changes may occur in the cells, or what problems may be encountered on scale up. The economics of these systems cannot yet be compared with serum-containing systems since cell yields and growth rates have not yet been accurately determined. It is clear however that low cost formulations based upon commercially available dry powder media supplemented with bactopeptone will support the growth of many cell lines and if the conditions are carefully selected can be converted to totally autoclavable systems for cell and virus production.

ACKNOWLEDGMENTS

The dedicated technical assistance of Mr. Fred Anderson is gratefully acknowledged. This work was supported by the Human Cell Biology Program of the National Science Foundation (Grants #GB 38657 and BMS 73-07013-A01).

REFERENCES

Birch, J. R. and Pirt, S. J. (1970) *J. Cell Sc.* 7, 661-670.

Blaker, G. J., Birch, J. R. and Pirt, S. J. (1971). *J. Cell Sc.* 9, 529-537.

Boone, C. W., Mantel, N., Caruso, T. D., Kazam, E., and Stevenson, R. E. (1972). *In Vitro* 7, 174-189.

Bryant, J. C., Evans, V. J., Schilling, E. L., and Earle, W. R. (1961). *J. Natl. Cancer Inst. 26,* 239-252.

Eagle, H. (1955a). *Science 122,* 43-46.

Eagle, H. (1955b), *J. Expt. Med. 102,* 37-48.

Eagle, H. (1955c). *J. Expt. Med. 102,* 595-600.

Eagle, H., and Piez, K. (1962). *J. Expt. Med. 116,* 29-43.

Fedoroff, S., Evans, V. J., Hopps, H. E., Sanford, K. K. and Boone, C. W. (1972). *In Vitro 7,* 161-167.

Fioramonte, M. C., Evans, V. J. and Earle, W. R. (1958). *J. Natl. Cancer Inst. 21,* 579-583.

Ginsberg, H. S., Gold, E. and Jordan, W. S. (1955). *Proc. Soc. Expt. Biol. Med. 89,* 66-71.

Ham, R. G. (1963). *Expt. Cell Res. 29,* 515-526.

Ham, R. G. (1968). *Proc. Nat. Acad. Sci. 53,* 288-293.

Ham, R. G. (1974). *In Vitro 10,* 119-129.

Healy, G. M. and MacMorine, H. G. (1972). *Progress in Immunobiol. 45,* 202-208.

Healy, G. M., Fisher, D. C. and Parker, C. (1954). *Canad. J. Biochem. Physiol. 32,* 327-337.

Healy, G. M., Fisher, D. C. and Parker, C. (1955). *Proc. Soc. Expt. Biol. Med. 89,* 71-77.

Higuchi, K. (1963). *J. Infect. Dis. 112,* 213-220.

Higuchi, K. (1970). *J. Cell Physiol. 75,* 65-72.

Higuchi, M. (1973). *Adv. Appl. Microbiol. 16,* 111-136.

Higuchi, K. and Robinson, R. C. (1973). *In Vitro, 9,* 114-121.

Holmes, R. and Wolfe, S. W. (1961). *J. Biophys. Biochem. Cytol., 10,* 389-401.

Hopps, H. E. (1974). *In Vitro 10,* 243-246.

Hsueh, H. W. and Moskowitz, M. (1973a). *Expt. Cell Res. 77,* 376-382.

Hsueh, H. W. and Moskowitz, M. (1973b). *Expt. Cell Res. 77,* 383-390.

Keay, L. (1975) *Biotechnol. Bioeng. 17,* 745-764.

Keay, L. (1976) *Biotechnol. Bioeng. 18,* 363-382.

Keay, L. and Schlesinger, S. (1974). *Biotechnol. Bioeng. 16,* 1025-1044.

Lasfargues, E. Y., Continho, W. G., Lasfargues, J. C. and Moore, D. H. (1973). *In Vitro 8,* 494-500.

Lieberman, I. and Ove, P. (1959). *J. Biol. Chem. 234,* 2754-2758.

Mayyasi, S. A. and Schuurmans, D. M. (1956). *Proc. Soc. Expt. Biol. Med. 93,* 207-210.

Molander, C. W., Kniazeff, A. J., Boone, C. W., Paley, A. and Imagawa, D. T. (1972). *In Vitro 7,* 168-173.

Morgan, J. F., Morton, H. J. and Parker, R. C. (1950). *Proc. Soc. Expt. Biol. Med. 73,* 1-8.

Morrison, S. J. and Jenkin, H. M. (1972). *In Vitro 9,* 94-100.

Nagle, S. C. (1968). *Appl. Microbiol. 16,* 53-55.

Nagle, S. C. (1969). *Appl. Microbiol. 17,* 318-319.

Nagle, S. C. and Brown, B. L. (1971). *J. Cell Physiol. 77,* 259-264.

Nagle, S. C., Tribble, H. R., Anderson, R. E. and Gary, N. D. (1963). *Proc. Soc. Expt. Biol. Med. 112,* 340-344.

Pierson, R. W. and Temin, H. M. (1972). *J. Cell Physiol. 79,* 319-330.

Pumper, R. W., Yamashiroya, H. M. and Molander, L. T. (1965). *Nature 207,* 662-663.

Stammers, C. P., Elicieri, G. L. and Green, H. (1971). *Nature New Biol. 230,* 52-53.

Takaoka, T. and Katsuta, H. (1971). *Expt. Cell Res. 67,* 295-304.

Taylor, W. G. (1974). *J. Natl. Cancer Inst. 53,* 1449-1457.

Taylor, W. G., Dworkin, R. A., Pumper, R. W. and Evans, V. J. (1972a) *Expt. Cell Res. 74,* 275-279.

Taylor, W. G., Taylor, M. J., Lewis, N. T. and Pumper, R. W. (1972b). *Proc. Soc. Expt. Biol. Med. 139,* 96-99.

Taylor, W. G., Evans, V. J. and Pumper, R. W. (1974). *In Vitro 9,* 278-286.

Temin, H. M. (1967). *J. Cell Physiol. 69,* 377-384.

Tribble, H. R. and Higuchi, K. (1963). *J. Infect. Dis. 112,* 221-225.

Waymouth, C. (1955). *Tex Reports Biol. Med. 13,* 522-536.

Waymouth, C. (1956). *J. Natl. Cancer Inst. 17,* 315-325.

Waymouth, C. (1959). *J. Natl. Cancer Inst. 22,* 1003-1017.

Yamane, I., Matsuya, Y. and Jimbo, K. (1968). *Proc. Soc. Expt. Biol. Med. 127,* 335-336.

QUALITATIVE AND QUANTITATIVE ANALYSIS OF NUTRITIONAL FACTORS FOR CELL GROWTH

RICHARD G. HAM
PATRICIA M. SULLIVAN

Department of Molecular, Cellular, and Developmental Biology
University of Colorado
Boulder, Colorado

I. DEVELOPMENT OF DEFINED MEDIA FOR NORMAL CELLS

A. Definitions of "Defined" Media

1. *Theoretical*

In theory a "defined" medium would be prepared from water and chemicals that are completely free of impurities. The final medium would contain only those substances listed in its formula. In practice, such a medium can never be fully attained.

2. *Rigorous*

The most precisely defined medium that is realistically possible would be prepared from water and chemicals of ultra high purity. The effects on cellular growth of any remaining impurities would be characterized as completely as possible. Fixed amounts would deliberately be added of those impurities that could not be removed completely in order to insure uniformity from batch to batch of the medium.

3. *Currently Used*

The best currently available media are prepared from "reagent" grade water and chemicals. They contain trace impurities that are not listed in their formulas. Such impurities are probably essential for cellular growth in most, if not all, "defined" media currently in use.

B. Principles Involved in Developing Media for Normal Cells

1. *Alter the Medium, Not the Cells*

Most currently available media were developed specifically for established cell lines that had undergone extensive adaptation and evolution in culture. Such media are not satisfactory for normal cells as they first come from the animal. If we are going to understand the nutrient requirements of such cells, we must devise media that satisfy their needs before they have undergone any nutritional adaptation *in vitro*.

2. *Be Alert for the Unexpected*

Research is accomplished primarily by asking questions. Many valuable insights can be obtained from unexpected answers. For example, unexpected growth stimulation by a saline "control" led to the discovery that the cysteine concentration in medium F12 was too high for clonal growth of WI-38 cells.

3. *Adjust the Nutrient Balance for the Cell Types*

Balance relationships among medium components are very important (Ham, 1974). Changing the amount of one nutrient in the medium can significantly change the amounts of others that are needed. Also, too much of a nutrient is often as bad as too little, and the optimum range is sometimes quite narrow. Serum proteins tend to bind nutrients, buffering their effects. When serum protein is reduced in concentration or replaced by partially purified protein fractions, the range of concentrations of specific small molecular nutrients that will support optimum growth may become significantly narrower. These effects are not the same for each type of cell studied (c.f. #9, below).

4. *Environmental Responses May be Different*

Different types of cells may respond very differently to environmental factors such as pH, osmolarity, carbon dioxide tension, oxygen tension, and nature of culture surface.

5. *Contaminants May be Essential Nutrients*

Growth in a defined medium may be dependent on contaminants not specified in the formula of the medium. Such growth may fail unexpectedly if chemical manufacturers quietly "improve" their products. We currently find that selenium salts must be added to medium F12 to support clonal growth of Chinese hamster cell lines that once grew in it without such a supplement.

6. *Nutrient Utilization by Normal Cells May be Inefficient*

Some established cell lines that have been adapted to growth in serum-free media have become very efficient at scavenging extreme traces of impurities from the culture medium. For example, WI-38 cells require more than 100 times as much selenium as Chinese hamster ovary cells that have been in culture for many years.

7. *Normal Cells are Intolerant of Deficiencies*

Established cell lines can compensate extensively for deficiencies in their culture medium. For example if no polyunsaturated fatty acids are available, they can make functional membranes without them. Normal cells appear not to have this adaptability.

8. *Special Growth Factors May be Needed by Normal Cells*

Many established cell lines can be grown in synthetic media that appear to contain only small molecular nutrients that are taken into the cells and utilized in substrate or catalytic roles. Preliminary evidence suggests that many normal cells have additional requirements for "regulatory" factors that may act only at the cell surface (e.g. nerve growth factor, epidermal growth factor, somatomedin, etc.).

9. *Different Cell Types Have Different Needs*

Classical established cell lines (mouse strain L, HeLa, etc.) have mostly undergone rather similar patterns of adaptation *in vitro,* and tend to exhibit rather similar nutritional requirements. Our current experience with diploid human, mouse, and chicken cells indicates that non-adapted cells do not exhibit a comparable similarity. For example, each cell type that we are currently working with requires a different pattern of relative and absolute concentrations of amino acids for optimum clonal growth, and each of these patterns is different from that required by typical established cell lines.

II. PROGRESS TOWARD DEFINED MEDIA FOR SPECIFIC TYPES OF CELLS

A. Diploid Human Fibroblast, WI-38

1. *Cysteine*

Medium F12 (Ham, 1965, 1972) supplemented with dialyzed fetal bovine serum has been used as the starting point for studying the nutrient requirements for clonal growth of WI-38 cells. Lowering the cysteine concentration to 9.0×10^{-5} *M* (from 2.0×10^{-4} *M*) significantly improves such growth. The optimum concentration range is extremely narrow. Essentially no growth can be obtained at either 3.0×10^{-5} *M* or 3.0×10^{-4} *M*. The entire growth response from deficiency to optimum to toxicity occurs within a tenfold concentration range.

2. *Glutamine*

Further improvement of clonal growth is obtained by increasing the glutamine concentration to 2.5×10^{-3} *M* (from 1.0×10^{-3} *M*). Medium F12 with 9.0×10^{-5} *M* cysteine, 2.5×10^{-3} *M* glutamine, and 2.8×10^{-2} *M* HEPES buffer has been designated MCDB 102. This medium supports excellent clonal growth of WI-38 cells when supplemented with 10% dialyzed fetal bovine serum.

3. *Selenium*

The amount of serum protein required for clonal growth can be significantly reduced when 3.0×10^{-8} *M* sodium selenite is added to the culture medium. Selenium appears to be functioning as an essential trace element for the cells. Medium MCDB 102 with $ZnSO_4$ reduced to 5.0×10^{-7} *M* and Na_2SeO_3 at 3.0×10^{-8} *M* has been designated MCDB 103. This medium will support good clonal growth with 1500 μg/ml fetal bovine serum protein (equivalent to 3% serum).

4. *Serum Protein*

The amount of serum protein required for clonal growth has been reduced significantly by adjusting the balance of nutrients already in the medium and by the inclusion of a trace element supplement. In the best current media, relatively good clonal growth can be obtained with 500 μg/ml serum protein. Preliminary fractionation studies indicate that at least two, and possibly more, growth promoting activities are being furnished by the serum protein.

B. Other Cell Types

1. *Chicken Embryo Fibroblasts*

Chicken embryo fibroblasts require higher concentrations of many amino acids than WI-38. They also require larger amounts of serum protein in the best media that we currently have. However, they appear not to require an elevated concentration of glutamine, and are not as sensitive to cysteine toxicity as WI-38.

2. *Mouse Embryo Fibroblasts*

Mouse embryo fibroblasts do not grow well in media based on F12. They do exhibit reasonably good plating efficiency in medium CMRL 1415 (Healy and Parker 1966) modified by eliminating cystine and reducing the cysteine concentration to 3.0×10^{-4} *M*. The macromolecular fractions from both fetal bovine serum and horse serum are needed. Preliminary studies suggest that the mouse cells require rather high concentrations of a number of nutrients. The optimum concentration of cysteine for the mouse cells is lethal to WI-38.

3. *Chinese Hamster Ovary and Lung Lines*

Chinese hamster ovary and lung lines that once grew in medium F12 without protein supplements (Ham 1965) no longer do so. We have recently found that the addition of sodium selenite will restore such growth. We suspect that adequate amounts of selenium were originally supplied to the cells as contaminants in the chemicals that were used 10-12 years ago for the original studies.

III. OTHER FACTORS INFLUENCING CELLULAR GROWTH

A. Petri Dishes

We have found that not all brands of tissue culture petri dishes support equivalent clonal growth of normal cells in media containing dialyzed serum, or of established lines in protein free media. The work described above was all done in Corning dishes, which in our experience have worked well. Other brands should be tested carefully before routine use.

B. Hydrogen Ion Concentration

The pH of the culture medium is critical for some types of cells (particularly WI-38). We routinely add 28 mM HEPES to most of our media, except the completely protein free medium for the Chinese hamster lines, where pH appears to be less critical and HEPES may be slightly toxic. Carbon dioxide is needed as a metabolic intermediate in a number of biosynthetic pathways. Under clonal conditions, a sufficient amount of metabolic carbon dioxide does not reliably accumulate. Therefore, a carbon dioxide-bicarbonate buffer is retained in the media in addition to HEPES.

ACKNOWLEDGMENTS

Much of this report is based on unpublished experiments prepared by Greg Hamilton, Susan Hammond, Billie Jean Lemmon, Kerstin McKeehan, Dr. Wallace McKeehan, and Gary Shipley. The research has been supported by contract 223-74-1156 from the Bureau of Biologics, Food and Drug Administration, Grant AG00310 from the National Institute on Aging, Grant HD08181 from the National Institute of Child Health and Human Development, and Grant CA 15305 from the National Cancer Institute.

REFERENCES

Ham, R. G. (1965). *Proc. Nat. Acad. Sci. USA 53,* 288-293.

Ham, R. G. (1972). *In* "Methods in Cell Physiology" (D. Prescott, ed.), Vol. 5, pp. 37-74. Academic Press, New York.

Ham, R. G. (1974). *In Vitro 10,* 119-129.

Healy, G. M. and Parker, R. C. (1966). *J. Cell Biol.* 30: 531-538.

Note Added in Proof: During the time since the symposium, the principles outlined above have proven to be extremely effective in reducing the amount of serum protein required for clonal growth of several types of normal cells. A number of papers describing various aspects of this work have been published or are in press, as follows:

Ham, R. G., Hammond, S. L., and Miller, L. L. (1977). Critical adjustment of cysteine and glutamine concentrations for improved clonal growth of WI-38 cells, *In Vitro* (in press).

McKeehan, W. L., Hamilton, W. G., and Ham, R. G. (1976). Selenium is an essential nutrient for growth of WI-38 human diploid fibroblasts in culture, *Proc. Natl. Acad. Sci. U.S.A. 73*: 2023-2027.

McKeehan, W. L., and Ham, R. G. (1976). Stimulation of clonal growth of normal fibroblasts with substrata coated with basic polymers, *J. Cell Biol. 71*: 727-734.

McKeehan, W. L., McKeehan, K. A., Hammond, S. L., and Ham, R. G. (1977). Improved medium for clonal growth of human diploid fibroblasts at low concentrations of serum protein, *In Vitro* (in press).

Hamilton, W. G. and Ham, R. G. (1977). Clonal growth of Chinese hamster cell lines in protein-free media, *In Vitro* (in press).

SALINE WASHES STIMULATE THE PROLIFERATION OF HUMAN FIBROBLASTS IN CULTURE

JAMES C. LACEY, JR.
JANNA D. STROBEL
DANIEL P. STEPHENS, JR.

Laboratory of Molecular Biology
University of Alabama in Birmingham
Birmingham, Alabama

I. SUMMARY

We found brief rinses with phosphate buffered saline or 0.9% NaCl stimulated the proliferation of density inhibited human fibroblasts. Additions of trypsin or pronase to phosphate buffered saline did not increase this effect. Systematic elimination of the components of phosphate buffered saline showed none of these components was essential to the overgrowth phenomenon. Part of the stimulation by phosphate buffered saline was probably due to a transient increase in pH. However, because unbuffered 0.9% NaCl, which assumes the pH of the cells, also caused stimulation, we believe other factors in addition to pH are involved. Since additions of dibutyryl cyclic AMP to phosphate buffered saline did not reduce the overgrowth effect, we believe a reduction in intracellular cyclic AMP levels is not responsible. The evidence does, however, suggest to us that a negative substance, perhaps a chalone or cell surface glycoprotein, is being extracted from the cell.

II. INTRODUCTION

Much experimental effort has been directed toward understanding the basic mechanisms controlling the proliferation of cells in multicellular organisms. Many experiments have been carried out using *in vitro* procedures involving cells in culture. Stationary cell cultures maintained under fixed conditions grow to a particular density and then their growth rate slows. Most researchers assume that the *in vitro* processes which lead to this reduction in growth are qualitatively the same as those operating in the intact creature. In addition to serum (Temin, 1971), a number of diverse factors have been reported to stimulate the overgrowth of such density inhibited cells. These factors include insulin (Temin, 1967), proteases (Burger, 1970; Burger *et al.*, 1972; Noonan and Burger, 1973; Rubin, 1970; Sefton and Rubin, 1970), guanosine 3'5' monophosphate (cyclic GMP) (Rudland *et al.*, 1974), low levels of cyclic AMP (Burger *et al.*, 1972), hyaluronidase, ribonuclease (Vasiliev *et al.*, 1970) and neuraminidase (Vaheri *et al.*, 1972). There is considerable controversy regarding the ability of several of these factors to reinitiate proliferation. We are particularly interested in a number of reports regarding the stimulation of overgrowth by various proteases. These include the reports from Burger's laboratory, using 3T3 cells (Burger, 1970; Burger *et al.*, 1972; Noonan and Burger, 1973) and Rubin's laboratory (Rubin, 1970; Sefton and Rubin, 1970), using chick embryo cells. There is ample reason to question whether 3T3 cells are normal since they are heteroploid and undergo a more rigorous density dependent inhibition than normal cells. Chick embryo cells, while they can be grown under conditions giving density dependent inhibition, might not be controlled by the same mechanisms as cells from either born chickens or mammals. Because of these considerations, we wanted to determine if proteases would cause the overgrowth of normal human fibroblasts in culture.

III. MATERIALS AND METHODS

A. Cell Cultures

Secondary cultures of normal human diploid foreskin fibroblasts (HF-6 and HF-7) from two different individuals* were grown at 37°C in 32 oz prescription flats (Brockway) containing 50 ml of medium. The medium consisted of MEM (50-50 Hanks'-Earle's salts) supplemented with 10% fetal calf serum, penicillin, streptomycin and fungizone purchased from GIBCO. Flats were innoculated at half confluency and required 5 days at 37°C to become confluent giving a cell density of about 90,000 cells/cm^2 or approximately 1.3×10^7 cells/flat. For experimental samples, flats were trypsinized with 0.25% trypsin (GIBCO). The detached cells were suspended in the same medium supplemented with 3%, instead of 10%, fetal calf serum. Borosilicate scintillation vials with rubber lined caps were innoculated with $2.0\text{-}2.5 \times 10^5$ cells in 2.0 ml of medium. Under these conditions, the cells reached a plateau level of about $4.0\text{-}4.5 \times 10^5$ cells/vial in 5 to 7 days. Passage numbers are given in the figure legends.

B. Cell Counts

Cell counts were made by removing the medium from vials by aspiration and adding 2.0 ml of 0.25% trypsin. After 20 min at 37°C, the cells were aspirated up and down in a disposable pipette 12 times to break up clumps. Eight ml of 0.9% NaCl were added

*See note added in proof.

and a Coulter Counter, Model B, probe was inserted directly into the vial. Routinely, triplicate counts were made on each of triplicate samples. When cell counts were approximately 20,000 counts/0.5 ml, the variation among a set of triplicate samples was about 1000 counts.

C. Washing Procedures

Our experimental procedure usually consisted of asceptically pouring off the medium from confluent vials at 00 time. The pH of the medium at 7 days was about 6.8. This conditioned medium was stored sterilly while not in use. Last traces of medium were carefully removed from the cells by aspiration using a sterile disposable pipette. One ml of the solution being tested (pH 7.2 except for the unbuffered 0.9% NaCl) was added to the cells and allowed to stand for 15 min. The solution was then carefully removed by aspiration and 2.0 ml of the conditioned medium was readded to each vial. Control samples were treated in the same manner only conditioned medium was used instead of a test solution. In the protease experiments, crystalline trypsin (Sigma) or pronase (Calbiochem) was dissolved in Dulbecco's phosphate buffered saline (PBS). When trypsin was used in an experiment, the cell layer was prewashed for 5 min with 1.0 ml of PBS buffer. This was removed and then the trypsin put on. The prewashing was done to remove trypsin inhibitors present in the medium. When pronase was used the prewash was omitted and replaced by a post-treatment wash because there is no pronase inhibitor present in the medium. After each washing experiment triplicate samples were counted to determine the amount of cell loss due to the washing (0 time). Subsequently cell counts were taken at intervals up to seven days.

IV. RESULTS

In studying the effects of proteases on cell proliferation, we ran two types of controls in parallel with the test samples; one set consisted of samples washed with PBS (Dulbecco's phosphate buffered saline) alone, the other set was exposed to the medium in which the cells were grown (conditioned medium). We were surprised to find that the control washes with PBS alone resulted in overgrowth above the levels observed with samples in which conditioned medium was used. Figure 1 shows that additions of up to 3 μg/ml of trypsin to the PBS did not cause additional overgrowth. In fact, 3 μg/ml resulted in disruption of the cell layer and early death of the cells. Higher levels of trypsin caused greater cell loss. Similarly, in Fig. 2, up to 5 μg pronase/ml did not stimulate overgrowth any more than PBS alone. Higher levels of pronase also caused cell loss.

Using the Student t Test for independent samples, we examined our samples for significant differences in cell numbers. In both protease experiments, the arithmetic means of PBS and the protease treated samples were significantly greater ($P < 0.05$) than those of the conditioned medium samples by the second day after washing. There were no significant differences between the PBS and protease treated samples ($P > 0.05$).

After observing overgrowth with PBS in these and many other experiments, (not shown) we wondered which component, if any, in the PBS might be responsible for the stimulation. We systematically eliminated every component in the PBS including the replacement of phosphate with HEPES (N-2-hydroxy-ethylpiperazine-N'-2-ethanesulfonic acid) buffer. We still got significant overgrowth in every case. Since the stimulation did not appear to depend on any factor in the PBS solution we then tried plain isotonic NaCl (0.9%), unbuffered, as shown in Fig. 3. There was still significant overgrowth, but reduced about 50% from that obtained with PBS. The pH of unbuffered 0.9% NaCl was 6.0 before putting it on

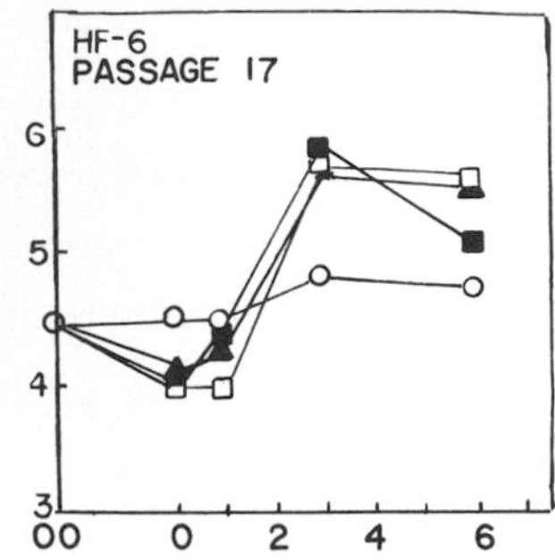

FIGURE 1 *Abscissa: days after washing; ordinate: cells/vial × 10^{-5}. Effect of trypsin on proliferation of confluent human fibroblasts (HF-6; Passage 17). At time 00, medium was removed from cells and saved. Last traces of medium were carefully aspirated. Samples to receive trypsin were washed with 1.0 ml of PBS for 5 min. PBS was carefully aspirated and then we added 1.0 ml of □, PBS; ▲, PBS with 1.5 μg/ml cryst. trypsin; ■, PBS with 3.0 μg/ml cryst. trypsin; for 15 min. Test solutions were aspirated and 2.0 ml of conditioned (cond.) medium readded to each sample. Control samples of cond. medium, o, were treated in an analogous manner with cond. medium. Cell counts were made at intervals shown.*

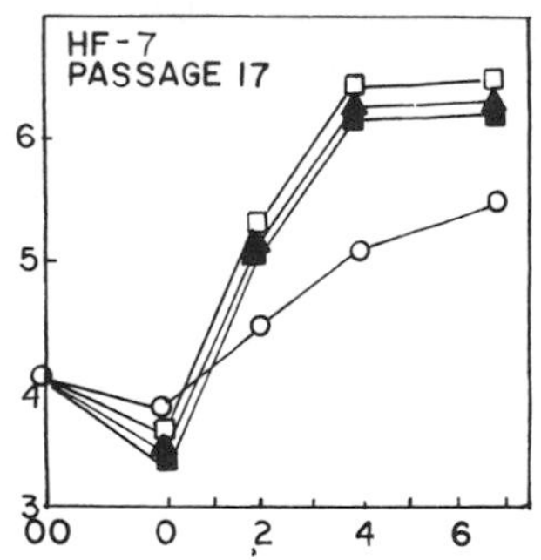

FIGURE 2 *Abscissa: days after washing; ordinate: cells/vial × 10^{-5}. Effect of pronase on proliferation of confluent human fibroblasts (HF-6; Passage 17). At time 00, medium was removed and saved. Last traces of medium were carefully aspirated. Cells were exposed for 15 min to 1.0 ml: □, PBS; ▲, PBS with 1 μg/ml pronase; ■, PBS with 5 μg/ml pronase; o, cond. medium control. All solutions were aspirated. Samples were washed with 1.0 ml of PBS for 5 min to remove pronase except controls which were exposed to cond. medium. All solutions were aspirated and cond. medium readded. Cell counts were made at intervals shown.*

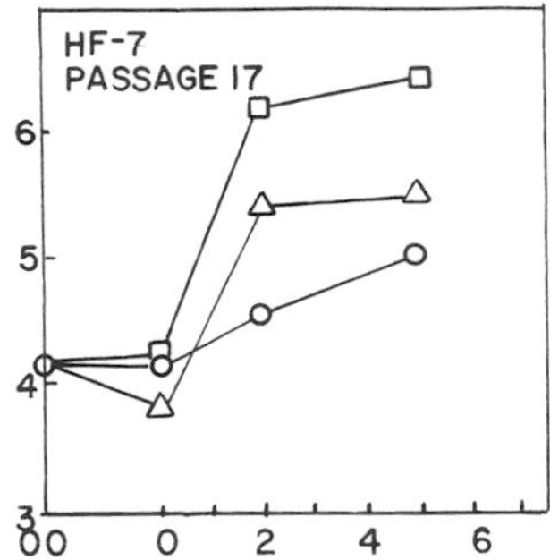

FIGURE 3 *Abscissa: days after washing; ordinate: cells/vial × 10^{-5}. Effect of PBS (pH 7.2) and 0.9% NaCl (unbuffered) on proliferation of confluent human fibroblasts (HF-7; Passage 17). At time 00, medium was removed from cells and saved. Last traces of medium were carefully aspirated and cells were exposed for 15 min to 1.0 ml of either □, PBS: Δ, 0.9% NaCl; or o, cond. medium. Solutions were carefully aspirated and cond. medium replaced. Cell counts were made at intervals shown.*

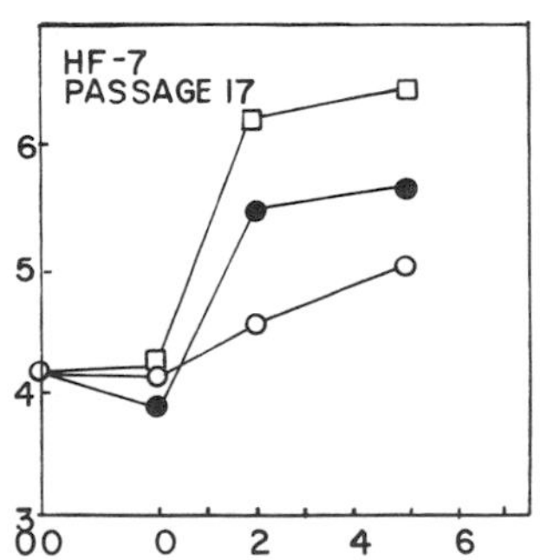

FIGURE 4 *Abscissa: days after washing; ordinate: cells/vial × 10^{-5}. Effect of small amount of residual medium on PBS stimulation of proliferation of confluent human fibroblasts (HF-7; Passage 17). At time 00, medium was poured off all vials. Last traces of medium were removed from 2 sets of vials by aspiration. A 3rd set of vials was not aspirated, leaving an estimated 0.1 ml of medium in each vial. 1.0 ml of PBS was added to the unaspirated vials, ●, and to one set of aspirated vials, □. 1.0 ml of cond. medium was added to the remaining set of aspirated vials, o. After 15 min all solutions were aspirated. 2.0 ml of cond. medium was added to each vial. Samples were counted at intervals shown.*

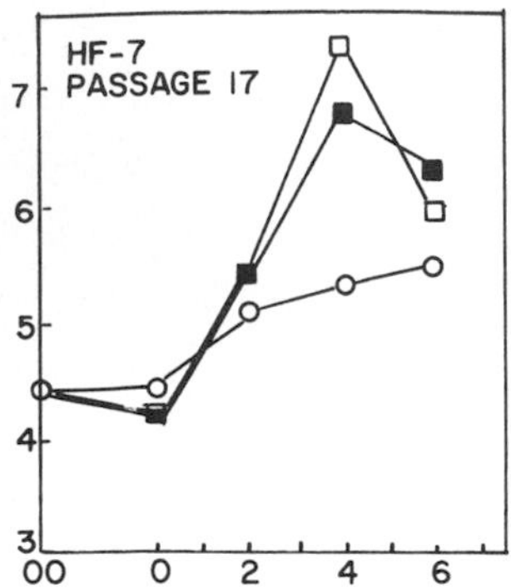

FIGURE 5 *Abscissa: days after washing; ordinate; cells/vial × 10^{-5}. Effect of dibutyryl cyclic AMP on PBS stimulation of proliferation of confluent human fibroblasts (HF-7; Passage 14). At time 00, medium was removed from cells and saved. Last traces of medium were carefully aspirated. Samples were exposed for 15 min to 1.0 ml: PBS with 1 × 10^{-5} M dibutyryl cyclic AMP, , PBS, ; and cond. medium, o. All solutions were aspirated and cond. medium readded. Cell counts were made at intervals shown.*

the cells, while the pH of PBS was 7.2. The NaCl solution probably assumed the pH of the medium (6.8) just removed from the cells.

In attempting to arrive at a mechanism which is consistent with our results, our first feeling was that the stimulation was caused by cell loss since washing the cell layers with either PBS or 0.9% NaCl caused a cell loss of up to 10 per cent. In some cases, however, there was no loss and stimulation still resulted. Multiple washes did not increase cell loss or growth. There was some time dependence, 15 min washes caused slightly more overgrowth than 5 min washes. However, simply putting PBS on the cells and aspirating immediately also resulted in overgrowth.

To our knowledge, stimulation of cell proliferation by simple saline extraction has not been previously reported. Burger (1970) washed 3T3 cells with PBS but reported no stimulation. We wondered if our careful removal of last traces of medium might have made a difference. In one experiment we simply poured the medium out of one set of vials, and did not aspirate the last traces of medium. This procedure left about 0.1 ml of medium in each vial.

The PBS was then added to this set of vials and also to a control set which had been carefully aspirated. The results are shown in Fig. 4. Residual medium caused about a 50% reduction in overgrowth.

A number of workers have shown that intracellular cyclic AMP levels are low in proliferating cells and high in confluent monolayers (Anderson *et al.*, 1973; Burger *et al.*, 1972). One possible explanation of our observed overgrowth phenomenon is that saline washing causes a drop in the level of cyclic AMP. To test this possibility we repeated the PBS wash experiment including a set of vials washed with a PBS solution containing 1×10^{-5} *M* dibutyrl cyclic AMP. As Fig. 5 shows, there was no statistically significant decrease in overgrowth.

V. DISCUSSION

These experimental results leave us with several important conclusions regarding the control of proliferation in culture:

(1) Washing of human fibroblasts with saline solutions (PBS or 0.9% NaCl) can induce proliferation in confluent cells and the induction does not seem dependent on any factor present in the solution. Washing confluent monolayers with saline solutions usually results in the loss of up to 10% of the cells. The stimulation may be partly dependent on a transient pH increase since unbuffered 0.9% NaCl caused less proliferation.

(2) We find no additional stimulation of proliferation by adding trypsin or pronase to the PBS washes. Our results do not support the reports for protease induced overgrowth in 3T3 cells and chick embryo fibroblasts. Glynn and associates (1973) also failed to find increases in cell number by exposing 3T3 cells to various pronase concentrations, although they did find that pronase treatment increases cell agglutinability with concanavalin A. However, we cannot rule out the possibility that in our

experiments the proteases may cause some overgrowth that is masked by the PBS effect.

Reich (1973, 1974) recently suggested that increased protease activity leading to the ability to lyse fibrin is a characteristic of transformed cells. This contention has been challenged by Mott and his co-workers (1974) who find that some normal cells have this ability and some transformed cells do not. Similarly, Chou *et al.*, (1974a) showed that suppression of fibrinolysin activity did not restore density dependent inhibition to SV40 transformed 3T3 cells. While some protease inhibitors do cause a reduction in growth of transformed cells, Chou and his associates (1974b) are of the opinion that this effect is due to an inhibition of protein synthesis rather than an inhibition of protease activity. On the other hand, since some growth factors (Jones and Ashwood-Smith, 1970; Greene *et al.*, 1971) have been shown to have proteolytic activity, it is probable that proteolysis is a part of the process leading to mitosis but is probably not the trigger.

Anderson and his collaborators (1973) found high levels of intracellular cyclic AMP in confluent monolayers but low levels during proliferation. These and other data (Burger *et al.*, 1972) suggest that low levels of cyclic AMP trigger cell proliferation. However, more recent experiments by Rudland *et al.* (1974) show high levels of cyclic GMP are more important than low cyclic AMP levels. Since the stimulatory effect of PBS in our experiments was not inhibited by exogenously added dibutyryl cyclic AMP, our inference is that cyclic AMP is not a part of the mechanism inducing proliferation by PBS.

Part of the stimulation by pH 7.2 PBS could be due to increased pH. As Ceccarini and Eagle (1971) have shown, pH plays an intimate role in controlling the growth of cells. Before washing with salt solutions our cells had a pH of 6.8-6.9, and 15 min exposure to pH 7.2 PBS stimulated considerable overgrowth. Unbuffered 0.9% NaCl, which would assume the pH of the cells, also caused overgrowth but only about 50% of that caused by PBS.

Therefore, the overgrowth due to the PBS wash probably results only in part from a pH effect. Furthermore, the pH elevation due to PBS was only brief and after replacing the conditioned medium on the cells there was no difference in the pH of the PBS washed cells and the control cells washed with conditioned medium. Consequently, whatever mechanisms were initiated during the brief exposure to the salt solutions were not reversed by the conditioned medium.

There are several other possible effects that could be adding to the pH effect. Ceccarini and Eagle (1971) showed that exposure of human fibroblasts to Earle's salt solution could increase uridine uptake, but they did not observe an increase in the synthesis of RNA, DNA or protein. More recently, Jimenez de Asua and his co-workers (1974) reported that additions of phosphate to the medium of cells in culture caused increased uridine transport and also that one of the primary results of serum addition to medium was an increase in phosphate uptake by cells. Increased phosphate uptake in turn resulted in an increased uridine uptake. These authors propose that an increase in phosphate uptake is a primary event in the initiation of the mitotic sequence. However, they did not report an increase in cell density. Our results with PBS are probably not related to this phenomenon since replacement of phosphate with HEPES still gave stimulation.

Another of the possibilities is that the saline wash is removing a negative factor from the cell layer. There are several possible candidates.

(a) Removal of cyclic GMP phosphodiesterase from the cell could cause elevation of cyclic GMP levels.

(b) Extraction of cell surface factors could decrease cell-cell adhesion.

(c) Removal of intercellular materials might act as a micro wound.

(d) Another possibility is that our saline washes are removing a fibroblast cell surface glycoprotein which is extractable in salt solutions (Yamada and Weston, 1974) or with 0.25% trypsin (Rouslahti and Vaheri, 1974) and is markedly diminshed or absent in transformed cells. However, it is not clear why removal of the glycoprotein would stimulate proliferation.

(e) Since fibroblast mitotic inhibitor (chalone) has been isolated and partially identified (Houck *et al.*, 1972; Houck *et al.*, 1973a; Houck *et al.*, 1973b), one of the most likely mechanisms is that the saline solution is extracting chalone from the cell surface. Chalone is found in conditioned medium from fibroblasts and retention of cells in a resting state probably requires a particular level of chalone in or on the cell. It is possible that chalone is equilibrium bound to the cell, and because PBS contains no chalone, PBS washing rapidly releases chalone which results in proliferation.

We are presently attempting to isolate fibroblast chalone from the PBS solutions which were used to wash cells. If we identify chalone in the PBS wash and can show that addition of exogenous chalone can inhibit PBS stimulation of overgrowth, we may conclude that the stimulation is due to chalone extraction.

ACKNOWLEDGMENTS

We thank Dr. Charles Alford for the gift of the fibroblasts and Dr. William Wingo for the use of his Coulter Counter. This work was supported by National Institute of Health General Research Support No. 5-501-RR-05300-10 and by Public Health Service Research Grant No. CA-13148 from the National Cancer Institute.

Note added in proof: The cell culture specimens HF-6 and HF-7 used in these experiments were obtained from another laboratory (Dr. Charles Alford of our university). We were under the impression that the cells had been derived from a single individual. Long after this manuscript was prepared, we found that the cultures were actually derived from multiple individuals, HF-7 from seven individuals. Furthermore, cells derived from a single individual required a double PBS wash to show the stimulation, and by preparing cultures from various numbers of individuals, we found the ease of stimulation generally increased with the number of individuals represented. The fact that cells derived from a single individual require a double wash for stimulation probably accounts for this phenomenon not being previously observed by other workers.

Furthermore, pH has been ruled out as being responsible for the PBS stimulation, since PBS at pH 6.8 stimulates the same as PBS at pH 7.2.

We have also shown, using SDS gel electrophoresis, that the PBS extract contains predominantly a protein with the same mobility as the so-called LETS protein and another of about 50,000 daltons, which is approximately equal to the molecular weight of the fibroblast chalone (Houck *et al.*, 1973a).

REFERENCES

Anderson, W. B., Russell, T. R., Carchman, R. A. and Pastan, I. (1973). *Proc. Nat. Acad. Sci. U.S.A. 70,* 3802-3805.

Burger, M. M. (1970). *Nature 227,* 170-171.

Burger, M. M., Bombik, B. M., Breckenridge, B. M. and Sheppard, J. R. (1972). *Nature New Biol. 239,* 161-163.

Ceccarini, C. and Eagle, H. (1971). *Proc. Nat. Acad. Sci. U.S.A. 68,* 229-233.

Chou, I., Black, P. H. and Roblin, R. O. (1974a). *Nature 250,* 739-741.

Chou, I., Black, P. H. and Roblin, R. O. (1974b). *In* "Control of Proliferation in Animal Cells." (B. Clarkson and R. Baserga, eds.) Vol. I, pp. 339-350. Cold Spring Harbor Laboratory.

Glynn, R. D., Thrash, C. R. and Cunningham, D. D. (1973). *Proc. Nat. Acad. Sci. U.S.A. 70,* 2676-2677.

Greene, L. A., Tomita, J. T. and Varon, S. (1971). *Exp. Cell Res. 64,* 387-395.

Houck, J. C., Weil, R. L. and Sharma, V. K. (1972). *Nature New Biol. 240,* 210-211.

Houck, J. C., Cheng, R. F. and Sharma, V. K. (1973a). *Nat. Cancer Inst. Mono. 38,* 161-170.

Houck, J. C., Sharma, V. K. and Cheng, R. F. (1973b). *Nature New Biol. 246,* 111-113.

Jimenez de Asua, L., Rosengurt, E. and Dulbecco, R. (1974). *Proc. Nat. Acad. Sci. U.S.A. 71,* 96-98.

Jones, R. O. and Ashwood-Smith, M. J. (1970). *Exp. Cell Res. 59,* 161-163.

Mott, D. M., Fabisch, P. H., Sani, B. P. and Sorof, S. (1974). *Biochem. Biophys. Res. Commun. 61,* 571-577.

Noonan, K. D. and Burger, M. M. (1973). *Exp. Cell Res. 80,* 405-414.

Reich, E. (1973). *Fed. Proc. 32,* 2174-2175.

Reich, E. (1974). *In* "Control of Proliferation in Animal Cells." (B. Clarkson and R. Baserga, eds.) Vol. I, pp. 351-355. Cold Spring Harbor Laboratory.

Rouslahti, E. and Vaheri, A. (1974). *Nature 248,* 790-791.

Rubin, H. (1970). *Science 167,* 1271-1272.

Rudland, P. S., Gospodarowicz, D. and Seifert, W. (1974). *Nature 250,* 741-742.

Sefton, B. M. and Rubin, H. (1970). *Nature 227,* 843-845.

Temin, H. M. (1967). *J. Cell. Physiol. 69,* 377-384.

Temin, H. M. (1971). *J. Cell. Physiol. 78,* 161-170.

Vaheri, A. E., Rouslahti, E. and Nordling, S. (1972). *Nature New Biol. 238,* 211-212.

Vasiliev, Ju. M., Gelfand, I. M., Guelstein, V. P. and Fetisova, E. K. (1970). *J. Cell. Physiol. 75,* 302-314.

Yamada, K. M. and Weston, J. A. (1974). *Proc. Nat. Acad. Sci. U.S.A. 71,* 3492-3496.

Workshop III

PROBLEMS OF LARGE-SCALE CELL CULTURE AND QUALITY CONTROL

TISSUE CULTURE REAGENT QUALITY ASSURANCE TESTING IN A LARGE-SCALE VIRUS PRODUCTION LABORATORY

CHARLES V. BENTON
ROGER W. JOHNSON
ALBERT PERRY
W. I. JONES
GEORGE P. SHIBLEY

National Cancer Institute
Frederick Cancer Research Center
Frederick, Maryland

I. INTRODUCTION

In the Viral Resources Laboratory at the Frederick Cancer Research Center, Frederick, Md. (FCRC), several oncogenic murine viruses are produced in large-scale tissue culture. These agents include (1) Rauscher leukemia virus, produced in JLS-V9 mouse bone marrow cells, (2) Gross leukemia virus, progapated in NIH-3T3 mouse fibroblasts, and (3) mouse mammary tumor virus, grown in $Mm5mt/c_1$ mouse mammary carcinoma cells. Approximately 600 liters per week of tissue culture fluids containing 10^7 - 10^{10} virus particles/ml, depending on the system, are produced. This effort involves, on a weekly basis, the use of (1) 750 l tissue culture media, (2) 75 l fetal bovine sera, and (3) 2 to 15 l each of penicillin-streptomycin solutions, L-glutamine, phosphate buffered saline and trypsin solutions.

The frequent turnover of such large volumes of tissue culture reagents necessitates an efficient and relatively rapid method of reagent quality assurance testing. Ideally, any system for achieving quality assurance will strike a balance of time, effort, cost, and value of the results obtained. Little is gained, for example, by enriching a bacteriological medium with expensive or unusual nutrients to detect a contaminating microbe that in most situations will not proliferate during routine tissue culture operations.

The present disscussion has been prepared primarily to summarize the results obtained at the FCRC with our current methods of reagent quality assurance. It should be kept in mind, however, that the methods employed are constantly being improved as more data is accumulated and analyzed.

II. SERUM

The cost and the nature of quality control in the processing of serum makes thorough quality assurance testing of each serum lot a necessity prior to its utilization in tissue culture.

A. Mycoplasma

With the use of routine antibiotic incorporation, which we find necessary in a large-scale system, mycoplasma screening receives the major emphasis of our serum quality assurance program. Data presented at a workshop on serum at the W. Alton Jones Cell Science Center (Fedoroff *et al.*, 1971) indicated that serum contaminated with bovine mycoplasma is a major source of mycoplasma contamination of tissue culture. The most common species found in bovine serum are *Mycoplasma arginini* and *Acholeplasma laidlawii* (Barile, 1975). Barile (1973) and others (Rodwell, 1969;

Stanbridge *et al.*, 1969; Markov *et al.*, 1969) have demonstrated that mycoplasma may alter the karyology and metabolism of cultured cells as well as affect the production of virus in certain cell systems.

Our testing consists of the inoculation of a total of 25 ml of serum into several types of liquid and solid media: (1) BBL medium (Vera, 1974), (2) Medium No. 243 (Hayflick, 1965), (3) M96 medium (Frey, 1973), and (4) U.S. Department of Agriculture medium. We have found that excluding any one of these media could cause the occasional lack of detection of certain species of strains at low concentrations. Samples are incubated aerobically and anaerobically at 37°C with both positive and negative controls. Broth culture material is used as a source of material for retesting and identification of corresponding positive agar plates. Observations are made at 7, 14, and 21 days post-inoculation. Results indicate a relatively low incidence of mycoplasma in the 38 commercial serum lots that we have screened (Table I).

Currently, biochemical and physical methods of mycoplasma detection are being evaluated to determine the relative sensitivities as compared to the biological culture approach. Schneider *et al.* (1973) have reported increased sensitivity using the detection of 16S and 23S mycoplasma ribosomal RNA (Markov *et al.*, 1969; Harley *et al.*, 1970; Levine *et al.*, 1968) isolated by polyacrylamide gel electrophoresis (Grossbach and Weinstein, 1968). Todaro *et al.* (1971) demonstrated that these entities could be physically separated based on their characteristic density of 1.22-1.24 g/cm^3. Mycoplasma species are also noted for their characteristic cytoplasmic labeling in autoradiographs and for their ability to deplete arginine from tissue culture growth medium (Schneider *et al.*, 1973; Barile, 1973).

TABLE I Quality Assurance Testing of Commercial Serum Lots

A. Contamination

Agent tested for	No. lots tested	No. positive	Percent positive
Mycoplasma	38	3	7.9
Bacteriophage	45	12	26.7
Bacteria	38	0	0
Fungi	38	2	5.3
Endotoxins	3	1	33.5
EM[a]	14	5[b]	35.7

B. Growth Support

	Cell lines tested		
	JLS-V9	NIH-3T3	$Mm5mt/c_1$
No. lots accepted	24	3	3
No. lots tested	27	3	3
Percent lots accepted	88.9	100	100

[a]Electron microscopic examination for all agents.
[b]Positives were: 1 bacteriophage, 1 C-type virus-like, 1 mycoplasma and 2 lots bacteria.

B. Bacteriophage

The second group of biological contaminants for which serum is routinely tested before purchasing is bacteriophage. Bacteriophage testing is relatively simple and indicates the existence or former presence of other contaminants, i.e., the homologous host bacterium or bacterial endotoxins. Also, bacteriophage Lambda has been demonstrated to alter the growth of human fibroblasts in tissue culture (Merril *et al.*, 1971).

Briefly, the test employs 50–100 ml of serum concentrated 50-fold by pelleting at 100,000 × g for 4 hr. This concentrate is mixed with 2 ml agar at 45°C and a suspension of *Escherchia coli* C, employed because of its wide range of bacteriophage susceptibility, is added and mixed. This inoculum is then added to a warm tryptone agar plate and the culture is incubated 24 hr at 37°C with positive and negative controls. Results (Table I) indicate that a high perce ntage of serum lots are contaminated with bacteriophage. Generally, the level of contamination is low, in the range of 1 plaque-forming unit (PFU) per ml; however, samples with as high as 2,000 PFU/ml have been encountered.

C. Animal Viruses

The commercial serum sources that we utilize routinely screen their product for several animal viruses. The viruses most often encountered are (1) bovine enteroviruses, (2) bovine viral diarrhea virus, (3) bovine herpes viruses, (4) parainfluenza type 3, and (5) adenoviruses (Fedoroff *et al.*, 1972; Molander *et al.*, 1972; Boone *et al.*, 1972). At the present time, no testing of serum for possible viral contamination is performed at FCRC. However, as the current picture continues to develop concerning the possible origin and/or tropism of some oncogenic viruses, increasing concern has been generated over the possibility of the introduction of oncogenic viruses, viral components, or even antisera to these agents into our tissue culture systems. Since the logistics and, in some cases, the technology of screening for such agents is overwhelming, the problem, if it is real, might best be circumvented by the use of serum free or chemically defined media. This approach obviously would also reduce sources for other types of contamination, not to mention the probable economic advantage.

D. Bacteria and Fungi

Our bacteriological and fungal screening involves the routine use of 90 ml of the following media inoculated with 10 ml of serum: (1) yeast-mold broth, acidified to reduce bacterial growth, (2) Sabouraud's dextrose broth, primarily for culturing fungi, (3) tryptose phosphate broth for streptococci and pneumococci, (4) tryptic soy broth for pathogenic and saphrophytic microbes, and (5) brain heart infusion broth for staphylococci and other fastidious organisms. In addition, blood agar plates are inoculated. All media are incubated both at 26° and 37°C and observed on days 7, 14, and 21 post-inoculation. Such a battery of media is required to insure detection of the bulk of the contaminants routinely encountered (Boone *et al.*, 1972). Our screening has revealed a minimal amount of bacterial and fungal contamination (Table I).

E. Endotoxins

Recently, testing of serum lots for endotoxins has been initiated by an independent source utilizing the method of Levine and Bang (1964), which relies on the clotting of *Limulus* amebocyte lysate in the presence of endotoxin. Although the metabolic effect, if any, of endotoxins on the growth and virus production of tissue culture cells is unknown, the presence of these lipopolysaccharide components of the outer bacterial cell wall does nevertheless signal the former or concurrent presence of gram-negative bacteria. For this reason lots of serum that have detectable, picagram amounts of endotoxin are rejected. To date, endotoxin has been detected in one of three serum lots tested (Table I).

F. Electron Microscopy

Our final screening of serum for contamination, a procedure also recently established, involves the electron microscopic examination of serum concentrated 100-fold by ultracentrifugation.

The observations are used as collaborative data to substantiate the results of biological testing. That is, unless the micrographs are overwhelmingly conclusive on their own merit, a serum lot will not be rejected unless another test concurs with the electron microscopic finding. Interestingly, bacterial entities were detected in two instances with the electron microscope that were not detected biologically - perhaps due to lack of viability (Table I).

G. Growth Support

Before final acceptance, a serum lot must be tested for its growth-supporting abilities. Basically, the test involves seeding in duplicate 10^4 cells/cm^2 into two sets of 75 cm^2 flasks - one with control serum previously demonstrated to be adequate for growth support and one with test serum. All other medium constituents are the same in each group. Cells are fed every 72 hr and passed when confluent, usually after 5-7 days; this process is repeated for three passages. Observations are made daily and include (1) relative plating efficiency, (2) degree of spindling, (3) refractility and granularity, (4) membrane integrity, (5) degree of floating cells or cellular debris, (6) percent confluency, and (7) cell counts at each passage. At the last passage the tissue culture fluid is assayed for virus particles and viral polymerases. Table I demonstrates that most lots were acceptable by these criteria.

III. MEDIUM

The tissue culture reagent that receives the next largest amount of quality assurance testing is the medium itself. At the FCRC this is a two-step process due to the fact that we prepare our media from commercial powders. The first step involves the initial testing of a representative reconstituted sample of the powder lot, which if purchased will usually be large enough to make 2,000-5,000 l of medium, primarily for cellular growth support and virus production and, secondarily, for contamination. The basic procedures for testing media are similar to those outlined for serum. Few commercial powders fail to pass these quality assurance tests (Table III).

A commercial powder is reconstituted to make multiple 50-200 l lots. During membrane filtration, 500 ml samples for sterility testing are taken at the beginning, in the middle and at the end of the aliquoting of each lot. The bulk of the lot is held at room temperature for 3 weeks on a quarantine status. A small portion of each lot is incubated at 37°C until that particular lot is either rejected or utilized in the production unit. At the end of the first week of the quarantine period, the 500 ml samples are tested for sterility by inoculation into broths and agar plates and for ability to support cellular growth as previously described. Bacterial and fungal contamination has represented the only major problem in our processing of liquid medium lots (Table II). Once these lots pass all quality assurance tests they are incubated for 72 hr at 37°C just prior to use to allow for detection of any overt contamination heretofore missed.

IV. MISCELLANEOUS REAGENTS

The other tissue culture reagents utilized in our production laboratories are (1) L-glutamine, (2) penicillin-streptomycin solutions, (3) trypsin solutions, and (4) phosphate buffered saline. Bacterial and fungal screening is performed on these reagents. Even though mycoplasma have not been isolated from trypsin solutions, the source of the raw material for this enzyme makes it suspect for possible mycoplasma contamination (Barile, 1975); we therefore routinely screen these solutions for mycoplasma (Table III).

TABLE II Quality Assurance Testing of Tissue Culture Media

A. Commercial Powders[a]

No. lots tested	Cell growth acceptable lots/lots tested			Virus production acceptable lots/lots tested			No. lots contaminated
	JLS-V9	NIH-3T3	$Mm5mt/c_1$	RLV	GLV	MMTV	
15	10/10	4/4	1/1	10/10	4/4	1/1	2[b]

B. Liquid Lots[c]

No. lots tested	Growth support[d] acceptable lots/lots tested			Contamination testing No. lots positive		
	JLS-V9	NIH-3T3	Mm5mt/cl	Mycoplasma[e]	Bacteria/fungi	Encotoxin[f]
88	55/58	14/15	15/15	3	7	1

[a]Reconstituted prior to testing. One lot = 2000-5000 liter equivalents. [b]One lot mycoplasma positive. [c]One lot = 50-200 liters. [d]Cellular growth support and virus production. [e]Tested randomly (not every lot tested). [f]Six lots tested.

TABLE III Quality Assurance Testing of Tissue Culture Reagents

Reagent	No. lots tested[a]	No. positive	Percent positive
L-glutamine	19	1	5.3
P/S[b]	22	1	4.5
Trypsin	10	0	0
PBS[c]	13	2	15.4

[a]Tested for bacteria and fungi. Trypsin also tested for mycoplasma.
[b]Penicillin-streptomycin solutions.
[c]Phosphate buffered saline.

V. DISCUSSION

The final criterion on which any quality assurance program is based is the integrity of the product produced. At present, 7-10% of all virus-containing tissue culture fluids that we produce are rejected due to contamination or, infrequently, poor quality. The determination of the origin of these contaminants has, in virtually every instance, demonstrated that the reagents used were not the source. Moreover, it is estimated that if we were to eliminate our reagent quality assurance program, the rejection level of the viral fluids would approach 50% - an estimate that clearly justifies the continued monitoring of our reagents.

ACKNOWLEDGMENTS

Appreciation is extended to Drs. J. Gruber and H. J. Hearn, Jr. of the National Cancer Institute for their support of this program. The technical assistance of Mrs. Ruth Herring and the FCRC Virus Production staff is gratefully acknowledged.

Research was sponsored by the National Cancer Institute under Public Health Service Contract NO1-CO-25423 MOD 27 with Litton Bionetics, Inc.

REFERENCES

Barile, M. F. (1973). *In* "Contamination of Tissue Cultures" (J. Fogh, ed.) p. 136. Academic Press, New York.

Barile, M. F. (1975). These proceedings.

Boone, C. W., Mantel, N., Caruso, T. D., Kazam, E., and Stevenson, R. E. (1972). *In Vitro 7,* 174-189.

Fedoroff, S., Evans, V. J., Hopps, H. E., Sanford, K. K., and Boone, C. W. (1972). *In Vitro 7,* 161-167.

Frey, M. L. (1973). *In* "Standardization Methods in Veterinary Microbiology" Nat. Acad. Sci., Washington, D. C..

Grossbach, U., and Weinstein, I. B. (1968). *Anal. Biochem. 22,* 311-320.

Harley, E. H., Rees, K. R., and Cohen, A. (1970). *Biochim. Biophys. Acta 213,* 171-182.

Hayflick, L. (1965). *Texas Rpts. Biol. Med. 23,* 285-303.

Levin, J., and Bang, F. B. (1964). *Bull. Johns Hopkins Hosp. 115,* 265-274.

Levine, E. M., Thomas, L., McGregor, D., Hayflick, L., and Eagle, H. (1968). *Proc. Nat. Acad. Sci. U.S.A. 60,* 583-586.

Markov, G. G., Bradvarova, I., Mintcheva, A., Petrov, P., Shishkov, N., and Tsaneu, R. G. (1969). *Exptl. Cell Res. 57,* 374-384.

Merril, C. P., Geier, M. R., and Petricciani, J. C. (1971). *Nature (London) 233,* 398-400.

Molander, C. W., Kniazeff, A. J., Boone, C. W., Paley, A., and Imagawa, D. T. (1972). *In Vitro 7,* 168-173.

Rodwell, A. (1969). *In* "The Mycoplasmatales and the L Phase of Bacteria" (L. Hayflick, ed.) p. 413. Appleton-Century-Crofts,

New York.

Schneider, E. L., Epstein, C. J., Epstein, W. L., Betlach, M., and Halbasch, G. A. (1973). *Exptl. Cell Res. 79*, 343-349.

Stanbridge, E., Onen, M., Perkins, F. T., and Hayflick, L. (1969). *Exptl. Cell Res. 57*, 397-410.

Todaro, G. J., Aaronson, S. A., and Rands, E. (1971). *Exptl. Cell Res. 65*, 256-257.

Vera, H. D. (1974). Personal communication.

CONTAMINATION MONITORING AND CONTROL IN A LARGE SCALE TISSUE CULTURE VIRUS PRODUCTION LABORATORY

ROGER W. JOHNSON
CHARLES V. BENTON
ALBERT PERRY
JOHN HATGI
GEORGE P. SHIBLEY

National Cancer Institute
Frederick Cancer Research Center
Frederick, Maryland

I. INTRODUCTION

Those involved in the use of tissue culture on a large scale basis are well aware of the consequences of contamination. The growth of bacteria in several hundred roller bottle cultures represents not only an economic loss, but a loss of time and effort. The failure of cells to grow in culture because of endotoxin contamination of the medium or detergent not rinsed from glassware results in similar losses. It is the task of our laboratory to provide virus for investigators within the National Cancer Institute's Virus Cancer Program for biochemical, immunological and genetic studies. Therefore, it is essential that a steady supply of virus, free of contaminants be produced and be made available to these investigators at all times. To inadvertantly distribute a contaminated product results in the loss of respect and reputation within the scientific community, and leads to wasted research efforts by the investigators receiving the product.

II. TYPES OF CONTAMINATION ENCOUNTERED IN A LARGE SCALE VIRUS PRODUCTION LABORATORY

Contaminants which one may encounter in a tissue culture virus production operation are shown in Table I.

A. Microbial

The type of contamination encountered most frequently in the tissue culture laboratory is microbial.

1. Bacteria, Fungi, and Yeasts

Bacteria are the most common offenders in this group, followed by fungi and yeast. While antibiotics may completely supress some contaminants, they will only mask others and still other organisms will proliferate at a remarkable rate in the tissue culture medium. In our experience gram negative rods, generally of the genus Flavobacter, are encountered most frequently. Tumilowicz (1974) recently described the contamination of tissue cultures with leptospira; the source was possibly a deionized water system used to prepare the tissue culture media.

2. Mycoplasma

Several years ago mycoplasma were described as the crab-grass of tissue culture. Ten years later the description still holds. Although there are numerous descriptions in the literature of methods for freeing cell cultures of mycoplasma contamination (Barile, 1975), it is evident that these organisms are merely being supressed rather than eliminated.

TABLE I Types of Contamination Encountered in the Large-Scale Virus Production Laboratory

A. Microbial
   1. Bacteria
   2. Fungi and yeasts
   3. Mycoplasma
   4. Viruses
      a. Animal
      b. Bacterial
   5. Endotoxins
B. Foreign cells
C. Inert matter
   1. Chemicals
   2. Viral antigens

Although mycoplasma contamination of cultures is difficult to detect (Hayflick, 1965, Barile, 1975), it is easier to prevent then other microbial contaminants, since the sources of potential mycoplasma contamination are limited. The serum used in tissue culture media, or cell stocks obtained from other laboratories represent the primary sources of contamination. Another common source is mouth pipetting, still practiced in some laboratories. This practice results in the introduction of mycoplasma into culture from the oropharynx of the technician (Crawford and Kraybill, 1967, Barile, 1975).

3. Viruses

Adventitious viruses represent one of the least recognized forms of tissue culture contamination. Bovine viral diarrhea virus, bovine herpes virus, and para-influenza type-3 virus (Molander *et al.*, 1972) all may be isolated from serum. If murine tissues are used, contamination can occur with a variety of adventitious murine viruses (Collins and Parker, 1972). When several viruses are under investigation in a laboratory in which common reagents, equipment and technicians are used, contamination of cell lines with virus may and often does occur. Merrill *et al.*

(1972) and others (Petriccian *et al.*, 1973; Cho *et al.*, 1973 and Moody *et al.*, 1975) have demonstrated that serum may be contaminated with bacteriophages. Although the bacteriophage will not replicate in tissue culture, their presence none the less, constitutes a foreign antigen in concentrated culture fluids.

4. *Endotoxins*

Endotoxin contamination of serum (Moody *et al.*, 1975) or medium, detectected by the *Limulus* amebocyte lysate procedure (Wildfeuer *et al.*, 1974; Jorgenson and Smith, 1973) occurs when gram negative bacteria were allowed to proliferate prior to filtration of these reagents.

5. *Viral Antigens*

Autoclaved or chemically decontaminated equipment which contained virus material must be adequately washed and rinsed to remove all traces of coagulated protein and virus. Residual virus, although inactivated, represents a potential source of foreign antigen. Certain pieces of equipment used in the harvesting and centrifugation process, e.g., rubber and tygon tubing, continuous flow rotors, present special problems of cleaning.

B. Foreign Cells

Not every cell line one receives from another laboratory is necessarily what it is labelled to be, or entirely of one species of origin (Nelson-Rees *et al.*, 1974). Only by karyotyping and fluorescent staining of cells (Stulberg and Simpson, 1973) is it possible to determine correctly the species of origin of a cell line.

C. Chemicals

Improperly washed and rinsed glassware or equipment may be contaminated with detergents or disinfectants used in the glassware preparation area or elsewhere in the laboratory. This results in failure of cells to grow well, or failure to grow over portions of the glass surface of the culture bottle. Phenol and aldehyde groups are often involved, as are the detergents used in washing the glassware.

III. SOURCES OF CONTAMINATION

Potential sources of contamination are summarized in Table II.

A. Cell Seed Stocks

Cell stocks received from other laboratories or even developed within one's own laboratory are potential sources of contamination with mycoplasma (Giradi, *et al.*, 1965, Macpherson, 1966), viruses or foreign cells (Nelson-Rees, 1974). Mycoplasma may be spread by aerosols from infected to uninfected cultures (O'Connel *et al.*, 1964). In a similar manner, cells from one line may be introduced into cultures of another line as a result of improperly labelled flasks or by carry over through the use of common bottles of reagent. We encountered a first-hand example of this problem recently when a putative human breast tumor cell line was sent to us for study. Upon investigation, by karyotyping and fluorescent staining testing, this line was found to be a mixture of 95% rat cells and 5% HeLa cells.

B. Tissue Culture Reagents

Tissue culture reagents are an obvious source of microbial contamination, albeit a controllable one. Bacterial or fungal contamination of medium usually can be detected with no testing other than observation. However, only by testing can endotoxin contamination be detected.

TABLE II Sources of Contaminants

A.	Cell seed stocks
B.	Tissue culture reagents
C.	Failure to maintain aseptic technique
D.	Environmental overload
E.	Inadequate washing, rinsing and sterilization of equipment

Low (1974) reported the isolation of mycoplasma from a commercial lot of serum-free medium. This is an isolated and probably unusual instance of such contamination. Low suggested possible aerosol exposure of the medium to contaminated bovine serum during manufacture.

Serum is a potential source of more contaminants than any other reagent (Fedoroff *et al.*, 1972; Molender *et al.*, 1972; Boone *et al.*, 1972; and Merrill *et al.*, 1972). Serum can be contaminated with bacteria or fungi, as well as with mycoplasma, animal and bacterial viruses and endotoxins.

Trypsin, although a likely source of mycoplasma contamination has apparently never been proven to be an actual source (Barile, 1973).

C. Failure to Maintain Aseptic Technique

Failure to exercise proper aseptic technique while carrying out tissue culture procedures is a common source of contamination. Proper aseptic technique is developed only through training and repeated stressing of the importance of care being exercised during culture manipulations and the importance of each individual's participation in the program.

D. Environmental Overload

Under certain conditions, the laboratory environment may become overloaded with microorganisms. Dr. Lloyd Herman, Environmental Serivices, NIH, Bethesda, Md. (Personal Communication), has described this situation to have occurred when floor drains are left uncovered or when large amounts of moisture are allowed to collect in the tissue culture areas. Environmental overload also occurs when technicians fail to exercise personal hygiene (Voxakis *et al.*, 1974), or when the integrity of laminar flow hoods is repeatedly broken with surface contaminated items.

Breakdown in aseptic technique, e.g., allowing gloved hands to come in contact with the lips of culture vessels and environmental over-load, account for most of the microbial contamination we experience in our laboratory.

E. Inadequate Washing, Rinsing and Sterilization of Equipment

Inadequately washed, rinsed or sterilized glassware and other equipment are obvious sources of inert or microbial contamination. Unfortunately, this contamination is recognized only after the fact. Strict adherance to glassware processing protocols is the only way such contamination can be prevented.

IV. SCALE-UP AND PRODUCTION

In our laboratory each production lot of virus is initiated from an ampoule of tested and proven working cell seed. These cells, which are chronically infected with the desired RNA C- or B-type tumor viruses, are scaled up to production level within 4 to 5 transfers (Fig. 1). At each one of these transfers, and at each intermediate feeding of the cultures with fresh medium, samles of culture fluid are taken for sterility tests. These sterility checks are examined before the next step in the scale-up sequence is carried out.

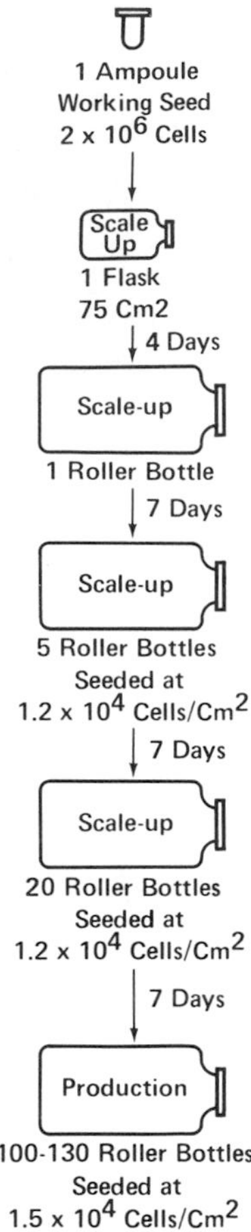

FIGURE 1 *Typical scale-up procedure for large-scale virus production.*

V. PRODUCTION PROCEDURE

Each week the Virus Production Laboratory seeds from 300 to 800 production roller bottle cultures. Currently, we are producing 300 l/week of oncogenic RNA virus. This material is obtained from chronically infected cells by multiple harvests of the cultures at 24- to 72-hour intervals. A single production lot of cultures may be at production level for as long as 10 to 14 days, depending on the cell line, and as many as seven harvests may be obtained. For example, in the production of Gross leukemia virus, six harvests are obtained from 400 production-level cultures. Therefore, 2800 manipulations are made during the production sequence, from the seeding of the production level cultures through the sixth harvest. At the end of the production period, cells from working seed stock are phased into a new production lot.

All culture procedures are carried out in laminar flow biological safety cabinets. Virus containing fluids are harvested from the cultures by vacuum. The fluids are pumped through stainless steel coils imbedded in ice, into plastic 20 liter carboys, then through a continuous flow centrifuge rotor to remove cellular debris. The fluids are stored at 4C until the virus can be concentrated and purified by continuous flow zonal centrifugation in a sucrose gradient in a K-II ultracentrifuge.

VI MASTER AND WORKING CELL SEED STOCK DEVELOPMENT AND TESTING

The first step in our contamination control system is the preparation and testing of master and working seed stocks (Table III).

A. Quarantine and Testing

Upon receipt, starter cultures or cells are placed in quarantine. All culture manipulations are carried out in a laminar flow biological safety cabinet in a laboratory isolated both physically and by air balance from the remainder of the production area. Portions of the cells and culture supernatants from the starter cultures are submitted for mycoplasma testing, karyotyping and fluorescent antibody staining. In the case of cells which are murine in origin, a culture is submitted for mouse antibody

TABLE III Cell Seed Stock Development and Testing

A. Quarantine and testing
   1. Mycoplasma
   2. Karyotyping
   3. Cytotoxicity
   4. Mouse antibody production testing
B. Cell preparation for freezing
   1. Antibiotic free culture
   2. Tested reagents
   3. Contamination checks at each culture feeding and transfer
   4. Testing of approximately 10% of cell pool
C. Testing of frozen seed stock
   1. Antibiotic-free culture
   2. Mycoplasma
   3. Standard contamination testing
   4. Product testing
      a. Viral antigens
      b. Enzyme activity
      c. In vitro or in vivo activity

production testing (Collins and Parker, 1972) to assure freedom from a number of murine virus contaminants.

When the described testing indicates that the cultures are free of mycoplasma and are of the appropriate karyotype, the cells are amplified in antibiotic-free, pretested medium to yield enough cells to make a frozen master seed stock. The fluids removed from these cultures, when they are transferred or when intermediately fed fresh medium are filtered thru 0.22 μ membrane filters, and the filters incubated on blood agar at 37 C. Portions of the cells at transfer are tested on blood agar and in broths at both 37C and room temperature. When the final pool of cells is prepared in the cryoprotective medium for ampouling, 10% is set aside for conventional bacteria sterility testing. Duplicate broth cultures of brain heart infusion, tryptose phosphate, trypticase soy, yeast and mold, and sabouraud dextrose, along with blood agar plates are inoculated and incubated at 37C and room temperature. Test cultures are held for 21 days before the sterility testing is terminated.

Twenty-four to 48 hours after the cell seed stock is frozen, three ampoules are recovered in antibiotic-free medium. Before confluency, the cells are scraped into the medium and the material submitted for mycoplasma and bacterial contamination testing and karyotyping.

An ampoule of these cells is then expanded to prepare a working seed stock in the same manner and with the same testing methods used to prepare the master seed stock. When this testing is completed, an ampoule of the working seed stock is scaled up to production level and the purified and concentrated virus product is tested by immunological and biochemical methods to assure freedom from contamination with other C-type viruses.

VII. TISSUE CULTURE REAGENT TESTING

A. Medium

Medium (Benton *et al.*, 1975) is generally produced in 200 l lots without glutamine. Each lot of medium (Table IV) is held in quarantine at room temperature for 3 weeks prior to transfer to storage at 4C. During the course of filtration, 500-ml samples are taken for additional sterility testing. These random samples are held at room temperature for 7 days and then submitted for sterility testing at 37C and room temperature using five broth media and blood agar plates. These sterility tests are held for 21 days. Medium is usually released for use 28 days after its preparation.

Medium is not routinely submitted for endotoxin testing. However, it has been our experience that medium found to contain endotoxin fails to support cell growth.

B. Serum

Prior to purchase of a serum lot, two 500-ml samples are obtained from the manufacturer. These samples are tested (Table IV)

TABLE IV Tissue Culture Reagent Testing

A. Medium
   1. Three week quarantine
   2. Large volume testing
   3. Endotoxin
B. Serum
   1. Mycoplasma
   2. Bacteria/fungi
   3. Phage
   4. Endotoxin
C. Trypsin
   1. Mycoplasma
   2. Bacteria/fungi
D. Glutamine, penicillin/streptomycin

for mycoplasma, bacteria, fungi, and bacteriophage. Several samples have recently been tested for endotoxin. Lots of serum have been rejected due to contamination with one or more of each contaminant mentioned with the exception of bacterial contamination (Benton *et al.*, 1975).

C. Trypsin

Trypsin is tested for mycoplasma and bacterial and fungal contamination when it is dispensed from the manufacturer's bottle into smaller aliquots and refrozen (Table IV).

D. Glutamine and Penicillin/Streptomycin

Glutamine and penicillin/streptomycin solutions are frozen as soon as they are filtered and random samples are submitted for bacteriological testing (Table IV). Glutamine is tested by inoculation of duplicate sets of broths and blood agar plates. Penicillin/streptomycin solution is filtered and the filters are placed in bottles of broths and on blood agar plates.

VIII. Contamination Monitoring During Scale-up, Production, and Final Product Testing

A. Scale-Up

Tissue culture media (Table V) used for scale up are incubated 72 hr at 37C prior to addition of serum, glutamine, and penicillin/streptomycin. Complete media are then held at room temperature until used. As previously described, culture supernatants, at the time of transfer or at intermediate feedings are pooled and 100-ml amounts filtered. The filters are put on blood agar plates and incubated. Cell pools, obtained when transfers are made, are plated in blood agar and inoculated into trypcase soy broth. Portions of the complete cell culture medium are also incubated at 37C.

As the scale up cultures increase in number, the lot is divided into groups of cultures and sterility checks are made on these groups rather than individual cultures.

Cells used in the final scale-up step are submitted for mycoplasma testing. This allows for retrospective testing on the production lot.

B. Production Cultures

Prior to harvesting and refeeding of production level cultures, each bottle is examined before it is processed. Before the harvested culture fluids are clarified of cellular debris, a 500-ml sample of the pool of unclarified fluid is obtained for contamination testing. After centrifugation, 500-ml samples are obtained from each of several carboys containing the clarified material. These samples are filtered and the filters incubated on blood agar plates. Before the clarification step, the centrifuge hose lines and rotor are flushed with 20 l of sterile water. Five hundred-ml amounts of the water are obtained both before and after flushing and are filtered and incubated to determine that the water, rotor and all the hose lines were sterile.

TABLE V Contamination Monitoring During Scale-up, Production, and on the Final Product

A. Scale-up cultures
   1. Reagents
   2. Culture supernatants
   3. Seed pools
B. Production level cultures
   1. Reagents
   2. Harvested culture fluids
C. Products
   1. Microbial testing
   2. Antigens

The medium, in 15-l carboys, used to feed the production cultures is incubated at 37C for 3 days before it is used. The incubation serves the dual purpose of enhancing growth of microbial agents to a visible level and warming the medium to 37C prior to use. Samples of the medium and the medium plus the individual constituents, i.e., glutamine, serum and penicillin/streptomycin are obtained in 500-ml amounts and incubated at 37C (Table V).

If the sterility results indicate freedom from microbial contamination, the harvested tissue culture fluids are then processed by continuous flow zonal centrifugation to concentrate and purify the virus. The final viral product is tested for freedom from bacterial and fungal contaminants, for antigen content, biochemical and immunological characterization, and *in vitro* and, in some cases, *in vivo* bioassays.

The production level represents the highest level of monetary investment in the production process. Each harvest averages around 50 l of material at a cost of several hundreds of dollars, exclusive of the cost of the glassware preparation.

The procedures outlined above include greater than 100 tests on each production lot. Although cumbersome, these procedures have precluded the use of contaminated reagents, prevented the further processing of contaminated material and provided us with clues as to the probable sources of contamination.

ACKNOWLEDGMENTS

We express our appreciation to Drs. J. Gruber and H. J. Hearn, Jr. of the National Cancer Institute, for their continued support of this program, and to Dr. L. Herman, Environmental Services, National Institutes of Health, for his help and suggestions in enviornmental contamination control and monitoring. We wish to acknowledge Dr. W. Nelson-Rees for the karyology of all our cells and Dr. C. Stulberg for the fluorescent antibody staining of them. We are indebted to Mr. O. R. Robinson, Jr. for his valuable suggestions on testing and to Mr. W. I. Jones and Mrs. R. Herring for reagent and product microbial contamination testing. We also wish to acknowledge the Frederick Cancer Research Center, Departments of Virus Production and Virus Purification staffs for their technical assistance.

Research was sponsored by the National Cancer Institute under Public Health Service Contract NO1-CO-25423 MOD 27 with Litton Bionetics, Inc.

REFERENCES

Barile, M. F. (1973). *In* "Tissue Culture Methods and Applications" (P. F. Kruse, Jr. and M. K. Patterson, Jr., eds.), pp. 729-735. Academic Press, New York.

Barile, M. F. (1975). In this volume.

Benton, C. V., Johnson, R. W., Perry, A., Jones, W. I. and Shibley, G. P. (1975). In this volume.

Boone, C. W., Mantel, H., Caruso, T. D., Jr., Kazam, E. and Stevenson, R. E. (1972). *In Vitro 7*, 174-189.

Chu, F. C., Johnson, J. B., Orr, H. C., Probst, P. G. and Petricciani, J. C. (1973). *In Vitro 9*, 31-34.

Collins, M. J., Jr. and Parker, J. C. (1972). *J. Nat. Cancer Inst. 49*, 1139-1143.

Crawford, Y. E. and Kraybill, W. H. (1967). *Ann. N.Y. Acad. Sci. 143*, 411-421.

Federoff, S., Evans, V. J., Hopps, H. E., Sanford, K. I. and Boone, C. W. (1972). *In Vitro 7*, 161-167.

Giradi, A. J., Hamparian, V. V., Somerson, N. L. and Hayflick, L. (1965). *Proc. Soc. Exptl. Biol. Med. 120*, 760-771.

Hayflick, L. (1965). Texas Reports. Biol. Med. *23*, 285-303.

Jorgensen, J. H. and Smith, R. F. (1973). *Appl. Microbiol. 26*, 43-48.

Low, J. E. (1974). *Appl. Microbiol. 27*, 1046-1052.

Macpherson, I. (1966). *J. Cell Sci. 1*, 145-168.

Merrill, C. R., Friedman, T. B., Attalluh, A. F. M., Greier, M. R., Krell, K. and Yarkin, R. (1972). *In Vitro 8*, 91-93.

Molander, C. W., Kniazeff, A. J., Boone, C. W., Paley, A. and Imagawa, D. T. (1972). *In Vitro 7*, 168-173.

Moody, E. E. M., Trousdale, M. D., Jargensen, J. H. and Shelokov, A. (1975). *J. Inf. Dis. 131*, 588-591.

Nelson-Rees, W. A., Flandermeyer, R. A. and Hawthorne, P. K. (1974). *Science 184*, 1093-1096.

O'Connell, R., Witten, R. G. and Faber, J. E. (1964). *Appl. Microbiol. 12*, 337-342.

Petricciani, J. C., Chu, F. C. and Johnson, J. B. (1973). *Proc. Soc. Exptl. Biol. Med. 144*, 789-792.

Stulberg, C. S., Coriell, L. L., Kniazeff, A. J. and Shannon, J. E. (1970). *In Vitro 5*, 1-16.

Tumilowicz, J. J., Alexander, A. D. and Stafford, K. (1974). *In Vitro 10*, 238-242.

Voxakis, A. C., Lomy, P. P. and Herman, L. G. (1974). *Formula Management 12*, 14-18.

Wildfeuer, H. B., Schleifer, K. H. and Haferkamp, O. (1974). *Appl. Microbiol. 28*, 867-871.

MASS CELL CULTURES IN A CONTROLLED ENVIRONMENT

MONA D. JENSEN

Instrumentation Laboratory, Inc.
Lexington, Massachusetts

I. INTRODUCTION

The importance of environmental control in tissue culture has become increasingly evident over the last few years. Ideally, culture methods should provide the environment cells experience in vivo; but there is increasing evidence that cells grown by popular methods create their own micro-environment in which such important factors as pO_2, pCO_2, and pH can be far from physiological values (McLimans, *et al.*, 1968; Rubin, 1971; Jensen, *et al.*, 1976). One documented example of the effect on cells of culturing under nonphysiological conditions is the fact that established cell lines respire at much lower rates than freshly isolated cells of the same type (McLimans, *et al.*, 1968). An obvious corollary of this effect is increased non-volatile acid production through glycolysis. Another general effect of classical culture methods on cells is contact inhibition in cultures of "normal" cells; living organisms do not grow in monolayers.

Several methods have been developed which increase environmental control in small scale cultures; but many of these methods have been extremely difficult to scale up for production of cells in quantity. I would like to describe a recent development in cell culture which allows large scale production of cells in layered culture and also provides controls on the cells' environment which are not possible with other methods. This instrument is the 410 Cell Culture System made by Instrumentation Laboratory, Inc.

II. THE IL 410 SYSTEM

A. Theory

The major inovation in this system is the use of a gas-permeable, nonporous plastic film (FEP - teflon)[1] as the growth surface. This plastic is permeable to gases, water vapor, and some small molecules such as methanol, but will not allow passage of salts or other tissue culture medium constituents. The O_2 and CO_2 permeabilities for this film compare favorably with those of other films which have been used in tissue culture (Table I). Due to its strength, FEP - teflon can be used as a thin film (2.5×10^{-3} cm thickness). Since the flow rate of gas across a diffusion barrier is proportional to P/t, where P is the gas permeability in cm^3 gas, cm barrier/sec, cm^2 barrier, atm and t is the barrier thickness in cm, flow across FEP will be faster than across the other diffusion barriers in Table I. For instance, P/t for oxygen is 2.6×10^{-4} across FEP and 1.4×10^{-5} across a typical culture medium thickness of 1 mm.

[1]Fluoroethylene - propylene copolymer. "TEFLON" is a registered trademark of DuPont.

TABLE I Permeabilities of Various Diffusion Barriers[a]

	O_2[b]	CO_2[b]
FEP - Teflon	6.48×10^{-7}	8.16×10^{-7}
Polyethylene	2.23×10^{-8}	6.75×10^{-8}
Polypropylene	7.3×10^{-9}	2.48×10^{-8}
Medium	1.4×10^{-6}	1.6×10^{-5}

[a]From Jensen, *et al.* (1976)
[b]Permeabilities in cm^3 gas, cm film/sec, cm^2 film, atm.

Gas used or generated on one side of the FEP film rapidly equilibrates with the atmosphere on the other side of the film. Therefore, cells attached to the film grow at essentially the atmospheric partial pressures of the incubator and do not create steep gas concentration gradients such as those which are formed by cells grown on a gas-impermeable surface under a layer of medium (McLimans, *et al.*, 1968). The steady state pericellular partial pressure of a gas can be calculated for confluent cultures growing on gas-permeable and impermeable supports (Jensen, *et al.*, 1976). Table II gives some sample results for the pO_2 of several cell-types grown on various surfaces. Depletion of pericellular oxygen is drastic on all but the FEP surface. Similarly, the build-up of pericellular CO_2 due to cell respiration, and the resultant drop in pericellular pH, is much greater on surfaces other than FEP.

Since the pO_2 and pCO_2 of cultures grown on FEP are essentially those of the incubator atmosphere, measurement and control of these gases in the atmosphere results in control of the pericellular values. Medium pH must be measured directly because it is a function of CO_2 and acid metabolic product concentration.

TABLE II Steady State Pericellular pO_2[a]

Cell type	O_2 Consumption (moles/cm^2, sec)	pO_2[b] (mm Hg) FEP-teflon[c]	Polyethylene[c]	Polypropylene[c]	Medium[d]
Liver (in vitro)	6.25×10^{-12}	132.4	85.0	0	83.8
Liver (in vivo)	2.03×10^{-11}	126.4	0	0	0
Alveolar macrophage	1.11×10^{-10}	88.2	0	0	0

[a]From Jensen, *et al*. (1976).
[b]Atmospheric pO_2 = 135.
[c]Oxygen diffusion through plastic film 2.5×10^{-3} cm thick.
[d]Oxygen diffusion through medium layer 1 mm thick.

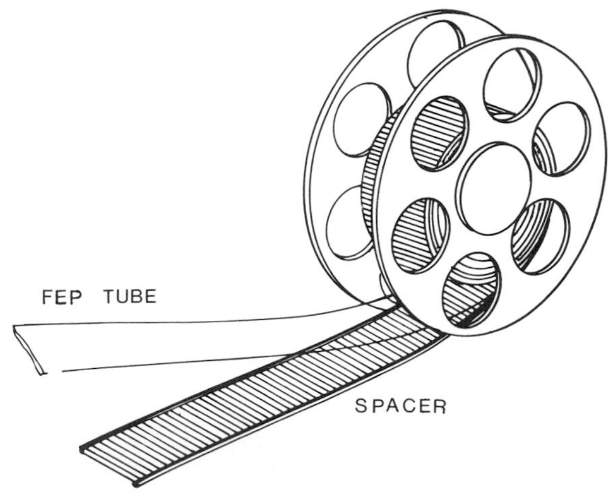

FIGURE 1 *The IL 410 culture reel.*

B. Instrumentation

In this system, cells grow inside containers made of the FEP - teflon film. This film is formed as a long tube which is then wound spirally with spacers to allow air contact to the membrane (Fig. 1). The resulting configuration is a culture reel which packs considerable growing surface area into a relatively small volume. Figure 2 shows a cross-section of the culture tube and its dimensions. A reel with 8 feet of tubing contains 2.6×10^3 cm^2 of growing surface. Production reels, for large scale culture, have 2.5×10^4 cm^2. Since the tube is completely filled with medium, cells grow on both upper and lower interior surfaces of the vessel.

Inlet and outlet ports in the culture reel allow medium circulation. Medium pH is continuously monitored by an autoclavable pH electrode. Figure 3 shows the system's medium flow. The medium is pumped past the cells in the culture tube(s), past the pH electrode, and then back to the culture reel(s). The pH is controlled to a selected set-point by addition or dilution of CO_2 in the incubator atmosphere.

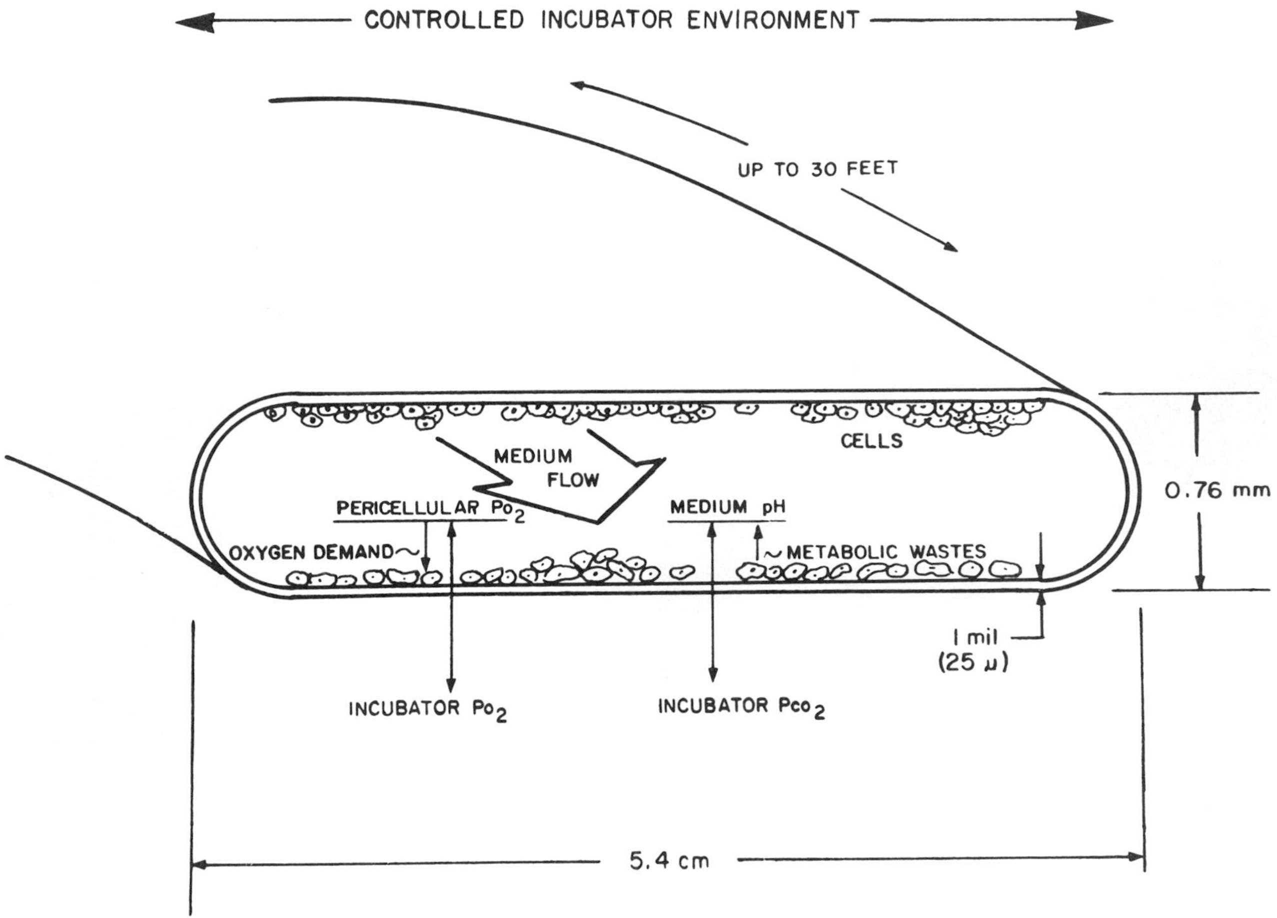

FIGURE 2 *Cross section of the FEP - teflon culture tube.*

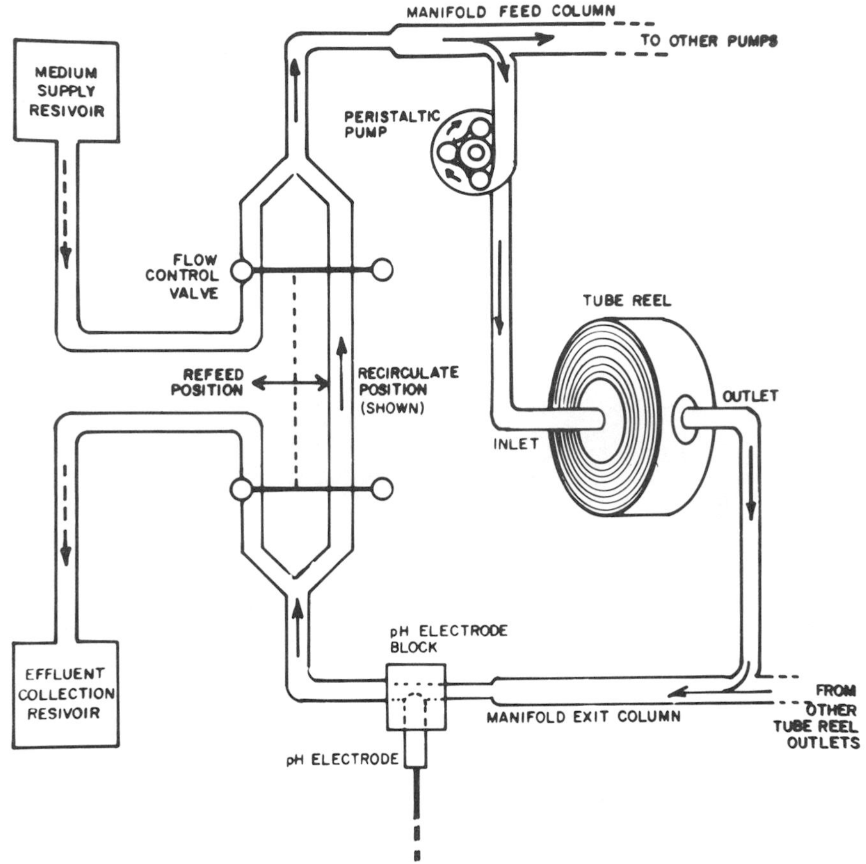

FIGURE 3 *Medium flow in the IL 410 System.*

The instrument will automatically refeed the cultures and collect the effluent medium by operating the flow control valve. When this valve is in the refeed position, spent medium flows into a collection reservoir and is replenished by fresh medium from the medium reservoir. Figure 4 shows typical fluxuations in pH and pCO_2. The upper graph shows a culture grown at constant pCO_2; pH drops until the culture is refed. The lower graph gives the fluxuations in the 410. As non-volatile metabolic cell products accumulate in the medium, the 410 lowers incubator pCO_2 un-

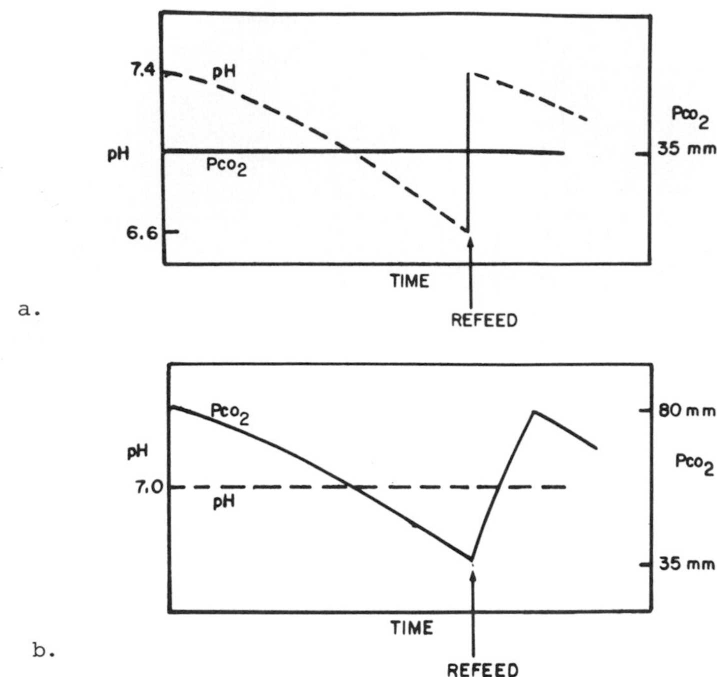

FIGURE 4 *Typical pH - pCO_2 fluctuations of cultures grown (a) on a non-permeable support, (b) in the IL 410.*

til a predetermined value is reached, in this case 35 mm. The cultures are then automatically refed.

The IL 410 continuously monitors the pO_2 of the incubator atmosphere using a Clark-type electrode, and controls the oxygen level to a set point selected by the operator by adding oxygen or nitrogen. Incubator pCO_2 is continuously monitored as a measure of cell lactate production, and the value is used to operate the refeed mechanism.

The instrument has two sections, the incubator and the electronic console (Figure 5). Fully loaded, the incubator will hold 200,000 cm^2 of growth surface. That is the equivalent of 129 large roller bottles. Reels, tubing, pH electrodes, and medium and collection reservoirs can be set up on a wheeled transport

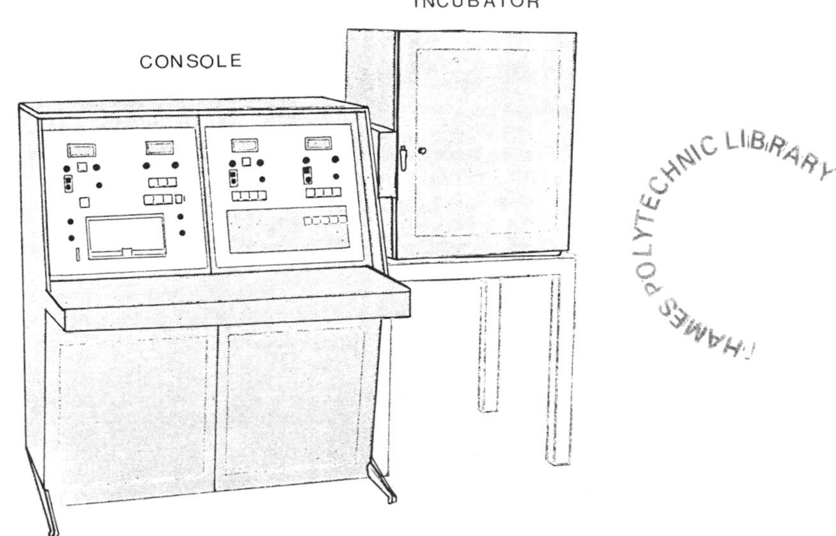

FIGURE 5 *The IL 410 Cell Culture System.*

which fits inside the incubator. The entire assembly can be autoclaved together; as a result, maintaining sterility is fairly simple. The incubator also contains the pO_2, pCO_2, and temperature sensors, humidifier, and pump drive mechanism.

The incubator environment is controlled by a console unit. This console displays and controls temperature, pCO_2, pO_2, and pH, provides a real-time chart record of the 4 parameters, and displays a variety of alarm messages.

III. CULTURE STUDIES

Instruments have been evaluated in several laboratories where they have been used for production of animal viruses and various cellular products, as well as for large scale cell production. Cells grown in the 410 include WI-38 bovine kidney and mammary epithelial primaries, chick embryo primaries, several human tumor lines and a variety of common normal lines. For normal cell

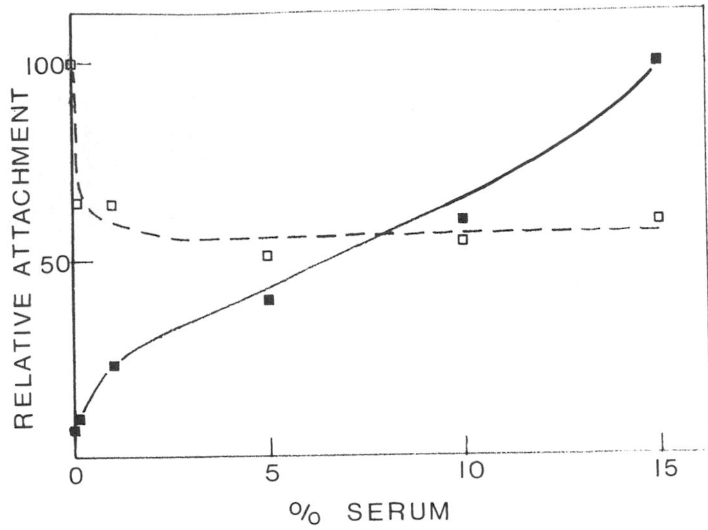

FIGURE 6 *Relative attachment of NBC-6 cells to FEP - teflon, □, and Falcon plastic, □, as a function of fetal calf serum concentration. Maximum cells attached per cm² in the range of serum concentration tested is given as 100%.*

lines run under controlled optimal conditions, the IL 410 at maximum cm^2 capacity will produce in the vicinity of 200 ml packed cell volume per culture run.

Most of the studies in our lab have been with VERO cells (African green monkey kidney fibroblasts). Growth properties of this system seem to be fairly typical of all the cell types grown so far. Cultures multilayer readily, typically showing 3 or 4 layers. Up to 7 layers have been observed. When multilayered cells are returned to glass or plastic flasks, the resulting cultures show typical monolayer, "contact inhibited," growth. Cultures have been grown at serum levels as low as 0.1% with the pH controlled at the optimum of 7.2 (Jensen, *et al.*, 1974a). Cells can be removed non-enzymatically by flexing the FEP membrane. Cells pop off easily and retain their long microvilli.

Cell attachment studies have been made using NBC-6, a bovine leukemic lymphocyte line (Lin, *et al.*, 1973). A maximum of about

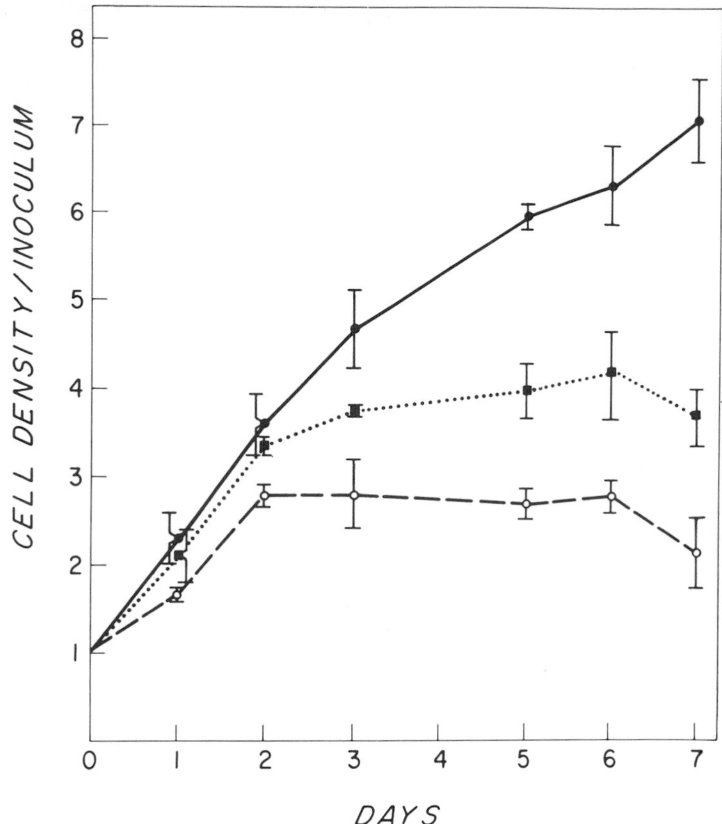

FIGURE 7 *Growth of VERO with and without diffusion across the FEP film.* ●, *FEP - teflon;* ■, *Falcon plastic;* o, *FEP - teflon attached to a pitri dish to prevent gas diffusion across the film.*

5% of these cells in a suspension culture will adhere to and spread out on the culture vessel; spread cells per a given area are easily counted microscopically. Figure 6 shows the relative attachment of these cells to FEP and to Falcon plastic as a function of medium serum content. Attachment to Falcon plastic is maximum in the absence of serum and drops to roughly 60% of the maximum at 0.1% serum or more. Attachment to FEP shows an essentially linear increase with serum content over the range tested.

Culture multilayering is probably a function of several variables, including oxygen availability. VERO cells grown on a sheet of FEP attached to a petri dish to prevent gas diffusion through the teflon divide at a slightly reduced rate for 2 days and then go into stationary phase at a relatively low cell density. (Figure 7) Cells on Falcon plastic reach a higher density in the same manner, while cells on FEP with gas diffusion through the plastic continue to devide for 5 more days. Electron micrographs of thin sections of multilayered DON cultures on FEP, made by Dr. W. H. G. Douglas of the W. Alton Jones Cell Science Center, show unusually high densities of well-defined mitochondria and endoplasmic reticula. The multiple effects of cell dependance on oxydative phosphorylation as opposed to glycolysis may well determine the layering characteristics of a culture.

Some preliminary work has been done with primary cultures, particularly cell types which are difficult to grow or maintain in other systems. Normal mouse lymphocytes have been maintained in culture for up to 10 days and good primary immune response can be elicited throughout this culture period. (Munder, *et al.*, 1973) If these lymphocytes are cultured with a macrophage layer attached to the FEP - teflon, the lymphocyte population doubles in 4 days (Munder, *et al.*, 1971).

Epithelial primaries do very well in this system. Normal bovine mammary cultures form epithelial nests. (Munder, *et al.*, 1971) Primary cultures from decapitated 10-day chick embryos are roughly 40% fibroblast and 40% epithelial cells at confluence if the cultures are grown at 130 mm O_2. The other 20% is a variety of cell types, such as lymphocytes and yoke cells. (Jensen, *et al.*, 1974b) We have maintained chick cultures at 130 mm O_2 for 6 months, or about 13 passages. During this time, the fibroblasts did not dominate the cultures, and the cell types maintained their histochemical properties. Also, many epithelial clumps formed vacuoles with histochemical staining properties different from the surrounding cells, which may indicate cellular secretions. If the cultures are grown at 100 mm O_2 or if cultures are passaged into non-gas permeable surfaces, fibroblasts dominate by the next passage.

IV. SUMMARY

Since both the instrument and the environmental control it provides are quite recent developments, there is still much to be learned about the culture of specific cell types for specific purposes, as well as the effects of culture conditions on those cells. Evidence now available does indicate that cells grown by this method more closely resemble cell morphology in vivo. Successful culturing of such difficult cells as normal lymphocytes suggests that the method can be used to increase the number of cell types which can be cultured. We believe that these factors, plus the capacity for culturing large volumes of cells under controlled conditions will greatly expand cell culture capabilities and their uses.

REFERENCES

Jensen, M. D., Wallach, D. F. H. and Lin, P-S. (1974a) *Exptl. Cell Res.*, *84*, 271-281.

Jensen, M. D., Wallach, D. F. H. and Lin, P-S. (1974b) *In Vitro*, *9*, 384.

Jensen, M. D., Wallach, D. F. H. and Sherwood, P. (1976) *J. Theor. Biol.*, *56*, 443-458.

Lin, P-S, Jensen, M. D., Wallach, D. F. and Tsai, S. (1973) *In Vitro*, *8*, 405.

McLimans, W. F., Blumeson, L. E. and Tunnah, K. V. (1968) *Biotech. Bioeng.*, *10*, 741.

Munder, P. G., Modolell, M., and Wallach, D. F. H. (1971) *FEBS Letters*, *15*, 191-196.

Munder, P. G., Modolell, M., Raetz, W. and Luckenbach, G. A. (1973) *Eur. J. Immunol*, *3*, 454-457.

Rubin, H. (1973) *J. Cell Physiol.*, *82*, 231.

GROWTH OF BHK CELLS FOR THE PRODUCTION AND PURIFICATION OF FOOT-AND-MOUTH DISEASE VIRUS

J. POLATNICK
H. L. BACHRACH

Plum Island Animal Disease Center
Agricultural Research Service
U.S. Department of Agriculture
Greenport, New York

I. INTRODUCTION

The primary function of the Plum Island Animal Disease Center is to study animal diseases not presently endemic to the United States of America. Foot-and-mouth disease is the disease under most intensive study. To meet the needs for the production of foot-and-mouth disease virus (FMDV), we have developed and organized facilities for the production of baby hamster kidney (BHK) cells in large quantities in roller bottles. These cells are infected with FMDV, and the crude virus harvests are subsequently processed to yield pure virus of good quality and in satisfactory quantity for biochemical and immunological studies. Previous papers during the past decade (Polatnick and Bachrach, 1964; Bachrach and Polatnick, 1968; Polatnick and Bachrach, 1972) and the following discussion present details of the procedures used.

II. EQUIPMENT, CELLS, AND MEDIA

We are currently producing about 3×10^{11} BHK cells weekly, distributed among some 400 2-liter round Baxter bottles. Sets of nineteen of these bottles are held in place in cylindrical wire racks and rotated on a three-tiered roller mill. The bottles remain enclosed in the racks throughout all cycles of washing, dry heat sterilization, cell seeding, virus production, and decontamination. Figure 1 shows the roller mill with bottle-containing racks in place and the movable overhead electric hoist that is

FIGURE 1 *Three-tiered roller mill for cylindrical wire racks each holding 19 Baxter bottle cultures.*

FIGURE 2 *Partially exploded view of cylindrical rack for holding 19 Baxter bottles showing metal lid with polyurethane foam bonded to underside, aluminum foil, and two-piece wire rack - upper section is joined to lower section by three wing nuts.*

used to maneuver the racks. Figure 2 provides an exploded view of a rack, showing the sheet of heavy-weight aluminum foil that serves to cover the bottles instead of individual rubber stoppering. The aluminum foil is pressed tightly in place by polyurethane foam bonded to the underside of the lid which in turn clips onto the upper rim of the rack.

Uncloned BHK cells were originally obtained from McPherson and Stoker. Later, BHK clone 13 from passage 21 cells were obtained from the American Type Culture Collection, Rockville, Md. Early passages of these cells were dispersed in a dimethysulfoxide-containing medium, frozen at a controlled rate, and stored in liquid nitrogen. These frozen cells are drawn upon whenever cells currently in production require replacement.

The original McPherson-Stoker (1962) growth medium called for the use of twice the usual concentration of purified amino acids in a bicarbonate-buffered modified Eagle's basal medium, 80%,

TABLE I Growth Medium for Production of BHK Cells

Component	Percentage in medium
Eagle's basal medium, modified to contain 0.02 M Tris and 0.5% lactalbumin hydrolysate	80
Tryptose phosphate broth (Difco)	10
Bovine serum	10
Penicillin	100 units/ml
Streptomycin	100 μg/ml
Fungizone	2 μg/ml

containing 10% tryptose phosphate, and 10% calf serum. For economic reasons, we replaced the costly purified amino acids with lactalbumin hydrolyzate fortified with histidine. For ease of handling and for preventing large swings in pH, the bicarbonate buffer was essentially replaced by 0.02M Tris, and an adjustment was made in the sodium chloride concentration to maintain isotonicity (Polatnick and Bachrach, 1964). Table I outlines the composition of the growth medium.

The media (minus serum, glutamine, vitamins, and antibiotics, which are prepared separately and stored frozen) is sterilized by passage through 220 millimicron Millipore filters and usually stored for at least one week at 37° and another two weeks at room temperature. Just prior to use, the serum and other temperature-sensitive components are added. The bottles are washed in a specially constructed unit servicing 4 loaded racks at a time, as shown in Fig. 3, after which they are sterilized in a dry-heat room-sized oven as shown in Fig. 4 (Bachrach and Polatnick, 1968).

FIGURE 3 *Glassware washer with special header to accommodate four racks of Baxter bottles.*

FIGURE 4 *Dry heat sterilizing chamber measuring 4 feet 10 inches by 9 feet 5 inches.*

III. CELL PASSAGE

Cells grown in the presence of serum screened for adventitious viruses and mycoplasma are reserved for use in seeding new cultures. One hundred ml of growth medium containing 20-25 $\times 10^6$ of these cells are delivered into each production bottle, and the cells are incubated for 4 days at 37°. The fluid is changed and incubation is continued for another 2 days. At this time, growth has reached 600-800 $\times 10^6$ cells, which are now used for virus production.

IV. VIRUS PRODUCTION

FIGURE 5 *Tilting apparatus for discarding spent media directly to a waste line and also for collecting virus fluids into a chilled vessel. The figure shows equipment for the latter operation.*

The spent growth medium from the 6-day-old BHK cells is discarded by means of the tilting apparatus shown in Fig. 5, and the drained cultures are infected with virus in 60 ml of an equivolume mixture of the modified Eagle's salts and 0.16 M Tris buffer in the absence of any serum. Seed virus is maintained by passage only through primary calf kidney culture cells. For virus production, the seed virus is used to infect a few 5-day-old cultures of BHK cells to obtain virus for use in infecting all the major virus-production cells on the following day. The multiplicity of infection is of the order of several tenths of a plaque-forming unit (PFU) per cell, and the infectious fluid is harvested when the cells start sloughing off the glass. The time required for maximum virus yield varies with the virus type, but it is usually 20-24 hours.

V. VIRUS PURIFICATION

The crude virus harvests, with titers averaging 2-3 × 10^8 PFU per ml, are freed of cell debris by passage through a Sharples centrifuge. Polyethylene glycol (PEG) of 20,000 daltons is added to a clarified harvest to make a 6% solution (Wagner *et al.*, 1970). After standing for a minimum of one hour, or over a weekend at 4° if the work schedule so requires, the PEG precipitate is collected in Spinco 15 rotor tubes by spinning for 20 minutes at 15,000 rpm. The precipitate is resuspended in 0.16 M Tris buffer containing 1.0% EDTA (Tris-EDTA) to one-tenth of the original harvest volume. The PEG precipitation cycle is repeated once more, yielding a hundred-fold concentrate of partially purified virus with titers ranging from 1-3 × 10^{10} PFU per ml.

This virus concentrate is then layered over CsCl and centrifuged in a Spinco SW 25.2 rotor for 200 minutes at 25,000 rpm at 20°. The procedure, modified from Bachrach *et al.* (1964), has in each centrifuge tube, from the bottom up, 0.5 ml of ρ1.6 CsCl in

Tris-EDTA, 7 ml of ρ1.42 CsCl in Tris-EDTA, 7 ml of 80% solution of ρ1.42 CsCl in Tris-EDTA, and 42 ml of the twice PEG-precipitated viral concentrate. The viral zone is detectable after centrifugation at ρ1.43 by its strong light-scattering properties. The tubes are punctured, and the viral zone in each tube is collected dropwise directly in a dialysis bag. The purified virus is dialyzed against several changes of 0.05 M potassium phosphate, pH 7.5, containing 0.2 M potassium chloride, and is then stored in plastic vials at 4° for future investigative work. The virus preparations, with titers ranging from 5×10^{10} to 5×10^{11} PFU/ml, meet several different standards of purity which will be discussed in the following section. There may be occasions when additional purification is required, and this is accomplished by an additional centrifugation of the virus through an organic liquid into CsCl (Bachrach and Polatnick, 1968).

The concentration of virus is determined from optical density measurements at 259 mμ, at which wavelength FMDV has an extinction coefficient of 76.0 (Bachrach *et al.*, 1964). Yields of pure virus have reached as high as 18 mg per rack of BHK cells, or approximately 1 mg per bottle, depending on virus type. A more usual yield is from 5 to 6 mg per rack.

VI. PARAMETERS OF VIRUS QUALITY

Pure FMDV has icosahedral symmetry with an estimated 32 capsomeric subunits (Breese *et al.*, 1965). Figure 6 shows a negatively stained electron micrograph in which some details of the fine structure of the virus can be seen.

The procedures in standard use to establish virus purity and integrity are the absorbance-temperature (A-T) profile (Bachrach, 1964), and SDS disc polyacrylamide gel electrophoresis (PAGE). In determining the A-T profile, virus is first diluted in 0.01 M potassium phosphate, pH 7.5, to an optical density value of about

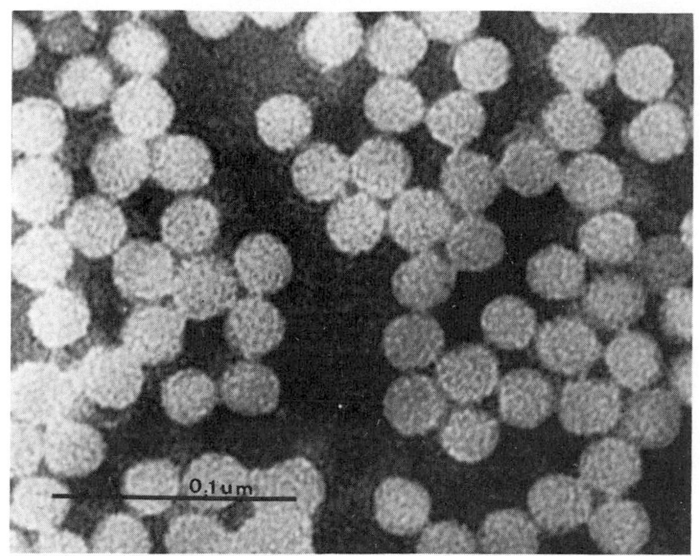

FIGURE 6 *Negatively stained (phosphotungstic acid) electron micrograph of foot-and-mouth disease virus, type A; magnification 450,000 X. Picture courtesy of S. S. Breese, Jr. of this laboratory.*

0.3. Readings are taken in a Beckman DK2 ratio-recording spectrophotometer, using electrically heated sealed cuvettes. As seen in Fig. 7, the relative absorbance of intact purified virus remains unchanged from 0 to about 52° on curve 1, after which the viral RNA is released and undergoes reversible heat denaturation (hyperchromic section of curve 1 and all of curve 2). The exact temperature of virus degradation (T_d) is a function of the concentration and valence of the cation, the virus type and its passage history. Virus from animals, which has had only a few passages in tissue culture, degrades at temperatures of 40° or lower, producing a transient hypochromic dip in the A-T profile of the released RNA. The RNA has a melting temperature (T_m = midpoint of O.D. rise) of about 56° and a hyperchromicity of 1.24. Partially degraded virus manifests itself by the appear-

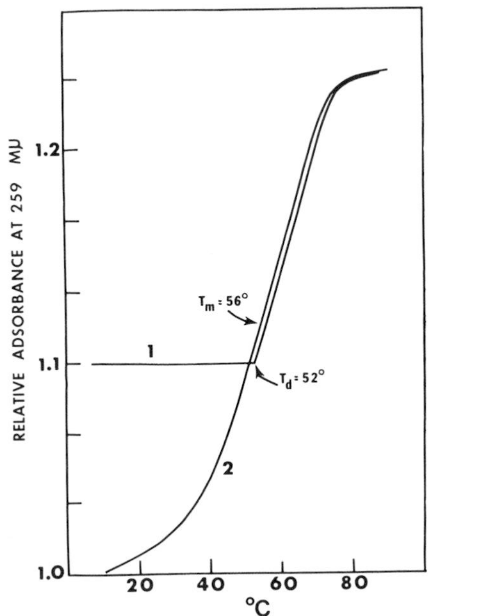

FIGURE 7 *Absorbance-temperature profile and T_m at potassium ion strength of 0.01, pH 7.5, of foot-and-mouth disease virus. Curves 1 and 2 represent first and second heatings, respectively.*

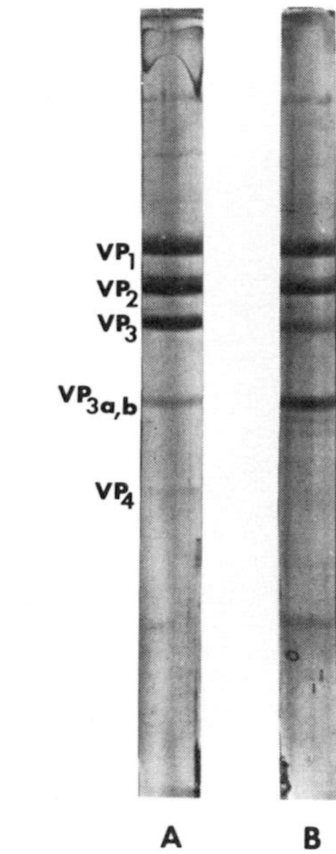

FIGURE 8 *Polyacrylamide gel electrophoresis runs on 12.5% gels containing 8 M urea. Four virus proteins, VP_{1-4}, are seen, together with small (gel A) and large (gel B) amounts of $VP_{3a,b}$ formed from VP_3 by trypsin or similar enzyme digestion.*

TABLE II Parameters Used in the Routine Assessment of the Quality of Production Runs of Purified and Concentrated Foot-and-Mouth Disease Virus, Types A, O, and C

Parameter	Values
1. Spectral diagram and concentration	O.D. minimum at 239°, maximum at 259°. Concentration determined from $E^{1\%}_{259\ nm} = 76$.
2. Absorbance-temperature profile	No tilt to initial plateau in curve. Degradation temperature usually about 52° for high passage tissue culture virus.
3. Sedimentation constant	140 S
4. Polyacrylamide gel electrophoresis	Four proteins, VP_{1-4}
5. Infectivity titer	5×10^{10} to 5×10^{11} PFU/ml
6. Complement fixation	Determinations of virus type

ance at low temperatures of a positively sloped line in place of the absorbance plateau shown by intact virus.

PAGE runs on 12.5% gels containing 8 M urea reveal three major proteins of FMDV, VP_1, VP_2, and VP_3 of molecular weights in thousand daltons of about 33, 30, and 27, respectively. A fourth virus protein, VP_4, is frequently seen at about the 13,000 dalton zone. Virus protein 3 is sensitive to tryptic digestion, yielding a $VP_{3a,\ 3b}$ doublet difficult to resolve photographically. Virus preparations from BHK cells sometimes contain $VP_{3a,\ 3b}$, presumably produced by cellular or serum trypsin or by other enzymes with similar specificity. Figure 8 shows typical PAGE patterns of virus containing small (A) and large (B) amounts of $VP_{3a,\ 3b}$.

When considered necessary, analytical ultracentrifugation and complement fixation tests are conducted on purified virus for confirmation of homogeneity and immunological type, respectively. Table II shows the parameters used in assessing the quality of production runs of purified virus.

VII. DISCUSSION AND CONCLUSION

The current weekly output of the virus production unit from 400 Baxter bottle cultures of BHK cells is 100 to 200 mg of purified virus. This production represents only a portion of our capacity of 2,052 cultures. All phases of production at the present level, including cleaning and decontaminating the working area, are carried out by three technicians. It is estimated that full production could be attained without an appreciable increase in staff, but such production would necessitate more serum for cell growth and zonal centrifuges for purifying the virus.

As in most cell production facilities, our system is not without its problems. Difficulties are at times encountered with the cells, virus, serum and other components of the system even when these components are pretested. Problems appear to occur most frequently during the humid weeks of late summer and have been attributed in the past to mold contamination or to the seemingly poor growth-promoting properties of serum collected from our cattle during that season. For purposes of decontamination, the whole working area can be fumigated by the introduction of formaldehyde through the air inlet ducts.

The use of bottle cultures rather than large suspension tanks provides flexibility for the concommitant preparation of different cell types. Of perhaps more importance, cells used in the roller cultures can be and have been of primary tissue origin, in contrast to the heteroploid or tumorigenic cells required for growth in suspension. A further advantage of the roller bottle system lies in its adaptability to researchers with various needs. Racks of cells are delivered routinely to several laboratory research sections where they are infected with any one of the 7 different types (at least 61 subtypes) of FMDV or other viruses being investigated.

REFERENCES

Bachrach, H. L., (1964), *J. Mol. Biol. 8,* 348-358.

Bachrach, H. L., Trautman, R., Breese, S. S., (1964), *Am. J. Vet. Res. 25,* 333-342.

Bachrach, H. L., Polatnick, J., (1968), *Biotech. and Bioeng. 10,* 589-599.

Breese, S. S., Jr., Trautman, R., Bachrach, H. L., (1965), *Science 150,* 1303-1305.

MacPherson, I., Stoker, M., (1962), *Virology 16,* 147-151.

Polatnick, J., Bachrach, H. L., (1964), *Appl. Microbiol. 12,* 368-373.

Polatnick, J., Bachrach, H. L., (1972), *Growth 36,* 247-253.

Wagner, G. G., Card, J. L., Cowan, K., (1970), *Archiv. ges. Virusforsch 30,* 343-352.

SIGNIFICANCE OF OXIDATION-REDUCTION POTENTIAL IN MAMMALIAN CELL CULTURES

G. M. Toth

Fermentation Design, Inc.
Division of New Brunswick Scientific Co., Inc.
Bethlehem, Pennsylvania

I. INTRODUCTION

The role of electron distribution in animal cell membrane constituents and the ion dependency of electrical coupling of cells was recognized in recent years (DeRobertis *et al.*, 1970). These seem to explain some chemical changes accompanying cellular interactions and cell differentiation. Also, the effect of surfactants and chelating agents (e.g. EDTA) masking the cell membrane's receptor structure (Hakomori *et al.*, 1974) indicates the importance of the molecular charge conditions on the regulation of animal cell membrane activity.

With the development of complex cell culture techniques, the question is raised: "Is there a sensor system by which changes in the culture's charge conditions can be monitored?" According to the data accumulated so far, the Oxidation-Reduction Potential (ORP) can be used as an indicator of cellular metabolism initiated

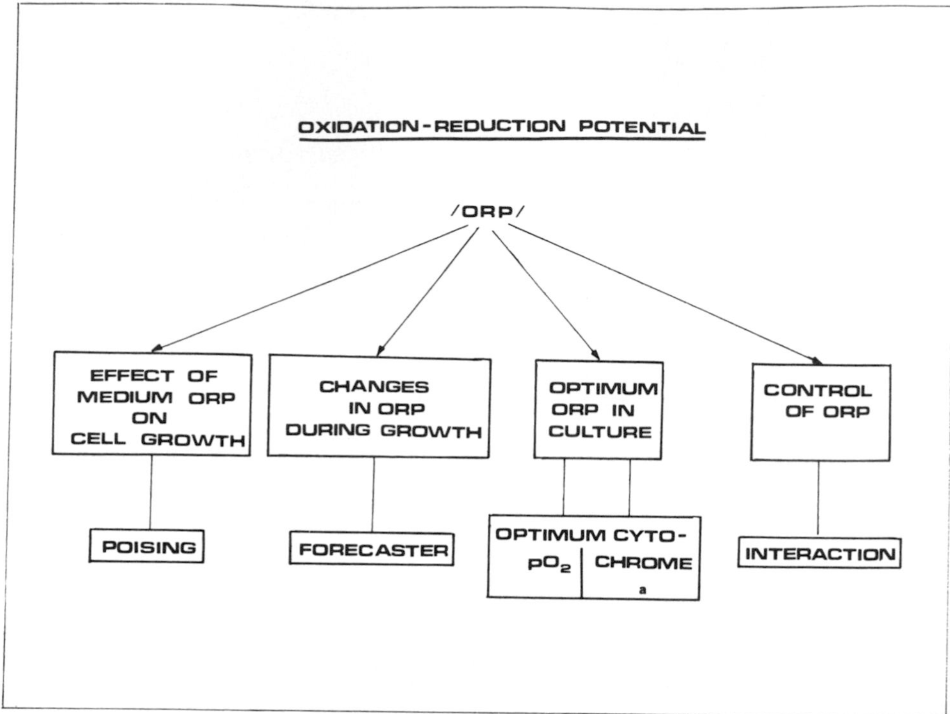

FIGURE 1 *Key points in the known role of ORP in eucaryotic cell cultures.*

changes in charge conditions. This was observed both in the microbial biochemistry (Hewitt, 1950; Hongo, 1958; Tengerdy, 1961; Rabotnova, 1963) and in animal cell cultures (Cater, 1960; Loewenstein and Kanno, 1964; Potter *et al.*, 1960). All of the data indicated that a correlation can be established between changes in ORP and the biochemical events during the cell growth cycle. In view of the most recent achievements in automatic control of cell culture environment (Nyiri, 1972) the utilization of a direct sensor of such importance is considered to be necessary for both process monitoring and control purposes.

With this need in mind, the objective of this paper is to give a brief overview of the present status of ORP in animal cell cultures.

The topic is divided into several major parts (Fig. 1) which, in fact, attempt to cover our present knowledge of the role of ORP in the eucaryotic cell growth.

II. CHANGING ORP VALUES IN THE MEDIUM PRIOR TO INOCULATION

It was observed by Daniels and his coworkers (1970a; 1970b) that the ORP of sterile culture medium undergoes drastic changes if it is left unattended during an incubation period at 37°C (Fig. 2). The initial ORP value of a double strength Eagle's essential medium falls to -230mV from -100mV then rises again to about +75- +100mV in three days. This latter ORP level seems to be the optimum to start growing Earle's L cells.

Wiles and Smith (1969) introduced 95% N_2 + 5% CO_2 mixture to lower the ORP to -320mV then switched to 95% air + 5% CO_2 to reach +75 - 100mV. This method was adapted by us for the treatment of RPMI 1640 medium in a 600 ml flask (Fig. 3) between the ORP values of +150 and -20mV. Technically speaking, this latter method has advantages in reduction of medium treating time which is of importance in large scale systems.

The effects of introducing N_2 or CO_2 containing air on the ORP value suggests that some pO_2 related charge condition changes take place in the medium constituents during the treatment. The same effect can be achieved by processing the medium under nitrogen to prevent the oxidation of some of its components (Mohberg and Johnson 1963).

The treatment is called *poising*. The advantages of this operation are in higher initial ORP levels, higher viable cell counts and longer growth periods.

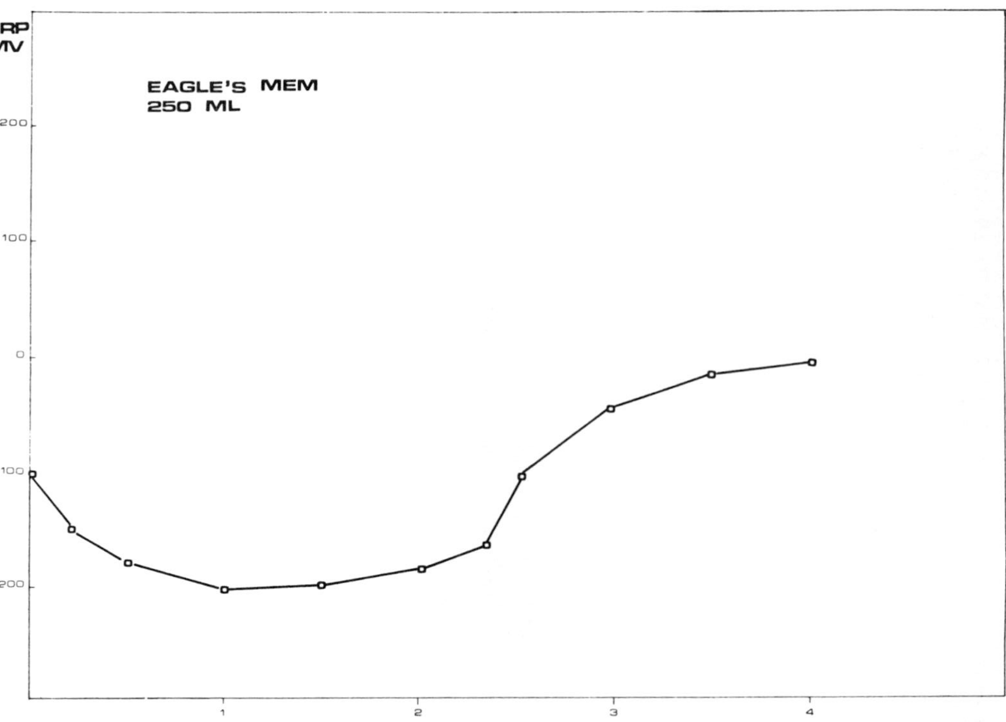

FIGURE 2 *Changes in ORP of Eagle's MEM during a 3 day incubation period.*

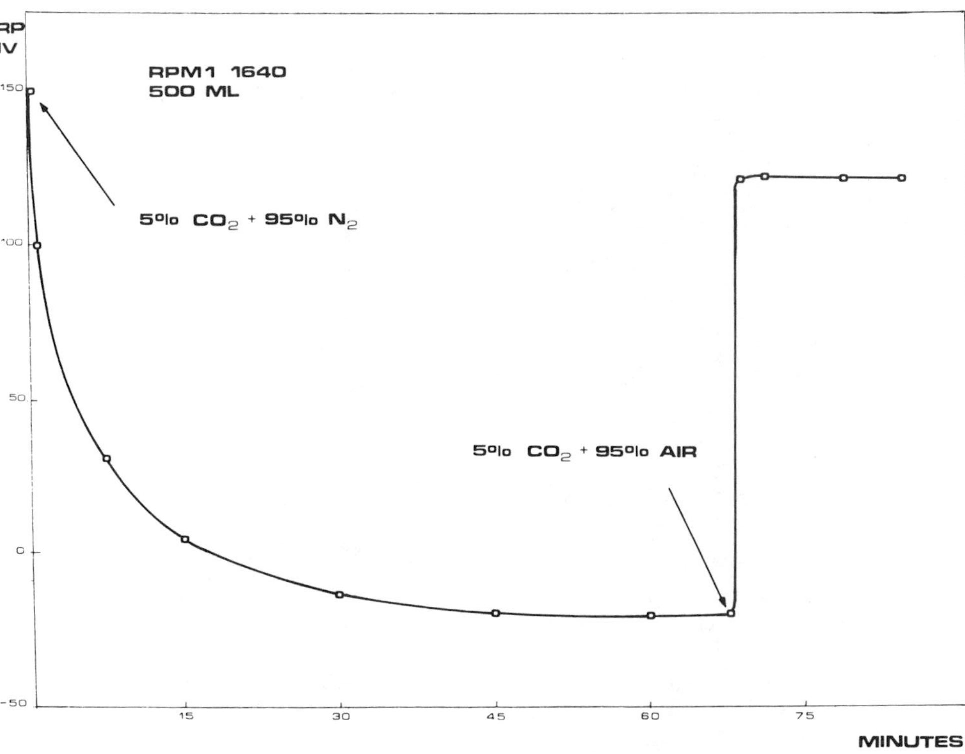

FIGURE 3 *Poising RPMI - 1640 medium with gas introduction.*

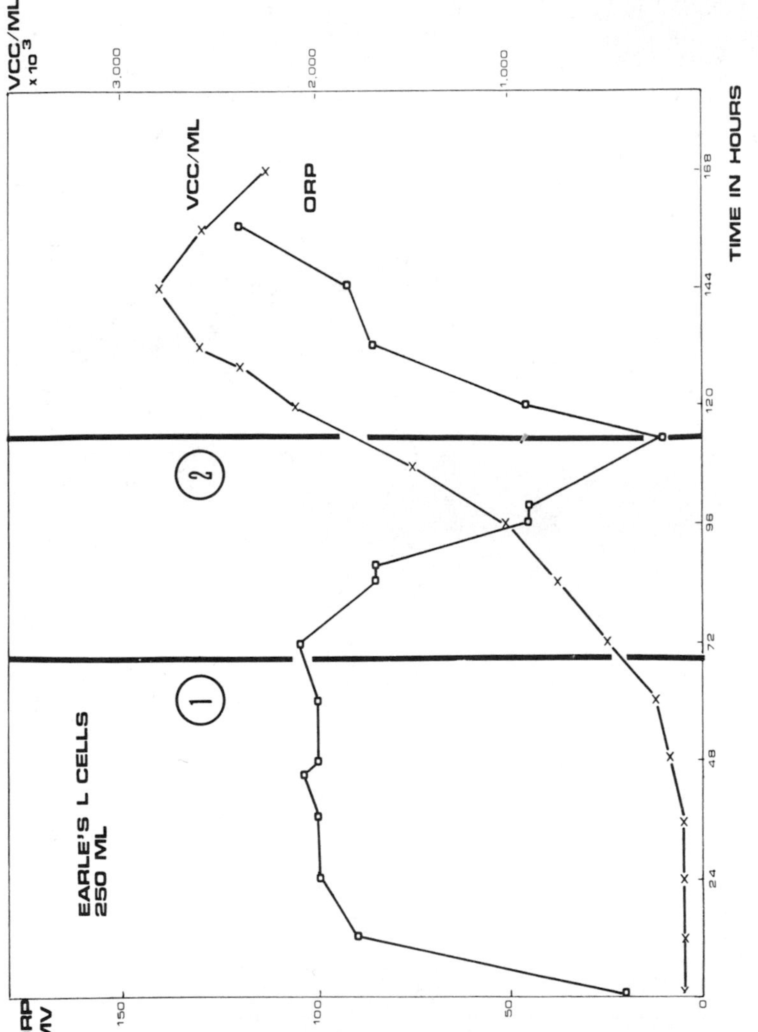

FIGURE 4 *Relationship of ORP and VCC in a culture of Earle's L cells.*

III. CHANGES IN ORP DURING GROWTH CYCLE

In Fig. 4 a typical growth curve is shown for Earle's L cells (Wiles and Smith, 1969) with corresponding ORP changes during culture.

In this experiment both the ORP and the cell count behave as forecasters for the sequence of events during the culture (Nyiri *et al.*, 1976). In the first 70 hours the ORP is around +100mV and the cell count indicates a lag in growth. In the 60th hour the increase in cell number forecasts a fall in the ORP level (Fig. 4, section 1) which will drop to a minimum of +15mV during the logarithmic state of growth. From this point on the subsequent increase in the ORP level (Fig. 4, section 2) forecasts the peak in cell count which follows the ORP minimum within 44 hours.

IV. OPTIMUM ORP IN ANIMAL CELL CULTURE

The observations discussed above indicate the significance of ORP both as an environmental factor and as a means for process control implementation. Both of them require the determination of optimum ORP levels in cell cultures.

The optimum ORP level for Earle's L cells is about +75mV (or +290mV against H_2 electrode) and was established independently by two research groups (Wiles and Smith, 1969; Taylor *et al.*, 1971) (Fig. 5). This potential is the same as the standard potential for cytochrome oxidase. The control of the ORP at or near this potential value probably keeps the cell cytochrome oxidase in the optimum condition to carry out tetravalent reduction of oxygen to water.

On the other hand Taylor and his co-workers pointed out (1971) that +75mV ORP value corresponds to about 100mmHg pO_2. This oxygen tension exists in the *in vivo* growing tissues.

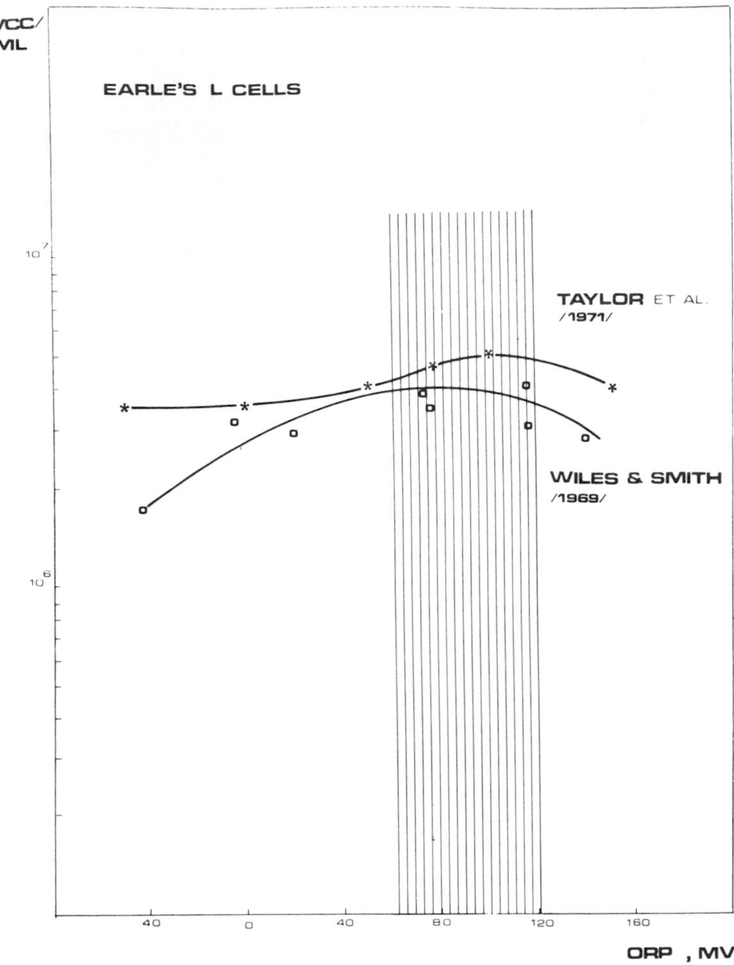

FIGURE 5 *Optimum ORP levels for growing Earle's L cells.*

V. CONTROL OF ORP

The information concerning the changes in pH, ORP and DO patterns during a cell culture give the key to control of ORP in mammalian cell cultures.

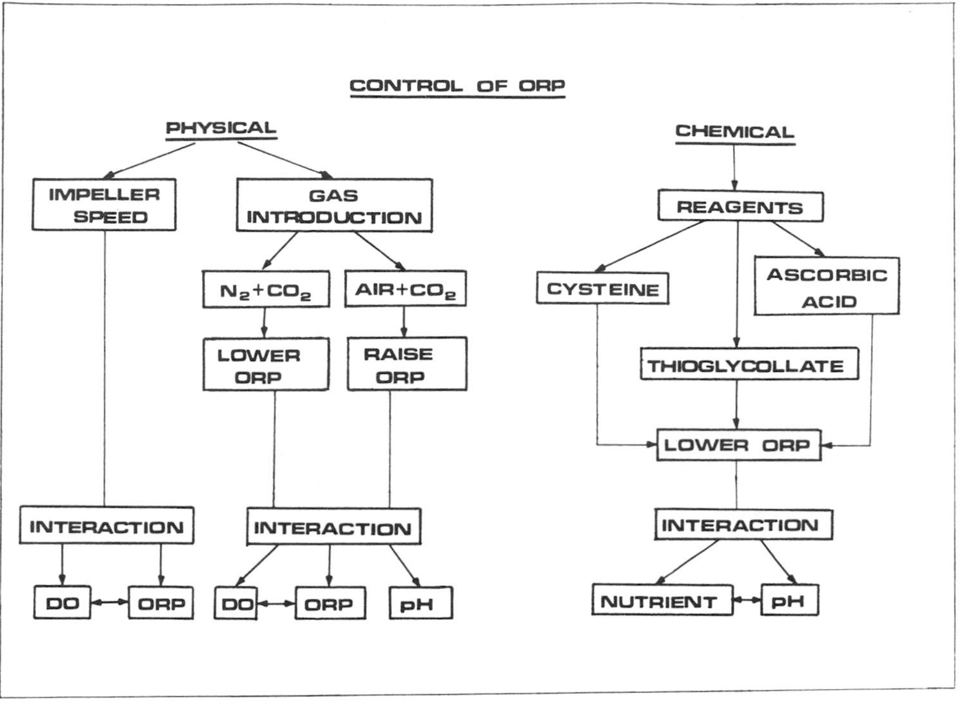

FIGURE 6 *Control of ORP in eucaryotic cell cultures.*

There are different approaches reported so far (Fig. 6): the control can be implemented either physically by changing the impeller speed and/or sparging with gases (N_2, CO_2, Air) (Daniels and Browning, 1962; Daniels *et al.*, 1965) or chemically by adding cysteine, ascorbic acid, sodium thioglycollate or a mixture of cysteine and ascorbic acid (Daniels *et al.*, 1965). The time of response on the ORP level is different in each case. The fastest response was obtained in the case of chemically influencing the ORP (2 minutes) and the slowest was in the case of changing the impeller speed with fixed air/gas flow rate (1½ hour).

The control of ORP through changing agitation speed and gas introduction reflects the relationship between ORP and pO_2 as well as emphasises the necessity of proper control of dissolved oxygen

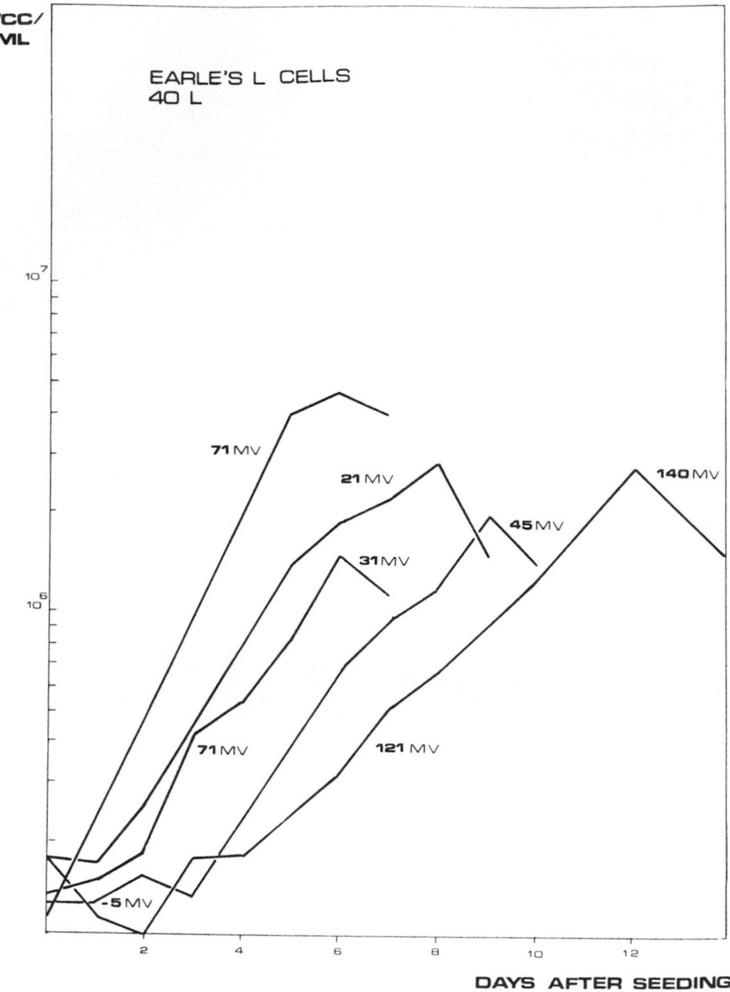

FIGURE 7 *Growth curves for Earle's L cells influenced by different ORP levels during culture.*

concentration as expressed in the following oxygen balance equation:

$$-QO_2 \ X = K_La \quad (C^* - \overline{C}) \qquad /1/$$

The value of K_La can be improved by increasing the impeller speed (Aiba *et al.*, 1973).

The control of ORP can also be carried out by introducing a standard mixture of different gases as the culture conditions require the lowering or raising of the ORP level. The gas mixture can comprise N_2, CO_2 and/or air as specified earlier. The nature of this control is strictly interactive. Since the introduction of N_2 lowers pO_2 and pCO_2 (so influencing ORP and pH) it is suggested to apply a N_2/CO_2 mixture instead of pure N_2. However, because of the difference in the solubility of CO_2 and O_2 in the culture liquid, the ratio of the N_2 to CO_2 must be defined by the changes in the pH which is in turn, defined by pCO_2.

The control of ORP level in the culture by adding chemical compounds to it is applied the least because of the inconvenience of steril addition of the compounds and because of their extensive effect on the pH and ORP levels.

The advantage of controlling ORP in mammalian cell cultures was demonstrated by Wiles and Smith (1969). If the optimal ORP level - which is about +75mV - was maintained throughout the run, the peak cell count and the generation time were at their maximum (Fig. 7). If, however, the ORP was changed significantly during the run, the maximum count was significantly reduced.

VI. STANDARDIZATION OF ORP ELECTRODES

The key problem in the proper control of redox potential is the reliability of the ORP sensor system.

There has been different approaches in attempting to standardize redox electrodes both in bacterial and in mammalian cell cultures. These experiments were either external or internal procedures according to the nature of the standardization, e.g. using the culture itself or taking a sample. Biological cultures are very weakly poised in the sense that very little redox agent is needed to cause very large ORP changes. This makes sampling the culture for purposes of checking externally the ORP difficult,

TABLE I Standardization of Redox Electrodes

Culture	Author	Electrode	Mode	System
Bacteria and fungi	Tengerdy, 1961	Pt-Calomel	External	Quinhydrone
	Tabatabai & Walker, 1970	Pt-Calomel	External	Quinhydrone + phtalate
	Oblinger & Kraft, 1973	Pt-Calomel	External	Quinhydrone + phtalate
	Jacob, 1974	Pt	Internal	Bacterial culture
Mammalian cells	Wiles & Smith, 1969	Pt-$AgCl_2$	Internal	Culture medium
	Taylor, 1971	Pt-$AgCl_2$	External	Electronic
	Toth, 1975	Pt	External	Chemical

because in transferring the sample to an external measuring system, the absorption of small amounts of air causes the ORP to increase sufficiently to make the result meaningless. Therefore, the ORP system can not be standardized the same way with an external standard as is done with the pH of the culture liquid.

Table I compiles the standardization techniques for ORP sensors applied in bacterial and mammalian cell cultures.

Concerning the latter one, Wiles and Smith (1969) and Toth's (1975) method will be discussed.

Wiles and Smith (1969) used the medium itself to standardize redox electrodes from batch to batch. The principle of their measurement is that the medium contains a significant amount of redox groups that have equivalence points in the range of ORP's encountered in poising. These points can be used as internal standards to check the accuracy of the ORP electrodes.

They developed the method empirically for a Beckman Pt and L & N Ag-AgCl reference electrode by comparing poising data from several runs. Freshly mixed Eagle's MEM normally had a negative ORP between -80 - 100mV. The low point for their medium composition was -322mV.

The method consists of stripping the medium of all oxygen by sparging with 95% N_2 + 5% CO_2 until a minimum ORP was constant for at least 1/2 hour and then by raising the ORP to the desired level by sparging with 95% air + 5% CO_2. They obtained a curve similar to the potentiometric titration curves with two break points. The ORP electrode system was calibrated by using a mean value for the break point as an internal standard. The method for locating the break point was developed empirically. With successive runs for the double-strength Eagle's minimum essential medium they found -79 + 20mV for the break point with a 99% confidence level.

It has to be emphasized that the values of the characteristic break points are valid only for the particular medium composition. For other media the poising behavior must be studied to determine its characteristics before trying to apply this technique. On the other hand this method does not establish a reference point for the electrode by which the validity of readings can be checked.

The method developed by Toth (1975) for a steam sterilizable Ingold Platinum combined redox electrode is an external procedure using chemical means for establishing a standard ORP value for electrode calibration.

The basic principle of the method is to use systems where the redox potential is dependent only upon the concentration of the reduced form of certain compounds so by changing the absolute amount of the substance, the electrode potential will change proportionally (Hewitt 1950). Such a system is the cystine-cysteine, where the potential depends only upon the concentration of cysteine (RSH) and the hydrogen ions and not upon the concentration of the oxidized form, cystine (RSSR):

$$2\ RSH \longrightarrow RSSR + H_2$$

The reaction is irreversible.

VII. APPARATUS FOR CALIBRATION OF ORP ELECTRODES

The apparatus used for the calibration is a 600 ml reaction vessel with an airtight removable top secured with a cover clamp (Thomas Co., Philadelphia, Pa) (Fig. 8). The top has four immersion holes which provide for pH, DO and redox probes and the N_2 sparging line. These variables are monitored and recorded during the measurement. The vessel stands on a hot plate - magnetic stirrer which allows the temperature control and constant agitation of the solution.

Before starting the probe's calibration, a known electromotive force is impressed on the input of the redox meter making the redox measurements in a range of -1000mV to +1000mV possible.

The procedure starts with the making of a 600 ml phosphate buffer mixture in the reaction vessel. Before installing the redox Pt electrode the surface of it is cleaned with CeO_2. After the installation of all the electrodes and adjustment of the temperature the pH of this solution should set to 7.0. At this time the deoxygenation of the solution commences by passing N_2 through it at a rate of 2000 ml/min so deoxygenation is achieved in 40 minutes. At this point the ORP value read on the meter is zero and serves to check out the measuring system and the amplifier.

The compounds used for chemical calibration are cysteine .H Cl and glutathione in the concentration range of $0.2\text{-}1.4 \times 10^{-3}$ Mole. The corresponding amounts of the compounds are measured out and added to the reaction vessel while maintaining the parameters of pH = 7.0, DO = Ø, T = 30°. The obtained relationship between reagent concentration and redox readings is shown in Fig. 9.

After the amplifier and the Pt electrode has been checked out and calibrated, the Pt electrode is removed, its surface is cleaned again by the same procedure mentioned earlier and it can be installed on the culture vessel.

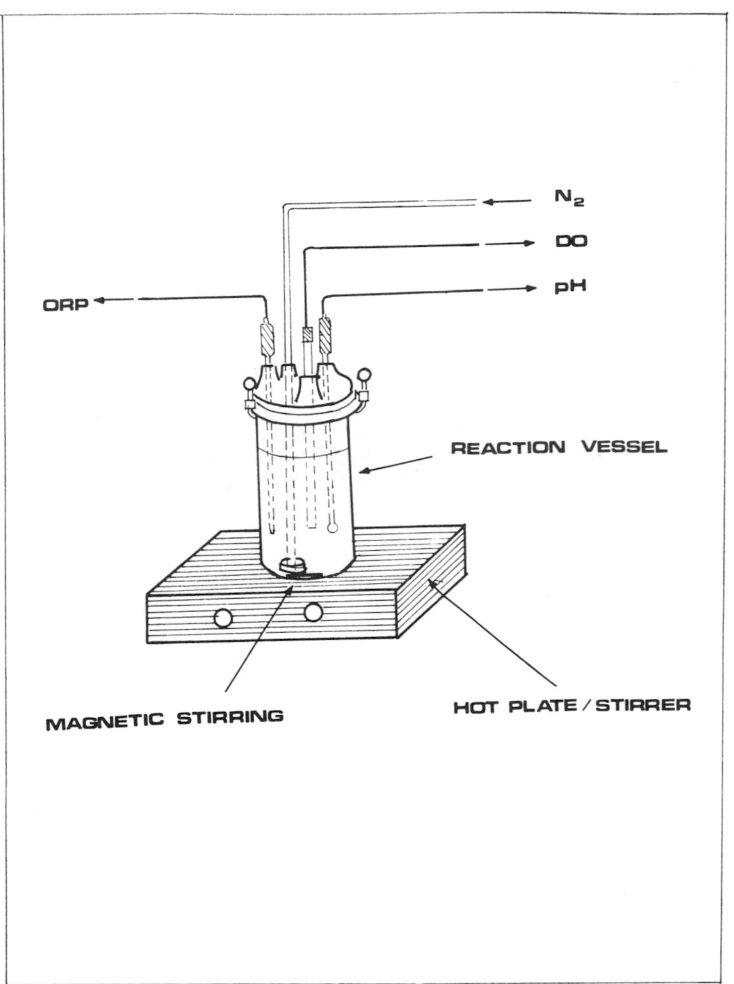

FIGURE 8 *Apparatus for external calibration of steam sterilizable Pt redox electrodes.*

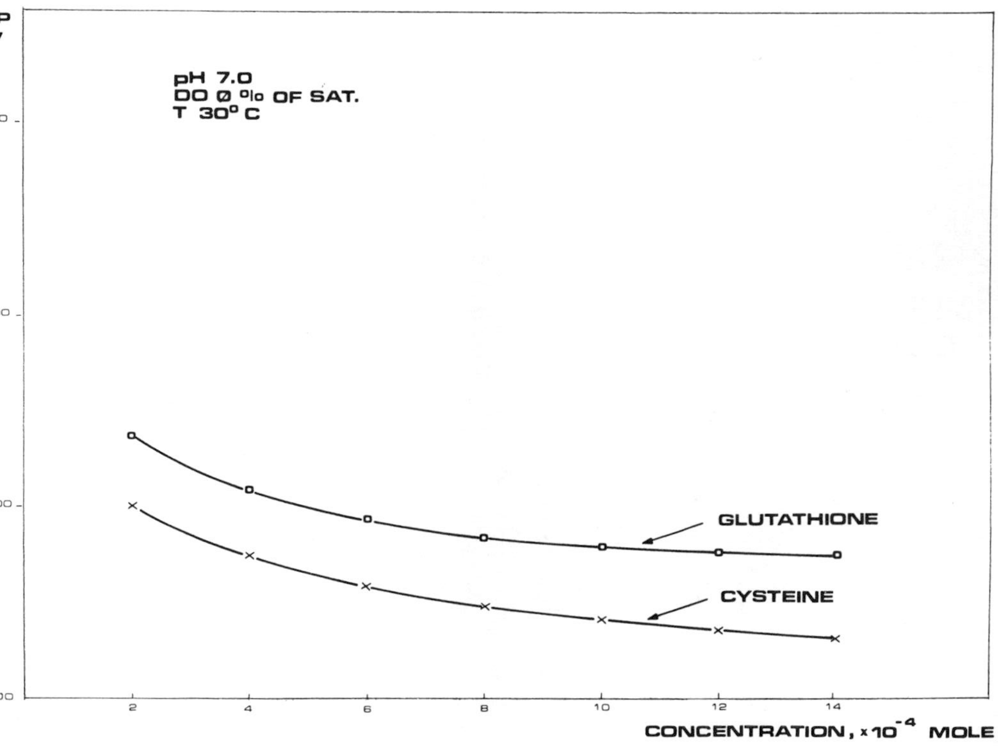

FIGURE 9 *Concentration - ORP relations for sulphydryl compounds.*

VIII. CONCLUSIONS

The significance of oxidation-reduction potential in animal cell cultures was recognized in the early thirties and was verified *in vitro* studies almost a decade ago.

Recently the effect of poising of culture medium, the optimum ORP as well as its control is well documented. These facts indicate the ORP's correlation with charge condition changes during the mammalian cell's life cycle.

The most recent discoveries with regard to the changes within the cell-environment and cell-cell interactions as well as the changes in the cell membrane structure during differentiation and malignant transformation amplify the importance of studies related to charge conditions in animal cell cultures.

Although a direct sensor (redox electrode) is available to implement such studies the literature indicates few efforts in this area. With the development of proper calibration methods of redox electrodes and with their extended application in mammalian cell cultures it is anticipated that correlation studies will be conducted revealing the role of ORP in the cells' life cycle. Such studies ultimately will lead to improved environmental control of the *in vitro* cultures which is the prerequisite of the proliferation of mammalian cells.

NOTATIONS

C^* Concentration of dissolved oxygen which is in equilibrium with partial pressure (pO_2) in bulk gas phase (g mole/cm^3).

$\overline{C}$ Actual concentration of dissolved oxygen in bulk liquid (g mole/cm^3).

K_La Oxygen mass transfer coefficient in aqueous phase (l/hr).

QO_2 Specific rate of oxygen uptake by an animal cell in suspension culture at $\overline{C}$ (g mole/cell/hr).

X Number of viable cells in suspension (cells/m^3).

REFERENCES

1. Aiba, S., Humphrey, A. E., Millis, N. F., (1973), *in* "Biochemical Engineering," Second Ed., pp. 179-181, University of Tokyo Press.
2. Balakireva, L. M., Kantere, V. M., Rabotnova, I. L., (1974), *Biotechnol. Bioeng. Symp. 4,* 769-780.
3. Cater, D. B., (1960), *Proc. in Biophys. and Biophys. Chem. 10,* 154-193.
4. Daniels, W. F., Browning, E. W., (1962), *Biotechnol. Bioeng. 4,* 79-86.
5. Daniels, W. F., Parker, D. A., Johnson, R. W., (1965), *Biotechnol. Bioeng. 7,* 529-553.
6. Daniels, W. F., Garcia, L. H., Rosensteel, J. F., (1970a), *Biotechnol. Bioeng. 12,* 409-417.
7. Daniels, W. F., Garcia, L. H., Rosensteel, J. F., (1970b), *Biotechnol. Bioeng. 12,* 419-428.
8. DeRobertis, E. D. P., Nowinski, W. W., Saez, F. A., (1970), *in* "Cell Biology," Fifth Edition, pp. 429-432, W. B. Saunders Co.
9. Garcia, L. H., Daniels, W. F., Rosensteel, J. F., (1967), *Biotechnol. Bioeng. 9,* 626-629.
10. Hakomori, S., Gahmberg, C. G., Laine, R., Kijimoto, S., (1974), Cold Spring Harbor Conference on Cell Prolif. 1, 461-471, Cold Spring Harbor Laboratory.
11. Hewitt, L. F., (1950) *in* "Oxidation-Reduction Potentials in Bacteriology and Biochemistry," Sixth Ed., pp. 45-48, E. & S. Livingstone Ltd., Edinburgh.
12. Hongo, M., (1958), *Nippon Nogeikagaku Kaishi, 32A,* 101-113.
13. Jacob, H. E., (1974), *Biotechnol. Bioeng. Symp. No. 4,* 781-788.
14. Loewenstein, W. R., Kanno, Y., (1964), *J. Cell. Biol. 22,* 565-570.

15. Mohberg, J., Johnson, M. J., (1963), *J. Nat. Cancer Inst. 31(3)*, 603-610.

16. Nyiri, L. K., (1972), *in* "Advances in Biochemical Engineering," (Ghose, T. K., Fiechter, A. and Blakebrough, N. eds.), Vol. 2., pp. 49-95, Springer Verlag, Berlin.

17. Nyiri, L. K., Toth, G. M., Charles, M., (to be published) *Biotechnol. Bioeng. 18*.

18. Oblinger, F. L., Kraft, A. A., (1973), *J. Food Science, 38*, 1108-1112.

19. Potter, D. D., Furshpan, E. J., Lennox, E. S., (1966), *Proc. Nat. Acad. Sci. U.S.A., 55*, 328-332.

20. Rabotnova, I. L., (1963), *in* "Die Bedeutung physikalisch-chemischer Factoren (pH and rH_2) für die Lebenstatigkeit der Mikroorganismen," VEB Gustav Fischer Verlag, Jena.

21. Tabatabai, L. B., Walker, H. W., (1970), *Appl. Microbiol. 20(3)*, 441-446.

22. Taylor, G. W., Kondig, J. P., Nagle, S. C., Jr., Higuchi, K., (1971), *Appl. Microbiol. 21(5)*, 928-933.

23. Tengerdy, R. P., (1961), *J. Biochem. Microbiol. Tech. Eng. 3(3)*, 241-253.

24. Toth, G. M., (1975), unpublished communication.

25. Wiles, C. C., Smith, V., (1969), *Amer. Inst. Chem. Eng. Symp. Bioeng. Technol. November*, 85-93.

HUMAN CELL CULTURES AS SENTINEL SYSTEMS FOR MONITORING AIR QUALITY

ERNEST V. ORSI
PETER LEVINS
GREGORY AMES
LOIS BUONINCONTRI
NESTOR HOLYK
RYTIS BALCIUNAS

Biology and Chemistry Departments
Seton Hall University
South Orange, New Jersey

I. AIR POLLUTION AND BIOLOGICAL STRESS

A. The Cellular Level Approach

The response of biological systems to air pollutants has become an increasingly important focal point for experimentation and discussion (Ayres & Buehler, 1970, Emik *et al.*, 1971, Coffin & Gardner 1972, Kilburn, 1974). In contrast to the voluminous amount of information at the organism level, reports of studies utilizing cell culture systems have been few (Rhim *et al.*, 1972, Gorden *et al.*, 1973, Roszman & Rogers, 1973). In addition these studies did not use a system in which cells replicated during continued exposure to naturally existing highly polluted urban air. Consequently a series of experiments was started employing replicate cultures from the same parent population which were taken and maintained with appropriate control procedures in regions of very different patterns of air quality.

II. MATERIALS AND METHODS

A. Cell Cultures

1. Cell Line

BS-C-1 (African Green Monkey, continuous, kidney) HEp-2 (Human, continuous, laryngeal epidermoid carcinoma) and HEL-299 (Human, diploid, embryonic lung) were obtained from the American Type Culture Collection.

2. Cell Medium

Stock cultures were grown in Eagle's Minimal Essential Medium (Eagle, 1959) in Earle's Balanced Salt Solution (Earle, 1943) supplemented with 10% heat inactivated calf serum and with each liter containing 1.9 g sodium bicarbonate, 100,000 units penicillin, 20 mg streptomycin, 2 mg Fungizone, and 50 mg Aureomycin. The medium used for the "open" and "closed" test cultures was L-15 (Leibovitz, 1963) with the same serum and antibiotic supplements. Both media were made from powder components from the same manufacturer and tested at the base laboratory at Seton Hall University (SHU) before distribution under controlled conditions to other test sites. In addition, all other compounds and vessels were prepared or obtained from the base laboratory.

3. Cell Counts

Inocula were adjusted and cell yield assayed by conventional trypsinization procedures using 0.25% trypsin-saline and trypan blue exclusion for viability (Merchant *et al.*, 1964).

B. Atmospheric Sampling

1. *Particle Counts*

The small particle detector, patented by Rich of General Electric (manufactured by Gardner Associates, Schenectady, N. Y.) was used to measure pollution in terms of condensation nuclei both inside and outside each of the test sites. Counting was carried out with the chamber set for a minimum particle radius of 10^{-7} cm. This allowed passage of some material through the bacteriologic barriers used to maintain sterility of the "open" culture vessels. These portable instruments were checked with the continuously recording modules at the Atmospheric Research Center (ASRC), State University of New York (SUNY) at Albany, Wilmington, N.Y.

2. *Ozone and Other Oxidants*

Analysis was carried out by the "buffered alkaline" method (American Pub. Heal. Assoc., 1972). In this method a measured amount of air was drawn for 30 minutes through an impinger containing a 1% solution potassium iodide buffered at pH 6.8 +/- 0.2. The iodine liberated by the oxidants was measured by the absorption of triiodide ion at 352 nm.

3. *Sulfur Dioxide*

Analysis was carried out in a similar manner using the West-Gaeke colorimetric method (American Pub. Heal. Assoc., 1972). In this case the impingers contained potassium or sodium tetrachloromercurate which produced a dichlorosulfitomercurate complex. A series of stabilizing and color producing reactions followed with formation of pararosaniline methyl sulfonic acid which was measured at 548 or 575 nm.

4. *Barometric Pressure and Relative Humidity*

At sites other than ASRC a simple continuous recorder and dial hygrometer were used respectively.

C. General Experimental Procedure

The basic idea behind all the experiments was as follows. If the air quality differed radically in one region from another and if some of the various environmental factors could reach the cell milieu, then a cell response might occur. Published reports (Falconer, 1970) had indicated a much greater particle count in the New York metropolitan area than the Whiteface Mountain region in Wilmington, N. Y. as seen in Table I. This suggested that a fairly simple procedure utilizing short term cultures involving few and relatively easily measured parameters ought to prove productive.

The two sites SHU and ASRC also differ in that the field station is about 300 miles due north at an elevation of about 2,000 feet situated among state forest in which conifers predominate. SHU on the other hand is about 20 miles from New York City at an elevation of about 100 feet and the nearby parks consist mainly of oaks and maples. Consequently one had to consider the role of differences in gravity, barometric pressure and electromagnetic spectrum radiation in addition to vegetation.

The device of comparing differences between "open" and "closed" cultures at each site was utilized not only to obviate most if not all ecologic variables but also variations in cell culturing ability among participants and slight temperature variations of the incubators monitored by strip chart recorders.

Additional controls were provided by starting each experiment in the following manner. Stock bottles of cells were prepared at S.H.U. as "closed" cultures. When confluency was reached, two randomly selected groups were placed in picnic coolers with frozen containers. Measurements with a recording thermometer showed that a temperature of 8.0 ± 0.5°C was maintained. The bottles in both coolers were arranged in the same manner and location not only to prevent undue chilling but also to provide uniform splashing of the medium over the cells during travel. Bottles were placed up-

TABLE I Particle Count Patterns in Urban and Rural Regions

Month	Condensation nuclei × 10^3/CC	
	New York metropolitan area	Whiteface Mtn Wilmington, NY
February	30.9	6.6
March	62.8	3.9
April	61.1	4.6
May	73.0	4.4
November	53.9	2.7
December	49.3	1.9
January	57.3	3.5

right with the side opposite the cells touching a thin piece of insulation next to the frozen containers. An arrow on the cooler cover indicated the same orientation of bottles in both coolers to the direction of travel in their respective vehicles. One set of cultures was taken to ASRC and the other set driven for the equivalent length of time and returned to SHU. These control trips were needed since earlier attempts to simulate travel with shakers at 4°C showed differences in both the color of the medium and the number of cells on the surface of the closed vessels of the ASRC and SHU groups.

After a convenient time of incubation at 37°C following arrival at the respective sites, test cultures were started in the following manner. An equal number of stock bottles were harvested at both sites and "open" and "closed" cultures started with the same number of cells in 1.0 ml in 16 × 150 mm tubes provided with solid rubber stoppers or either stoppers with cotton plugs or Morton closures. At appropriate intervals duplicate cultures from groups at each site were harvested for viable cell count.

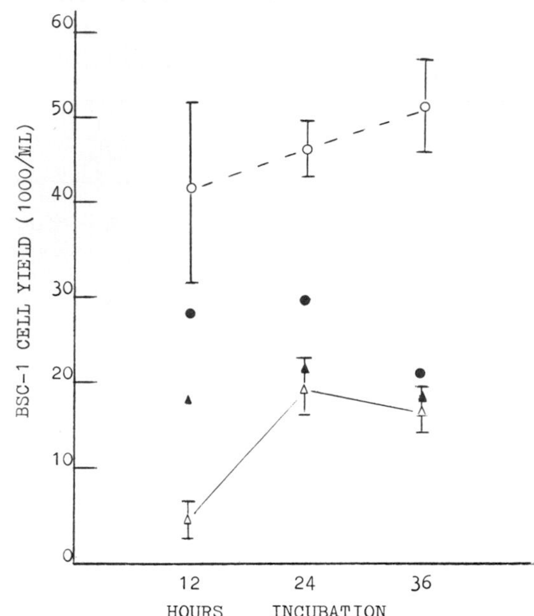

FIGURE 1 o - *open cultures - SHU,* ● - *closed cultures - SHU,* Δ - *open cultures - ASRC,* ▲ - *closed cultures - ASRC,* I - *standard error.*

Rate of travel, initiation of incubation after arrival, start of passage, selection of harvest intervals, and any other aspect were coordinated by telephone communication from the senior member.

III. RESULTS

A. Response of Continuous Cell Lines

The first series of experiments with BS-C-1 cells showed consistently higher yields of cells in the "open" system at SHU in contrast to the "open" cultures at ASRC. As seen in Figure 1 there was no significant difference between cultures grown in a "closed" system at both installations. Similar results were obtained with HEp-2 cells. During the course of these experiments, the media in "open" cultures at SHU were noticeably more acidic than those at ASRC. Measurements of the growth medium pH at the time of cell counting seen in Table II are typical of the results

TABLE II Regional Air Quality and pH of Open and Closed Systems[a]

Hours incubation	Open		Closed	
	SHU	ASRC	SHU	ASRC
12	7.10	7.37	7.37	7.36
	7.25	7.40	7.38	7.38
	7.17	7.39	7.37	7.37
24	7.21	7.43	7.35	7.35
	7.22	7.45	7.41	7.40
	7.21	7.44	7.38	7.37
36	7.20	7.35	7.38	7.38
	7.20	7.48	7.40	7.41
	7.20	7.42	7.39	7.39

[a]pH values are of growth media of BS-C-1 cultures removed at time of cell counting.

obtained with either BS-C-1 or HEp-2 cells. The tendency for the medium to be more alakaline in regions of low particle counts was also demonstrated with media incubated in culture tubes without cells. This is shown in Table III which presents the results of an experiment in which daily particle counts and pH measurements of "open" and "closed" cultures were made at both sites for 12 days.

This evidence for medium-atmosphere interaction prompted us to see if exposing medium to ASRC environment would produce ASRC cell yields at SHU. Replicate samples of L-15 medium capped by Morton closures were exposed to the air in a refrigerator at both sites for two weeks. No obvious volatile substances were noticed in either unit. Both units were opened and shut at least once daily, but no door opening schedule was followed. After exposure the medium at ASRC was returned to SHU and a control trip for the SHU exposed medium was carried out at the same time in the usual manner. Stock HEp-2 cultures which had been routinely propagated

TABLE III Particle Counts and pH of Growth Media Stored in Open and Closed Tubes[a]

	SHU (South Orange, NJ)			ASRC (Wilmington, NY)		
	pH			pH		
	Open	Closed	C.N./CC[b]	Open	Closed	C.N./CC[b]
	7.6	7.6	53,000	7.8	7.7	12,000
	7.5	7.6	59,000	7.5	7.5	12,000
	8.3	8.5	24,000	8.8	8.8	9,000
	8.6	8.3	57,000	9.1	8.7	22,000
	7.9	7.9	52,000	8.3	8.2	15,000
	7.9	7.9	28,000	8.6	8.5	25,000
	7.9	7.8	15,000	8.7	8.7	28,000
	6.5	6.5	28,000	8.7	8.7	20,000
	7.9	7.9	19,000	8.0	8.0	6,000
	7.7	7.7	30,000	7.9	7.9	2,000
	7.7	7.7	30,000	7.8	7.9	13,000
	7.8	7.7	43,000	7.9	7.9	20,000
X=	7.8	7.7	36,500	8.3	8.2	15,333

[a]pH values are of growth media incubated in tubes without the use of cells. P = 0.05 that SHU open media are more acidic than ASRC; P = 0.05 that SHU closed media are more acidic than ASRC; P = 0.001 that SHU particle counts are greater than ASRC.

[b]Particle counts of laboratory air near incubator.

at SHU provided cells for growing replicate samples of cells growing in the different media but all at one location (SHU). The results seen in Fig. 2 demonstrate that exposing medium to ASRC conditions produced the typical lowered cell yield but now at SHU.

B. Response of Diploid Cells

Since all our work had involved non-diploid cell lines, a series of experiments was carried out during summer, fall, and winter employing HEp-2 and HEL-299 cultures. The WI-38, Human, diploid lung line had also been selected, but in our hands it would not grow satisfactorily in L-15 medium. Another deficiency was the fact that particle counts had been the only measurements made of pollutant levels in the areas of the incubators. The as-

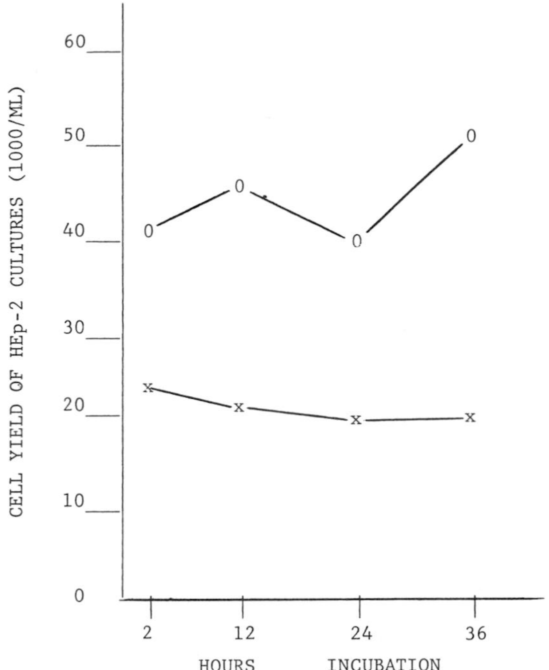

FIGURE 2 0 - *medium exposed - SHU*, X - *medium exposed - ASRC. Plots are midpoints of two samples.*

sumption that the indoor air quality was reflecting the known patterns of the region was probably valid since the interior quality will indeed often reflect that of the exterior (Schaefer *et al.*, 1972). Nevertheless, this relation had to be verified for our specific sites.

The summer phase utilized two installations in the South Orange-Newark, N. J. area and two in the Wilmington, N. Y. area. The fall and winter experiments were restricted to the Beaver-Wil and SHU locations. The same pattern was obtained regardless of the time of the year. The lowest and highest values obtained for particle counts, and SO_2, and total oxidants levels adjacent to the incubators are shown in Table IV. The air quality values were obviously more similar at both regional stations than between regions. The Beaver-Wil highest inside count was recorded during

TABLE IV Pattern of Condensation Nuclei Counts and Air Quality Parameters

	Wilmington, N.Y.[b]		So. Orange-Newark, N.J.[c]	
Lowest values	Beaver-Wil	ASRC	SHU	CM & DNJ
Particle count (inside)	2,100	10,300	16,000	26,000
Particle count (outside)[a]	1,350	2,800	66,000	56,000
SO_2 (PPB) (inside)	0.99	1.13	7.88	15.60
Oxidants (PPB) (inside)	8.37	4.22	8.69	9.10
Highest values				
Particle count (inside)	46,000	20,700	53,000	118,000
Particle count (outside)[a]	2,500	4,800	148,000	96,000
SO_2 (PPB) (inside)	0.92	0.19	27.50	18.30
Oxidants (PPB) (inside)	8.79	5.32	13.82	20.64

[a]Outside counts were selected according to the lowest and highest inside particle count readings for each site during a six week period.

[b]Beaver-Wil, private laboratory in tree farm area, el. 950 ft; ASRC, field station, SUNY at Albany in forest, el. 2000 ft.

[c]SHU, Seton Hall University; CM & DNJ, College of Med. & Dent. of N.J.; five miles apart, 20 miles from N.Y.C., el. 100 ft.

construction in another part of the building, but usually these counts paralleled those at ASRC about four miles distant. It is noteworthy that a correlation between higher particle counts and increased SO_2 and oxidants levels was found only in the urban area.

The HEp-2 cultures once again showed the highest cell counts in "open" systems in the region of greater air pollution. The HEL-299 cells always failed to show this since the average cell

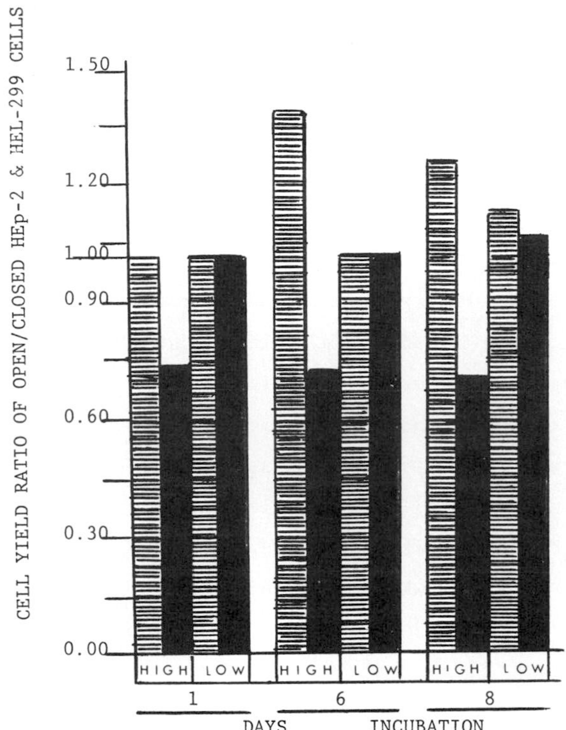

FIGURE 3 □ *HEp-2 cultures,* ■ *HEL-299 cultures,* □ *air pollution levels; high = SHU, low = ASRC.*

count was consistently below the average for the "closed" cultures in areas of higher pollution. At the lower pollution sites there was no significant difference in cell yield of HEL-299 cultures regardless of the system. This is demonstrated in Figure 3 which presents the ratio of "open"/"closed" cultures cell yield obtained from averages of the two sites at both regions.

IV. DISCUSSION

The ambient air measurements obtained at the four installations demonstrated once again the striking extremes in air quality between the Adirondack Mountains and the New York metropolitan area. The closer values at the lower range of oxidant levels between the two regions is not surprising in view of other similar reports (Stasiuk & Coffey, 1974). It should be pointed out that urban ozone levels are closer to those of rural regions usually during the day, and most of the air samples in our study were collected in the daytime.

The results with the cells were admittedly surprising since an inhibitory effect had been anticipated assuming any effect at all. The fact that enhancement of continuous lines was repeated and that the media exposure experiment corrected for barometric pressure differences supported the idea that the response reflected air quality differences. Finding that media in higher air pollution regions tended to be more acidic suggested that the higher levels of SO_2 in urban areas compensated for the tendency for media to drift particularly in an "open" system. This is of course a possibility. Experiments still in progress utilizing water through which ambient air has been bubbled at SHU and ASRC have produced similar results as exposing complete media. This is noteworthy since this water was then used to prepare complete media at SHU and utilized in "closed" systems.

Evidence that medium pH changes may not be the only mechanism involved was the selective inhibition of the diploid HEL-299 with the simultaneous enhancement of the heteroploid HEp-2 cell line.

V. SUMMARY

Controlled passages of monkey and human cell lines showed cell replication responses that reflected the air quality of the laboratory and the degree of exposure of the medium to the environment both prior to passage and during cultivation. Heteroploid monkey and human cell line yields were enhanced in areas of higher air pollution, but a human diploid line showed inhibition in this environment. These results indicate that continuous exposure of cell cultures would not only provide a suitable system for studying cell-air toxicity mechanisms but these same cultures could serve as sensitive sentinels for air quality changes of biologic relevancy.

ACKNOWLEDGMENTS

We would like to express our thanks to the following individuals: Nesrine Baturay and Thomas Genova were most helpful in the day to day operation of the cell culture laboratory in addition to their regular biology graduate program. Thanks are also extended to Leon Pirak (SHU), Barbara Parsons (SHU), Richard Deininger (SUNY), Ben Novograd (SUNY) and Robert Sheridan (Upsala College, N. J.) for their participation in the summer four sites project.

The stimulating discussions with Vincent Schaefer and Raymond Falconer of the Atmospheric Sciences Dept. State University of New York at Albany as well as the courtesies extended to us at the field station have been most appreciated.

Finally a heartfelt and vigorous nod of thanks to Douglas Wolfe of the ASRC, Wilmington, N. Y. for his valuable assistance ranging from introducing a biologist to weather recording instruments to rigging an emergency power line following a destructive bolt of lightning.

Supported in part by the Office of Naval Research #N00014-68-A-0340-0001 and National Science Foundation, Summer 1974 SOS.

REFERENCES

American Public Health Association (1972) "Methods of Air Sampling and Analysis," Washington, D. C.

Ayers, S. M. and Buehler, M. E. (1970). *Clin. Pharmacol. Ther. 11*, 337-371.

Coffin, D. L. and Gardner, D. E. (1972). *Ann. Occup. Hyg. 15*, 219-

235.

Emik, L. O., Plata, R. L., Campbell, K. I. and Clark, G. L. (1971). *Arch. Environ. Health. 23*, 335-342.

Falconer, R. E. (1970). "Third Ann. Rept. of Atmos. Sci. Res. Cen." State Univ. of New York.

Gorden, R. J., Bryan, R. J., Rhim, J. S., Demoise, C., Wolford, R. G., Freeman, A. E. and Huebner, R. J. (1973). *Int. J. Cancer 12*, 223-232.

Kilburn, K. H., (1974). *Ann. N. Y. Acad. Sci. 221*, 386-390.

Merchant, D. J., Kuckler, R. J. and Munyon, W. H. (1964) "Handbook of Cell and Organ Culture." Burgess Pub. Co., Minneapolis.

Rhim, J. S., Cho, H. Y., Rabstein, L., Gorden, R. J., Bryon, R. J., Gardner, M. B. and Huebner, R. J. (1972). *Nature 239*, 103-107.

Roszman, T. L. and Rogers, A. S. (1973). *Ann. Rev. Resp. Dis., 108*, 1158-1163.

Schaefer, V. J., Mohnen, V. A. and Veirs, V. R. (1972). *Science, 175*, 173-175.

Stasiuk, W. N. and Coffey, P. E. (1974). *J. Air Poll. Con. Assoc. 24*, 564-568.

Workshop IV

CELL CULTURE FOR THE STUDY OF DISEASE

FAT BODY CULTURE SYSTEMS IN THE STUDY OF FROG DISEASES

ROBERT L. AMBORSKI
DEPARTMENT OF MICROBIOLOGY
GRACE F. AMBORSKI
DEPARTMENT OF VETERINARY MICROBIOLOGY AND PARASITOLOGY

Louisiana State University
Baton Rouge, Louisiana

I. INTRODUCTION

A. Significance of the Problem

The decline of amphibian species is a world wide problem. The pressures responsible for this decline can be identified in the natural environment as well as in the conditions of maintenance after the desired species have been collected. Thus Prestt *er al*. (1974) have indicated that loss of habitat, predation, road mortality, complications associated with hibernation, over collection, pollution and diseases all play roles in reducing the number of frogs in the natural environment. At the other end of the spectrum, Gibbs *et al*. (1971) suggested that once frogs have been collected and placed in water holding pens factors including overcrowding, lack of environmental temperature control, low oxygen level and lack of food contribute to poor health and increased susceptibility to bacterial disease. It was noted in this report that in holding pens containing 20,000 to 30,000 frogs each, over half the frogs developed bacterial disease and died during one winter. The control of such diseases would make a significant contribution to the current frog problem.

B. The Major Bacterial Disease of Frogs

The most serious microbial threat to a frog's health is an epizootic form of bacterial septicemia usually accompanied by a cutaneous ulcerative condition, designated by Emerson and Norris (1905) as Red-leg disease. This disease has destroyed both pond and laboratory populations of a variety of frog species. The disease in the adult frog is insidious sometimes lasting as long as six months before death. However the same disease in the tadpole often results in massive overnight mortalities. Figure 1 presents a comparison between healthy and diseased tadpoles. The clinical signs of the disease include edema, hemorrhaging of the dermis from the mandible to the toes and the distension of the lymph sacs with a blood-tinged serous fluid, giving a bloated appearance. In the case of the tadpole this latter effect is especially pronounced in the femoral lymph sacs of the hind legs.

Numerous studies have been directed toward the identification of the bacteria associated with this disease. These efforts are summarized in Table I and indicate that at least thirteen gram negative species of six different genera can cause Red-leg disease. As an approach to the study of the possible interactions between these bacteria we attempted to determine whether or not any of the bacterial isolates in Table I were invasive or demonstrated any tissue affinities. Using standard histological techniques and a tissue gram stain, tissues from diseased and healthy animals were screened for bacterial involvement. In no instance could gram negative bacteria be shown to be invasive in any of the frog tissues. However quite unexpectedly, bacteria were

FIGURE 1 *Bullfrog, Rana catesbeiana, tadpoles. Normal tadpole (A) and diseased tadpoles (B).*

found to invade the fat body, and in every observed case these bacteria were gram positive. Figure 2 shows a histological section of frog fat body which demonstrates the presence of the gram positive bacterium. This organism has been isolated and through the determinations of substrate utilization, molar percent Guanine/Cytosine content of the DNA and cell wall analysis, this bacterium appears to be a species of *Corynebacterium*. As of yet all attempts to demonstrate the pathogenicity of this isolate have been negative. A major problem with this system lies in the fact that the *Corynebacterium* is a slow growing organism. Furthermore in those frogs which have died, the microbiological studies have been complicated by the presence of one or more of the gram negative species listed in Table I.

The inability to demonstrate the pathogenicity of the *Corynebacterium* or determine its contribution to the disease syndrome (if any) led us on a search for other ways to look at this prob-

TABLE I History of the Attempts to Define the Etiology of Red-leg Disease

Date	Investigator	Bacterial isolates
1898	Russell	*Bacillus hydrophilus fuscus* (*Aeromonas hydrophila*)
1905	Emerson and Norris	*Bacillus hydrophilus fuscus* (*Aeromonas hydrophila*)
1942	Kulp and Borden	*Bacillus hydrophilus fuscus* (*Aeromonas hydrophila*)
1950	Miles	*Bacterium alkaligenes* (*Alkaligenes faecalis*)
1953	Kaplan	*Pseudomonas hydrophila* (*Aeromonas hydrophila*)
1963	Gibbs	*Aeromonas hydrophila* *Citrobacter freundii*
1966	Gibbs *et al.*	*Aeromonas hydrophila* *Mima sp.* (*Acinetobacter sp.*)
1974	Glorioso *et al.*	*Aeromonas hydrophila* *Aeromonas shigelloides* *Citrobacter freundii* *Flavobacterium sp.* *Mima polymorpha* (*Acinetobacter sp.*) *Proteus mirabilis* *Proteus morganii* *Proteus retgerii* *Proteus vulgaris* *Pseudomonas aeruginosa* *Pseudomonas fluorescens* *Pseudomonas putida*

lem. This report represents our attempts to develop a cell culture system which would be applicable to further studies on the *Corynebacterium*.

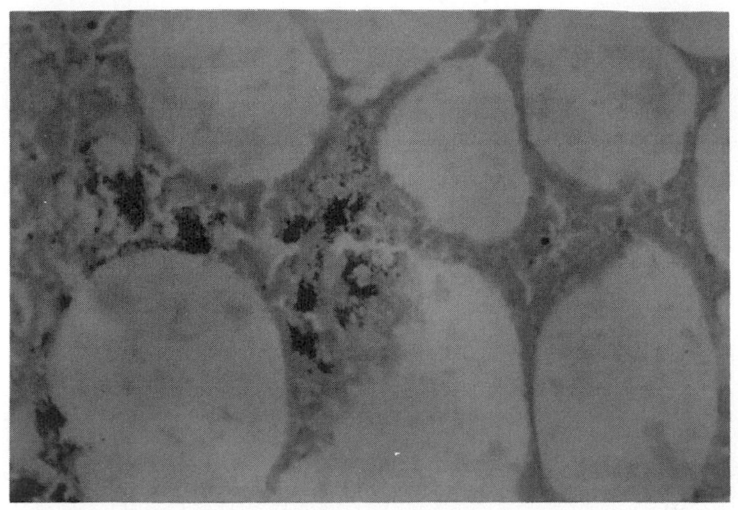

FIGURE 2 *Gram stained histological section of a fat body from a diseased frog. Arrows indicate foci of gram positive bacteria. 1720X.*

II. MATERIALS AND METHODS

A. Bacteria

The gram positive organism was isolated as previously described and maintained on Ordal Earp agar (Glorioso *et al.*, 1974). Quantitative determinations were made by serial dilution and the spreader plate technique.

B. Frogs.

All animals were obtained from the Louisiana State University fisheries ponds in Ben Hur farms. The animals were maintained in flow through systems using dechlorinated tap water. Larvae (tadpoles) were maintained on a diet of wheat bran. Adults were fed on a diet of mosquitofish (*Gambusia afinis*) and crayfish (*Procambarus clarkii*). The water temperature was regulated at 24 C and a photoperiod of 12L:12D was maintained. Blood samples were collected weekly by heart puncture and screened for the presence of bacteria. All confirmed or suspect cases of bacteremia or Red-leg disease were removed from the study.

C. Organ culture

Animals were removed from their chambers and sacrificed by pithing. The ventral surface of each animal was washed with 10 percent formalin, and standard aseptic techniques were used to expose the peritoneal cavity. The fat bodies, corpora adiposa, are attached to the anterior ends of both the ovaries and the testes, and are easily recognizable by their yellow color and their division into a number of finger-like bodies.

Fat bodies were aseptically removed and placed in sterile frog Ringer's solution without antibiotics. Fat bodies were then fragmented and the fragments were allowed to soak for 30 minutes in the frog Ringer's solution to help remove blood cells. One half of the fragments were randomly selected and removed for bacteriological determinations, and the remaining fragments were used to set up organ cultures. Each frog was processed separately and tissue samples were not mixed. The organ cultures were prepared by placing individual fragments measuring 5 to 10 mm in length on grids in plastic organ culture dishes (Bioquest, California) or by allowing fragments to float on the surface of the media. Fragments were removed at intervals and histological sections were prepared by standard techniques and stained for bacteria (Luna, 1968). All incubations were carried out at 24C.

D. Media

Eagle's minimal essential medium (MEM), modified MEM equivalent to 30, 50, 70 and 90 percent MEM and Wolf and Quimby amphibian tissue culture medium was used in this study. Media and sera were purchased from Grand Island Biological (New York). Each medium was buffered with 14 mM HEPES (N-2-hydroxy-ethyl-piperazine-N'-ethane sulfonic acid) and did not contain antibiotics.

III. RESULTS

In a preliminary report, Amborski *et al.* (1974) indicated that frog fat bodies could be maintained in organ culture for periods up to 14 days before central necrosis was observed. These observations have now been extended by using media with a reduced osmotic pressure as shown in Table II. The structural integrity of the organ culture system was maintained for periods of up to 30 days under the described conditions. Furthermore the histological patterns indicated that the free floating fragments gave results comparable to those observed in the fragments maintained in the organ culture dishes. The concentration of the serum (1 to 10 percent) did not appear to be critical, and the source of the sera did not affect the results as fetal calf, bovine, equine, porcine and human sera gave comparable responses. However one of the problems encountered during the long term incubations was microbial contamination as many frogs are parasitized with bacteria, yeasts, molds and ciliates. In addition, 30 percent of the fat bodies were discarded when the *Corynebacterium* was demonstrated in the tissue fragments of the apparently healthy donors. Figure 3 shows a typical fragment under floating conditions.

Fragments similar to those shown in Fig. 3 were exposed to 10^6 each of the *Corynebacterium sp.* by adding the bacteria directly to the growth medium. After 24 hours the medium was removed and then each fragment was washed 3X with fresh medium and then incubated. Fragments were harvested daily and histological sections were prepared and gram stained. In the presence of precipitated stain, it was difficult to identify the *Corynebacterium* during the first four days of incubation, but after eight days of

TABLE II Morphological Integrity of Fat Body Organ Cultures

Medium[a]	Number of days in culture		
	10	20	30
Wolf and Quimby	0/8[b]	1/6	2/5
Eagle's MEM	1/9	4/8	7/8
90% MEM	0/7	3/7	4/5
70% MEM	0/9	1/8	2/8
50% MEM	1/9	3/6	4/6
30% MEM	3/7	4/6	6/6

[a]Media supplemented with 10 percent fetal calf serum.
[b]Ratio represents number of organ cultures showing central necrosis/total number of organ cultures observed.

incubation foci of bacteria were obvious. A histological section of an uninoculated culture is shown in Fig. 4, and a histological section prepared from an inoculated culture is shown in Fig. 5. The clumping of the bacteria made it difficult to make any quantitative measurements on these cultures, but the *Corynebacterium* was readily isolated after eight days of culture. The amount of tissue destruction was apparently light, and it would appear that the action of the bacterium in the animal and in the culture system does not result in gross destruction of the fat body tissue.

As a result of some of our original observations, fragments of fat bodies were also set up as classical explant cultures. After an incubation period of 6 to 10 days, cells were observed around the periphery of the explants. However the continued outgrowth was only observed in media containing 70 to 80 percent MEM and in the WQ medium, and under these conditions monolayers were observed to form after two weeks in culture. However it must be pointed out that the outgrowth did not present a uniform pattern, and even in cultures in media with a reduced osmolality, the expansion of the initial outgrowth did not always occur. Typical

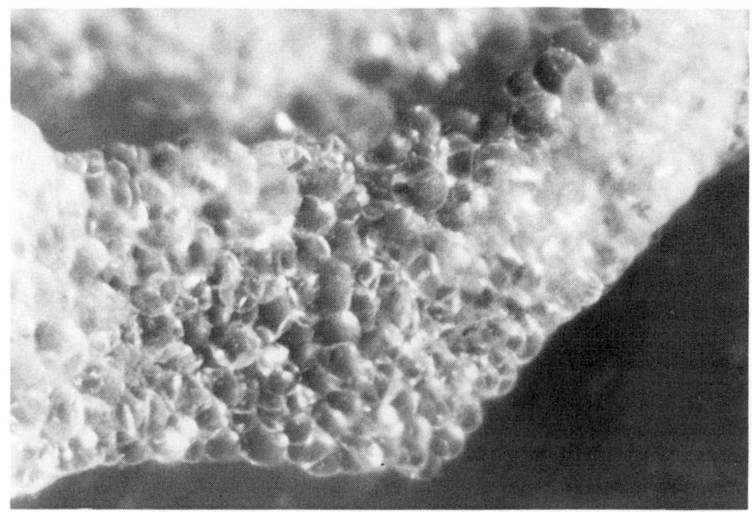

FIGURE 3 *Frog fat body in organ culture. 50X.*

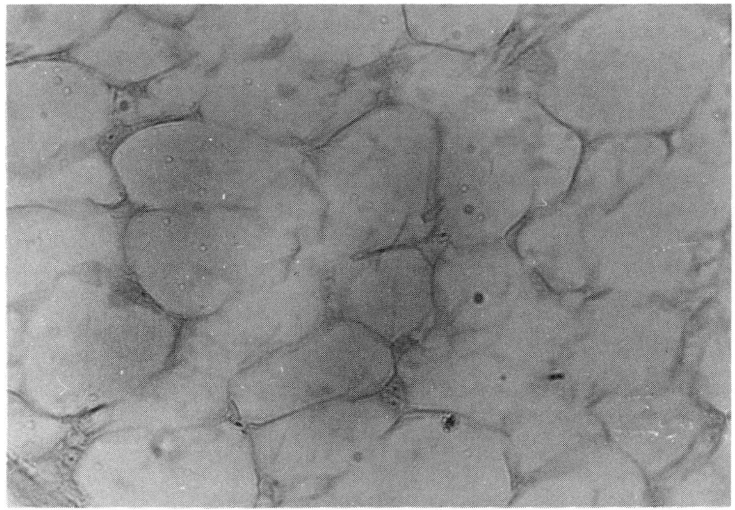

FIGURE 4 *Gram stained histological section of frog fat body organ culture after 13 days incubation. No bacteria added. 800X.*

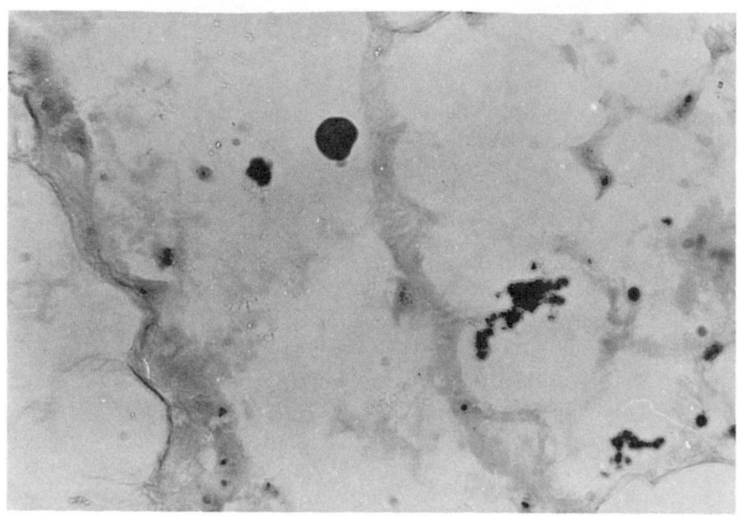

FIGURE 5 *Gram stained histological section of frog fat body organ culture after 13 days incubation. Arrows indicate foci of gram positive bacteria. Bacteria were added five days after the organ culture was initiated. 800X.*

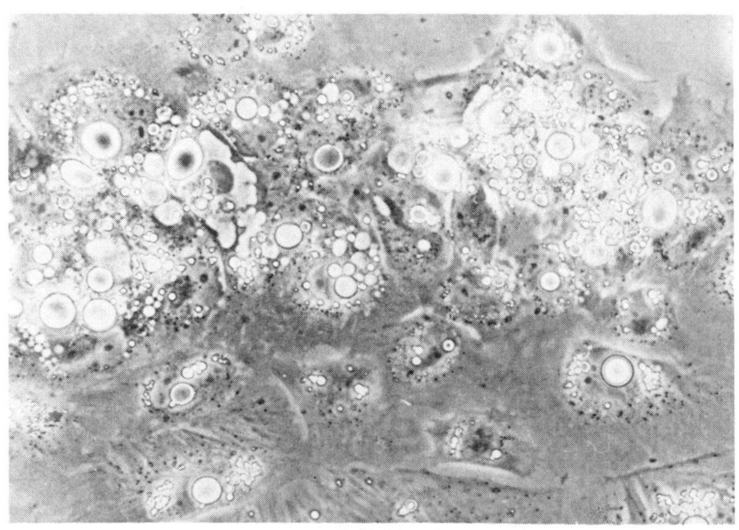

FIGURE 6 *Adipocytes from explant culture after eight days in culture. 700X.*

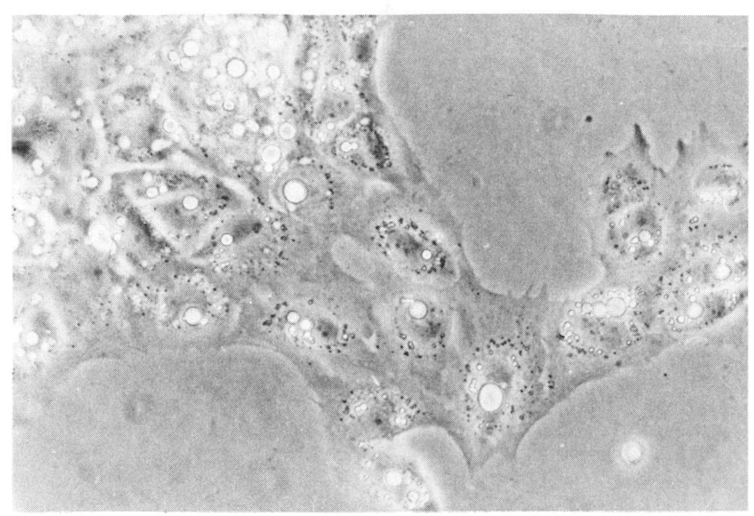

FIGURE 7 *Adipocytes from explant culture after 12 days in culture. 700X.*

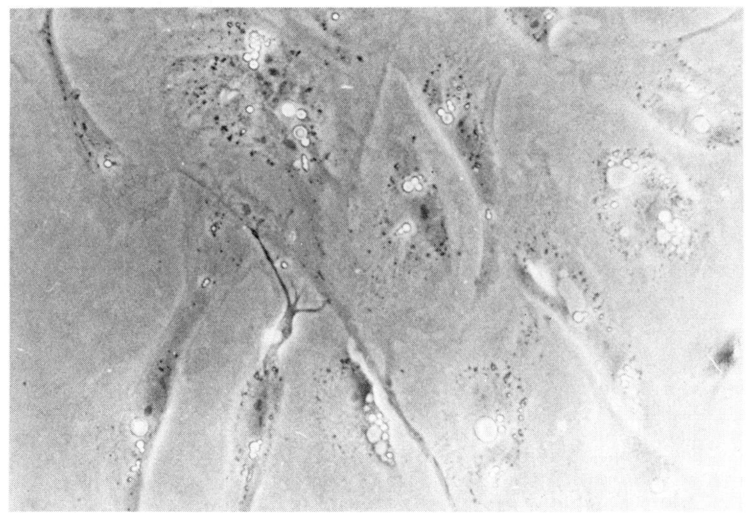

FIGURE 8 *Adipocytes from explant culture after 18 days in culture. 700X.*

patterns of outgrowth are shown in Figs. 6, 7 and 8. Figure 6 shows the outgrowth immediately adjacent to the explant after 8 days in culture. Light refractile fat inclusions of varying size can be seen. It would appear that the loss of fat might involve the disruption of the large single inclusion into a number of smaller droplets. Even after 12 days in culture as shown in Fig. 7, the cells are still stellate and contain substantial amounts of fat. Finally as shown in Fig. 8, after 18 days in culture, the cells are more fibroblastic in shape, an intact nucleus can be observed and there is almost complete replacement of the fat by cytoplasm.

IV. DISCUSSION

A major problem in the study of Red-leg disease in frogs has been the large number of different bacteria which have been shown to contribute to the disease syndrome. This complexity has effectively limited the studies on this syndrome and may reflect the same interactions which are involved in human infections caused by opportunistic pathogens. However in the case of the frog disease, the observation by Van der Waaij *et al.* (1974) that the microbial flora of the frog could be substantially reduced by hibernating temperatures, and the present observations that the *Corynebacterium sp.* can infect fat in an organ culture system offers some new experimental approaches to understanding Red-leg disease.

The limits on the use of the frog fat body organ culture system appear to be the variability that we have observed in the cultural response of that tissue. Amphibians are essentially lean animals and do not have stores of subcutaneous fatty tissue as found in birds and mammals. The fat stored in the fat bodies are apparently necessary for maintaining the health and normal development of the gonads. There is thus a cyclic response in the deposition of fat and some of the variability observed in our studies may reflect differences in the stages of fat body development in the donor animal.

In general our results agree with the known response of amphibian cells in culture (Monnickendam and Balls, 1973), and more specifically there is agreement with the known response of adipose cells in culture (Adebonojo, 1975).

ACKNOWLEDGMENTS

This work was supported by NIH Grant No. 5 P06 RR00635-04AR, The Louisiana State University Agriculture Experiment Station and The Department of Microbiology.

REFERENCES

Amborski, R. L., Glorioso, J. C. and Amborski, G. F. (1974). *In* "Proc. Gulf Coast Regional Symp. Diseases of Aquatic Animals." (R. L. Amborski, M. A. Hood a d R. R. Miller, eds.), pp. 217-224. Center for Wetland Resources Louisiana State University, Baton Rouge, La.

Abebonojo, F. O. (1975). *The Yale J. of Biol. and Med. 48,* 9-16.

Emmerson, H. and Norris, C. (1905). *J. Exptl. Med. 7,* 32-58.

Gibbs, E. L., Gibbs, T. J. and Van Dyck, P. C. (1966). *Lab. Anim. Care 16,* 142-160.

Gibbs, E. L., Nace, G. W. and Emmons, M. B. (1971). *Bioscience 21,* 1027-1034.

Gibbs, E. L. (1973). *Am. Zool. 13,* 781-783.

Glorioso, J. C., Amborski, R. L., Amborski, G. F. and Culley, D. D. (1974). *Am. J. Vet. Res. 35,* 1241-1245.

Kulp, W. L. and Borden, D. G. (1942). *J. Bact. 44,* 673-679.

Luna, L. G. (1968). "Manual of Histological Staining Methods of

The Armed Forces Institute of Pathology." McGraw-Hill Book Co., New York.

Miles, E. M. (1950). *J. Gen. Microbiol. 4,* 434-436.

Monnickendam, M. A. and Balls, M. (1973). *Experientia 29,* 1-17.

Prestt, I., Cooke, A. S. and Corbett, K. F. (1974). *In* "The Changing Flora and Fauna of Britain" (D. L. Haksworth, ed.) pp. 229-254. Academic Press, New York.

Russel, F. H. (1898). *J. Am. Med. A. 30,* 1442-1449.

Van der Waaij, D., Cohen, B. J. and Nace, G. W. (1974). *Lab. Anim. Sci. 24,* 307-317.

MICROASSAY OF BLUETONGUE VIRUS FROM CULICOIDES VARIIPENNIS*

B. M. BANDO
R. H. JONES

U. S. Department of Agriculture
Agricultural Research Service
Arthropod-borne Animal Disease Research Laboratory
Denver, Colorado

I. INTRODUCTION

Bluetongue is an arthropod-borne disease of domestic and wild ruminants and is biologically transmitted by the gnat, *Culicoides variipennis*. BTV has been isolated from a variety of animal tissues and insects by inoculation of ECE (Foster and Luedke, 1968). The EAS is convenient and economical and has been used routinely by researchers at the Arthropod-borne Animal Disease Research Laboratory. However, studies of the genetic factors affecting oral susceptibility of *C. variipennis* to infection with BTV have created a need to assay large numbers of individual gnats; the EAS presently in use does not satisfy this increased demand.

Cell culture systems are not widely used to isolate BTV (Howell *et al.*, 1970). However, a MAS that economizes on space,

*Abbreviations used: BTV, bluetongue virus; CPE, cytopathic effect; EAS, egg assay system; EBME, Eagle's basal minimal essential medium in Earle's balanced salt solution; ECE, embroyonating chicken eggs; ELD_{50}, virus dose required to infect 50% of embryos inoculated; FBS, fetal bovine serum; IHM, insect-holding medium; MAS, microassay system; $TCID_{50}$, virus dose required to infect 50% of cultures inoculated.

reagents, and personnel would be an ideal replacement for the EAS.

The purpose of this study was to determine whether a cell culture MAS would be as efficient in detecting BTV-infected gnats as the EAS and to compare the infection rate[1] of gnats artificially provided a blood meal containing cell culture adapted BTV to the infection rate of gnats fed a blood meal containing ECE adapted BTV.

II. MATERIALS AND METHODS

A. *C. variipennis*

Gnats were from two colonies of *C. variipennis* maintained at this laboratory. The Sonora (Texas) strain (000 line) has been maintained by Jones since 1957 without the addition of wild flies (Jones *et al.*, 1969). The Bruneau (Idaho) strain (036 line) was established in 1973.

B. Cell Cultures

BHK-21 (Clone 13)[2] was subcultured twice weekly and was grown in 10% FBS, 10% tryptose phosphate broth, and 80% EBME at a pH of 7.2. Suspensions for microplates contained 100,000 cells/ml.

VERO[3], African green monkey kidney, was subcultured once weekly and was grown in 10% FBS and 90% 199E at a pH of 7.0. Suspensions for microplates contained 150,000 cells/ml. Antibiotics were added to growth media to obtain a final concentration of

[1]Infection rate, a measure of oral susceptibility, was determined for each group of gnats and is the percentage of BTV-infected gnats per group assayed.

[2]American Type Culture Collection, 12301 Park Lawn Drive, Rockville, Maryland 20852.

[3]Gorgas Memorial Institute - Middle America Research Unit, Box 2011, Balboa Heights, Canal Zone.

gentamicin 50 mcg/ml, penicillin 200 units/ml, and Fungizone[1] 2 mcg/ml.

C. Egg Assay System

ECE were used to detect BTV in gnat samples and to prepare ECE adapted virus. Inoculations were made into the yolk sac of 11-day old ECE. Death pattern in eggs was the principal criterion used to determine whether gnats became infected with BTV.

D. Virus

BTV (62-45S)[2] was subpassaged twice in BHK-21; final titer was $10^{7.5}TCID_{50}$/ml in BHK-21 MAS. To prepare ECE adapted BTV, we subpassaged 62-45S 8 times in ECE; final titer was $10^{7.5}ELD_{50}$/ml.

E. Artificial BTV-Blood Meal

Gnats, 1-4 days old, were provided a blood meal that consisted of 1 part cell culture or ECE adapted BTV and 9 parts defibrinated sheep blood (Jones and Foster, 1971). Blood-fed female gnats (only the female is blood-sucking) were maintained on a sugar and water diet for 14 days to allow time for adequate virus multiplication; they were then killed and stored at -70°C until sample preparation. Individual gnats were disrupted by sonification in IHM (pH 7.0) that was 20% FBS and 80% EBME; gentamicin, penicillin, and Fungizone were added to obtain a final concentration of 100 mcg/ml, 200 units/ml, and 4 mcg/ml, respectively. Prepared gnat samples were held at 4°C until assayed.

F. Microassay System

Rigid, plastic, 96-well, flat-bottomed microplates[3] treated for cell culture were used. Each well received 50 μL of cell

[1] E. R. Squibb and Sons, Inc., Princeton, New Jersey 07540.
[2] Thomas L. Barber, USDA-ARS, Denver, Colorado 80225.
[3] Linbro Chemical Co., Los Angeles, California 90020.

TABLE I Infection Rate of BTV[a] Positive *C. variipennis*, Sonora Strain (000 Line), Assayed in MAS and EAS

Test	Virus adapted to	Number assayed	MAS[e]				EAS[b]	
			BHK-21		VERO			
			#Pos	IR[d]	#Pos	IR[d]	#Pos	IR[d]
1	ECE[c]	30	7	23%	6	20%	5	17%
	BHK-21	30	6	20%	5	20%	6	20%
2	ECE[c]	23	4	17%	2	9%	N.T.[f]	
	BHK-21	23	5	22%	4	17%	N.T.[f]	

[a]BTV: Bluetongue virus.
[b]EAS: egg assay system.
[c]ECE: embryonating chicken eggs.
[d]IR: infection rate, number positive/number assayed X 100
[e]MAS: microassay system.
[f]N.T.: not tested.

TABLE II Infection Rate of BTV[a] Positive *C. variipennis*, Bruneau Strain (036 Line), Assayed in MAS and EAS

			MAS[e]				EAS[b]	
			BHK-21		VERO			
Test	Virus adapted to	Number assayed	#Pos	IR[d]	#Pos	IR[d]	#Pos	IR[d]
1	ECE[c]	30	8	27%	6	20%	10	33%
	BHK-21	30	14	47%	12	40%	13	43%
2	ECE[c]	23	10	44%	9	39%	N.T.[f]	
	BHK-21	23	12	52%	11	48%	N.T.[f]	

[a]BTV: bluetongue virus.
[b]EAS: egg assay system.
[c]ECE: embryonating chicken eggs
[d]IR: infection rate, number positive/number assayed X 100
[e]MAS: microassay system
[f]N.T.: not tested.

suspension and 50 μL of the appropriate growth medium; 4 wells of BHK-21 and VERO cell culture received 25 μL of sonified gnat sample. To serve as a control, gnat-free IHM was added to 4 wells of each culture. Microplates were incubated at 35°C in a humidified, 2% CO_2 in air atmosphere. Monolayers were examined microscopically on the 5th, 6th, and 7th days for the development of CPE. Monolayers with 75% or greater CPE were scraped with a pipette, aspirated, and placed in a equal volume of buffered lactose peptone[1]. Cell cultures with CPE were confirmed as BTV-positive by an indirect immunofluorescent staining technique (Jochim *et al.*, 1974).

III. RESULTS AND DISCUSSION

Data in Table I, Test 1 showed that infection rates were similar for each assay system. Test 2 supported the results of Test 1 for the BHK-21 and VERO MAS. In all tests (Tables I and II) the BHK-21 MAS appeared to be able to detect more BTV-infected gnats than the VERO MAS regardless of the source of the blood meal.

Data in Table II, Test 1 confirmed that the infection rates obtained with each assay system were similar; however, the infection rates for the 036 line gnats were typically higher than those for the 000 line. Unpublished laboratory data suggested that the 036 line had a greater oral susceptibility than the 000 line. Therefore, we expected the 036 line to have a higher infection rate than the 33% actually obtained when gnats were fed ECE adapted BTV and were assayed in EAS. An infection rate of 47% was obtained for the 036 line fed BHK-21 adapted BTV and assayed in BHK-21 MAS.

[1]Each liter of buffered lactose peptone (pH 7.2) contains 220 ml 0.15 M $NaH_2PO_4 \cdot H_2O$, 780 ml 0.15 M Na_2HPO_4, 100 g lactose, and 2 g peptone. Final pH is 7.2.

Our results indicated that the source of virus may have influenced the difference in the infection rates between the 000 and the 036 lines. When the infection rates for the 000 and 036 line gnats fed ECE adapted BTV are compared in all assay systems (ECE, Test 1, Table I), differences are 4%, 0%, and 16%. However, when infection rates for the 000 and 036 line gnats fed BHK-21 adapted BTV are compared in all assay systems (BHK-21, Test 1, Table I), differences are 27%, 23%, and 23%. This comparison of differences between Tables I and II for Test 1 suggests that the source of the BTV will affect the infection rate obtained in all the assay systems used. But differences in infection rates for the 000 line fed ECE or BHK-21 adapted BTV are 3% (Table I, Test 1) whereas differences for the 036 line were 10-20% (Table II, Test 1). Apparently, the source of virus affected the infection rates for the 036 line but not for the 000 line. The methods of virus preparation may also have influenced the number of gnats that are infected; the ECE adapted BTV was a whole embryo extract containing extraneous protein but the BHK-21 adapted BTV contained very little extraneous protein.

In summary, the MAS apparently will be a suitable substitute for the EAS. Also the use of the BHK-21 MAS and BHK-21 adapted BTV infective blood meals may result in higher and more consistent infection rates.

ACKNOWLEDGMENTS

The authors gratefully acknowledge the technical assistance of Marlin Larson, Howard Rhodes, and Regina DeHerrera.

REFERENCES

Foster, N. M. and Luedke, A. J. (1968). *American J. Vet. Res. 29,* 749-753.

Howell, P. G., Kumm, N. A. and Botha, M. J. (1970). *Onderstpoort J. Vet. Res. 37,* 59-66.

Jochim, M. M., Barber, T. L. and Bando, B. M. (1974). Proceedings of American Association of Veterinary Laboratory Diagnosticians, 17th Annual Meeting, 91-103.

Jones, R. H. and Foster, N. M. (1971). *J. Med. Ent. 8,* 499-501.

Jones, R. H., Potter, H. W. and Baker, S. K. (1969). *J. Econ. Ent. 62,* 1483-1486.

HUMAN COLORECTAL CELL LINES AS A SOURCE FOR CARCINOEMBRYONIC ANTIGEN (CEA)

ALBERT LEIBOVITZ
WILLIAM B. MCCOMBS, III
CAMERON E. MCCOY
KENNETH C. MAZUR
NANCY D. MABRY
JAMES C. STINSON

Scott and White Clinic
Temple, Texas

I. INTRODUCTION

Cancer of the large bowel represents this country's most common malignant neoplasm (excluding those of skin), with an overall five-year survival rate of only about 40% (Sherlock, 1974). The detection by Gold and Freedman (1965) of a tumor-specific antigen, carcinoembryonic antigen (CEA), in colorectal carcinomas stimulated intensive research to develop suitable immunologic tests. Most published studies on the isolation and purification of CEA are based on the use of human colorectal solid tumors as the starting material; metastatic tumors to the liver are also rich sources of CEA (Krupey *et al.*, 1972). However, as noted by

Goldenberg and Hansen (1972), such tumor material is difficult to procure and the results obtained are variable.

The establishment of permanent cell lines from human colorectal adenocarcinomas may offer an alternate route for the isolation and purification of CEA. Although relatively few permanent cell lines have been established from human colorectal adenocarcinomas, several of these have been shown to secrete CEA in moderate amounts (Egan and Todd, 1972; Tompkins *et al.*, 1974; Drewinko *et al.*, 1976; Tom *et al.*, 1976; McCombs *et al.*, 1976). During the past few years, we have established 11 human colorectal cell lines. These cell lines were classified into three groups based on morphology, modal chromosome number, and ability to synthesize CEA (Leibovitz *et al.*, 1976). Group 1 and group 2 cell lines were low to moderate producers of CEA (8 to 214 ng/10^6 cells), whereas group 3 cell lines produced CEA in relatively large amounts (1200 to 7500 ng/10^6 cells).

II. MATERIALS AND METHODS

A. Processing of Specimens

All colon, rectal, and involved lymph node tissues studied were from surgical specimens submitted for pathologic diagnosis and grading. Portions of those specimens known to represent adenocarcinoma were placed in sterile Petri dishes and sent to the tissue culture laboratory. These specimens were immediately covered with a complete growth medium containing antibiotics. Specimens received in the morning were processed the same day. Those received in the afternoon were stored at 4° and processed on the following morning.

Except in the early days of our studies (McCombs *et al.*, 1976), nonenzymatic methods were used in the processing of tumor specimens for tissue culture, namely, the spillout techniques (Lasfargues and Ozzello, 1958; Leibovitz *et al.*, 1976).

1. *Spillout Technique*

After removal of normal tissue, necrotic areas, and blood clots from the tumor tissue specimen, it was placed in a sterile plastic Petri dish containing about 15 ml of growth medium. In this dish it was sliced into 1-mm cubes with crossed Bard-Parker no. 11 blades. The supernatant fluid was harvested with a 5-ml pipet fitted with a rubber bulb (Flow Laboratories, Rockville, Maryland) and transferred to 15-ml sterile plastic screw-capped centrifuge tubes (BioQuest, Cockeysville, Maryland). The tubes were placed in a rack for several minutes to permit the fine minces to settle. Then the supernate was removed to fresh tubes (second supernate). The fine minces were suspended in 2 ml of growth medium, dispersed into 25-cm^2 plastic flasks, and incubated at 37° for at least three days (without being disturbed) to permit adherence and initial outgrowth.

The second supernate was centrifuged at 500*g* for 5 min to settle clusters of cancer cells in suspension. This yield was washed at least three times with fresh growth medium to remove debris and toxic products being given off by dead or dying cells. After resuspension in growth medium, the viable cells were counted by the trypan blue exclusion technique. This suspension was inoculated into 25-cm^2 flasks at 10^6 viable cells/flask; if less than 10^6 viable cells were present, the entire yield was inoculated into one flask.

2. *Spinner Spillout Technique*

The minces remaining after the initial removal of supernatant fluid were transferred to a 50-ml spinner flask (Bellco Glass, Vineland, New Jersey) in 30 ml of complete growth medium and rotated at 200 rpm for 30 min. The supernate of this spinner-spillout technique was processed in the same manner as the second supernate.

After overnight incubation at 37°, all flasks except those containing fine mince were examined for the presence of viable cancer cells. The supernates were removed, pooled, and centrifuged at 500*g* for 5 min. The yield was washed three times with growth medium and inoculated into a 25-cm^2 flask in 5 ml of growth medium. The original flasks were refed with 5 ml of medium. The flasks containing fine mince were examined after three days and fed with 5 ml of growth medium. From then on, all flasks were fed once per week by complete removal of the spent medium and replacement with 5 ml of fresh medium.

B. Growth Media

Our initial growth medium, L-15-CI (Leibovitz *et al.*, 1973), was Leibovitz L-15 medium (Leibovitz, 1963) supplemented with 10% unheated fetal calf serum (KC Biological, Lenexa, Kansas), insulin (Eli Lilly, Indianapolis, Indiana), 0.01 units/ml, and cortisol (Solu-Cortef, Upjohn, Kalamazoo, Michigan), 10 μg/ml. Additional ingredients were added as detoxifiers, as growth stimulants, or to enable the tumor cells to compete with the more rapidly growing stromal cells. Our most recently developed medium, L-15-D (Leibovitz *et al.*, 1976), which is used for detoxification, is described in Table I. This medium also contains α-mercaptopropionylglycine for its lathyrogenlike (Karnovsky and Karnovsky, 1961) action in inhibiting the growth of most collagen-producing stromal fibroblasts. All cultures were maintained in this medium until a cell line was established. Once the cells were adapted to the *in vitro* environment, this complex medium was no longer required and they were maintained on medium L-15, 10% foetal calf serum.

C. Screen for CEA Synthesis

Three 25-cm^2 plastic tissue culture flasks were inoculated with 10^6 cells in 10 ml of L-15-CI medium and incubated at 37° for 21 days without refeeding. The supernate from each flask was

TABLE I Medium L-15-D (Detoxification Medium) for Establishment of Human Cancer Cells *in Vitro*

To 765 ml of pyrogen-free distilled water, add the following:

L-15 powder medium[a]	1 liter pkg.
Polyvinylpyrrolidine,[b] GMW 360,000, 10% solution	10.0 ml
Methylcellulose,[c] 15 cps, 2% solution	100.0 ml
Catalase,[d] 50,000 units/ml	1.0 ml
Insulin,[e] 40 units/ml	0.25 ml
Cortisol,[f] (Solu-Cortef), 10 mg/ml	1.0 ml
Heparin sodium,[g] 0.5%	0.1 ml
Sodium polypectate,[h] 0.25% in 10% sucrose	10.0 ml
Yeastolate,[i] 5%	10.0 ml
α-Mercaptopropionylglycine,[g] 10 mg/ml	1.0 ml
Polyestradiol phosphate (Estradurin),[j] 100 μg/ 100 ml	1.0 ml
Fetal calf serum,[k]	100.0 ml

Sterilize by ultrafiltration (Millipore)

[a]Grand Island Biological Co., Grand Island, New York.
[b]General Biochemicals, Chagrin Falls, Ohio.
[c]Dow Chemical Co., Midland, Michigan
[d]Worthington Biochemical Co., Freehold, New Jersey.
[e]Eli Lilly, Indianapolis, Indiana.
[f]Upjohn Co., Kalamazoo, Michigan.
[g]Calbiochem, LaJolla, California
[h]ICN Nutritional Biochemicals, Cleveland, Ohio
[i]Difco Laboratories, Detroit, Michigan
[j]Ayerst, New York, New York.
[k]KC Biological Co., Lenexa, Kansas

transferred to a 15-ml centrifuge tube, spun at 500*g* for 10 min to remove cells, and then transferred to sterile screw-capped test tubes for storage at -70° until it could be assayed for CEA. All assays were done within two weeks by radioimmunoassay using the Roche kit and procedure manual (Hoffman-La Roche, Nutley, New Jersey). The cells remaining in each flask were harvested by the trypsin-EDTA technique, and the final population of viable cells was determined by the trypan blue exclusion method. CEA synthesis was calculated per 10^6 cells to relate our findings to those of other investigators (Drewinko *et al.*, 1976).

III. RESULTS

A. Establishment of Cell Lines

Of the 163 specimens processed from the middle of 1971 to the end of 1975, 50 were lost to contamination (usually by saprophytic fungi), 88 failed to develop into a cell line, 11 cell lines became established, and 14 are still in progress. The cell line establishment rate increased significantly with the use of complex media designed to neutralize toxic substances released by the dead and dying cells and to enable the tumor cells to grow competitively with the stromal cells. A complex medium was no longer required after the cells became established.

The spinner-spillout technique usually yielded two- to five-fold more cancer cell clusters than did the spillout technique, but these cultures were more grossly contaminated with stromal cells. The supernate harvest pool yielded about as many clusters as did the spinner-spillout supernate. The fine mince bottles usually showed a mixture of cancer cells and stromal cells. The use of all four methods enhanced the chances of a successful isolation of a permanent cell line.

Viability counts of cancer cells, obtained by the spillout techniques, by the trypan blue exclusion method ranged from less than 1 to 50%, with the majority of specimens having 10 to 20% viable cells. Although such counts indicated that from about 15,000 to several million viable cells were present (mean, 5×10^5), there was no correlation between cell count and yield of cancer cells capable of proliferating into monolayers. Regardless of the initial count, from 0 to 100 islands of epitheliumlike cells were evident (commonly 10 to 20 in flasks containing viable cancer cells). The cancer cells were readily recognized as tight clusters of islands of epitheliumlike cells; some became firmly attached to the flask wall within 24 hours of explantation and others were floating in the medium. The clusters of cancer cells appeared to lie dormant for variable periods ranging from several weeks to about six months before obvious growth was noted.

This long lag phase permitted the stromal cells, which were relatively few in number initially, to proliferate. At first, the collagenase method of Lasfargues and Moore (1971) was used to prevent the fibroblastlike cells from smothering the cancer cells. However, some fibroblastlike cells were not deterred by collagenase and they completely filled the flask. In one instance, fibroblastlike cells grew under the cancer cell clusters and popped them into the supernatant fluid; the cancer cell clusters were recovered from the supernatant fluid and established as a stroma-cell-free cell line (McCombs *et al.*, 1976). Recently, we noted that a mercaptan, α-mercaptopropionylglycine (Calbiochem, La Jolla, California), functioned like a lathyrogen and permitted the cancer cells and stromal cells to grow as co-cultures. As with collagenase, this compound does not deter the growth of all fibroblastlike cells. However, the mercaptan also enhanced cancer cell growth and was incorporated, at 0.1 μg/ml, as part of the detoxification growth medium (L-15-D).

The clusters or islands of epitheliumlike cells slowly expanded but, in the initial flasks, would rarely expand sufficiently to form a complete monolayer. Attempts to pass the cells, either by the trypsin-EDTA method or by scraping, before they became well established in their *in vitro* environment usually were disastrous. When the clusters showed definite evidence of doming, they usually could be passed by either method. One of our lines was maintained in the original flask for more than a year before it could be passed successfully.

B. Classification of Cell Lines

The 11 cell lines were placed into one of three groups based on their cytogenetics, morphologic features, and ability to synthesize CEA (Leibovitz *et al.*, 1976). These characteristics are summarized in Table II.

TABLE II Grouping of Human Colorectal Cell Lines by Morphology Cytogenetics, and Ability to Synthesize CEA

Cell line	Passage	CEA synthesis ($ng/10^6$ cells)[a]	Micro-vesicular bodies	Modal chromosome number
Group 1				
SW-48	63	8	0	47
SW-707	12	30	0	47
SW-802	7	215	0	47
Group 2				
SW-480	73	21	0	55
SW-620	51	11	0	54
Group 3				
SW-403	53	7500	+	66
SW-742	9	2000	+	54
SW-837	5	1200	+	42 (85%) 80 (15%)
SW-948	4	2000	+	76
SW-1083	6	5500	+	42 (50%) 80 (50%)
SW-1116	7	7000	+	63

[a]Mean of triplicate analyses by radioimmunoassay (Roche kit) of supernate from 10^6 cells grown in 10 ml of media from three weeks without refeeding; results corrected for number of cells in final population.

These cytogenetic studies were limited to determination of modal chromosome numbers. As noted by Drewinko *et al.* (1976), modern banding techniques would be essential for proper analysis of the karyograms. CEA content in the medium was measured after 21 days of incubation because growth studies in our laboratory had determined that maximal CEA synthesis starts after the cells progress through the log phase of growth and enter the stationary phase.

Group 1 cells had a modal chromosome number of about 47. By light microscopy, these cells were similar morphologically to those reported in the literature. Ultrastructurally, two distinct cell types were apparent. The cells in the middle of the colonies were isodiametric and loosely arranged with the most notable form of adhesion being desmosomes. Fasciculated filaments were prominent in the cytoplasm. The cells on the periphery of the islands were more columnar and often aligned in a pattern resembling that of normal absorptive epithelium. The free surface of these peripheral cells often formed microvilli. A definite glycocalyx was evident in cell line SW-802. CEA synthesis in the group 1 cell lines was low to moderate, ranging from 8 to 214 ng/10^6 cells.

Both group 2 cell lines were derived from the same patient. SW-480 was isolated from the primary adenocarcinoma arising in the colon, whereas SW-620 was isolated from a lymph node when the malignancy recurred with widespread metastasis. SW-480 cells grew as a mixture of small islands of epithelial cells and individual bipolar cells. On electron microscopy, the SW-480 cells were polygonal and often had microvilli on their free surfaces. SW-620 had fewer islands, and most of the cells appeared as a mixture of small individual spherical cells and bipolar cells. The SW-620 cells seemed to be further dedifferentiated because all cells were isodiametric and there was no evidence of microvilli on their cell surfaces. Both cell lines were hyperdiploid and both were low producers of CEA, from 11 to 21 ng/10^6 cells.

The group 3 cell lines were not significantly different from the group 1 cell lines by light microscopy. Ultrastructurally, SW-403 and SW-948 had a demonstrable glycocalyx, as did SW-802 (a group 1 cell line). Two of the group 3 cell lines, SW-403 and SW-1083, had a prominent Golgi apparatus, which was not noted in any of the other cell lines. SW-837 had unique cytoplasmic structures, resembling stacked lamina, which were often associated with lipid; a thin section of a single cell could contain as many as five to ten of these lamina, which were often associated with lipid; a thin section of a single cell could contain as many as five to ten of these laminar structures. Multivesicular bodies were present in all group 3 cell lines in the cytoplasm and along the brush borders. Multivesicular bodies have been reported in solid tumor tissue from the surface of neoplastic glandular epithelium, especially in the digestive tract and in mammary glands (Dalton and Haguenau, 1973). Two cell lines, SW-837 and SW-1083, had bimodal populations. The other cell lines in this group were hyperdiploid to hypertriploid. All cell lines in group 3 synthesized relatively large amounts of CEA, from 1200 to 7500 ng/10^6 cells.

IV. DISCUSSION

Establishment of permanent cell lines from human colorectal adenocarcinomas requires isolation of viable clusters of cancer cells from a milieu of dead and dying cells as well as bacteria and fungi and protection of these isolated cells from the competitive outgrowth of normal stromal cells while they are adapting to the *in vitro* environment. In our hands, the spillout and spinner-spillout methods have been more satisfactory than trypsinization methods in regard to yield of viable clusters of cancer cells with minimal stromal contamination. The relatively small yield of viable clusters of cancer cells in relation to the initial count of viable cells (trypan blue exclusion method) suggested that most of the cells were dying and that use of a detoxification medium might enable the surviving cells to withstand the powerful proteases and peroxides released by the dying cells. Although the tumor cells did not grow any better in medium L-15-D than in medium L-15-CI, the success rate increased from about 3 to 16% when the medium with detoxification ingredients was used.

To combat contamination, we now use the following antibiotics: gentamicin, 40 μg/ml; streptomycin sulfate, 40 μg/ml; and amphotericin B, 1.25 μg/ml. No evidence of contamination by Mycoplasma organisms has been noted by electron microscopy screening in our laboratory or by a battery of tests in Dr. Fogh's laboratory (personal communication).

The addition of α-mercaptopropionylglycine enabled the cancer cells to compete with the more rapidly growing stromal cells for survival. Often, they grew as cocultures. When the colonies of cancer cells grew sufficiently to start doming (three-dimensional growth), the monolayers could be trypsinized and passed; the original flask was always refed. The cancer cells seemed to be extremely sensitive to trypsinization before they started to grow three-dimensionally and often would be destroyed if passed prematurely. The stromal cells eventually died out or were overgrown by the cancer cells. The detoxification medium may also be important at this stage to neutralize the toxins released by the dying stromal cells. When the cancer cells were readily subculturable and in relatively pure culture, the complex detoxification medium was no longer required.

Previous investigators (Drewinko *et al.*, 1976; McCombs *et al.*, 1976; Tom *et al.*, 1976; Tompkins *et al.*, 1974) have reported that human colorectal adenocarcinoma cells have a modal chromosome number of 47 to 49. These synthesize low to moderate amounts of CEA and resemble our group 1 cell lines. HT29, isolated by Fogh and Trempe (1975), had a biomodal population, resembling some of our group 3 cell lines, but synthesized relatively low amounts of CEA (Egan and Todd, 1972; Tom *et al.*, 1976). Our report (Leibovitz *et al.*, 1976) was the first, to our knowledge, to record the group 2 and group 3 cell lines.

The group 3 cell lines offer a controlled source of obtaining CEA in relatively large quantities. The cells retain this capability over multiple passages. Our most extensively studied cell line, SW-403, has often been split as high as 1:100, and after 53 splits still yield over 7000 ng CEA per 10^6 cells. Whether these cells will eventually dedifferentiate is not known at this time. However, ample stocks can be stored in the liquid nitrogen refrigerator to ensure a continuous supply of high CEA producing cells. The relatively high synthesis of CEA by the group 3 cell lines appears to be related to differentiating cells that have multiple copies of the chromosome having the loci for CEA synthesis. A collaborative study is now in progress with the Wistar Institute to test this hypothesis.

V. SUMMARY

Eleven human colorectal adenocarcinoma cell lines established in this laboratory were classified into three groups based on morphologic features (light and electron microscopy), modal chromosome number, and ability to synthesize carcinoembryonic antigen (CEA). Group 1 cell lines contained both dedifferentiated and differentiating cells growing in tight clusters or islands of epitheliumlike cells; their modal chromosome number was about 47 and they synthesized small to moderate amounts of CEA, from 8 to 214 ng/10^6 cells. Group 2 cell lines were more dedifferentiated, were hyperdiploid, and synthesized small amounts of CEA, from 11 to 21 ng/10^6 cells. Group 3 cell lines were morphologically similar to group 1 by light microscopy. They differed ultrastructurally by containing multivesicular bodies; the modal chromosome number varied from hyperdiploid to hypertriploid, or they had bimodal populations of hypodiploid and hypertriploid cells, and they synthesized relatively large amounts of CEA, from 1200 to 7500 ng/10^6 cells.

The relatively high production of CEA by the group 3 cell lines appears to be related to differentiating cells that have multiple copies of the chromosome having the loci for CEA synthesis. These cell lines offer a constant source of CEA in significant quantities over multiple passages.

REFERENCES

Dalton, A. J., Haguenau, F. (1973). "Ultrastructure of Animal Viruses and Bacteriophages: An Atlas," pp. 392-397. Academic Press, New York.

Drewinko, B., Rohmsdahl, M. M., Yang, L. Y., Ahearn, M. J., Trujillo, J. M. (1976). *Cancer Research 36*: 467-475.

Egan, M. L., Todd, C. W. (1972). *J. Nat. Cancer Inst. 49*: 887-889.

Fogh, J., Trempe, G. (1975). *In* "Human Tumor Cells In Vitro" (J. Fogh, ed.), pp. 115-160. Plenum Press, New York.

Gold, P., Freedman, S. O. (1965). *J. Exp. Medi. 121*: 439-462.

Goldenberg, D. H., Hansen, H. J. (1972). *Science 175*: 1117-1118.

Krupey, J., Wilson, T., Freedman, S. O., Gold P. (1972). *Immunochemistry 9*: 617-622.

Karnovsky, M. J., Karnovsky, M. L. (1961). *J. Exp. Medi. 113*: 381-403.

Lasfargues, E. Y., Ozzello, L. (1958). *J. Nat. Cancer Inst. 7*: 21-25.

Leibovitz, A. (1963). *Am. J. Hygiene 78*: 173-180.

Leibovitz, A. (1975). *In* "Human Tumor Cells In Vitro" (J. Fogh, ed.), pp. 23-50. Plenum Press, New York.

Leibovitz, A., McCombs, W. B. III, Johnston, D., McCoy, C. E., Stinson, J. C. (1973). *J. Natl. Cancer Inst. 51*: 691-697.

Leibovitz, A., Stinson, J. C., McCombs, W. B., III, McCoy, C. E., Mazur, K. C., Mabry, N. D. (1976). *Cancer Res. 36*: 4562-69.

McCombs, W. B. III, Leibovitz, A., McCoy, C. E., Stinson, J. C., Berlin, J. D. (1976). *Cancer 38*: 2316-2327.

Sherlock, P. (1974). *Am. J. Dig. Dis. 19*: 933-934.

Tom, B. H., Rutzky, L. P., Jakstys, M. M., Oyasu, R., Kaye, C. I., Kahan, B. D. (1976). *In Vitro 12*: 180-191.

Tompkins, W. A. F., Watrach, A. M., Schmale, J. D., Schultz, R. M., Harris, J. A. (1974). *J. Natl. Cancer Inst. 52*: 1101-1110.

THE ELABORATION OF GLYCOPROTEINS INTO CULTURE MEDIUM BY HUMAN OVARIAN TUMOR CELLS

H. J. ALLEN
J. J. BARLOW
E. A. Z. JOHNSON

Roswell Park Memorial Institute
Department of Gynecology
Buffalo, New York

I. INTRODUCTION

Ovarian cancer is the fourth most frequently fatal cancer in women in the United States. There are generally no early symptoms for this disease and diagnosis is usually made at an advanced stage when prognosis is poor (Lingeman, 1974). Epithelial cancers of the ovary are the predominant types of which serous tumors account for 40-60%; mucinous tumors account for 10-15%; undifferentiated tumors account for 16%; and endometrioid tumors account for 14% (Lingeman, 1974; Barber *et al.*, 1975; Curling *et al.*, 1975; and Young *et al.*, 1974).

Some of the clinical features of this disease include (1) high cure rate (75-90%) for early cancer; (2) low cure rate (<10%) for advanced cancer; (3) tumor implantation on the peritoneal surface; (4) plugging of the lymphatics draining the peritoneal cavity; (5) generation of large volumes of ascites with occasionally large numbers of free tumor islets present; and (6) occasional generation of pleural effusions due to metastatic spread.

As part of a program to elucidate the role of cell surface glycoproteins in carcinogenic and metastatic processes, we have chosen ovarian cancer as one of our model systems. It is well known that a variety of cells elaborate glycoproteins into their growth medium and that at least some of these glycoproteins are derived from the cell surface (Molnar *et al.*, 1965; Cooper *et al.*, 1974; Ruoslahti *et al.*, 1975; Kraemer, 1967). It has also been demonstrated that some cells derived from human cancer can elaborate tumor-specific (associated) products into their culture medium (Hickok *et al.*, 1974; Breborowicz *et al.*, 1975). We have therefore made the assumption that ovarian tumor cells elaborate glycoproteins into their growth medium *in vivo*. Since ascites may result from the movement of fluids into the peritoneal cavity but not from the cavity (Feldman *et al.*, 1974; and Feldman, 1975), the tumor-derived glycoproteins might accumulate in this fluid. The amount of tumor-derived glycoproteins present would depend upon many factors, including tumor load, but is expected to be small.

In order to develop a method for detecting all tumor-derived glycoproteins present in pleural and peritoneal effusions independent of antigenicity, we have carried out studies on the biosynthesis and secretion of glycoproteins *in vitro* by human ovarian tumor cells. With the use of radioactive glycoprotein precursors, it was hoped to obtain radiolabeled glycoproteins from *in vitro* cultures which could be used as markers for the isolation of tumor-derived glycoproteins present in the patient effusions.

II. EXPERIMENTAL

A. Materials

D-[6-^{3}H] glucosamine (Sp. act. 10.13 Ci/mM, L-[6-^{3}H] fucose (Sp. act. 12.066 Ci/mM) and Aquasol were obtained from New England Nuclear, Boston, Mass. Sephadex and Sepharose were purchased from Pharmacia Fine Chemicals, Piscataway, N. J. RPMI 1640 culture medium with HEPES buffer was obtained from Grand Island Biological Co., Grand Island, N. Y.

B. Methods

1. Cell Work-up from Patient Effusions

Fresh pleural and peritoneal effusions obtained from ovarian cancer patients were centrifuged for 15 minutes, 500 xg at 4°C. Cell pellets were examined microscopically and slide smears were prepared.

The red blood cells in excessively bloody pellets were lysed by suspending the cells in 20 volumes of 0.83% ammonium chloride and incubating for 10 minutes at 4°C. The cells were washed 2 times with 20 volumes of cold PBS (0.15 M NaCl, 0.01 M phosphate, pH 7.2). For some cell preparations, the tumor cell population was enriched by incubating 10 ml portions of a PBS cell suspension in 4 oz Blake bottles at 37°C for 30 minutes. The tumor cells were poured from the bottles while the normal cells remained attached to the glass. The cells were suspended in 10 volumes PBS and the cell concentration was estimated by counting the cells in a hemocytometer. Cell viability was determined by the Trypan Blue method.

2. Cell Culture Conditions

Large cultures were set up in 4 oz glass Blake bottles containing 30 × 10^6 cells in 10 ml medium consisting of 90% RPMI

1640 medium with HEPES buffer, 10% heat inactivated autologous patient effusion, 0.3 mg streptomycin and 0.5 mg penicillin. To each were added 25 μCi of D-[6-^{3}H] glucosamine or L-[6-^{3}H] fucose.

The bottles were capped tightly and laid flat in a 37°C incubator for 15 to 40 hours.

For time course experiments, 1 ml cultures were set up in 20 × 40 mm serum vials with one-tenth the quantities used for the large cultures.

3. *Work-up of Cell Cultures*

The culture medium from large cultures was separated from the cells by centrifugation, dialyzed against deionized water and lyophilized.

At given time intervals, time course cultures were centrifuged. To 0.7 ml culture medium were added 2.5 ml ice cold 12.5% TCA. TCA precipitates were washed 4 times with cold 10% TCA, dissolved in 0.5 ml of 0.5 M KOH, bleached with 50 μl of 50% H_2O_2 and transferred to polyethylene Mini-vials with 5 ml Aquasol for tritium determination.

TCA supernatants were dialyzed exhaustively against deionized water and lyophilized in glass scintillation vials. To the vials were added 10 ml Aquasol.

The cells were washed 4 times with cold PBS and suspended in 0.7 ml of 0.5% bovine serum albumin solution. From this stage, TCA-precipitable and soluble fractions of the cells were processed as described for the culture medium. All radioactivity determinations were carried out in a Packard 3375 Tri-Carb scintillation spectrometer.

4. *Gel Filtration Chromatography*

Sephadex and Sepharose chromatography was carried out with a buffer consisting of 0.15 M NaCl, 0.015 M phosphate, pH = 7.0, 0.05% NaN_3. A constant ascending flow was maintained with a Gilson Mini-pulse pump. Column temperature was maintained at 18°C. Load samples were dissolved in 5 ml column buffer and the solutions were clarified by centrifugation before applying to gel columns. Column effluents were monitored for U.V. absorbing material. Aliquots of effluent fractions were placed in polyethylene Mini-vials with 4 ml Aquasol and their radioactivity was measured.

III. RESULTS

A. Cell Preparations

The biosynthesis and secretion of glycoproteins by four different cell populations were investigated. The cells shown in Fig. 1 were obtained from a peritoneal effusion of a patient with a serous cystadenocarcinoma of the ovary. About 97% of the cell population were non-malignant cells which included some leukocytes, mesothelial cells and a large number of activated histiocytes. Many of the latter had phagocytized erythrocytes. Since there were a very small number (less than 3%) of malignant cells present in this preparation, it was used in these studies as a "normal" cell population and has served as a control for comparative purposes.

In Fig. 2 are shown the cells from a pleural effusion of a patient with a primary Krukenberg tumor of the ovary. A minimum of 75% of these cells were malignant. The malignant cells were characteristically large, single cells with a highly vacuolated cytoplasm and a large, irregular nucleus.

Tumor cells obtained from peritoneal effusions of patients with serous and mucinous cystadenocarcinoma of the ovary are shown in Figs. 3 and 4, respectively. Tumor cells in the effusions of these two types of ovarian cancer were usually present

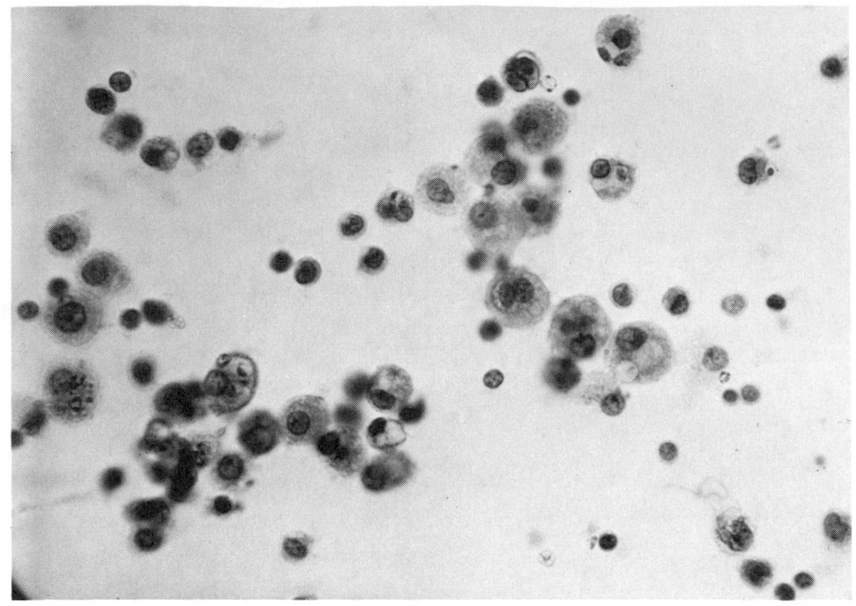

FIGURE 1 *Non-malignant cells obtained from the peritoneal effusion of a patient with serous cystadenocarcinoma of the ovary. Papanicolaou staining. Magnification 630 X.*

as clusters which had from several to about 50 cells. The cells contained large vacuoles, large irregular nuclei and often, prominent nucleoli were observed.

It was found that the tumor cell clusters did not adhere to glass as did the viable normal cells. This difference was used to enrich the tumor cell population to more than 90% in these preparations as described in Methods B. 1.

From viability determinations it was observed that 90% or more of the tumor cells and of normal cells which adhered to glass were viable. Invariably, a very low viability was observed for the normal cells which did not attach to glass surfaces.

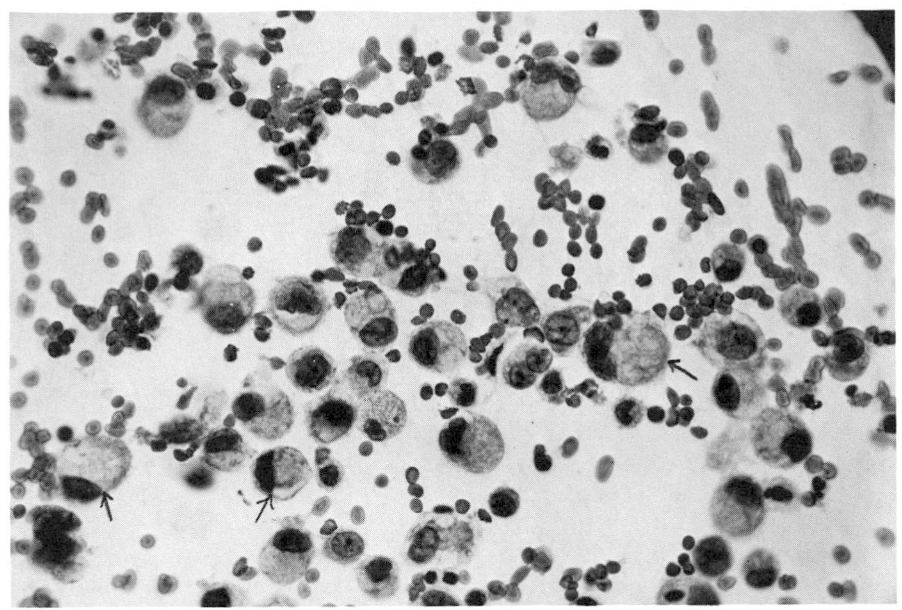

FIGURE 2 *Mixture of malignant and non-malignant cells obtained from pleural effusion of patient with primary Krukenberg tumor of the ovary. Several malignant cells are indicated by arrows. Papanicolaou staining. Magnification 630 X.*

B. Time Course Incorporation of D-[6-^{3}H] Glucosamine

Time course experiments were carried out to examine the ability of cell preparations to incorporate radiolabeled glycoprotein precursor into cellular components and macromolecules elaborated into the culture medium.

The radioactivity present in the TCA-precipitable and TCA-soluble non-dialyzable fractions from cells cultured with tritiated glucosamine has been plotted as a function of time in Fig. 5. Very little radioactivity was incorporated into the TCA-soluble non-dialyzable fraction from normal cells; whereas, there was a significant amount of radioactivity in corresponding fractions from both types of tumor cells.

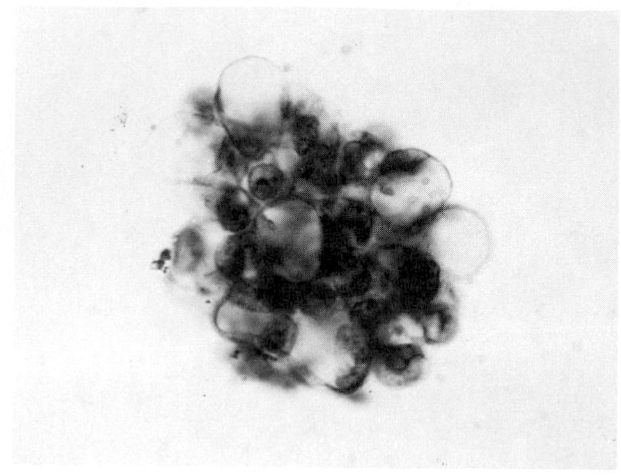

FIGURE 3 *Cluster of malignant cells obtained from peritoneal effusion of patient with serous cystadenocarcinoma of the ovary. Papanicolaou staining. Magnification 630 X.*

The incorporation of tritiated glucosamine into TCA-precipitable cellular glycoproteins continues for at least 120 hours with respect to the normal cell population. In contrast, for the serous tumor cells, the incorporation of tritiated glucosamine into corresponding fractions reaches a plateau after 50-60 hours. For the Krukenberg tumor cell population a plateau was reached after 30-40 hours. The experiment with the Krukenberg tumor cells was repeated on a different date and similar results were obtained. The decline in cellular TCA-precipitable glycoproteins for the Krukenberg tumor cells appeared to be due to cellular leakage during the washing of the cultured cells. The time course of elaboration of glycoproteins into the culture medium is shown in Fig. 6. As was the case for the cellular glycoproteins, the normal cells incorporated no tritiated glucosamine into the TCA-soluble non-dialyzable fraction of the culture medium. However, a significant portion of the radiolabeled glycoproteins elaborated

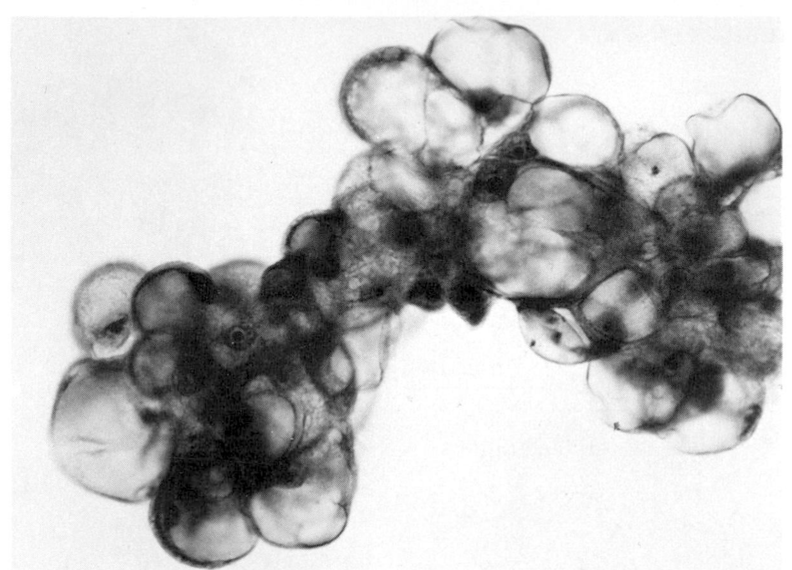

FIGURE 4 *Cluster of malignant cells obtained from peritoneal effusion of patient with mucinous cystadenocarcinoma of the ovary. Papanicolaou staining. Magnification 630 X.*

by the tumor cells into the culture medium were in the TCA soluble non-dialyzable fraction.

The incorporation of label into the acid-precipitable extracellular glycoproteins plateaued after about 120 hours for the serous tumor cells, plateaued after 60-80 hours for the Krukenberg tumor cells whereas no plateau was evident up to 120 hours for the normal cell population.

C. Sephadex Gel Filtration of Radiolabeled Glycoproteins in Culture Medium

From the time course experiments it was apparent that the cultured cells elaborated glycoproteins into the culture medium. To determine the molecular weight distribution of the radiolabeled glycoproteins, the dialyzed and lyophilized culture media were subjected to gel filtration on columns of Sephadex G-150 or G-200. The elution profiles for $[^3H]$ glucosamine-labeled glycoproteins are shown for four different cell types in Figs. 7, 8, 9 and 10. Essentially identical profiles were obtained for all culture media examined. Virtually all of the radioactivity was

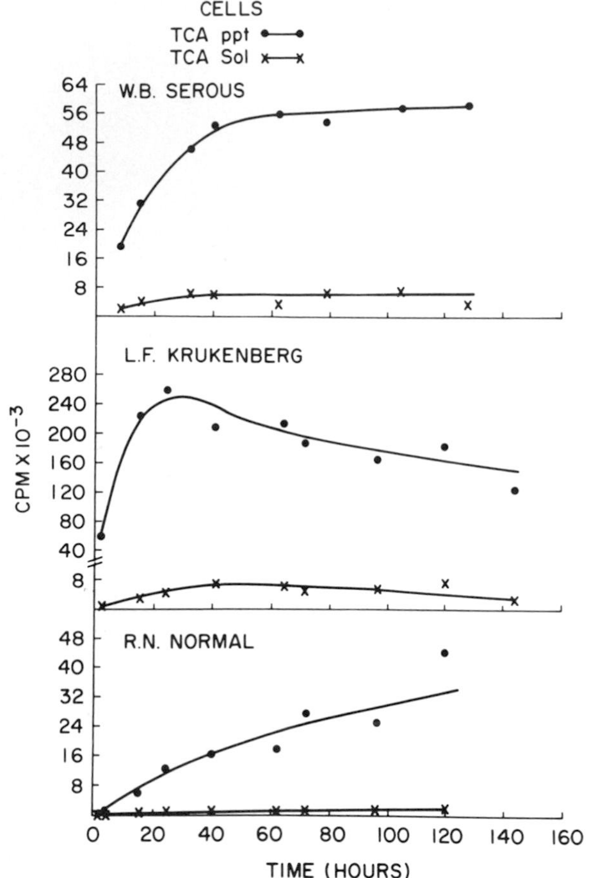

FIGURE 5 *Time course incorporation of [^3H] glucosamine into TCA-precipitable (●) and TCA-soluble non-dialyzable (x) fractions of cells incubated for 0 to 140 hours.*

eluted in the excluded fraction while only a small amount of radioactivity was distributed in the retarded fractions. No difference was observed in the radioactivity elution profile when 5 mg/ml cold D-glucosamine was included in the column load.

The protein elution profiles were also determined for these experiments. Protein profiles for all culture media were similar and the protein peaks corresponded to the three major peaks ob-

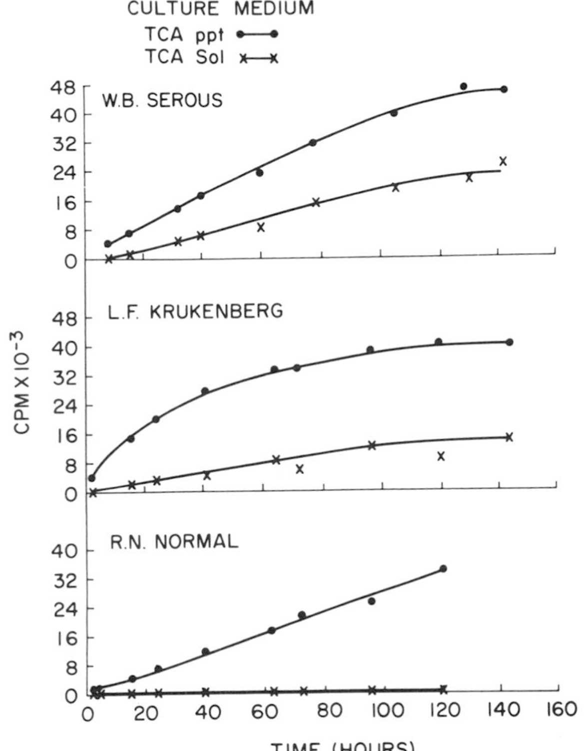

FIGURE 6 *Time course incorporation of [^{3}H] glucosamine into TCA-precipitable (●) and TCA-soluble non-dialyzable (x) fractions of culture medium from cells incubated for 0 to 140 hours.*

served when either normal human serum or plasma was chromatographed on the same columns.

In Fig. 11 is shown a comparison of the elution profiles for the [^{3}H] glucosamine-labeled and [^{3}H] fucose-labeled glycoproteins present in the culture media obtained from serous tumor cell cultures. The profile for the [^{3}H] fucose-labeled glycoproteins was identical to that obtained for the [^{3}H] glucosamine-labeled glycoproteins with the exception that a somewhat greater proportion of the [^{3}H] fucose label was eluted in retarded fractions. This difference in elution profiles has been consistently observed for all culture media studied.

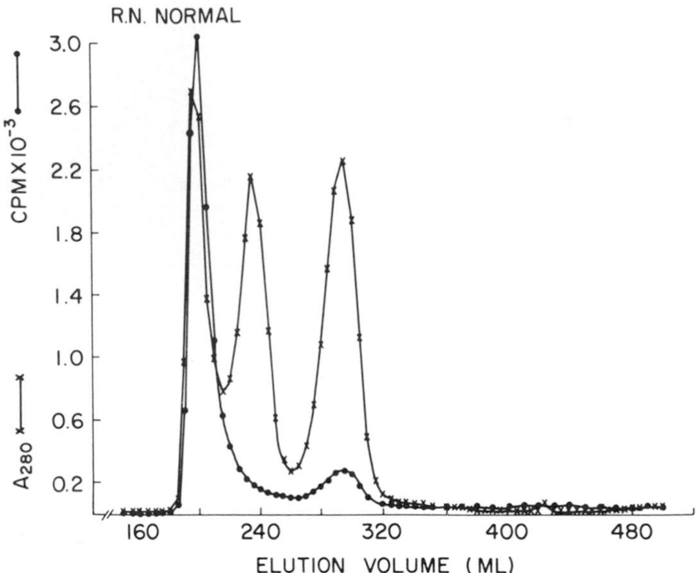

FIGURE 7 *Sephadex G-150 chromatography of culture medium from normal cell cultures incubated with [^{3}H] glucosamine. The culture medium from five 10 ml cultures incubated for 40 hours was prepared for chromatography as described in Methods B.4. Column: 2.6 × 95.5 cm. (Vo = 201.6 ml); flow rate = 17.1 ml/hr; fraction folume = 5 ml. Radioactivity in 50 μl aliquots from fractions was determined.*

Sephadex gel filtration profiles for [^{3}H] glucosamine and [^{3}H] fucose labeled glycoproteins have been obtained for the culture media derived from cultures of (1) cells collected from the same patient on different dates; (2) cells collected from several patients with serous cystadenocarcinoma of ovary; and (3) cells collected from patients with different histological types of ovarian cancer. All of these profiles for the [^{3}H] glucosamine-labeled glycoproteins were similar to those shown in Figs. 7-11. As was noted previously, all of the elution profiles for the [^{3}H] fucose-labeled glycoproteins were similar to each other with a consistently greater proportion of the radioactivity present in

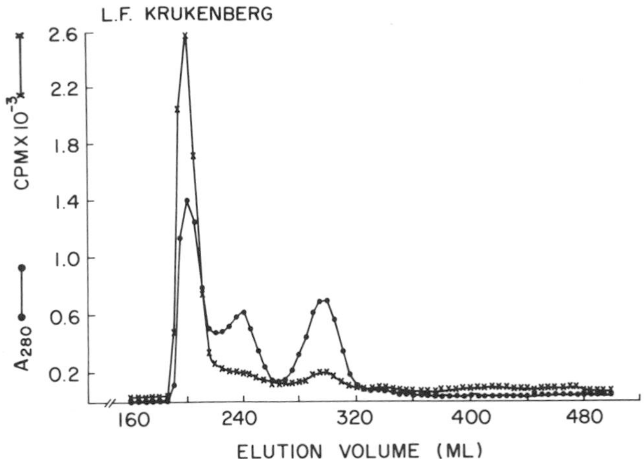

FIGURE 8 *Sephadex G-150 chromatography of culture medium from cultures of Krukenberg tumor cells. Culture medium from five 10 ml cultures incubated with [^{3}H] glucosamine for 18 hours was prepared for chromatography as described in Methods B.4. Column: 2.6 × 95.5 cm (Vo = 201.6 ml); flow rate = 17.0 ml/hr; fraction volume = 5 ml. Radioactivity in 100 μl aliquots from fractions was determined.*

the retarded fractions as compared to the [^{3}H] glucosamine-labeled glycoproteins.

D. Sepharose 6B Gel Filtration of the Excluded Fractions from Sephadex Columns

To further resolve the high molecular weight radiolabeled glycoproteins elaborated by the cultured cells, the excluded fraction from Sephadex chromatography of the four different culture media were chromatographed on columns of Sepharose 6B. As shown in Figs. 12-14, the radioactivity and protein of the 3 tumor cell-derived fractions were both resolved into an excluded fraction and a broad retarded fraction in which the radioactivity was somewhat displaced from the protein toward the high molecular weight region. In Fig. 15 is shown the protein and radioactivity profile obtained for the normal cell-derived fraction. In contrast to the tumor cell-derived fractions, neither protein nor radioactivi-

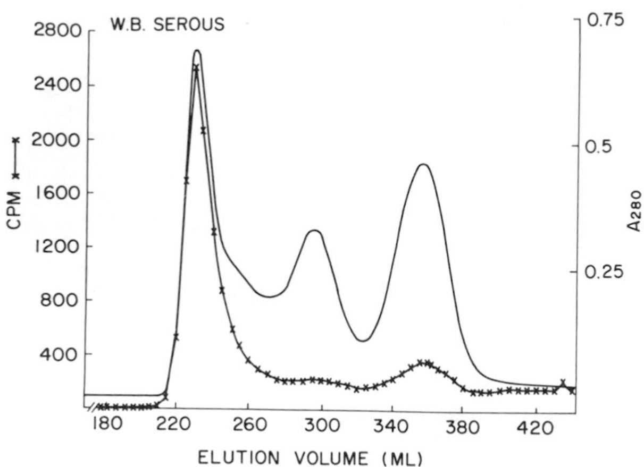

FIGURE 9 *Sephadex G-200 chromatography of culture medium from serous cystadenocarcinoma tumor cell cultures. Half of the culture medium from five 10 ml cultures incubated with [^3H] glucosamine for 14.5 hours was prepared as described in Methods B.4. Column: 2.6 × 94.5 cm (Vo = 218 ml); flow rate= 17.0 ml/hr; fraction volume = 5 ml. Radioactivity in 300 μl aliquots from fractions was determined.*

ty were eluted in the excluded fraction from the Sepharose 6B column. A similar displacement of retarded radioactivity in relation to retarded protein was observed but the radioactivity appeared to be less heterogenous than that obtained from the tumor cell-derived fractions.

In Fig. 16 is shown a comparison of the Sepharose 6B elution profiles for the [^{3}H] glucosamine-labeled and [^{3}H] fucose-labeled glycoproteins of the excluded fractions from the Sephadex chromatography shown in Fig. 11. No significant differences in the elution profiles were observed.

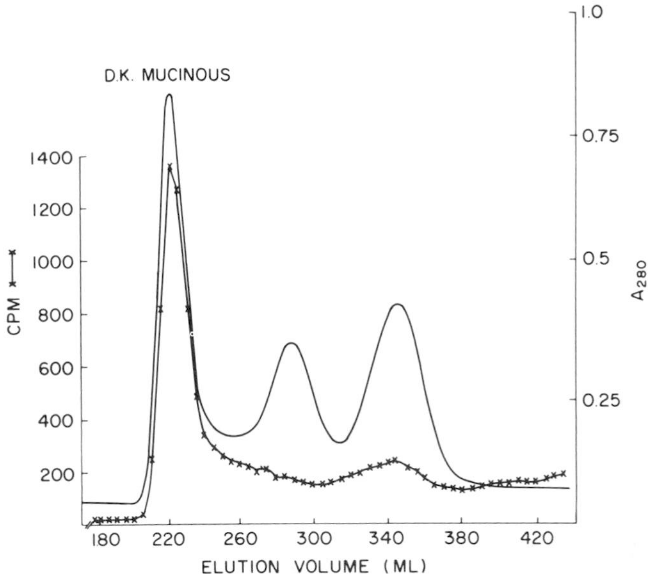

FIGURE 10 *Sephadex G-200 chromatography of culture medium from cultures of mucinous cystadenocarcinoma tumor cells. Culture medium from ten 10 ml cultures incubated with [^{3}H] glucosamine for 24 hours was prepared for chromatography as described in Methods B.4. Column: 2.6 × 94.5 cm (Vo = 218 ml); flow rate = 16.9 ml/hr; fraction volume 5.0 ml. Radioactivity in 50 μl aliquots from fractions was determined.*

IV. DISCUSSION

For the characterization of ovarian tumor glycoproteins, it has been proposed that pleural and peritoneal effusions may serve as a source of tumor-derived glycoproteins. This followed from the postulate that tumor cells present in the effusions and/or implanted on the pleura or peritoneum may elaborate glycoproteins into the effusion and that these glycoproteins will accumulate there since the drainage of the effusion from the pleural and peritoneal cavities has been severely restricted or completely blocked.

In the studies reported here, viable cells were collected from effusions and cultured in media containing radiolabeled gly-

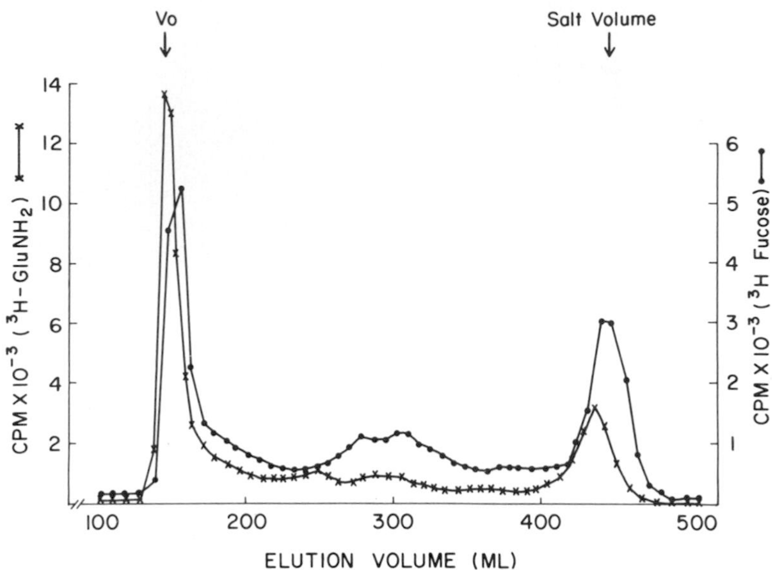

FIGURE 11 *Comparison of radioactivity profiles from Sephadex G-150 chromatography of culture medium from cultures of serous cystadenocarcinoma tumor cells. Culture medium from four 10 ml cultures incubated with [^3H] glucosamine for 15 hours and another nine 10 ml cultures incubated with [^3H] fucose for 15 hours were prepared separately for chromatography as described in Methods B.4. Sample load volume = 10 ml; Column: 2.5 × 89.2 cm (Vo = 140 ml); flow rate = 7.0 ml/hr; fraction volume = 4 ml and 7.75 ml for [^3H] glucosamine and [^3H] fucose labeled culture medium, respectively. Radioactivity in 100 μl aliquots from fractions was determined.*

coprotein precursors. The results of the time course experiments shown in Figs. 5 and 6 indicate that both normal cells (i.e. non-malignant cells) and tumor cells are quite capable of synthesizing and elaborating glycoproteins into the culture medium for a period of 20-120 hours. When comparing different cell populations, differences were observed between the normal and tumor cells with respect to the amount and kinetics of radiolabel appearing in the TCA-precipitable and TCA-soluble non-dialyzable fractions from both cells and culture medium. However, since we do not know what affect the chemotherapy administered to the donor

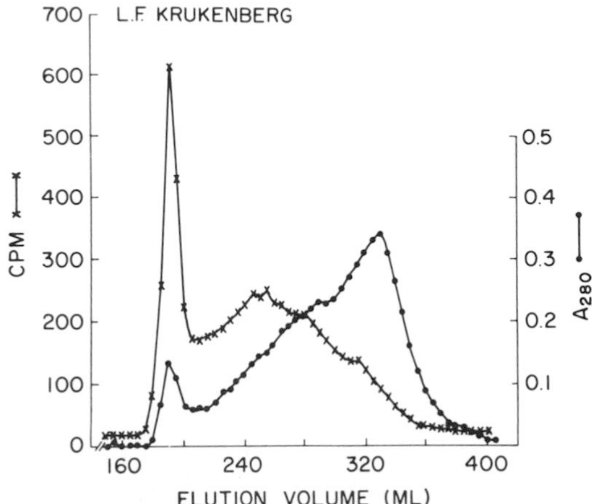

FIGURE 12 *Sepharose 6B chromatography of excluded fraction from Sephadex chromatography of culture medium from Krukenberg tumor cell culture. Column: 2.6 × 93.1 cm (Vo = 186.2 ml); flow rate = 16.4 ml/hr; fraction volume = 5 ml. Radioactivity in 100 μl aliquots from fractions was determined.*

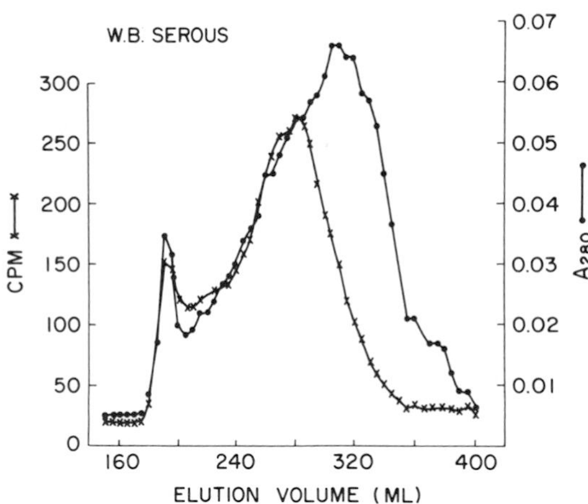

FIGURE 13 *Sepharose 6B chromatography of excluded fraction from Sephadex chromatography of culture medium from serous cystadenocarcinoma tumor cell cultures. Chromatography conditions as described for Fig. 12. Radioactivity in 300 μl aliquots from fractions was determined.*

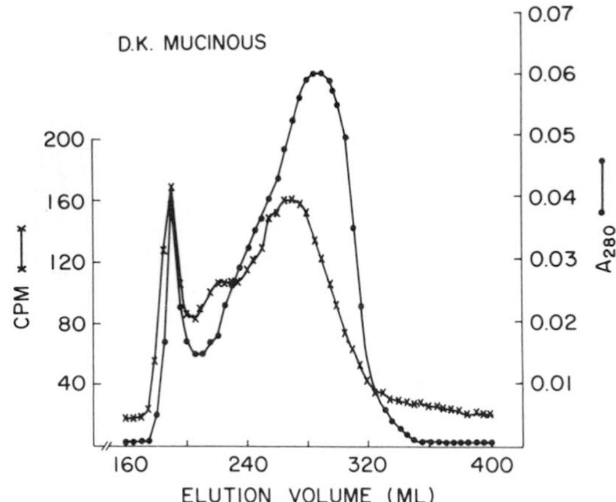

FIGURE 14 *Sepharose 6B chromatography of excluded fraction from Sephadex chromatography of culture medium from mucinous cystadenocarcinoma tumor cell cultures. Chromatography conditions as described for Fig. 12. Radioactivity in 100 μl aliquots from fractions was determined.*

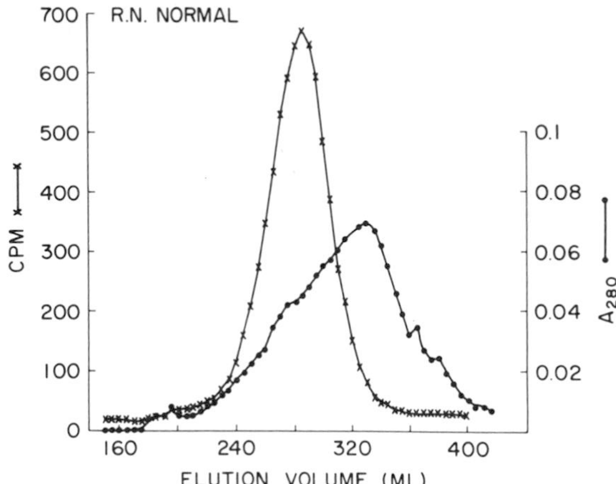

FIGURE 15 *Sepharose 6B chromatography of excluded fraction from Sephadex chromatography of culture medium from normal cell cultures. Chromatography conditions as described for Fig. 12. Radioactivity in 100 μl aliquots from fractions was determined.*

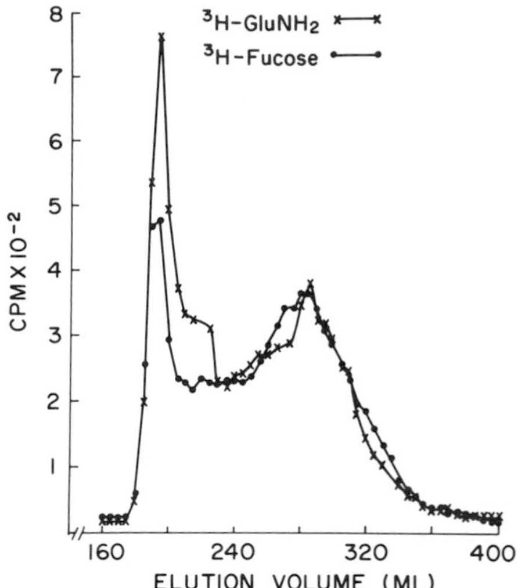

FIGURE 16 *Comparison of radioactivity profiles from Sepharose 6B chromatography of excluded fractions from Sephadex chromatography described for Fig. 11. Sepharose chromatography conditions the same as described for Fig. 12. Radioactivity in 50 μl aliquots from fractions was determined.*

patient may have on the cells, no definitive conclusions may be made at this time about differences observed in the kinetics of [^{3}H] glucosamine incorporation.

To assess the ability of cells to replicate in the culture system described, several time course experiments were carried out in which the incorporation of [^{3}H] thymidine into DNA was determined. The results indicated that some DNA synthesis occurred during the first several hours in culture.

Sephadex gel filtration of the culture media indicated that most of the radiolabeled precursor was incorporated into high molecular weight glycoproteins as evidenced by their elution in the excluded fractions from the columns. Since more than two-thirds of the *unlabeled* protein was eluted in the retarded fraction, a significant enrichment of the high molecular weight radiolabeled glycoprotein was achieved.

It should be made note of that when the dialyzed and lyophilized culture medium was reconstituted with column buffer, usually a small portion of the lyophilizate was insoluble. This was also true of the dialyzed and lyophilized high molecular weight fractions obtained from the Sephadex chromatography of culture media. The nature of these fractions has not been studied.

Sepharose 6B chromatography of the high molecular weight fractions from Sephadex elution of the culture media showed the presence of a radioactive component(s) excluded by the gel and a retarded radioactive fraction. The elution profile of the latter indicated the presence of molecular weight heterogeneity. The displacement of the retarded peak of radioactivity from the retarded protein peak suggested that the radiolabeled glycoproteins are distinctly different from the bulk protein in the effusions and are characteristic of the cultured cells. The absence of an excluded radiolabeled fraction and the symmetrical retarded peak of radioactivity obtained from the Sepharose 6B chromatography of the high molecular weight fraction from the normal cell culture medium was in contrast to the elution profiles obtained for the Sepharose chromatography of the high molecular weight fraction from the tumor cell culture media. The significance of these differences awaits further study.

The isolation and charact erization of the radiolabeled glycoproteins are now in progress.

ACKNOWLEDGMENTS

We are grateful to Dr. Marie Gamarra of the Department of Cytology for her valuable assistance in classifying cell populations. We thank Miss Jessica Pawlowicz for her technical assistance.

This work was supported by Grant DRG-1221 awarded by the Damon Runyon Memorial Fund for Cancer Research Incorporated and by NIH Grant 1R01 CA 14854-01A1 from the National Cancer Institute.

REFERENCES

Barber, H. R. K., Sommers, S. C., Snyder, R. and Kwon, T. H. (1975). *Am. J. Obstet. Gynec. 121:* 795-807.

Breborowicz, J., Easty, G. C., Birbeck, M., Robertson, D., Nery, R. and Neville, A. M. (1975). *Br. J. Cancer 31,* 559-569.

Cooper, A. G., Codington, J. F. and Brown, M. (1974). *Proc. Nat. Acad. Sci. U.S.A. 71,* 1224-1228.

Curling, O. M. and Hudson, C. N. (1975). *Br. J. Obstet. Gynaec. 82,* 405-411.

Feldman, G. B. (1975). *Cancer Res. 35,* 325-332.

Feldman, G. B. and Knapp, R. C. (1974). *Am. J. Obstet. Gynec. 119:* 991-994.

Hickok, D. F. and Miller, L. (1974). *In Vitro. 10,* 157-166.

Kraemer, P. M. (1967). *J. Cell Physiol. 69,* 199-208.

Lingeman, C. H. (1974). *J. Nat. Cancer Inst. 53,* 1603-1618.

Molnar, J., Teegarden, D. W. and Winzler, R. J. (1965). *Cancer Res. 25,* 1860-1866.

Ruoslahti, E. and Vaheri, A. (1975). *J. Exptl. Med. 141,* 497-501.

Young, R. C. and DeVita, V. T. (1974). *Am. J. Obstet. Gynec. 120,* 1012-1024.

ESTABLISHMENT AND CHARACTERIZATION OF EPITHELIAL CELL CULTURES FROM TUMORS OF THE HUMAN URINARY TRACT

A. Y. ELLIOTT
D. L. BRONSON
J. CERVENKA
P. H. CLEVELAND
E. E. FRALEY

Department of Urologic Surgery
University of Minnesota Health Sciences Center
Minneapolis, Minnesota

SUMMARY

Fifty-four surgical specimens of transitional cell cancer (TCC) of the human urinary tract were placed in culture. Epithelium-like cells grew out from explants of 38 (70%) of the tumors cultured, and 11 (22%) have been maintained in culture for more than 18 months. Characterization studies performed on six of these lines show the cells to (1) be small, (2) have a rapid doubling time, (3) exhibit multilayering, (4) produce tumors in the cheek pouches of immune-depressed hamsters, (5) form colonies in soft agar, (6) possess a hyperdiploid karyotype, and (7) contain ultrastructural features common to epithelial cells. The established cultures grow rapidly in roller bottles and therefore can be produced in large quantities. These cells also remain viable after storage for 4 years in liquid nitrogen (LN_2).

I. INTRODUCTION

Since the initial work of Burrows, *et al.* (1917), many laboratories have attempted to establish long-term cell lines from transitional cell cancers (TCC) of the human urinary tract. Although there are presently available to investigators three long-term lines: RT_4, established by Rigby and Franks (1970); T_{24}, established by Bubenik *et al.* (1973); and 253J established by Elliott *et al.* (1974); no single laboratory has been successful in establishing more than one long-term epithelial cell line. Reports by Walker *et al.* (1965), Jones (1967), Yajima (1968), and Bubenik *et al.* (1970) emphasize the problems involved in establishing epithelial-cell cultures from human tumors.

Previously, Elliott *et al.* (1974) reported the establishment and characterization of a cell line (253J) derived from a metastic lymph node from a patient with multiple TCC of the urinary tract. Utilizing the technique employed to establish the 253J cell line, as well as another method described by Elliott *et al.* (1976), we have established ten additional long-term TCC cell lines.

II. MATERIALS AND METHODS

A. Tissue Culture Procedures

1. *Establishment of Cell Cultures*

Tumor tissues obtained at surgery were brought directly to the tissue culture laboratory in RPMI-1640 medium (Gibco, Grand Island, New York) containing 20 percent heat-inactivated fetal bovine serum (Armour Co., Kankakee, Illinois), 10 percent tryptose phosphate broth (Difco Laboratories Inc., Detroit, Michigan), and 100 U of penicillin and 100 ug of streptomycin (Gibco) per milliliter.

In most cases, processing of the tumors began within 15 minutes of the time they were removed from the patient. The tissue was minced very fine (1 mm) with sharp scissors in a sterile Petri dish. The minced tissue was floated in an appropriate amount of RPMI outgrowth medium in a Petri dish, and the small fragments and the medium were drawn up into a 10-ml plastic syringe without a needle. Approximately 1 ml of outgrowth medium containing tumor fragments was placed in each of several 25-cm^2 plastic flasks (Falcon Plastics, Oxnard, California), and the flasks were rotated to distribute the fragments evenly over the surface. The flasks were then incubated at 37°C without CO_2 for 24 hours. It is very important to use as little medium as possible, or many of the tissue fragments will not attach to the surface of the flask. After growth began, 2 to 3 ml of outgrowth medium were added carefully so as not to disturb the attached pieces.

2. *Subculture*

When the primary cultures reached a monolayer, the cells were removed from the flask by the technique previously described by Elliott *et al.* (1974).

3. Freezing and Storage

Cells were removed from the flasks by trypsinization, washed several times in outgrowth medium, counted, and diluted in storage medium (RPMI-1640 with 30 percent fetal bovine serum, 10 percent dimethyl sulfoxide, penicillin, and streptomycin) to a concentration of 5.0×10^6 cells/ml. Three ml of the cell suspension were placed in a 5-ml blue-line break ampoule (Kimble Glass Products, Vineland, N. J.) and heat-sealed. The ampoules were placed at 4^oC for one hour and then put into the vapor phase compartment of a liquid nitrogen refrigerator (Minnesota Valley Engineering Co., New Prague, Minnesota).

To revive the stored cells, the ampoules were removed from the refrigerator and thawed at 40^oC. The cells were washed several times in outgrowth medium and seeded at a density of 5.0×10^6 per ml of outgrowth medium in 75-cm^2 Falcon plastic flasks.

4. Large Volume Production

Glass roller bottles (#7000 Bellco disposable bottles, Bellco Glass Co., Vineland, N. J.) were used to produce large volumes of tumor cells. Before the cells were planted, the surface of each bottle was conditioned by adding 100 ml of outgrowth medium and rotating the bottles at 37^oC for 30 minutes. The medium was then removed from the bottles and replaced with 100 ml of fresh outgrowth medium containing 1×10^7 tumor cells. The bottles were placed in a Bellco roller bottle apparatus, incubated at 37^oC, and rotated at 0.5 rpm.

B. Malignancy Studies

1. Hamster Inoculation

The procedure employed for inoculation of TCC into the cheek pouches of immune-depressed hamsters have been previously described by Elliott *et al.* (1974).

2. *Cell Culture in Agar*

Cultures were prepared according to the procedure described by McAllister *et al.* (1967). The cell concentrations employed, as well as the medium and cultural requirements, were previously described by Elliott *et al.* (1974).

3. *Electron Microscopy*

A portion of the original tumor tissue from each TCC was fixed in 2 percent glutaraldehyde (Electron Microscopy Sciences, Fort Washington, Pa.) as previous described by Elliott *et al.* (1974). Tissue culture cells were fixed, dehydrated, and embedded in Epon 812 (Electron Microscopy Sciences) as monolayers while still in the culture flasks by a method described by Elliott *et al.* (1973).

4. *Chromosome Studies*

Near-monolayer cultures of TCC cells were harvested by trypsinization after incubation with 0.05 mg colcemid/ml (Ciba Pharmaceutical Products Inc., Summit, N. J.) for 5 to 10 hours. The remainder of the procedure was previously described by Elliott *et al.* (1974).

III. RESULTS

A. Cells in Culture

1. *Cell Outgrowth*

Cultures were observed daily for evidence of cell growth from the tumor explants. In 48 of 54 tumors (89%), epithelial-like cells grew out from the tumor within 24 to 48 hours after planting (Fig. 1). In 12 tumors, the epithelial cells continued to grow for 1 to 4 weeks, then began to round up and detach from the sur-

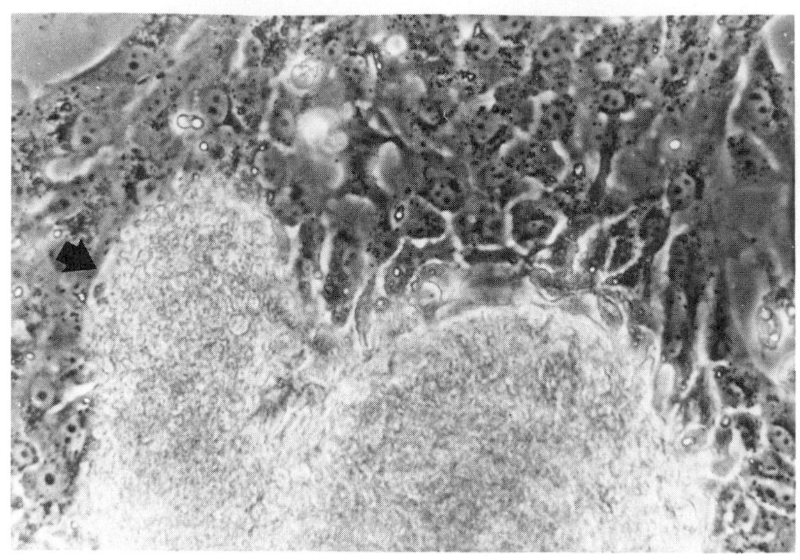

FIGURE 1 *Phase contrast photograph of epithelial cells growing out from tumor (292W) piece (at arrow). X 400.*

face of the flask. Three cultures exhibiting this spontaneous degeneration have been described in detail by Elliott *et al.* (1973).

Each of these cultures showed complete destruction of the cells within 7 days, and attempts to subculture the detached, floating cells proved futile. Extensive culture and ultrastructure studies failed to detect bacteria, mold, yeast, mycoplasma, or ubiquitous viruses.

Cells from 38 tumors (70%) grew rapidly and were subcultured. However, 27 of the epithelium-cell cultures could not be kept in culture longer than 12 months. At various times between 2 and 12 months after planting, the cells would either slow their growth rate or cease to grow altogether. The cells that remained on the flask usually were much larger than the characteristically small TCC cells and were frequently multinucleated. All attempts to keep these cells in culture by increasing the serum concentration,

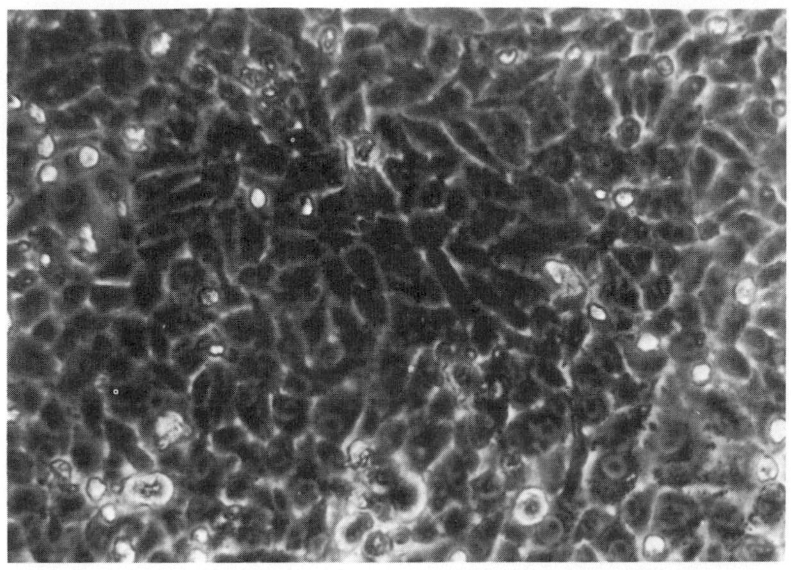

FIGURE 2 *Phase contrast photograph of monolayer culture of 292W cells in passage #30. X 400.*

by adding L-glutamine, nonessential amino acids, insulin, cortisone, or testosterone, or by using Eagle's MEM, Medium 199, or Earle's MEM instead of our usual medium, did not prevent the death of these cells. Even when earlier passage levels of the same cells were revived from storage in LN_2 and placed in culture, the cells stopped growing after a short time.

Eleven long-term cell lines have been in culture for more than a year. Six of these lines (253J, 292W, 192B, 647V, 639V and 486P) have been subcultured more than 50 times and continue to grow at a constant rate. The cells are small (averaging 10 μ in diameter), epitheloid, and, usually, uninucleate. They grow rapidly in culture, exhibit multilayering, and can be subcultured at a 1-to-3 split at weekly intervals (Fig. 2). For practical reasons, the other 5 lines, after more than 20 passages, have been stored in LN_2 for future study.

Cells retrieved from LN_2 storage settled onto the surface of the flasks and grew within 1 hour after planting, often forming a monolayer in 4 to 7 days. Cells have been revived from LN_2 storage of early passages of each of the 11 long-term cultures with no apparent damage or change in growth pattern.

After the tumor cells had undergone approximately 12 passages in the RPMI outgrowth medium, it was possible to switch to Eagle's MEM (GIBCO) containing 10 percent heat-inactivated fetal bovine serum, penicillin, and streptomycin as both outgrowth and maintenance medium.

B. Large Volume Cell Production

TCC cells grew better on plastic flasks than on disposable glass roller bottles, but after several passages of the cells on plastic flasks, tumor cells seeded in roller bottles at a high cell density (1×10^7 cells/bottle) would grow and form a heavy multilayered cell sheet. Good cell attachment and outgrowth were not obtained if the surface of the bottles was not exposed to outgrowth medium before the cells were seeded. The cells grew out more slowly on glass, requiring 10 to 14 days to form a monolayer. When confluent, the cells continued to grow in multilayers; and after 3 to 4 weeks in culture, 1×10^9 cells could be recovered from a single bottle. This procedure has allowed us to produce larger numbers of tumor cells in early passages and to store them for future studies.

Culture and ultrastructure studies in our laboratory and by HEM Corporation (Rockville, Maryland) have failed to detect bacteria, yeast, molds, or mycoplasma in these cell lines.

1. *Cell Growth in Agar*

Passage-10 cells from each of the TCC lines listed in Table I produced multilayered colonies in soft agar. Microscopic examination of the agar plates 24 hours after the cells were planted re-

TABLE I Properties of TCC Cells in Long-Term Culture

Tumor number	Site of tumor	Cell morph.	Doubling time	Modal[a] number	Tumor prod in hamsters	Growth in agar	Lack of contact inhibition	Passage number
253 J	M-TCC[b]	Epith.	48 hrs.	60-67	+	+	+	90
192 B	B[c]	Epith.	20 hrs.	52-54	+	+	+	68
292 W	RP[d]	Epith.	21 hrs.	50-55	+	+	+	64
639 V	U[e]	Epith.	24 hrs.	56-60	+	+	+	51
647 V	B	Epith.	42 hrs.	66-70	+	+	+	58
486 P	B	Epith.	35 hrs.	76-80	+	+	+	52

[a]Modal number represents the modal chromosome number for each cell line.
[b]M-TCC = multiple transitional cell cancers.
[c]B = bladder.
[d]RP = renal pelvis.
[e]U = ureter.

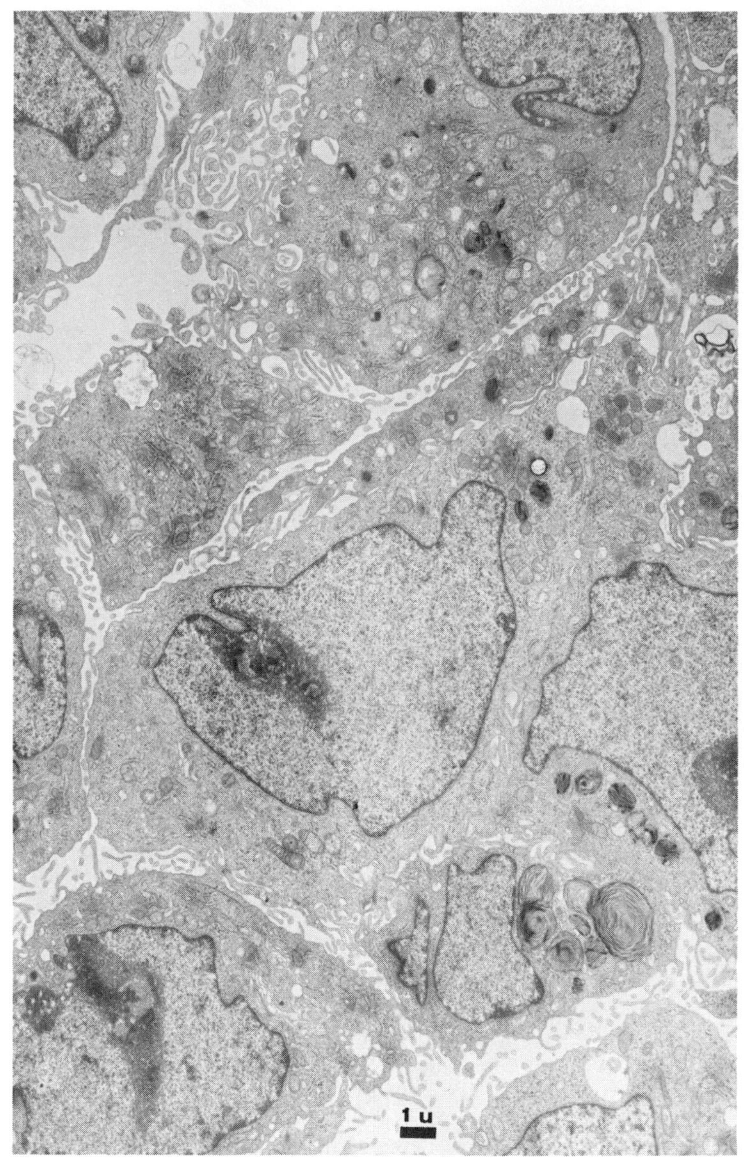

FIGURE 3 *Electron micrograph of 192B cells in passage #34. Numerous microvilli are shown. X 6000.*

vealed single cells well dispersed throughout the top agar layer; areas could be marked and the colonies from these single cells observed. Except in the case of 639V and 486P, colonies consisting of a few cells were visible several days after inoculation onto soft agar. At the end of 21 days in culture, the colonies consisted of tightly packed, multilayered cell sheets. The size of the colonies again depended on the growth rate of the cells in culture, with the largest colonies seen in 192B cells and the smallest in 486P cells. Approximately 65 percent of the 192B cells and 10 percent of the 486P cells planted in agar formed colonies. At this time, clones were established by transplanting individual colonies to Petri dishes under liquid medium. These colonies attached to the surface of the dish and, except for the 192B cells, grew slowly; the cells were usually a monolayer within 3 weeks (see Fig. 3).

C. Electron Microscopy

Sections from both the original surgical specimen and from passaged TCC cells revealed structures which are considered to be consistently found in epithelial cells. Most of the cells, regardless of the passage level, contained numerous microvilli (Fig. 4). Mitochondria were abundant, as were the ribosomes and polysomes. Golgi bodies were found in the cells, although they were not well developed. Residual bodies, vacuoles, and, to lesser extent, lysosomes and phagosomes were also observed. Multivesicular bodies were found in abundance, especially in the 253J and 192B cell lines. Desmosomes and bundles of tonofibrils were scarce in most of the cell lines but readily seen in the 647V and 486P cell lines (Fig. 5). No remarkable ultrastructural differences were noted between the early and the late passages of any of the cell lines.

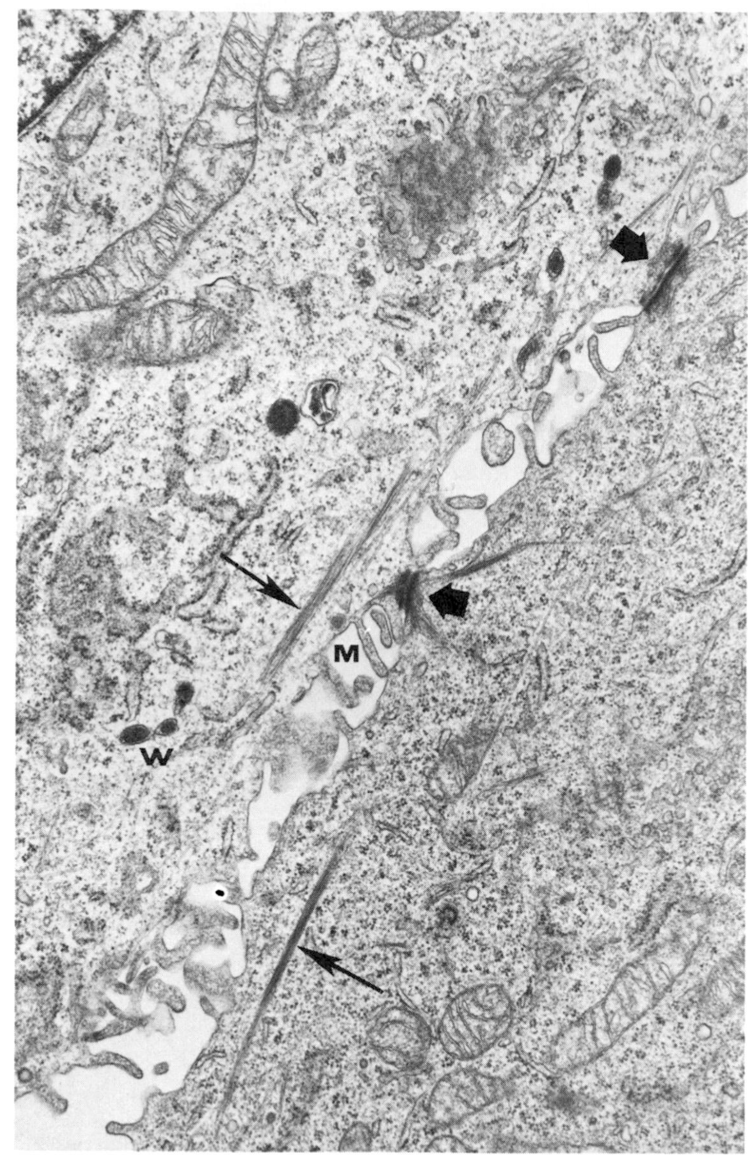

FIGURE 4 *Electron micrograph of 647V cells in passage #19. Desmosomes (large arrows), bundles of tonofibrils (small arrows), microvilli (M), and Weibel Palade bodies (W) are shown. X 18,000.*

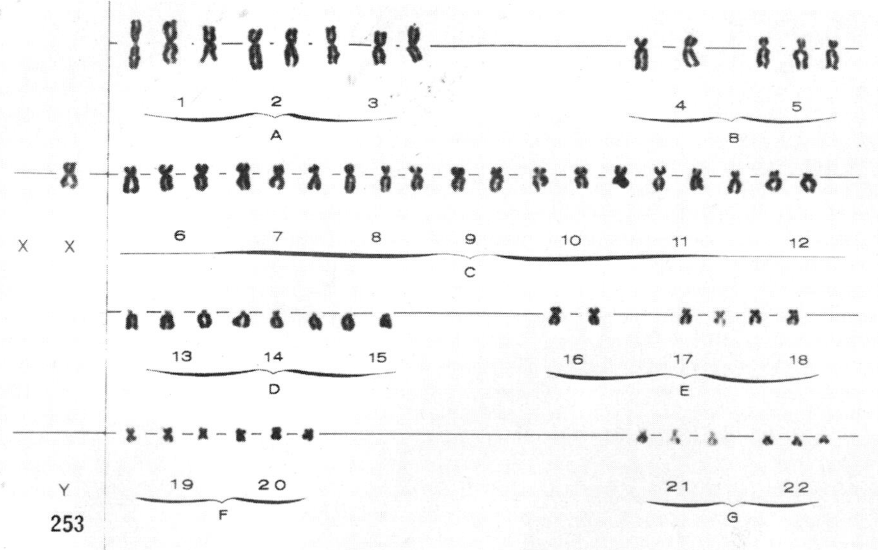

FIGURE 5 *Karyotype of 253J cells in passage #76. X 1500.*

D. Chromosome Studies

Chromosome analysis was performed on each cell line at passage 10 and at every tenth passage thereafter. Although there was some shifting and chromosomal rearrangement the basic karyotype of each cell line did not change drastically with time. The karyotype of passage 76 253J cells is shown in Fig. 5. This cell line has been in culture since April 1972 and has undergone more than 90 subcultures. The basic karyotype has not changed since passage 10.

The karyotype of passage-1 192B cells is shown in Fig. 6. This cell line is the fastest growing and has the smallest number of chromosomes, the modal number for this line being 52 to 54. In general, all of the cell lines possessed hyperdiploid karyotypes with numerous minutes and breaks. Although no definite chromosome abnormality or rearrangement seems consistent with TCC cells, all of the lines: (1) are definitely of human origin, (2) possess karyotypes consistent with malignant cells, (3) can be distinguished from one another on the basis of distinctive chro-

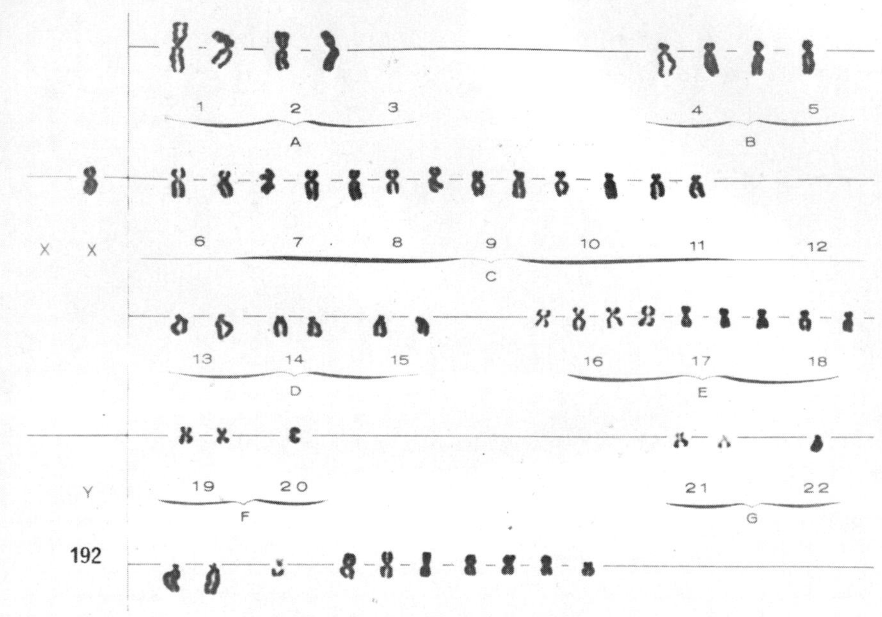

FIGURE 6 *Karyotype of 192B cells in passage #1. X 1500.*

mosome patterns, and (4) do not have karyotypes in common with either HeLa or any other characterized human cell line.

IV. DISCUSSION

Using the procedure with which we established the 253J cell line, and a second method, we have been able to establish 10 additional long-term TCC cell lines. Six of these have been studied in detail, and characterization studies have been performed. Properties common to all six cell lines, as listed in Table I, are (1) epithelial cell morphology, (2) multilayering of cells in culture, (3) hyperdiploid chromosome profiles, (4) production of tumors in immune-depressed hamsters, (5) growth in soft agar, (6) rapid doubling times, and (7) presence of ultrastructural features common to epithelial cells.

The difficulties encountered by other investigators in their attempts to establish epithelial cell lines from TCC are also reflected in our work. As shown in Table II, there was initial cell outgrowth from 38 (70%) of the 54 tumors placed in culture, but cells from 27 of these tumors could be maintained in culture for no longer than six months or four subcultures, and, as previously mentioned, all attempts to keep these cells in culture by the addition of growth-promoting hormones or other nutrients proved futile. No similar "crisis period" was noted in the eleven cell lines which have been carried for at least 20 passages.

Employing the roller bottle technique described in this report, it was possible to produce large volumes of cells from each of the 11 TCC cell lines. This technique provided large numbers of cells from both early and late tissue culture passages so that we could compare the results of cytogenetic, immunologic, biochemical, ultrastructural, oncogenicity, and growth studies. It was also possible to store large numbers of these cells in LN_2 and later revive them for tissue culture passage and study.

Not only has it been difficult to establish long-term cell cultures from human transitional cancers of the urinary tract, but similar problems have been encountered by investigators attempting to culture epithelium-like cells from other solid human tumors. For example, Giard *et al.* (1973) summarized the results of their attempts to establish cell lines from 200 human tumors. Having obtained only a 6 percent success rate, the authors concluded that, in contrast to human fibroblasts, epithelium-like cell cultures from human tumors were indeed difficult to develop into long-term lines.

However, the studies reported herein suggest that a reasonable success rate can be achieved with TCC if a certain protocol is followed. First, tumors should be placed in culture as soon as possible. We believe that it is important to process the tumors and place them in culture within an hour of their removal from the patient. Second, although TCC may be obtained by many surgi-

TABLE II Growth of Tumor Cells from Tissue Culture Explants[a]

Site of primary tumor	Number of specimens studied	Outgrowth of cells from tumor	Established long term cultures
Renal pelvis	17	12	4
Ureter	29	19	4
Bladder	7	6	2
Urethra	1	1	1
Totals	54	38 (70%)	11 (22%)

[a]Only epithelial-like cells are counted in the 70% figure listed under outgrowth of cells from tumor. Fibroblasts were disregarded. Each of the 11 cell lines listed in the last column has been in culture for over 18 months and has undergone at least 50 subcultures.

cal techniques, we have had our best results with specimens removed mechanically through the resectoscope with biopsy forceps. Specimens obtained by transurethral resection with an electrically activated cutting loop often are so badly damaged by the electrical current that they are not satisfactory for cultivation. Third, it is imperative that the outgrowth medium provides the proper nutrients for rapid outgrowth of cells from tumor explants. The medium described herein appears to meet this requirement for TCC cells.

Cell lines established from human TCC have been used in various immunologic and biochemical studies. It was previously shown by Hakala *et al.* (1974a) that humoral complement-dependent cytotoxic to this cell line were obtained more frequently from patients. Hakala *et al.* (1974b) also showed that lymphocytes cytotoxic to this cell line were obtained more frequently from patients with TCC than from patients without TCC.

Transitional cell cancer of the urinary tract has been studied in the laboratory as comprehensively as any other solid human tumor. There has been special interest in both the immunobiology and virology of these tumors. Almost all of these studies have utilized an established TCC cell line, usually either T_{24} or RT_4. If studies on the basic biology of TCC are to continue and to be expanded, additional cell lines will be required in order to establish or refute the accuracy of data gathered on TCC in the above assays.

ACKNOWLEDGMENTS

We thank H. Asmussen, S. Dombrovskis, and N. Stein for excellent technical assistance.

Supported in part by Public Health Service Training Grant AM 05514-10 from the National Institute of Arthritis and Metabolic Diseases, and by grants CA 13095-03 and CA 15551-02 from the National Cancer Institute.

REFERENCES

1. Bubenik, J., Baresova, M., Viklick, V., Jakoubkova, J., Sainerova, H. and Donner, J. (1973). *Int. J. Cancer 11,* 765-773.
2. Bubenik, J., Perlmann, P., Helmstein, K. and Moberger, G. (1970). *Int. J. Cancer 5,* 39-46.
3. Burrows, M. T., Burns, J. E. and Suzuki, Y. (1917). *Johns Hopkins Hosp. Bull. 28,* 178-179.
4. Elliott, A. Y., Bronson, D. L., Stein, N., and Fraley, E. E. (1976). *Cancer Res. 36,* 365-369.
5. Elliott, A. Y., Cleveland, P., Cervenka, J., Castro, A. E., Stein, N., Hakala, T. R. and Fraley, E. E. (1974). *J. Natl. Cancer Inst. 53,* 1341-1349.
6. Elliott, A. Y., Fraley, E. E., Cleveland, P., Castro, A. E., and Stein, N. (1973). *Science 179,* 393-395.

7. Giard, D. J., Aaronson, S. A., Todaro, G. J., Arnstein, P., Kersey, J. H., Dosik, H., and Parks, W. P. (1973). *J. Natl. Cancer Inst.* *51*, 1417-1423.

8. Hakala, T. R., Castro, A. E., Elliott, A. Y. and Fraley, E. E. (1974a). *J. Urol.* *111*, 382-385.

9. Hakala, T. R., Lange, P. H., Castro, A. E., Elliott, A. Y. and Fraley, E. E. (1974b). *Cancer* *34*, 1929-1934.

10. Jones, G. W. (1967). *Cancer* *20*, 1893-1898.

11. McAllister, R., Reed, G., and Huebner, R. (1967). *J. Natl. Cancer Inst.* *39*, 43-53.

12. Rigby, C., and Franks, L. M. (1970). *Brit. J. Cancer* *24*, 746-754.

13. Walker, D. G., Lyons, M. M., and Wright, J. C. (1965). *Eur. J. Cancer* *!*, 265-273.

14. Yajima, T. (1968). *Jap. J. Urol.* *59*, 128-134.

CARDIAC MUSCLE CELL CULTURE AS A PHARMACOLOGICAL TOOL: THE EFFECT OF CORONARY DILATORS ON THE METABOLISM OF ADENOSINE

S. JAMAL MUSTAFA

Department of Pharmacology
University of South Alabama
College of Medicine
Mobile, Alabama

I. INTRODUCTION

During the last decade a number of efforts have been made to characterize the contractile phenomena of the mammalian myocardium and to study the biochemistry of its physiological response to nutritional and pharmacological agents (16). Investigators have related their findings to the physiological status of the myocardium using the spontaneous, rhythmic contractibility typical of the mammalian heart as a parameter of its function. However, studies of the *in vivo* heart are difficult to assess because of the complex and sometimes confusing influence of humoral and neuronal interactions which effect the performance and response of the myocardium (42). *In vitro* or isolated heart preparations are useful only for short term experiments but undesirable for long term experiments because of their inability to maintain stable function for more than a few hours due to the development of ischemia and necrosis (42). Similarly, the organ culture technique developed by Wildenthal (42) for pre-natal rat hearts is not useful in maintaining long term myocardial function at physiological levels. Subsequently methods were developed for separating heart tissue into its component cells and culturing them for long periods of time (8, 14, 15, 17, 19, 38). Cultured heart cells offer some advantages over other heart preparations, and some of these advantages are:

(a) The direct effect of drugs or other chemical agents can be studied without equivocation since the cells are denervated;

(b) Pure myocardial muscle cell cultures can be prepared; thus, studies can be made without contamination by other cell types present in the myocardium, such as endothelial or vascular smooth muscle cells;

(c) The problem of diffusion lag in the interstitial fluid space of the intact heart is eliminated;

(d) Cytochemistry, autoradiography, and flurorescence microscopy are facilitated, because sectioning is not necessary, and the cells can be examined under living conditions; and

(c) cultured cells can be maintained for long-term experiments.

In the present communication, cultured chick embryonic cardiac muscle cells were chosen as a model for mammalian myocardium (there are no significant differences between the adult and embryonic cardiac cells) in order to answer some of the criticisms of the adenosine hypothesis for the regulation of coronary blood flow. According to this hypothesis, first proposed by Berne (3) in 1963, a reduction in myocardial oxygen tension associated with a negative oxygen balance in response to conditions such as hypoxia, reduced coronary blood flow, or increased myocardial oxygen demand leads to the breakdown of adenine nucleotides to adenosine. This adenosine may enter the interstitial fluid space and produce anteriolar dilation thereby increasing coronary blood flow to match the oxygen needs of the heart. Adenosine has been shown to be one of the factors responsible for this type of vasodilatory response (30).

Most of the studies concerning adenosine hypothesis for the regulation of coronary blood flow (3, 19, 22, 30, 32) have been conducted with either *in situ* hearts or isolated perfused hearts. These approaches are open to certain criticisms since; (a) it cannot be determined with certainty that adenosine is released by muscle cells or other cell types present in the myocardium; (b) the observed amounts of adenosine released into the perfusates of the isolated perfused hearts may not reflect the *in situ* amounts since during its passage across the capillary membrane some of the adenosine is degraded to inosine and hypoxanthine (33); and (c) the extracellular concentrations of adenosine cannot be directly measured. Thus, the first objective of this study was to use cultured chick embryonic cardiac muscle cells as a model in order to avoid some of these problems associated with the study of adenosine release due to hypoxia.

Drugs such as dipyridamole and aminophylline have been employed in attempts to elucidate the mechanism of coronary blood flow regulation. The interpretation of the results with these drugs is quite controversial with respect to the role played by adenosine. For example, the vasodilator effect of dipyridamole is primarily attributed to inhibition of cellular uptake of adenosine (4). However, Kubler and co-workers (24) have suggested that dipyridamole blocks the release of adenosine from myocardial cells, thus decreasing its extracellular concentration, a situation that would not be consistent with the adenosine hypothesis. Aminophylline attenuates the vasodilation produced by adenosine (2), but the exact mechanism of its action is not clear; therefore, a second object of the present investigation was to study the effects of dipyridamole and aminophylline on the metabolism of cultured cardiac cells.

II. MATERIALS AND METHODS

A. Preparation of Heart Cell Cultures

Cardiac cells were isolated from 16 day old chick embryonic hearts using essentially the same procedure as described by Mustafa and co-workers (31) with slight modification. The hearts were dissected with sterile techniques from six dozen chick embryos and placed in N-16 Pucks medium (obtained from Microbiological Associates, Bethesda, Md.). Each heart was cut into four pieces to facilitate removal of the intracardiac blood. The heart segments were left at room temperature in N-16 Pucks medium until all the hearts were harvested (about 1 hour). After one rinsing, the hearts were minced into pieces about 1-1.5 mm^3 while immersed in N-16 Pucks medium. These pieces were then transferred to a 50 ml sterile conical flask. About 10-12 ml of 0.1% trypsin solution (100 mg of trypsin dissolved in 50 ml of N-16 Pucks medium and 50 ml of normal modified Hanks medium containing 3.06 mM Ca^{++} and 0.81 mM Mg^{++}) was added to the flask and stirred for 5 minutes at low speed with a small magnetic stirring bar. In some experiments a mixture of 0.1% collagenase (Type I, Sigma Chemical Co., St. Louis, Missouri) and 0.2% hyaluronidase (Type I, Sigma Chemical Co., St. Louis, Missouri) was used to improve the yield of the cardiac cells. The first supernatant fraction which contained broken cells and blood was discarded. Fresh trypsin solution (maintained at 37°C) was added to the remaining tissue in the flask and agitated with a stirring bar for 10 minutes at low speed. When agitation was stopped, the supernatant fraction which contained cardiac cells was decanted into a 50 ml sterile precooled centrifuge tube. Immediately thereafter, an equal volume of growth medium having 15% horse serum (obtained from Microbiological Associates, Bethesda, Md.), 40% N-16 Pucks medium, 45% modified Hanks solution and 1% penicillin-streptomycin (obtained from Microbiological Associates, Bethesda, Md.) pH 7.2 was added to dilute the action of trypsin (or other proteolytic enzymes), and the tube was placed in ice. More fresh trypsin solution was added to the remaining tissue and this process was repeated 5-6 times to get a sufficient number of cardiac cells. The collected supernatant fractions were centrifuged at 250 g for 8 minutes at room temperature. The pellet was resuspended, washed once with growth medium, and resuspended again in about 80 ml of the growth medium. This suspension was transferred to 15 × 65 cm culture dishes (5 ml to each dish) and sealed under sterile conditions.

Some experiments were carried out on freshly isolated cardiac cells for comparison of ATP values with those of the cultured cells. In a few experiments the isolated cells were layered over a 3% Ficoll (obtained from Sigma Chemical Co., St. Louis, Mo.) solution according to the method of Glick and his co-workers (12) in order to separate the contracting cells (myocytes) from non-contracting cells. The cultures were then kept in a water jacketed CO_2 incubator, and the following day the growth medium replaced with fresh medium containing 1×10^{-4} M adenosine. Incubation was then carried out for about 24 hours to restore (Reference 36, and our observations) cell ATP values of the cultures to control values (hearts removed from chick embryos and immediately frozen with liquid nitrogen and analyzed).

The viability of the preparation was assessed by examining the monolayer cultures with a phase contrast microscope. The cultures, observed after 24 hours of plating, beat spontaneously, and about 85% of the cultured cells were beating cardiac muscle cells (26). Cultures older than 48 hours were not used because the ratio of fibroblasts to muscle cells increased as the culture aged.

B. Determination of ATP in Freshly Isolated and Cultured Cardiac Muscle Cells

The isolated cardiac muscle cells after separation (and also the cells isolated after Ficoll treatment) were thoroughly washed with modified Hanks solution to remove the broken cells and contaminating blood. The cells were then homogenized in 0.5 N perchloric acid, centrifuged, and processed for ATP and protein determination according to the methods described below in detail. Cultured cardiac muscle cells were treated in a similar manner.

The experimental protocol for the various experiments described further below is outlined on the next page.

C. Release of Adenosine from Cultured Cardiac Muscle Cells Due to Hypoxia

Cultures obtained after 2 days of incubation were washed several times with modified Hanks solution to remove all the adhering growth medium and cell debris without disturbing the thin film of cells attached to the bottom of the culture dishes. The assay was carried out in the culture dishes. The incubation mixture consisted of 1.0 ml of modified Hanks solution containing phosphate buffer but without glucose. The dishes were swirled several times to distribute the medium evenly over the cell layers, and the dishes incubated at $37^{o}C$ for 15 minutes in an atmosphere of 95% N_2 + 5% CO_2. A control culture with 95% O_2 + 5% CO_2 was run simultaneously. At the end of the incubation, the dishes were removed from the incubator and the thin layer of cultured cells scraped from the dishes with a plastic spatula. The contents of the dishes were immediately transferred to polyethylene centrifuge tubes and immersed in ice to stop the enzymatic reactions. Two dishes were used in each assay and each was washed once with 1 ml of Hanks solution, and the washings added to the original tubes. Immediately thereafter, the tubes were centrifuged at $0^{o}C$ at 10,000 g for 5 minutes. After centrifugation, as much supernatant as possible was decanted into a separate tube.

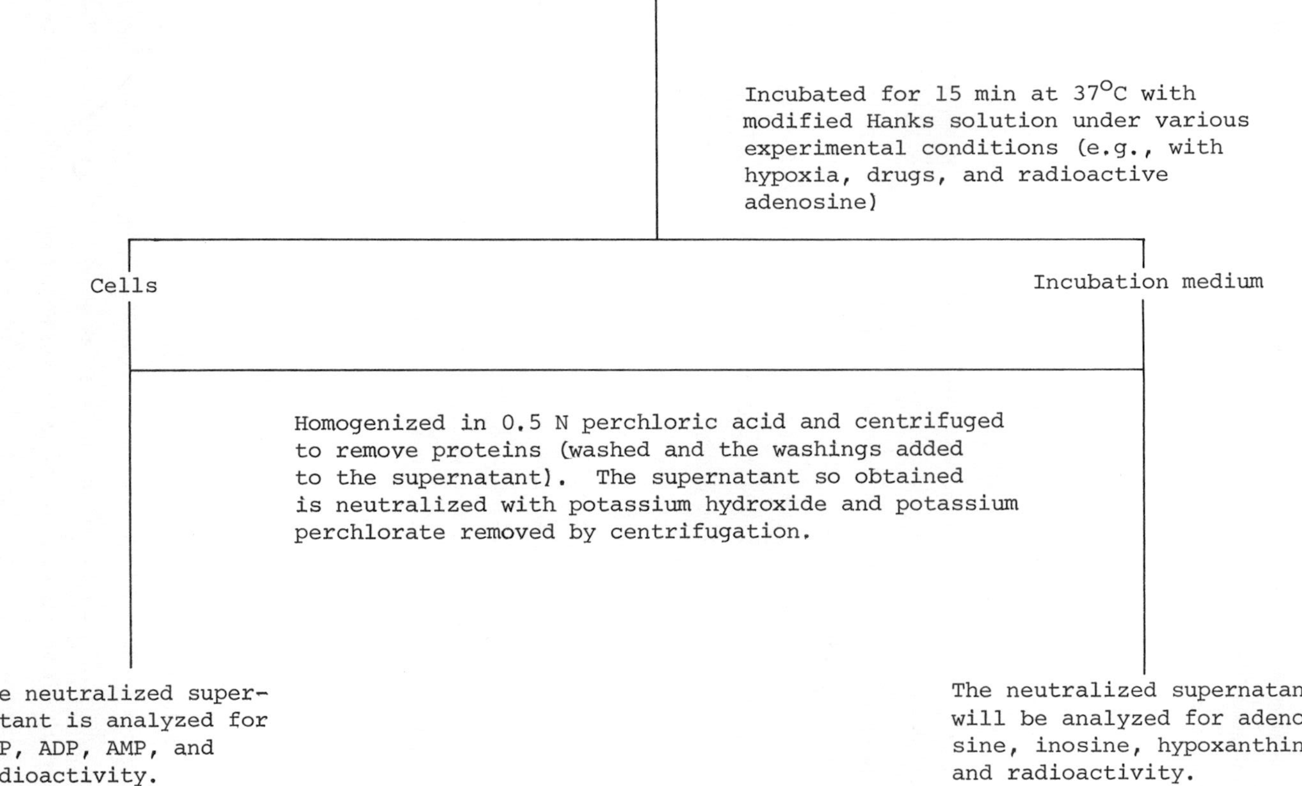
EXPERIMENTAL PROTOCOL
Cultured Cardiac Muscle Cells
Incubated for 15 min at 37°C with modified Hanks solution under various experimental conditions (e.g., with hypoxia, drugs, and radioactive adenosine)
Cells
Incubation medium
Homogenized in 0.5 N perchloric acid and centrifuged to remove proteins (washed and the washings added to the supernatant). The supernatant so obtained is neutralized with potassium hydroxide and potassium perchlorate removed by centrifugation.
The neutralized supernatant is analyzed for ATP, ADP, AMP, and radioactivity.
The neutralized supernatant will be analyzed for adenosine, inosine, hypoxanthine, and radioactivity.

The pellet was resuspended and washed once with 0.5 ml of Hanks solution, then centrifuged and again, the washings added to the supernatant fraction. The fraction will be referred to as "medium" and the pellet as "cells." Perchloric acid was added to both fractions in a final concentration of 0.5 N for precipitation of the proteins. The separated cells and medium were then immediately homogenized with a Polytron homogenizer for 90 seconds at a speed of 4000 rpm and then centrifuged for 10 min at 10,000 g in the cold ($4^{o}C$). The supernatant fractions were removed and the pellet of the protein precipitate was resuspended and washed once with 0.5 ml of 0.5 N perchloric acid, and the washings were added to the previous supernatant fraction. The protein precipitate obtained from the cell fraction was used for protein measurements by the method of Lowry and co-workers (28), and the results expressed on the basis of cellular protein concentrations.

The supernatant fractions were then brought to pH 7.0 by the addition of potassium hydroxide, placed in the refrigerator overnight, and the perchlorate precipitate was removed by centrifugation. The cell fractions were then analyzed for ATP by the luciferase method (13) and ADP and AMP by a spectrophotometric method (1). The medium fractions were analyzed for adenosine, inosine, and hypoxanthine by enzymatic assay (11). In two experiments the cells were analyzed for nucleosides, and the medium was analyzed for nucleotides, but none were detected.

D. Effect of Dipyridamole and Aminophylline on the Release of Radioactive Adenosine from Cultured Cardiac Muscle Cells

During the change of growth medium 24 hours after plating, 0.5 ml of U-C^{14}-adenosine (14.08 p moles; 15,000 CPM; obtained from Amersham/Searle Corp., Arlington Heights, Ill.) in addition to the unlabeled adenosine, was added to each culture dish in order to label the nucleotide pool(s) of the cells. After a total of 48 hours (or less) of growth these cultures were used for measuring the release of radioactivity under hypoxic conditions. The medium was decanted and the adhering cell layer washed several times with modified Hanks solution to remove cell debris and growth medium which also contained labeled adenosine and its metabolites. The assay mixture contained 0.1 ml drug (dipyridamole, obtained from Boehringer Ingelheim, Elmsford, New York, at a final concentration of 1×10^{-6} M or aminophylline, obtained from Sigma Chemical Co., St. Louis, Mo. at a final concentration of 1×10^{-5} M) and 0.9 ml modified Hanks solution. In the control experiments the drug was replaced by modified Hanks solution. The dishes were swirled to mix the contents thoroughly and then incubated at 37°C for 15 minutes in an incubator under 95% N_2 + 5% CO_2 atmosphere to accelerate adenine nucleotide degradation. The cells and medium fractions were isolated as described in the previous section and analyzed for adenine nucleotides and nucleosides. An aliquot from each fraction was removed for measurement of radioactivity in a Packard liquid scintillation counter. The scintillation cocktail consisted of 5.5 tm PPO, 0.1 gm POPOP, 333 ml Triton X-100, and 667 ml toluene.

E. Effect of Dipyridamole and Aminophylline on the Uptake of Radioactive Adenosine in Cultured Cardiac Muscle Cells

The cultures (grown for 48 hrs or less) were washed thoroughly without disturbing the cell layer in the manner described earlier. The assay was carried out in the culture dishes. The incubation mixture consisted of 0.05 ml U-C^{14}-adenosine (14.08 p moles; 15,000 CPM), 0.1 ml of drug (dipyridamole at a final concentration of 1×10^{-6} M or aminophylline at a final concentrations of 1×10^{-5} M) with the addition of modified Hanks solution containing phosphate buffer to a final volume of 1.0 ml. A control with vehicle (modified Hanks solution) in place of the drug was run simultaneously. The incubation mixture was added to the dishes and the latter were swirled several times to distribute the medium evenly over the cell layers. The dishes were then incubated at 37°C for 15 minutes in an atmosphere of 95% O_2 + 5%

CO_2. The cells and medium fractions were isolated and processed for nucleoside and adenine nucleotide analyses as described. An aliquot from each fraction was counted for radioactivity.

III. RESULTS

A. Levels of ATP in Various Preparations of Isolated Chick Embryonic Cardiac Muscle Cells

A common problem with the freshly isolated cardiac cells was the low concentration of ATP on the basis of per mg protein. The results of such an experiment are presented in Table I.

Therefore, freshly isolated cells were prepared according to the present method and according to the method of Glick and co-workers (12) in which the cells were isolated using conventional enzymatic methods and then layering them over a 3% Ficoll solution and at low centrifugation. Both isolation procedures resulted in low ATP values (together with low ADP and AMP). The present method did result in 65% higher ATP values than the method of Glick and co-workers (12). These isolated cells (with and without the Ficoll step) were placed in growth medium for 24 hours and then replaced with the same growth medium having 1×10^{-4} M adenosine and kept for another 24 hours in an incubator resulted in ATP values which are more comparable to control hearts (Table I). The ATP values of the cultured cardiac cells are 3-4 times higher than the freshly isolated cells (present method) from 16 day old chick embryonic hearts. Burns and Reddy (6) had reported higher ATP values for the freshly isolated cells from rat heart (using the Ficoll method) which we were unable to reproduce.

In another experiment with cultured cardiac cells, we found that inclusion of 1×10^{-4} M adenosine in the growth media for 24 hours increases the ATP values by 35% (unpublished observation). In our experience we have noticed no significant difference in

TABLE I Levels of ATP in Various Preparations of Isolated Cells from a 16 Day Old Chick Embryonic Heart (Values Expressed as N Moles/mg Cellular Protein)

Control hearts[a]	33.8 ± 2.23
Cultured cardiac cells[b]	23.59 ± 4.49
Freshly isolated cells	
(a) Present method[b]	7.32 ± 1.31
(b) According to Glick, *et al.* (12)[c]	4.43 ± 0.79

[a]Hearts removed from the embryo and homogenized immediately in 0.5 N PCA are taken as controls.

[b]A mixture of .2% hyaluronidase +.1% collagenase + 50% N-16 medium + 50% Hanks was used as a dispersion solution. The isolation procedure is described in the text.

[c]Cells were isolated (present method) and layered over 3% Ficoll Solution. The muscle cells were then isolated by low centrifugation and washed with Hanks solution.

the levels of ATP by using the earlier described methods where cells were isolated by trypsin or by a combination of collagenase and hyaluronidase (unpublished observation). Incubating the freshly isolated cells (present method) with growth media for 90 minutes at 37°C and then removing the media by thorough washings with modified Hanks solution did not seem to increase the ATP values. As a further note, we were not very successful using i-solated adult rat heart muscle when compared with chick embryonic cardiac muscle cells.

B. Levels of Nucleosides and Adenine Nucleotides in Cultured Cardiac Muscle Cells Due to Hypoxia

Hypoxia did not cause a significant change in the levels of ATP and ADP in the cultured cells compared to cells incubated in the presence of 95% O_2 + 5% CO_2 (Fig. 1). However, AMP in the cells and adenosine, inosine, and hypoxanthine in the medium were significantly increased by hypoxia and were 1.6, 2.2, 1.8 and 3.6 fold greater respectively than the corresponding controls.

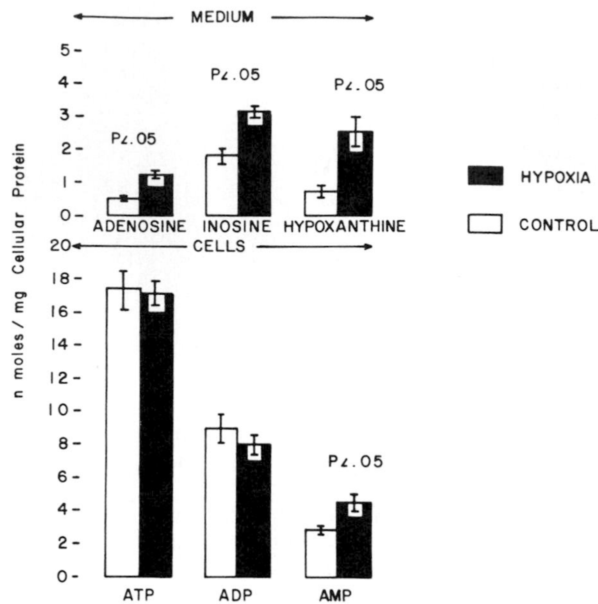

FIGURE 1 *The cells in the control and hypoxic groups were incubated with 95%* O_2 *+ 5%* CO_2 *and 95%* N_2 *+ 5%* CO_2 *respectively for 15 minutes at* $37^{o}C$ *in modified Hanks solution without glucose in a total volume of 1.0 ml.*

C. Effect of Dipyridamole and Aminophylline on the Release of Radio-active Adenosine and on the Levels of Adenine Nucleotides and Nucleosides in Cultured Cardiac Muscle Cells Due to Hypoxia

Neither dipyridamole nor aminophylline affected the amount of radio-activity released into the medium during hypoxia (Fig. 2). Furthermore, dipyridamole had no significant effect on the levels of ATP and ADP (Fig. 3), but AMP was significantly reduced. Dipyridamole did not alter the levels of hypoxanthine in the medium, but is significantly decreased the levels of inosine (Fig. 3).

Aminophylline caused (Fig. 3) no significant differences in the levels of ATP, ADP and AMP in the cells and also did not change the inosine and adenosine concentrations in the medium. However, there was a significant decrease in the level of hypoxanthine in the medium (Fig. 3).

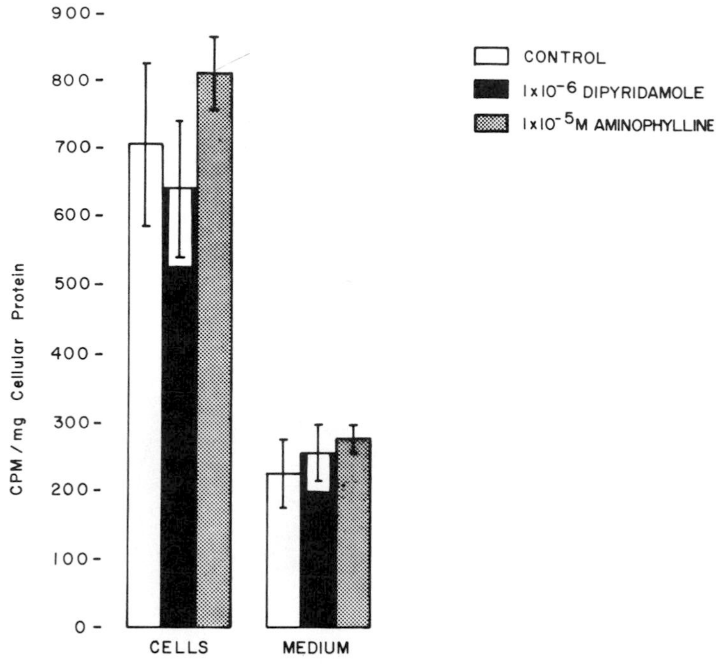

FIGURE 2 *The cells (prelabeled with U-C^{14}-adenosine) were incubated with 95% N_2 + 5% CO_2 for 15 minutes at 37°C in modified Hanks solution without glucose. The reaction mixture contained 0.1 ml drug (replaced with modified Hanks solution in control) and 0.9 ml modified Hanks solution.*

D. Effect of Dipyridamole and Aminophylline on the Uptake of Radio-active Adenosine and on the levels of Adenine Nucleotides and Nucleosides in Cultured Cardiac Muscle Cells

In the presence of dipyridamole, only 7% of the radio-activity was found in the cells; whereas, in the absence of dipyridamole, 42% of the radio-activity was present in the cells (Fig. 4). In terms of total cellular radio-activity, the control values were six fold greater than those of cells incubated in the presence of dipyridamole; hence, there was a marked inhibition of U-C^{14}-adenosine uptake by the cultured cardiac cells.

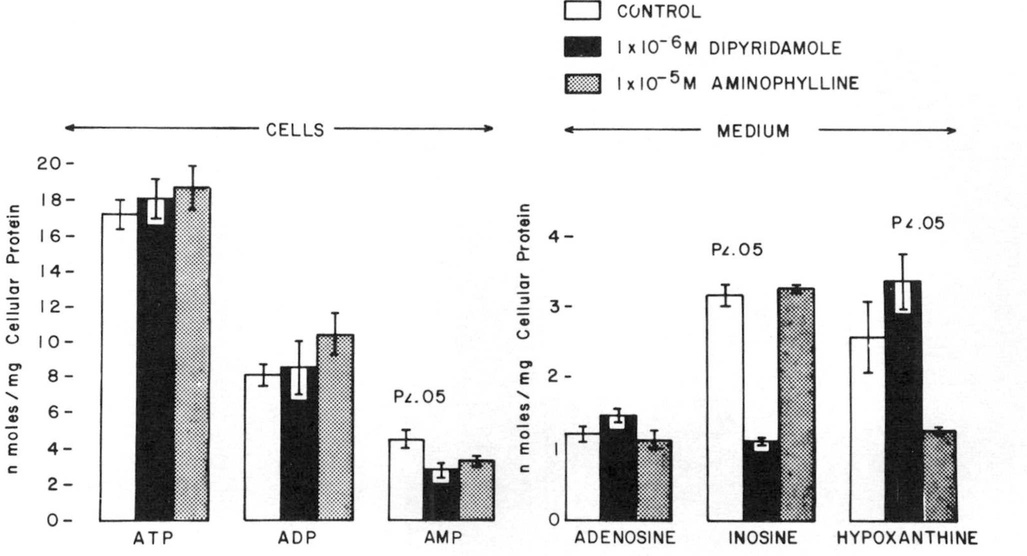

FIGURE 3 *The cells were incubated with 95%* N_2 *+ 5%* CO_2 *for 15 minutes at* $37^{o}C$ *in modified Hanks solution without glucose. The reaction mixture contained 0.1 ml drug (replaced with modified Hanks solution in control) and 0.9 ml modified Hanks solution.*

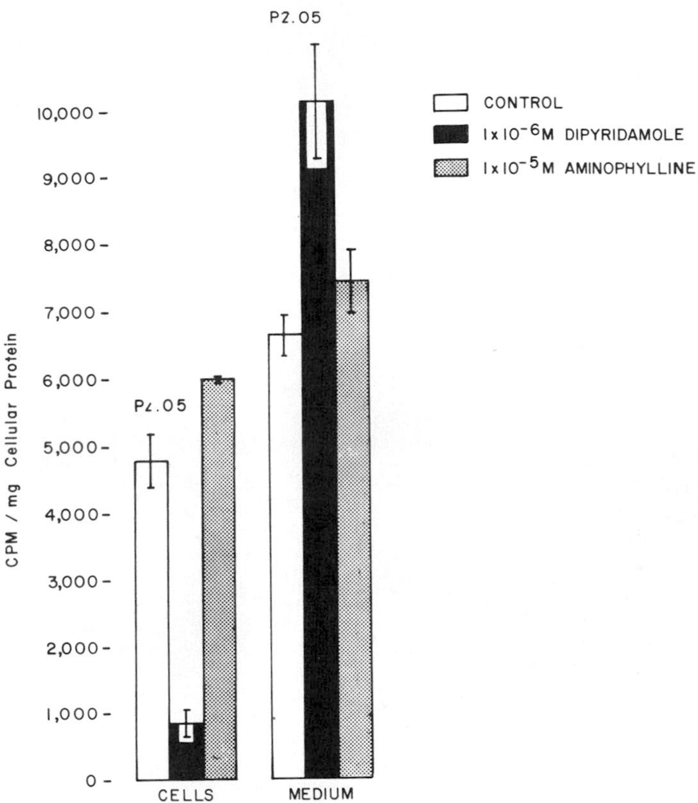

FIGURE 4 *The cells were incubated with 95% O_2 + 5% CO_2 for 15 minutes at 37°C in modified Hanks solution without glucose. The reaction mixture contained 0.05 ml U-C^{14}-adenosine + 0.1 ml of drug (replaced with modified Hanks solution in controls), and the volume made up to 1.0 ml with modified Hanks solution.*

Aminophylline produced no significant difference in the pattern of distribution of radioactivity from that of the controls (Fig. 4).

The radioactivity in the cells at $1X10^{-5}$ M aminophylline was slightly but not significantly higher than that of the controls.

In the presence of oxygen neither dipyridamole nor aminophylline significantly altered the levels of ATP, ADP, or AMP in the cells (Fig. 5). However, adenosine was significantly in-

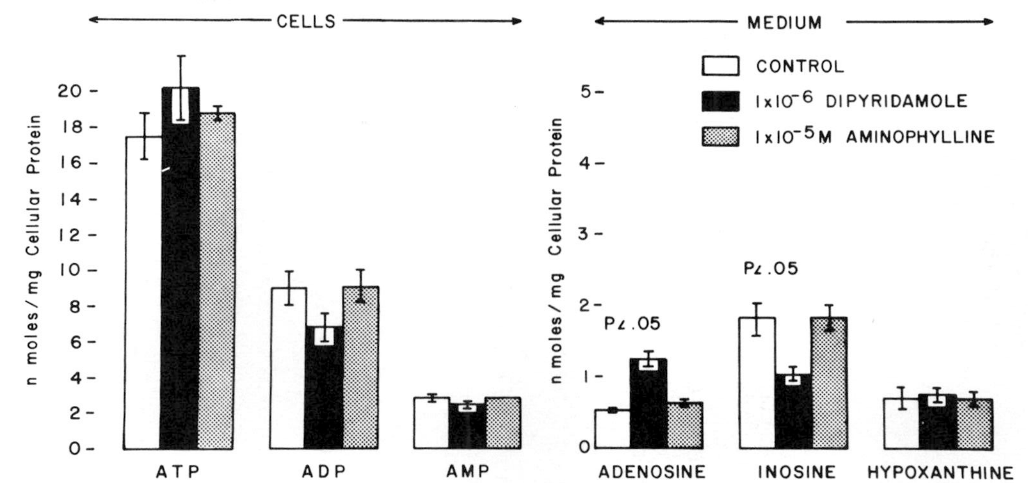

FIGURE 5 *The cells were incubated with 95%* O_2 *+ 5%* CO_2 *for 15 minutes at* 37^oC *in modified Hanks solution without glucose. The reaction mixture contained 0.05 ml U-*C^{14}*-adenosine + 0.1 ml of drug (replaced with modified Hanks solution in controls), and the volume made up to 1.0 ml with modified Hanks solution.*

creased in the medium in the presence of dipyridamole; while, inosine was significantly reduced (Fig. 5). Hypoxanthine levels in the medium were unaffected by dipyridamole, and adenosine, inosine, and hypoxanthine levels in the medium were not altered by aminophylline. Even a concentration of aminophylline as high as $1X10^{-3}$ M was without effect on cell and medium concentrations of the adenine nucleotides, nucleosides, or hypoxanthine.

IV. DISCUSSION

Adenosine triphosphate and other adenine nucleotides have been implicated as the major vehicles for energy coupling between the energy yielding and energy requiring sequences needed for the normal functioning of the cell. It is now becoming evident that many different cells, while they can survive major disturbances in their adenine nucleotide metabolism, show characteristic response patterns. The consequences to the cell of these fluctuation in high energy phosphates can result in disturbed physiological and metabolic states.

In the present study, efforts have been made to standardize a preparation of cardiac muscle cells, whose ATP values can be compared to the control hearts (*in vivo*). The ATP content of the cardiac muscle cells has been taken as one of the metabolic parameters (in addition to their spontaneous beating). Seraydarian and co-workers (35) have shown a definite correlation between intracellular levels of ATP and the maintenance of rhythmic contractions. The direct demonstration of ATP in the process of contraction has been demonstrated unequivocally by Cain and Davis (7); thus, the relationship between intracellular ATP concentration and the contraction of cardiac cells in culture follows. The ATP may be required for maintenance of the sodium pump, for proper membrane function, or for active transport of K^+ or Na^+.

Small changes in ATP may alter these functions. The mechanism of ATP involvement in spontaneous beating is not well understood.

The data presented in Table I definitely demonstrate that freshly isolated cardiac muscle cells have significantly lower ATP values (3-4 times) than cultured cardiac muscle cells; whereas, the ATP values of the cultured cardiac muscle cells are very close in ATP levels to those found in intact *in situ* hearts. Thus, the use of freshly isolated cardiac muscle cells for various metabolic and physiological studies should be done with extra caution because of their low ATP content which effects the correlation of the normal *in vivo* condition with the *in vitro* condition. The interpretation of higher ATP values in cultured cells compared with the freshly isolated cells is that the cells during harsh isolation procedures lose their adenine nucleotide pool (and perhaps other metabolites) due to altered membrane properties. Culturing the same cells in normal growth medium gives these cells a chance to regain their membrane properties and synthesize various metabolites to the extent of their original state. The growth medium has all the essential components required for various intermediary metabolic pathways.

A large number of workers in this area have used the dye exclusion method as an assessment of the viability of the preparation. Many of these dyes have acidic or toxic properties (25) that may be injurious to live cells in suspension and, hence, could possibly lead to variable results. The theory of vital dye staining of dead cells is related to the large increases in cell membrane permeability that occur following cell death, resulting into the passage of large molecules such as those of the dye into the cytoplasm. The small changes in membrane permeability which the dye exclusion technique may not show could result in the leakage of macromolecules such as enzymes from the cytoplasm into the extracellular space, changes in electrical resistance or impedence, and lack of other cofactors such as coenzymes or adenine nucleotide (21, 27- 37, 39, 40). With these factors in mind, the author devised conditions which would result in a higher ATP content of the cardiac muscle cells which could only be achieved by culturing the cells under the present situation. This constitutes one of the reasons for employing cell culture techniques. Based on this preparation of cardiac muscle cells the various experiments discussed in detail below were carried out.

The present data (Fig. 1) show a significant increase in the production of adenosine and its metabolic products, inosine and hypoxanthine when the cultured cardiac muscle cells were subjected to hypoxia. Although there were no significant differences in cellular ATP and ADP, AMP was significantly increased in the cells. The production of nucleosides under hypoxia is obviously by degradation of adenine nucleotides and the sum of adenine nucleotides and nucleosides in the control and hypoxic groups are comparable. These finding indicate that the adenosine released by hypoxia *in vivo* or isolated perfused hearts is primarily from the myocardial cells; however, the observations on isolated perfused or *in vivo* hearts to not exclude other sources of cardiac adenosine such as the vasculature and fibrous tissue.

Dipyridamole, a potent coronary vasodilator, enhances the vasodilator activity of exogenous adenosine by blocking its uptake and preventing its degradation by intracellular enzymes (18, 34). In agreement with this observation, the present investigation clearly demonstrates that the uptake of adenosine in cultured cardiac muscle cells is blocked by dipyridamole. These results are supported by the finding of higher absolute amounts of adenosine in the medium at the end of the incubation period in the dipyridamole treated cells. The release of adenosine was not influenced by dipyridamole, since the amount of radioactivity released into the medium from the hypoxic cardiac muscle cells (cell with prelabeled nucleotides) was the same in the absence or presence of the drug. The observation that the amount of adenosine in the medium with incubation in the absence of oxygen and in the presence of dipyridamole was slightly higher than with hypoxia alone can be explained on the basis that part of the adenosine released from hypoxic cells is not taken back into the cells because of the inhibition of uptake by dipyridamole. The decrease in the levels of AMP in the cells and inosine in the medium may be attributed to a reduction in substrate (adenosine) that is available to the adenosine kinase and adenosine deaminase of the cells respectively because of blockade of the uptake of adenosine by dipyridamole. To what extent AMP levels are altered by other intracellular enzymatic reactions cannot be determined from the present study.

These observations are not in agreement with the view put forward by Kubler and co-workers that dipyridamole blocks the release as well as the uptake of adenosine in myocardial cells (24). Kubler and co-workers (24) found that the adenosine concentration of the isolated hypoxic dog heart was greater in the presence of dipyridamole than in its absence. They concluded that dipyridamole blocked the release of adenosine from myocardial cells and, hence, adenosine could not be responsible for the coronary vasodilation observed in hypoxic hearts treated with dipyridamole. These experiments (24) were conducted on isolated heart preparations that have at least three compartments (vascular, interstitial fluid and intracellular) as well as non-myocardial cells and catecholamines. In an attempt to reconcile these observations with those of the present study we suggest that in the intact heart the adenosine may be sequestered in interstitial fluid located in the transverse tubules and intercalated discs and may not be completely washed out during coronary perfusion.

The mechanism(s) of action of dipyridamole is not clearly understood. It has been proposed (5, 10) that the drug inhibits adenosine deaminase and prevents endogenous or exogenous adenosine from being inactivated; other studies, on the other hand, have failed to detect a significant inhibitory effect on the enzymes involved in adenosine metabolism, namely adenosine deaminase (3, 23, 34) and adenosine kinase (34). In fact an activation of the latter has been demonstrated (9). How dipyridamole blocks the uptake of adenosine is not known. It is possible that the drug forms a complex with an adenosine carrier or that adenosine and dipyridamole compete for the same active site(s) on the carrier.

Aminophylline and theophylline (2, 20) attenuate the vasodilator action of exogenous adenosine and as in the case of dipyridamole, have been used to test the adenosine hypothesis for the regulation of coronary blood flow by studying their effect on myocardial reactive hyperemia. Our findings do not show a block in the uptake or release of adenosine in cultured cardiac cells in the presence of these agents. Additional support for these findings was obtained by quantifying the cell and medium fractions for adenosine contents together with other nucleosides and adenine nucleotides. The levels of adenosine released into the medium in the presence or absence of oxygen were not significantly changed by aminophylline. The mechanism whereby aminophylline attenuates the vasodilator action of adenosine is not known. It is possible that aminophylline may cause a direct release of bound Ca^{++} in the vascular smooth muscle in a manner similar to the effect of caffeine in skeletal muscle (41).

In summary, the use of monolayer cultures of beating cardiac muscle cells provides a useful method for studying many parameters of the heart, which can supplement and complement intact animal experiments and which are independent of many variables that plague *in vivo* studies of the beating heart. Cultured cardiac muscle cells (or other mammalian cells) can also serve as an excellent pharmacological tool for studying the direct effect of drugs on a specific cell type without the influence of neuronal and/or humoral factors.

ACKNOWLEDGMENTS

The technical assistance of Mrs. Kàren L. McGrain is greatly appreciated. This study was supported in part from an Intramural Research Grant of the University of South Alabama, College of Medicine, Mobile, Alabama. Part of this work was conducted at the Department of Physiology University of Virginia Medical School, Charlottesville, Virginia and published in Am. J. Physiology, 228: 1474-1478, 1975. This portion of the work was carried out during the tennure of a postdoctoral training fellowship from Public Health Service Grants NHLI HL-10384 and HL05815.

REFERENCES

Adams, H. (1965). *In* "Methods of Enzymatic Analysis" (H. V. Bergmeyer, ed.), pp. 573-577. Academic Press, New York.

Afonso, S. (1970). *Circulation. Res. 26,* 743-752.

Berne, R. M. (1963). *Am. J. Physiol. 204,* 317-322.

Bretschneider, H. J., Frank, A., Bernard, U., Kochspeik, K., Schler, R. (1959). *Arzneimittel. Forsch. 9,* 49-59.

Bunag, R. D., Douglas, C. R., Imai, S., Berne, R. M. (1964). *Circulation. Res. 15,* 83-88.

Burns, A. H., Reddy, W. J. (1975). *J. Mol. Cell. Cardiol. 7,* 553-561.

Cain, D. F., Davies, R. E. (1962). *Biochem. Biophys. Res. Commun. 8,* 361-366.

DeHaan, R. L. (1967). *In* "Factors Influencing Myocardial Contractility" (R. D. Tanz, F. Kavaler, J. Roberts, eds.), pp. 217-230. Academic Press, New York.

DeJong, J. W., Kalkman, C. (1973). *Biochem. Biophys. Acta. 320,* 388-396.

Deuticke, B., Gerlach, E. (1966). *Arch. Exptl. Pathol. Pharmakol. 225,* 107-119.

Dobson, J. G. Jr., Rubio, R., Berne, R. M. (1971). *Circulation. Res. 29*, 375-384.

Glick, M. R., Burns, A. H., Reddy, W. J. (1974). *Anal. Biochem. 61*, 32-42.

Greengard, P. (1965). *In* "Methods of Enzymatic Analysis" (H. V. Bergmeyer, ed.), pp. 551-555. Academic Press, New York.

Halle, W. (1967). *Morphologisches Jahrbuch III*, 3.

Halle, W., Wollenberger, A. (1968). *Zertschrift. Fur. Zellforschung. 87*, 292-314.

Harary, I., Farley, B. (1960). *Science. 131*, 1674-1675.

Harary, I., Farley, B. (1963). *Expt. Cell Res. 29*, 451-465.

Hashimoto, K., Kumakura, S., Tanemura, I. (1964). *Arzneimittel. Forschung. 14*, 1252-1254.

Imai, S., Riley, A. L., Berne, R. M. (1964). *Circulation. Res. 15*, 443-450.

Jurhan, W., Dietmann, K. (1970). *Pflugers. Arch. 315*, 105-109.

Kaltenbach, J. P., Kaltenbach, M. H., Lyons, W. B. (1958). *Exp. Cell. Res. 15*, 112-117.

Katori, M., Berne, R. M. (1966). *Circulation. Res. 19*, 420-425.

Kubler, W., Bretschneider, H. J. (1964). *Pflugers. Arch. 280*, 141-157.

Kubler, W., Spieckermann, P. G., Bretschneider, H. J. (1970). *J. Mol. Cell. Cardiol. 1*, 23-38.

Laiho, U. K., Trump, B. F. (1974). *Virchows. Arch. Abt. B. Zellpath. 15*, 267-277.

Lehmkuhl, D., Speralakis, N. (1967). *In* "Factors Influencing Myocardial Contractility" (R. D. Tanz, F. Kavaler, J. Roberts, eds.), pp. 245-278. Academic Press, New York.

Lepage, G. A. (1950). *Cancer. Res. 10*, 77-88.

Lowry, O. H., Rosebrough, H. J., Farr, A. L., Randall, R. J. (1951). *J. Biol. Chem. 193*, 265-275.

Mark, G. E., Strasser, F. F. (1966). *Expt. Cell. Res. 44*, 217-233.

Rubio, R., Berne, R. M. (1969). *Circulation. Res. 25,* 407-415.

Mustafa, S. J., Rubio, R., Berne, R. M. (1975). *Am. J. Physiol. 228,* 62-67.

Rubio, R., Berne, R. M., Katori, M. (1969). *Am. J. Physiol. 216,* 56-62.

Rubio, R., Wiedmeier, V. T., Berne, R. M. (1972). *Am. J. Physiol. 222,* 550-555.

Schrader, J., Berne, R. M., Rubio, R. (1972). *Am. J. Physiol. 233,* 159-166.

Seraydarian, M. W., Harary, I., Sato, E. (1968). *Biochem. Biophys. Acta. 162,* 414-423.

Seraydarian, M. W., Artaza, L., Abbott, B. C. (1972). *J. Mol. Cell. Cardiol. 4,* 477-484.

Slater, T. F. (1969). *In* "Lysosomes in Biology and Pathology" (J. T. Dingle and H. B. Amsterdam, eds.), pp. 467-492, North Holland Publishing Co.

Sperelakis, N. (1967). *In* "Electrophysiology and Ultrastructure of the Heart" (B. Taccardi, G. Marchetti, eds.), pp. 81. Bunkodo, Tokyo, Japan.

Trump, B. F., Ericsson, J. L. E. (1965). *In* "The Inflammatory Process" (B. W. Zweifach, L. Grant, and L. McCluskey, eds.), p. 35. Academic Press, New York.

Trump, B. F., Arstila, A. U. (1971). *In* "Principles of Pathobiology" (N. LaVia and R. Hill, eds.), pp. 9-96. Oxford University Press.

Weber, A., Herz, R. (1968). *J. Gen. Physiol. 52,* 750-772.

Wildenthal, K. (1971). *J. Applied. Physiol. 30,* 153-157.

A
B 7
C 8
D 9
E 0
F 1
G 2
H 3
I 4
J 5